W0225750

Mathematical Physics Studies

More information about this series at http://www.springer.com/series/6316

Jörg Teschner

Editor

New Dualities
of Supersymmetric Gauge
Theories

 Springer

Editor
Jörg Teschner
String Theory Group
DESY
Hamburg
Germany

ISSN 0921-3767   ISSN 2352-3905   (electronic)
Mathematical Physics Studies
ISBN 978-3-319-18768-6   ISBN 978-3-319-18769-3   (eBook)
DOI 10.1007/978-3-319-18769-3

Library of Congress Control Number: 2015941143

Springer Cham Heidelberg New York Dordrecht London

Printed on acid-free paper

Springer International Publishing AG Switzerland is part of Springer Science+Business Media (www.springer.com)

Contents

Exact Results on $\mathcal{N} = 2$ Supersymmetric Gauge Theories

Jörg Teschner

The following is meant to give an overview over our special volume. The first three Sects. 1–3 are intended to give a general overview over the physical motivations behind this direction of research, and some of the developments that initiated this project. These sections are written for a broad audience of readers with interest in quantum field theory, assuming only very basic knowledge of supersymmetric gauge theories and string theory. This will be followed in Sect. 4 by a brief overview over the different chapters collected in this volume, while Sect. 5 indicates some related developments that we were unfortunately not able to cover here.

Due to the large number of relevant papers the author felt forced to adopt a very restrictive citation policy. With the exception of very few original papers only review papers will be cited in Sects. 1 and 2. More references are given in later sections, but it still seems impossible to list all papers on the subjects mentioned there. The author apologises for any omission that results from this policy.

1 Background, History and Context

1.1 Strong Coupling Behavior of Gauge Theories

Gauge theories play a fundamental role in theoretical particle physics. They describe in particular the interactions that bind the quarks into hadrons. It is well understood how these interactions behave at high energies. This becomes possible due to the phenomenon of asymptotic freedom: The effective strength of the interactions

---

A citation of the form [V:x] refers to article number x in this volume.

J. Teschner (✉)
DESY Theory, Notkestr. 85, 22603 Hamburg, Germany
e-mail: teschner@mail.desy.de

© Springer International Publishing Switzerland 2016
J. Teschner (ed.), *New Dualities of Supersymmetric Gauge Theories*,
Mathematical Physics Studies, DOI 10.1007/978-3-319-18769-3_1

1

depends on the energy scale, and goes to zero for large energies. It is much less well understood how the interactions between quarks behave at low energies: The experimental evidence indicates that the interactions become strong enough to prevent complete separation of the quarks bound in a hadron (confinement). The theoretical understanding of this phenomenon has remained elusive.

When the interactions are weak one may approximate the resulting effects reasonably well using perturbation theory, as can be developed systematically using the existing Lagrangian formulations. However, the calculation of higher order effects in perturbation theory gets cumbersome very quickly. It is furthermore well-known that additional effects exist that can not be seen using perturbation theory. Exponentially suppressed contributions to the effective interactions are caused, for example, by the existence of nontrivial solutions to the Euclidean equations of motion called instantons. The task to understand the strong coupling behavior of gauge theories looks rather hopeless from this point of view: It would require having a complete resummation of all perturbative and non perturbative effects. Understanding the strong coupling behaviour of general gauge theories remains an important challenge for quantum field theory. However, there exist examples in which substantial progress has recently been made on this problem: Certain important physical quantities like expectation values of Wilson loop observables can even be calculated exactly. What makes these examples more tractable is the existence of supersymmetry. It describes relations between bosons and fermions which may imply that most quantum corrections from bosonic degrees of freedom cancel against similar contributions coming from the fermions. Whatever remains may be exactly calculable.

Even if supersymmetry has been crucial for getting exact results up to now, it seems likely that some of the lessons that can be learned by analysing supersymmetric field theories will hold in much larger generality. One may in particular hope to deepen our insights into the origin of quantum field theoretical duality phenomena by analysing supersymmetric field theories, as will be discussed in more detail below. As another example let us mention that it was expected for a long time that instantons play a key role for the behaviour of gauge theories at strong coupling. This can now be illustrated beautifully with the help of the new exact results to be discussed in this volume. We believe that the study of supersymmetric field theories offers a promising path to enter into the mostly unexplored world of non-perturbative phenomena in quantum field theory.

1.2 Electric-Magnetic Duality Conjectures

It is a hope going back to the early studies of gauge theories that there may exist asymptotic strong coupling regions in the gauge theory parameter space in which a conventional (perturbative) description is recovered using a suitable new set of field variables. This phenomenon is called a duality. Whenever this occurs, one may get access to highly nontrivial information about the gauge theory at strong coupling.

For future reference let us formulate a bit more precisely what it means to have a duality. Let us consider a family $\{\mathcal{F}_z; z \in \mathcal{M}\}$ of quantum theories having a moduli space $\mathcal{M}$ of parameters z. The quantum theory $\mathcal{F}_z$ is for each fixed value of z abstractly characterised by an algebra of observables $\mathcal{A}_z$ and a linear functional on $\mathcal{A}_z$ which assigns to each observable $\mathsf{O} \in \mathcal{A}_z$ its vacuum expectation value $\langle \mathsf{O} \rangle_z$. We say that $\{\mathcal{F}_z; z \in \mathcal{M}\}$ is a quantum field theory with fields Φ and action $S_\tau[\Phi]$ depending on certain parameters τ (like masses and coupling constants) if there exists a point z_0 in the boundary of the moduli space $\mathcal{M}$, a coordinate $\tau = \tau(z)$ in the vicinity of z_0, and a map $\mathcal{O}$ assigning to each $\mathsf{O} \in \mathcal{A}_z$ a functional $\mathcal{O}_{\mathsf{O},\tau}[\Phi]$ such that

$$\langle \mathsf{O} \rangle_z \simeq \int [\mathcal{D}\Phi] \, e^{-S_\tau[\Phi]} \, \mathcal{O}_{\mathsf{O},\tau}[\Phi], \tag{1.1}$$

where $\simeq$ means equality of asymptotic expansions around z_0 and the right hand side is defined in terms of the action $S[\Phi]$ using path integral methods.

We say that a theory with fields Φ, action $S_\tau[\Phi]$ and parameters τ is dual to a theory characterised by similar data $S'_{\tau'}[\Phi']$ if there exists a family of quantum theories $\{\mathcal{F}_z; z \in \mathcal{M}\}$ with moduli space $\mathcal{M}$ having boundary points z_0 and z'_0 such that the vacuum expectation values of $\mathcal{F}_z$ have an asymptotic expansion of the form (1.1) near z_0, and also an asymptotic expansion

$$\langle \mathsf{O} \rangle_z \simeq \int [\mathcal{D}\Phi']' \, e^{-S'_{\tau'}[\Phi']} \, \mathcal{O}'_{\mathsf{O},\tau'}[\Phi'], \tag{1.2}$$

near z'_0, with $\mathcal{O}'$ being a map assigning to each $\mathsf{O} \in \mathcal{A}_z$ a functional $\mathcal{O}'_{\mathsf{O},\tau'}[\Phi']$.

A class of long-standing conjectures concerning the strong coupling behavior of gauge theories are referred to as the electric-magnetic duality conjectures. Some of these conjectures concern the infrared (IR) physics as described in terms of low-energy effective actions, others are about the full ultraviolet (UV) descriptions of certain gauge theories. The main content of the first class of such conjectures is most easily described for theories having an effective description at low energies involving in particular an abelian gauge field A and some charged matter q. The effective action $S(A, q; \tau_{IR})$ will depend on an effective IR coupling constant τ_{IR}. The phenomenon of an electric magnetic duality would imply in particular that the strong coupling behavior of such a gauge theory can be represented using a dual action $S'(A', q'; \tau'_{IR})$ that depends on the dual abelian gauge field A' related to A simply as

$$F'_{\mu\nu} = \frac{1}{2} \epsilon_{\mu\nu\rho\sigma} F^{\rho\sigma}. \tag{1.3}$$

The relation between the dual coupling constant τ'_{IR} and τ_{IR} is also conjectured to be very simple,

$$\tau'_{IR} = -\frac{1}{\tau_{IR}}. \tag{1.4}$$

The relation expressing q' in terms of q and A may be very complicated, in general. In many cases one expects that q' is the field associated to solitons, localized particle-like excitations associated to classical solutions of the equations of motion of $S(A, q; \tau_{IR})$. Such solitons are usually very heavy at weak coupling but may become light at strong coupling where they may be identified with fundamental particle excitations of the theory with action $S'(A', q'; \tau'_{IR})$.

For certain theories there exist even deeper conjectures predicting dualities between different perturbative descriptions of the full *ultraviolet* quantum field theories. Such conjectures, often referred to as S-duality conjectures originated from the observations of Montonen and Olive [MoOl, GNO], and were subsequently refined in [WO, Os], leading to the conjecture of a duality between the $N = 4$ supersymmetric Yang-Mills theory with gauge group G and coupling τ one the one hand, and the $N = 4$ supersymmetric Yang-Mills theory with gauge group $^L G$ and coupling $-1/n_G \tau$ on the other hand. $^L G$ is the Langlands dual of a group G having as Cartan matrix the transpose of the Cartan matrix of G, and n_G is the lacing number[1] of the Lie algebra of G.

A given UV action S can be used to define such expectation values perturbatively, as well as certain non-perturbative corrections like the instantons. The question is whether all perturbative and non perturbative corrections can be resummed to get the cross-over to the perturbation theory defined using a different UV action S'.

A non-trivial strong-coupling check for the S-duality conjecture in the $N = 4$ supersymmetric Yang-Mills theory was performed in [VW].[2] Generalised S-duality conjectures have been formulated in [Ga09] (see [V:2] for a review) for a large class of $\mathcal{N} = 2$ supersymmetric gauge theories which are ultraviolet finite and therefore have well-defined bare UV coupling constants τ. It is of course a challenge to establish the validity of such conjectures in any nontrivial example.

1.3 Seiberg-Witten Theory

A breakthrough was initiated by the discovery of exact results for the low energy effective action of certain $\mathcal{N} = 2$ supersymmetric gauge theories by Seiberg and Witten [SW1, SW2]. There are several good reviews on the subject, see e.g. [Bi, Le, Pe97, DPh, Tac] containing further references.[3]

The constraints of $\mathcal{N} = 2$ supersymmetry restrict the low-energy physics considerably. As a typical example let us consider a gauge theory with $SU(M)$ gauge symmetry. The gauge field sits in a multiplet of $\mathcal{N} = 2$ supersymmetry containing a scalar field ϕ in the adjoint representation of $SU(M)$. $\mathcal{N} = 2$ supersymmetry allows

---

[1]The lacing number n_G is equal to 1 is the Lie-algebra of G is simply-laced, 2 if it is of type B_n, C_n and F_4, and 3 if it is of type G_2.

[2]The result of [Sen] furnishes a nontrivial check of a prediction following from the Montonen-Olive conjecture.

[3]A fairly extensive list of references to the early literature can be found e.g. in [Le].

parametric families of vacuum states. The vacuum states in the Coulomb branch can be parameterised by the vacuum expectation values of gauge-invariant functions of the scalars like $u^{(k)} := \langle \text{Tr}(\phi^k) \rangle$, $k = 2, \ldots M$. For generic values of these quantities one may describe the low-energy physics in terms of a Wilsonian effective action $S^{\text{eff}}[A]$ which is a functional of $d = M - 1$ vector multiplets A_k, $k = 1, \ldots, d$, having scalar components a_k and gauge group $U(1)_k$, respectively. The effective action $S^{\text{eff}}[A]$ turns out to be completely determined by a single holomorphic function $\mathcal{F}(a)$ of d variables $a = (a_1, \ldots, a_d)$ called the prepotential. It completely determines the (Wilsonian) low energy effective action as $S^{\text{eff}} = S^{\text{eff}}_{\text{bos}} + S^{\text{eff}}_{\text{fer}}$, where

$$S^{\text{eff}}_{\text{bos}} = \frac{1}{4\pi} \int d^4 x \left(\text{Im}(\tau^{kl}) \partial_\mu \bar{a}_k \partial^\mu a_l + \frac{1}{2} \text{Im}(\tau^{kl}) F_{k,\mu\nu} F_l^{\mu\nu} + \frac{1}{2} \text{Re}(\tau^{kl}) F_{k,\mu\nu} \tilde{F}_l^{\mu\nu} \right),$$
(1.5)

while $S^{\text{eff}}_{\text{fer}}$ is the sum of all terms containing fermionic fields, uniquely determined by $\mathcal{N} = 2$ supersymmetry. The a-dependent matrix $\tau^{kl}(a)$ in (1.5) is the matrix of second derivatives of the prepotential,

$$\tau^{kl}(a) := \partial_{a_k} \partial_{a_l} \mathcal{F}(a).$$
(1.6)

Based on physically motivated assumptions about the strong coupling behavior of the gauge theories under consideration, Seiberg and Witten proposed a precise mathematical definition of the relevant functions $\mathcal{F}(a)$ for $M = 2$. This type of description was subsequently generalised to large classes of $\mathcal{N} = 2$ supersymmetric gauge theories including the cases with $M > 2$.

The mathematics underlying the definition of $\mathcal{F}(a)$ is called special geometry. In many cases including the examples discussed above one may describe $\mathcal{F}(a)$ using an auxilliary Riemann surface Σ called the Seiberg-Witten curve which in suitable local coordinates can be described by a polynomial equation $P(x, y) = 0$. The polynomial $P(x, y)$ has coefficients determined by the mass parameters, the gauge coupling constants, and the values $u^{(k)}$ parameterising the vacua. Associated to Σ is the canonical one form $\lambda_{\text{SW}} = y dx$ on Σ. Picking a canonical basis for the first homology $H_1(\Sigma, \mathbb{Z})$ of Σ, represented by curves $\alpha_1, \ldots, \alpha_d$ and $\beta_1, \ldots, \beta_d$ with intersection index $\alpha_r \circ \beta_s = \delta_{rs}$ one may consider the periods

$$a_r = \int_{\alpha_r} \lambda_{\text{SW}}, \qquad a_r^D = \int_{\beta_r} \lambda_{\text{SW}}.$$
(1.7)

Both $a_r \equiv a_r(u)$ and $a_r^D \equiv a_r^D(u)$, $r = 1, \ldots, d$, represent sets of complex coordinates for the d-dimensional space of vacua, in our example parameterised by $u = (u^{(2)}, \ldots, u^{(M)})$. It must therefore be possible to express a^D in terms of a. It turns out that the relation can be expressed using a function $\mathcal{F}(a)$, $a = (a_1, \ldots, a_d)$, from which the coordinates a_r can be obtained via $a_r^D = \partial_{a_r} \mathcal{F}(a)$. It follows that $\mathcal{F}(a)$ is up to an additive constant defined by Σ and the choice of a basis for $H_1(\Sigma, \mathbb{Z})$.

The choice of the field coordinates a_k is not unique. Changing the basis $\alpha_1, \ldots, \alpha_d$ and $\beta_1, \ldots, \beta_d$ to $\alpha'_1, \ldots, \alpha'_d$ and $\beta'_1, \ldots, \beta'_d$ will produce new coordinates a'_r, a'^D_r, $k = 1, \ldots, d$ along with a new function $\mathcal{F}'(a')$ which is the prepotential determining a dual action $S'_{\text{eff}}[a']$. The actions $S_{\text{eff}}[a]$ and $S'_{\text{eff}}[a']$ give us equivalent descriptions of the low-energy physics. This gives an example for an IR duality.

1.4 Localization Calculations of SUSY Observables

Having unbroken SUSY opens the possibility to compute some important quantities exactly using a method called localization [W88]. This method forms the basis for much of the recent progress in this field.

Given a supersymmetry generator Q such that $Q^2 = P$, where P is the generator of a bosonic symmetry. Let $S = S[\Phi]$ be an action such that $QS = 0$. Let us furthermore introduce an auxiliary fermionic functional $V = V[\Phi]$ that satisfies $PV = 0$. We may then consider the path integral defined by deforming the action by the term tQV, with t being a real parameter. In many cases one can argue that expectation values of supersymmetric observables $\mathcal{O} \equiv \mathcal{O}[\Phi]$, $Q\mathcal{O} = 0$, defined by the deformed action, are in fact independent of t, as the following formal calculation indicates. Let us consider

$$\frac{d}{dt} \int [\mathcal{D}\Phi] \, e^{-S-tQV} \, \mathcal{O} = \int [\mathcal{D}\Phi] \, e^{-S-tQV} \, QV \, \mathcal{O}$$

$$= \int [\mathcal{D}\Phi] \, Q(e^{-S-tQV} \, V \, \mathcal{O}) = 0, \qquad (1.8)$$

if the path-integral measure is SUSY-invariant, $\int [\mathcal{D}\Phi] \, Q(\ldots) = 0$. This means that

$$\langle \mathcal{O} \rangle := \int [\mathcal{D}\Phi] \, e^{-S} \, \mathcal{O} = \lim_{t \to \infty} \int [\mathcal{D}\Phi] \, e^{-S-tQV} \, \mathcal{O}. \qquad (1.9)$$

If V is such that QV has positive semi-definite bosonic part, the only non-vanishing contributions are field configurations satisfying $QV = 0$. There are cases where the space $\mathcal{M}$ of solutions of $QV = 0$ is finite-dimensional.[4] The arguments above then imply that the expectation values can be expressed as an ordinary integral over the space $\mathcal{M}$ which may be calculable.

The reader should note that this argument bypasses the actual definition of the path integral in an interesting way. For the theories at hand, the definition of $\int [\mathcal{D}\Phi] \, e^{-S-tQV}$ represents a rather challenging task which is not yet done. What the argument underlying the localisation method shows is the following: If there is ultimately *any* definition of the theory that ensures unbroken supersymmetry in the

---

[4]In other cases $\mathcal{M}$ may a union of infinitely many finite-dimensional components of increasing dimensions, as happens in the cases discussed in Sect. 1.5.

sense that $\int [\mathcal{D}\Phi] \, Q(\ldots) = 0$, the argument (1.8) will be applicable, and may allow us to calculate *certain* expectation values exactly even if the precise definition of the full theory is unknown.

1.5 Instanton Calculus

The work of Seiberg and Witten was based on certain assumptions on the strong coupling behavior of the relevant gauge theories. It was therefore a major progress when it was shown in [N, NO03, NY, BE] that the mathematical description for the prepotential conjectured by Seiberg and Witten can be obtained by an honest calculation of the quantum corrections to a certain two-parameter deformation of the prepotential to all orders in the instanton expansion.

To this aim it turned out to be very useful to define a regularisation of certain IR divergences called Omega-deformation by adding terms to the action breaking Lorentz symmetry in such a way that a part of the supersymmetry is preserved [N],[5]

$$ S \rightarrow S_{\epsilon_1 \epsilon_2} = S + R_{\epsilon_1 \epsilon_2}. \tag{1.10} $$

One may then consider the partition function $\mathcal{Z}$ defined by means of the path integral defined by the action $S_{\epsilon_1 \epsilon_2}$. As an example let us again consider a theory with $SU(M)$ gauge group. This partition function $\mathcal{Z} = \mathcal{Z}(a, m, \tau; \epsilon_1, \epsilon_2)$ depends on the eigenvalues $a = (a_1, \ldots, a_{M-1})$ of the vector multiplet scalars at the infinity of $\mathbb{R}^4$, the collection m of all mass parameters of the theory, and the complexified gauge coupling τ formed out of the gauge coupling constant g and theta-angle θ as

$$ \tau = \frac{4\pi i}{g^2} + \frac{\theta}{2\pi}. \tag{1.11} $$

The unbroken supersymmetry can be used to apply the localisation method briefly described in Sect. 1.4, here leading to the conclusion that the path integral defining $\mathcal{Z}$ can be reduced to a sum of ordinary integrals over instanton moduli spaces. The culmination of a long series of works[6] were explicit formulae for the summands $\mathcal{Z}^{(k)}(a, m; \epsilon_1, \epsilon_2)$ that appear in the resulting infinite series[7] of instanton corrections

---

[5]The regularisation introduced in [N] provides a physical interpretation of a regularisation for integrals over instanton moduli spaces previously used in [LNS, MNS1].

[6]The results presented in [N, NO03] were based in particular on the previous work [LNS, MNS1, MNS2]. Similar results were presented in [FPS, Ho1, Ho2, FP, BFMT]; for a review see [V:4].

[7]The infinite series (1.12) are probably convergent. This was verified explicitly for the example of pure $SU(2)$ Super-Yang-Mills theory in [ILTy], and it is expected to follow for UV finite gauge theories from the relations with conformal field theory to be discussed in the next section.

$$\mathcal{Z}(a, m, \tau; \epsilon_1, \epsilon_2) = \mathcal{Z}^{\text{pert}}(a, m, \tau; \epsilon_1, \epsilon_2) \left(1 + \sum_{k=1}^{\infty} q^k \, \mathcal{Z}^{(k)}(a, m; \epsilon_1, \epsilon_2) \right), \quad (1.12)$$

with $q = e^{2\pi i \tau}$ in the ultraviolet finite cases, while it is related to the running effective scale Λ otherwise. The explicitly known prefactor $\mathcal{Z}^{\text{pert}}(a, m, \tau; \epsilon_1, \epsilon_2)$ is the product of the simple tree-level contribution with a one-loop determinant. The latter is independent of the coupling constants q_r, and can be expressed in terms of known special functions.

In order to complete the derivation of the prepotentials proposed by Seiberg and Witten it then remained to argue that $\mathcal{F}(a) \equiv \mathcal{F}(a, m, \tau)$ is related to the partition function $\mathcal{Z}$ as

$$\mathcal{F}(a, m; \tau) = - \lim_{\epsilon_1, \epsilon_2 \to 0} \epsilon_1 \epsilon_2 \mathcal{Z}(a, m; \tau; \epsilon_1, \epsilon_2), \qquad (1.13)$$

and to derive the mathematical definition of $\mathcal{F}(a)$ proposed by Seiberg and Witten from the exact results on $\mathcal{Z}(a, m; \tau; \epsilon_1, \epsilon_2)$ obtained in [N, NO03, NY, BE].

2 New Exact Results on $\mathcal{N} = 2$ Supersymmetric Field Theories

2.1 Localisation on Curved Backgrounds

Another useful way to regularise IR-divergences is to consider the quantum field theory on four-dimensional Euclidean space-times M^4 of finite volume. The finite-size effects encoded in the dependence of physical quantities with respect to the volume or other parameters of M^4 contain profound physical information. It has recently become possible to calculate some of the these quantities exactly. One may, for example, consider gauge theories on a four-sphere S^4 [Pe07], or more generally four-dimensional ellipsoids [HH],

$$S^4_{\epsilon_1, \epsilon_2} := \{ (x_0, \ldots, x_4) \mid x_0^2 + \epsilon_1^2(x_1^2 + x_2^2) + \epsilon_2^2(x_3^2 + x_4^2) = 1 \}. \qquad (2.1)$$

The spaces $S^4_{\epsilon_1, \epsilon_2}$ have sufficient symmetry for having an unbroken supersymmetry Q such that Q^2 is the sum of a space-time symmetry plus possibly an internal symmetry. Expectation values of supersymmetric observables on $S^4_{\epsilon_1, \epsilon_2}$ therefore represent candidates for quantities that may be calculated by the localisation method. Interesting physical quantities are the partition function on S^4, and the values of Wilson- and 't Hooft loop observables. Wilson loop observables can be defined as path-ordered exponentials of the general form $W_{r,i} := \text{Tr} \, \mathcal{P} \exp \left[\oint_C ds \, (i \dot{x}^\mu A_\mu^r + |\dot{x}| \phi^r) \right]$. The 't Hooft loop observables $T_{r,i}, i = 1, 2$, can be defined semiclassically by performing a path integral over field configurations with a specific singular behavior near a curve $\mathcal{C}$ describing the effect of parallel transport of a magnetically charged probe particle along $\mathcal{C}$. Choosing the support of the loop observables to be the circles $\mathcal{C}_1$ or $\mathcal{C}_2$ defined by $x_0 = x_3 = x_4 = 0$ or $x_0 = x_1 = x_2 = 0$, respectively, one gets operators commuting with part of the supersymmetries of the theory.

However, applying the localisation method to the field theories with $\mathcal{N} = 2$ supersymmetry is technically challenging [Pe07, HH, GOP]. A review of the necessary technology and of some subsequent developments in this direction can be found in the articles [V:6, V:7]. Appropriately modifying the action defining the theory under consideration on $\mathbb{R}^4$ gives a Q-invariant action S for the theory on $S^4_{\epsilon_1, \epsilon_2}$. A functional V is found in [Pe07] such that QV is positive definite. The field configurations solving $QV = 0$ have constant values of the scalar fields, and vanishing values of all other fields. This means that the path integral reduces to an ordinary integral over scalar zero modes. This phenomenon may be seen as a variant of the cancellations between contributions from fermionic and bosonic degrees of freedom that frequently occur in supersymmetric field theories, leaving behind only contributions from states of zero energy.

The results obtained by localisation [Pe07, GOP, HH] have the following structure:

- Partition functions:

$$\mathbf{Z}(m; \tau; \epsilon_1, \epsilon_2) := \langle 1 \rangle_{S^4} = \int da \; |\mathcal{Z}(a, m; \tau; \epsilon_1, \epsilon_2)|^2, \qquad (2.2)$$

where $\mathcal{Z}(a, m; \tau; \epsilon_1, \epsilon_2)$ are the instanton partition functions briefly discussed in Sect. 1.5. More details can be found in [V:6].

- Wilson or 't Hooft loop expectation values:

$$\langle \mathcal{L} \rangle_{S^4} = \int da \; (\mathcal{Z}(a, m; \tau; \epsilon_1, \epsilon_2))^* \, \mathcal{D}_{\mathcal{L}} \cdot \mathcal{Z}(a, m; \tau; \epsilon_1, \epsilon_2), \qquad (2.3)$$

where $\mathcal{Z}(a, m; \tau; \epsilon_1, \epsilon_2)$ are the instanton partition functions described in Sect. 1.5, and $\mathcal{D}_{\mathcal{L}}$ is a difference operator acting on the scalar zero mode variables collectively referred to by the notation a. The difference operators $\mathcal{D}$ are pure multiplication operators $\mathcal{D}_{\mathcal{L}} = 2 \cosh(2\pi a / \epsilon_i)$ if $\mathcal{L}$ is a Wilson loop supported on $\mathcal{C}_i$. These results are reviewed in [V:7].

The integral over a in (2.2), (2.3) is the integration over the scalar zero modes. One may interpret these results as reduction to an effective quantum mechanics of these zero modes. From this point of view one would interpret the instanton partition function $\mathcal{Z}(a, m; \tau; \epsilon_1, \epsilon_2)$ as the wave-function $\Psi_\tau(a)$ of a state $|\tau\rangle_0$ in the zero-mode sub-sector defined by the path integral over field configurations on the lower half-ellipsoid $S^{4,-}_{\epsilon_1, \epsilon_2} := \{ (x_0, \ldots, x_4) \in S^4_{\epsilon_1, \epsilon_2} ; x_0 < 0 \}$. The expectation value (2.3) can then be represented as

$$\langle \mathcal{L} \rangle_{S^4} = \langle \tau | \mathcal{L}_0 | \tau \rangle_0, \qquad (2.4)$$

where $\mathcal{L}_0$ denotes the projection of the operator representing $\mathcal{L}$ to the zero mode sub-sector. This point of view is further discussed in [V:12].

Although the dynamics of the zero mode sub-sector is protected by supersymmetry, it captures very important non-perturbative information about the rest of the theory. Dualities between different UV-descriptions of the gauge theory must be reflected in the zero mode dynamics, and can therefore be tested with the help of localisation calculations. But the definition of the full theory must be compatible with these results, which is ultimately a consequence of unbroken supersymmetry. One may view the zero-mode dynamics as a kind of skeleton of the SUSY field theory. Whatever the QFT-"flesh" may be, it must fit to the skeleton, and exhibit same dualities, for example.

The localisation method has furthermore recently been used to obtain exact results on some correlation functions in $\mathcal{N} = 2$ supersymmetric QCD [BNP].

2.2 Relation to Conformal Field Theory

In [AGT] is was observed that the results for partition functions of some four-dimensional supersymmetric gauge theories that can be calculated by the method of [Pe07] are in fact proportional to something known, namely the correlation functions of the two-dimensional quantum field theory known as Liouville theory. Such correlation functions are formally defined by the path integral using the action

$$S_b^{\text{Liou}} = \frac{1}{4\pi} \int d^2z \left[(\partial_a \phi)^2 + 4\pi \mu e^{2b\phi} \right]. \tag{2.5}$$

Liouville theory has been extensively studied in the past motivated by the relations to two-dimensional quantum gravity and noncritical string theory discovered by Polyakov. It is known to be conformally invariant, as suggested by the early investigation [CT], and established by the construction given in [Te01]. Conformal symmetry implies that the correlation functions can be represented in a holomorphically factorized form. As a typical example let us consider

$$\left(e^{2\alpha_4 \phi(\infty)} e^{2\alpha_3 \phi(1)} e^{2\alpha_2 \phi(q)} e^{2\alpha_1 \phi(0)} \right)_b^{\text{Liou}} = \int_{\mathbb{R}^+} \frac{dp}{2\pi} \, C_{21}(p) C_{43}(-p) \left| \mathcal{F}_p \left[\begin{smallmatrix} \alpha_3 \, \alpha_2 \\ \alpha_4 \, \alpha_1 \end{smallmatrix} \right](q) \right|^2, \tag{2.6}$$

where the conformal blocks $\mathcal{F}_p \left[\begin{smallmatrix} \alpha_3 \, \alpha_2 \\ \alpha_4 \, \alpha_1 \end{smallmatrix} \right](q)$ can be represented by power series of the form

$$\mathcal{F}_p \left[\begin{smallmatrix} \alpha_3 \, \alpha_2 \\ \alpha_4 \, \alpha_1 \end{smallmatrix} \right](q) = q^{\frac{Q^2}{4} + p^2 - \alpha_1 (Q - \alpha_1) - \alpha_2 (Q - \alpha_2)} \left(1 + \sum_{k=1}^{\infty} q^k \mathcal{F}_p^{(k)} \left[\begin{smallmatrix} \alpha_3 \, \alpha_2 \\ \alpha_4 \, \alpha_1 \end{smallmatrix} \right] \right), \tag{2.7}$$

having coefficients $\mathcal{F}_p^{(k)}\left[\begin{smallmatrix}\alpha_3\,\alpha_2\\\alpha_4\,\alpha_1\end{smallmatrix}\right]$ completely defined by conformal symmetry [BPZ].[8] Explicit formulae for the coefficient functions $C_{ij}(p) \equiv C(\alpha_i, \alpha_j, \frac{Q}{2} + ip)$, $Q = b + b^{-1}$, have been conjectured in [DOt, ZZ], and nontrivial checks for this conjecture were presented in [ZZ]. A derivation of all these results follows from the free-field construction of Liouville theory given in [Te01].

In order to describe an example for the relations discovered in [AGT] let us temporarily restrict attention to the $\mathcal{N} = 2$ supersymmetric gauge theory often referred to as $N_f = 4$-theory. This theory has field content consisting of an $SU(2)$-vector multiplet coupled to four massive hypermultiplets in the fundamental representation of the gauge group. The relation discovered in [AGT] can be written as

$$\mathcal{Z}(a, m; \tau; \epsilon_1, \epsilon_2) \propto N_{21}(p)N_{43}(p)\mathcal{F}_p\left[\begin{smallmatrix}\alpha_3\,\alpha_2\\\alpha_4\,\alpha_1\end{smallmatrix}\right](q), \tag{2.8}$$

where $|N_{ij}(p)|^2 = C_{ij}(p)$. The factors of proportionality dropped in (2.8) are explicitly known, and turn out to be inessential. The parameters are identified, respectively, as

$$b^2 = \frac{\epsilon_1}{\epsilon_2}, \qquad \hbar^2 = \epsilon_1\epsilon_2, \qquad q = e^{2\pi i\tau}, \tag{2.9a}$$

$$p = \frac{a}{\hbar}, \qquad \alpha_r = \frac{Q}{2} + i\frac{m_r}{\hbar}, \qquad Q := b + b^{-1}. \tag{2.9b}$$

In order to prove (2.8) one needs to show that the coefficients $\mathcal{Z}^{(k)}(a, m; \epsilon_1, \epsilon_2)$ in (1.12) are equal to $\mathcal{F}_p^{(k)}\left[\begin{smallmatrix}\alpha_3\,\alpha_2\\\alpha_4\,\alpha_1\end{smallmatrix}\right]$. This was done in [AGT] up order q^{11}. A proof of this equality for all values of k is now available [AFLT].

It furthermore follows easily from (2.8) that the partition function $\mathbf{Z}(m; \tau; \epsilon_1, \epsilon_2)$ defined in (2.2) can be represented up to multiplication with an inessential, explicitly known function as

$$\mathbf{Z}(m; \tau; \epsilon_1, \epsilon_2) \propto \left\langle e^{2\alpha_4\phi(\infty)} e^{2\alpha_3\phi(1)} e^{2\alpha_2\phi(q)} e^{2\alpha_1\phi(0)} \right\rangle_b^{\text{Liou}}. \tag{2.10}$$

The relations between certain $\mathcal{N} = 2$ supersymmetric gauge theories and Liouville theory are most clearly formulated in terms of the normalized expectation values of loop-observables

$$\langle\!\langle \mathcal{L} \rangle\!\rangle_{S^4} := \frac{\langle \mathcal{L} \rangle_{S^4}}{\langle 1 \rangle_{S^4}}. \tag{2.11}$$

To this aim let us note that the counterparts of the loop observables within Liouville theory will be certain nonlocal observables of the form

$$\mathsf{L}_\gamma := \text{tr}\left[\mathcal{P}\exp\left(\int_\gamma \mathcal{A}_y\right)\right], \tag{2.12}$$

---

[8] A concise description of the definition of the conformal blocks can be found in ([V:12], Sect. 2.5).

where γ is a simple closed curve on $\mathbb{C} \backslash \{0, q, 1\}$, and $\mathcal{A}$ is the flat connection

$$\mathcal{A} := \begin{pmatrix} -\frac{b}{2}\partial_z\phi & 0 \\ \mu e^{b\phi} & \frac{b}{2}\partial_z\phi \end{pmatrix} dz + \begin{pmatrix} \frac{b}{2}\partial_{\bar{z}}\phi & \mu e^{b\phi} \\ 0 & -\frac{b}{2}\partial_{\bar{z}}\phi \end{pmatrix} d\bar{z}. \tag{2.13}$$

Flatness of $\mathcal{A}$ follows from the equation of motion. Let us furthermore define normalized expectation values in Liouville theory schematically as

$$\langle\!\langle \mathcal{O} \rangle\!\rangle_b^{\text{Liou}} := \frac{\langle \mathcal{O} \, e^{2\alpha_4\phi(\infty)} e^{2\alpha_3\phi(1)} e^{2\alpha_2\phi(q)} e^{2\alpha_1\phi(0)} \rangle_b^{\text{Liou}}}{\langle e^{2\alpha_4\phi(\infty)} e^{2\alpha_3\phi(1)} e^{2\alpha_2\phi(q)} e^{2\alpha_1\phi(0)} \rangle_b^{\text{Liou}}}. \tag{2.14}$$

We then have

$$\langle\!\langle \mathcal{W} \rangle\!\rangle_{S^4} = \langle\!\langle \mathsf{L}_{\gamma_s} \rangle\!\rangle_b^{\text{Liou}}, \qquad \langle\!\langle \mathcal{T} \rangle\!\rangle_{S^4} = \langle\!\langle \mathsf{L}_{\gamma_t} \rangle\!\rangle_b^{\text{Liou}}, \tag{2.15}$$

where γ_s and γ_t are the simple closed curves encircling the pairs of points $0, q$ and $1, q$ on $\mathbb{C} \backslash \{0, q, 1\}$, respectively. A more detailed discussion can be found in [V:7, V:12].

2.3 Relation to Topological Quantum Field Theory

The localisation method is also applicable in the case when the manifold M^4 has the form $M^3 \times S^1$ with supersymmetric boundary conditions for the fermions on the S^1. In this case the partition function coincides with a quantity called index [Ro, KMMR], a trace $\text{tr}(-1)^F \prod_i \mu_i^{C_i} e^{-\beta\{Q, Q^\dagger\}}$ over the Hilbert space of the theory on $M^3 \times \mathbb{R}$, with F being the fermion number operator, Q being one of the supersymmetry generators, and C_i being operators commuting with Q. The index depends on parameters μ_i called fugacities. It has originally been used to perform nontrivial checks of existing duality conjectures on field theories with $\mathcal{N} = 1$ supersymmetry [DOs, SpV]. We will in the following restrict attention to cases where the field theories have $\mathcal{N} = 2$ supersymmetry which are more closely related to the rest of the material discussed in this special volume.

As before one may use the localisation method to express the path integral for such manifolds as an integral over the zero modes of certain fields, with integrands obtainable by simple one loop computations. This partition function can alternatively be computed by counting with signs and weights certain protected operators in a given theory. If, for example, one takes $M^3 = S^3$, the partition function of an $\mathcal{N} = 2$ gauge theory with gauge group G and N_f fundamental hypermultiplets takes the following form,

$$I(\mathbf{b}; p, q, t) = \oint [d\mathbf{a}]_G \, I_V(\mathbf{a}; p, q, t) \prod_{\ell=1}^{N_f} I_H(\mathbf{a}, \mathbf{b}; p, q, t), \tag{2.16}$$

where $[d\mathbf{a}]_G$ is the invariant Haar measure, we are using the notation $\{p, q, t, \mathbf{b}\}$ for the relevant fugacities, and I_V and I_H are contributions coming from free vector multiplets and hypermultiplets, respectively. The integral over $\mathbf{a}$ is roughly over the zero mode of the component of the gauge field in the S^1 direction. For more details see the article [V:9]. It is important to note that the supersymmetric partition functions on $M^3 \times S^1$ are independent of the coupling constants by an argument going back to [W88]. Nevertheless, they are in general intricate functions of the fugacities and encode a lot of information about the protected spectrum of the theory.

There exists a relationship between the supersymmetric partition function on $M^3 \times S^1$, the supersymmetric index, on the one hand, and a topological field theory in two dimensions on the other hand [GPRR] which is somewhat analogous to the relation of the S^4 partition function to Liouville theory discussed above. Let us consider the example discussed above, $\mathcal{N} = 2$ supersymmetric $SU(2)$ gauge theory with $N_f = 4$. The supersymmetric index of this theory can be represented in the form (2.16) noted above. In the particular case when the fugacities are chosen to satisfy $t = q$, this index is equal [GRRY11] to a four point correlation function in a topological quantum field theory (TQFT) which can be regarded as a one-parameter deformation of two-dimensional Yang-Mills theory with gauge grow $SU(2)$ [AOSV],

$$I(b_1, b_2, b_3, b_4; p, q, t = q) = \prod_{\ell=1}^{4} \mathcal{K}(b_\ell; q) \sum_{\mathcal{R}=0}^{\infty} C_{\mathcal{R}}^2 \prod_{\ell=1}^{4} \chi_{\mathcal{R}}(b_\ell). \qquad (2.17)$$

Here $\chi_{\mathcal{R}}(x)$ is the character of a representation $\mathcal{R}$ of $SU(2)$. The parameters b_i are fugacities for the $\prod_{i=1} SU(2)_i$ maximal subgroup of the $SO(8)$ flavor symmetry group of the theory. This relation can be generalized to a large class of $\mathcal{N} = 2$ theories and to indices depending on more general sets of fugacities [GRRY13, GRR].

3 What Are the Exact Results Good for?

In the following we will briefly describe a few applications of the results outlined above that have deepened our insights into supersymmetric field theories considerably.

3.1 Quantitative Verification of Electric-Magnetic Duality Conjectures

The verification of the conjectures of Seiberg and Witten by the works [N, NO03, NY, BE] leads in particular to a verification of the electric-magnetic duality conjectures about the low energy effective theories that were underlying the approach taken by Seiberg and Witten.[9]

Verification of UV duality relations like the Montonen-Olive duality seems hopeless in general (see, however, [VW]). However, in the cases where exact results on expectation values are available, as briefly described in Sect. 2.2, one can do better.

In the case of the $N_f = 4$ theory, for example, one expects to find weakly coupled Lagrangian descriptions when the UV gauge coupling q is near 0, 1 or infinity [SW2]. Let us denote the actions representing the expansions around these three values as S_s, S_t and S_u, respectively. A particularly important prediction of the S-duality conjectures is the exchange of the roles of Wilson- and 't Hooft loops,

$$\langle\!\langle \mathcal{W} \rangle\!\rangle_{S_s} = \langle\!\langle \mathcal{T} \rangle\!\rangle_{S_t}, \qquad \langle\!\langle \mathcal{T} \rangle\!\rangle_{S_s} = \langle\!\langle \mathcal{W} \rangle\!\rangle_{S_t}. \tag{3.1}$$

In order to check (3.1) we may combine the results (2.3) of the localisation computations with the relation (2.8) discovered in [AGT]. From the study of the Liouville theory one knows that the conformal blocks satisfy relations such as

$$\mathcal{F}_p\!\begin{bmatrix}\alpha_3\,\alpha_2\\\alpha_4\,\alpha_1\end{bmatrix}\!(q) = \int dp' \, F_{p,p'}\!\begin{bmatrix}\alpha_3\,\alpha_2\\\alpha_4\,\alpha_1\end{bmatrix}\mathcal{F}_{p'}\!\begin{bmatrix}\alpha_1\,\alpha_2\\\alpha_4\,\alpha_3\end{bmatrix}\!(1-q), \tag{3.2}$$

which had been established in [Te01]. These relations may now be re-interpreted as describing a resummation of the instanton expansion defined by action S_s (the left hand side of (3.2)) into an instanton expansion defined by the dual action S_t. This resummation gets represented as an integral transformation with kernel $F_{p,p'}$. Using (2.3), (2.8), (3.2) and certain identities satisfied by the kernel $F_{p,p'}\!\begin{bmatrix}\alpha_3\,\alpha_2\\\alpha_4\,\alpha_1\end{bmatrix}$ established in [TV13], one may now verify explicitly that the S-duality relations (3.1) are indeed satisfied.

In other words: Conformal field theory provides the techniques necessary to resum the instanton expansion defined from a given action in terms of the instanton expansions defined from a dual action. At least for the class of theories at hand, these results confirm in particular the long-standing expectations that the instantons play a crucial role for producing the cross-over between weakly-coupled descriptions related by electric-magnetic dualities.

The electric-magnetic dualities can be also checked using the supersymmetric index. Although the index does not depend on the coupling constants, in different duality frames it is given by different matrix integrals. Duality implies that these two matrix integrals evaluate to the same expression. In the relation of the index

---

[9]The IR duality conjectures can be used to describe the moduli space of vacua as manifold covered by charts with local coordinates a_r, a_r^D. The transition functions between different charts define a Riemann-Hilbert problem. The solution to this problem defines the function $\mathcal{F}(a)$. It was shown in [N, NO03, NY, BE] that the series expansion of $\mathcal{F}(a)$ around one of the singular points on the (Footnote 9 continued)
moduli space of vacua satisfies (1.13). Taken together, one obtains a highly nontrivial check of the IR-duality conjectures underlying Seiberg-Witten theory.

to TQFT discussed above, invariance under duality transformations in many cases follows from the associativity property of the TQFT.

3.2 Precision Tests of AdS-CFT Duality

Another famous set of duality conjectures concerns the behaviour of supersymmetric gauge theories in the limit where the rank of the gauge group(s) tends to infinity [Ma], see [AGMOO] for a review. In this limit one expects to find a dual description in terms of the perturbative expansion of string theory on a background that is equal or asymptotic to five-dimensional Anti-de Sitter space. This duality predicts in some cases representations for the leading strong-coupling behaviour of some gauge-theoretical observables in terms of geometric quantities in supergravity theories.

Some impressive quantitative checks of these duality conjectures are known in the case of maximal supersymmetry $\mathcal{N} = 4$ based on the (conjectured) integrability of the $\mathcal{N} = 4$ supersymmetric Yang-Mills theory with infinite rank of the gauge group [Bei]. Performing similar checks for theories with less supersymmetry is much harder. It is therefore worth noting that the localisation calculations of partition functions and Wilson loop expectation values described above have been used to carry out quantitative checks of AdS-CFT type duality conjectures for some gauge theories with $\mathcal{N} = 2$ supersymmetry [BRZ, BEFP].

It seems quite possible that the exact results described above can be used to carry out many further precision tests of the AdS-CFT duality for $\mathcal{N} = 2$ supersymmetric field theories. The relevant backgrounds for string theory are not always known, but when they are known, one may use the results obtained by localisation to check these generalised AdS-CFT duality conjectures. Another result in this direction was recently reported in [MP14a].

The exact results can furthermore be used to study the phase structure of these gauge theories in the planar limit as function of the 't Hooft coupling. A surprisingly rich structure is found in [RZ13a, RZ13b]. It seems that the full physical content of most of the data provided by the localisation calculations remains to be properly understood.

3.3 Evidence for the Existence of Six-Dimensional Theories with (2,0)-Supersymmetry

Low-energy limits of string theory can often be identified with conventional quantum field theories. The string theorist's toolkit contains a large choice of objects to play with, the most popular being compactifications and branes. One sometimes expects the existence of a low-energy limit with a certain amount of supersymmetry, but there is no known quantum field theory the limit could correspond to. Such a line of reasoning has led to the prediction that there exists a very interesting class of interacting quantum field theories with six-dimensional $(2, 0)$-superconformal invariance [W95a, St, W95b]. These theories have attracted a lot of attention over the last two decades, but not even the field content, not to speak of a Lagrangian, are known for these hypothetical theories, see [Sei, W09] for reviews of what is known.

Nevertheless, the mere existence of such theories leads to highly non-trivial predictions, many of which have been verified directly. One could, for example, study the six-dimensional $(2, 0)$-theories on manifolds of the form $M^4 \times C$, where C is a Riemann surface [Ga09, GMN2] (some aspects are reviewed in [V:2, V:3]). If C has small area, one expects that the theory has an effective description in terms of a quantum field theory on M^4. The resulting quantum field theory $\mathcal{G}_C$ is expected to depend only on the choice of a hyperbolic metric on C [ABBR], or equivalently (via the uniformisation theorem) on the choice of a complex structure on C. The $N_f = 4$-theory with four flavours mentioned above, for example, corresponds to $C = C_{0,4}$, which may be represented as Riemann sphere with four marked points at $0, 1, q, \infty$. One may use q as parameter for the complex structure of $C_{0,4}$. When q is near $0, 1, \infty$, respectively, it is natural to decompose $C_{0,4}$ into two pairs of pants by cutting along contours surrounding the pairs of marked points $(0, q)$, $(q, 1)$ and (q, ∞), respectively. It turns out that q can be identified with the function $e^{2\pi i \tau}$ of the complexified gauge coupling $\tau = \frac{4\pi i}{g^2} + \frac{\theta}{2\pi}$ of the four-dimensional theory. The limits where q approaches $0, 1$ and ∞ are geometrically very similar, but $q \to 0$ corresponds to small gauge coupling, while $q \to 1$ would correspond to a strong coupling limit. Note, on the other hand, that the marked points at 0 and 1, for example, can be exchanged by a conformal mapping. This already suggests that there might exist a dual description having a complexified gauge coupling τ' such that $q' = e^{2\pi i \tau'}$ vanishes when $q \to 1$. The results described above provide a rather non-trivial quantitative check for this prediction.

Playing with the choice of C, and with the choice of the Lie algebra $\mathfrak{g}$ one can generate a large class of interesting four-dimensional quantum field theories, and predict many non-trivial results about their physics [Ga09, GMN2]. The class of theories obtained in this way is often called class $\mathcal{S}$. Arguments of this type can be refined sufficiently to predict correspondences between four-dimensional gauge theories on M^4 and two-dimensional conformal field theories on C generalising the relations discovered in [AGT], see [Y12, CJ14]. In the resulting generalisations of the relation (2.10) one will find the correlation functions of the conformal Toda theory associated to $\mathfrak{g}$ on the Riemann surface C, in general. Considering the cases where $M^4 = M^3 \times S^1$, one may use similar arguments to predict that the partition functions are related to correlation functions in a TQFT on C, generalising the example noted in Sect. 2.3. Such correlation functions only depend on the topology of C, corresponding to the fact that the partition functions on $M^4 = M^3 \times S^1$ are independent of exactly marginal coupling constants. This will be discussed in more detail in [V:9].

Other compactifications are also interesting, like $M^3 \times C^3$ or $M^2 \times C^4$, where C^3 and C^4 are compact three- and four-dimensional manifolds. Compactifying on C^3 or C^4 one gets interesting quantum field theories on three- or two-dimensional manifolds M^3 or M^2, respectively. The origin from the six-dimensional $(2, 0)$-theory may again be used to predict various nontrivial properties of the resulting quantum field theories, including relations between three-dimensional field theories on M^3 and complex Chern-Simons theory on C^3 [Y13, LY, CJ13]. Such relations are further discussed in [V:11].

The six-dimensional $(2, 0)$-theories are for $d = 2, 3, 4$-dimensional quantum field theory therefore something like "Eierlegende Wollmilchsäue", mythical beasts capable of supplying us with eggs, wool, milk and meat at the same time. The steadily growing number of highly nontrivial checks that the predictions following from its existence have passed increase our confidence that such six-dimensional theories actually exist. Their existence supplies us with a vantage point from which we may get a better view on interesting parts of the landscape of supersymmetric quantum field theories.

3.4 Towards Understanding Non-Lagrangian Theories

There are many cases where strong-coupling limits of supersymmetric field theories are expected to exist and to have a quantum field-theoretical nature, but no Lagrangian description of the resulting theories is known [AD, APSW, AS]. The existence of non-Lagrangian theories is an interesting phenomenon by itself. Certain non-Lagrangian quantum field theories are expected to serve as elementary building blocks for the family of quantum field theories obtained by compactifying $(2, 0)$-theories [Ga09, CD].

The origin from a six-dimensional theory allows us to make quantitative predictions on some physical quantities of such non-Lagrangian theories including the prepotential giving us the low-energy effective action, and the supersymmetric index giving us the protected spectrum of the theory.

The results described in this special volume open the exciting perspective to go much further in the study of some non-Lagrangian theories. If the relation with two-dimensional conformal field theories continues to hold in the cases without known Lagrangian descriptions, one may, for example compute the partition functions and certain finite-volume expectation values of loop operators in such theories. First steps in this direction were made in [BMT, GT, KMST].

3.5 Interplay Between (topological) String Theory and Gauge Theory

Superstring theory compactified on Calabi-Yau manifolds has two "topological" relatives called the A- and the B-model respectively. The "topological" relatives are much simpler than the full superstring theories, but they capture important information about the full theories like the coefficients of certain terms in the corresponding space-time effective actions governing the low-energy physics. The A- and the B-model are not independent but related by mirror symmetry.

The definition of the B-model topological string theory can be extended to so-called local Calabi-Yau Y, defined (locally) by equations of the form

$$zw - P(u, v) = 0, \tag{3.3}$$

with $P(u, v)$ being a polynomial. Superstring theories on such local Calabi-Yau manifolds are expected to have decoupling limits in which they are effectively represented by four-dimensional gauge theories.[10] Describing four-dimensional gauge theory as decoupling limits of superstring theory is called geometric engineering [KLMVW, KKV, KMV]. String-theoretic arguments [N, LMN] predict that the instanton partition function (for $\epsilon_1 + \epsilon_2 = 0$) coincides with the topological string partition function $\mathcal{Z}^{top}$ of the B-model on Y, schematically

$$\mathcal{Z}^{inst} = \lim_{\beta \to 0} \mathcal{Z}^{top}, \tag{3.4}$$

where β is related to one of the parameters for the complex structures on Y. This prediction has been verified in various examples [IK02, IK03, EK, HIV]. It opens channels for the transport of information and insights from topological string theory to gauge theory and back. Interesting perspectives include:

- Results from topological string theory may help to understand 4d gauge theories even better, possibly including non-Lagrangian ones. To give an example, let us note that the topological vertex [AKMV, IKV] gives powerful tools for the calculation of topological string partition functions. These results give us predictions for the (yet undefined) instanton partition functions of non-Lagrangian theories, and may thereby provide a starting point for future studies of the physics of such theories. First steps in this direction were made in [KPW, BMPTY, HKN, MP14b].
- Exact results on supersymmetric gauge theories may feed back to topological string theory. As an example let us mention the development of the refined topological string, a one-parameter deformation of the usual topological string theory that appears to exist for certain local Calabi-Yau manifolds, capturing nontrivial additional information. The proposal was initially motivated by the observation that the instanton partition functions can be defined for more general values of the parameters ϵ_1, ϵ_2 than the case $\epsilon_1 + \epsilon_2 = 0$ corresponding to the usual topological string via geometric engineering [IKV, KW, HK, HKK]. There is growing evidence that such a deformation of the topological string has a world sheet realisation [AFHNZa, AFHNZb], and that the refinement fits well into the conjectured web of topological string/gauge theory dualities [AS12a, AS12b, CKK, NO14]. The relation with the holomorphic anomaly equation is reviewed in [V:14].

---

[10]This limit is easier to define in the A-model, but the definition can be translated to the B-model using mirror symmetry.

- As another interesting direction that deserves further investigations let us note that the topological string partition functions $\mathcal{Z}^{top}$ can be interpreted as particular wave-functions in the quantum theory obtained by the quantisation of the moduli space of complex structures on Calabi-Yau manifolds, as first pointed out in [W93], see [ST] and references therein for further developments along these lines. By using the holomorphic anomaly equation one may construct $\mathcal{Z}^{top}$ as formal series in the topological string coupling constant λ, identified with Planck's constant $\hbar$ in the quantisation of the moduli spaces of complex structures. However, it is not known how to define $\mathcal{Z}^{top}$ non-perturbatively in λ.

  On the other hand it was pointed out above that the instanton partition functions are naturally interpreted as wave-functions in some effective zero mode quantum mechanics to which the gauge theories in question can be reduced by the local-isation method. It seems likely that the effective zero mode quantum mechanics to which the gauge theories localise simply coincide with the quantum mechanics obtained from the quantisation of the moduli spaces of complex structures which appear in the geometric engineering of the gauge theories under considerations. These moduli spaces are closely related to the moduli spaces of flat connections on Riemann surfaces for the A_1 theories of class $\mathcal{S}$. The quantisation of these moduli spaces is understood for some range of values of ϵ_1, ϵ_2 [V:12]. Interpreting the results obtained thereby in terms of (refined) topological string theory may give us important insights on how to construct $\mathcal{Z}^{top}$ non-perturbatively, at least for many local Calabi-Yau-manifolds.

4 What Is Going to Be Discussed in This Volume?

Let us now give an overview of the material covered in this volume.

Chapter "Families of $N = 2$ Field Theories" [V:2] by Gaiotto describes how large families of field theories with $\mathcal{N} = 2$ supersymmetry can be described by means of Lagrangian formulations, or by compactification from the six-dimensional theory with $(2, 0)$ supersymmetry on spaces of the form $M^4 \times C$, with C being a Riemann surface. The class of theories that can be obtained in this way is called class $\mathcal{S}$. This description allows us to relate key aspects of the four-dimensional physics of class $\mathcal{S}$ theories to geometric structures on C.

The next chapter in our volume is titled "Hitchin Systems in $\mathcal{N} = 2$ Field Theory" by Neitzke [V:3]. The space of vacua of class $\mathcal{S}$ theories on $\mathbb{R}^3 \times S^1$ can be identified as the moduli space of solutions to the self-duality equations in two dimensions on Riemann surfaces studied by Hitchin. This fact plays a fundamental role for recent studies of the spectrum of BPS states in class $\mathcal{S}$ theories, and it is related to the integrable structure underlying Seiberg-Witten theory of theories of class $\mathcal{S}$. This article reviews important aspects of the role of the Hitchin system for the infrared physics of class $\mathcal{S}$ theories.

In Chapter "A Review on Instanton Counting and W-algebras" by Tachikawa [V:4], it is explained how to compute the instanton partition functions. The results can be written as sums over bases for the equivariant cohomology of instanton moduli spaces. The known results relating the symmetries of these spaces to the symmetries of conformal field theory are reviewed.

Chapter "β-Deformed Matrix Models and the 2d/4d Correspondence" by Maruyoshi [V:5] describes a very useful mathematical representation of the results of the localisation computations as integrals having a form familiar from the study of matrix models. Techniques from the study of matrix models can be employed to extract important information on the instanton partition functions in various limits and special cases.

Chapter "Localization for $\mathcal{N} = 2$ Supersymmetric Gauge Theories in Four Dimensions" by Pestun [V:6] describes the techniques necessary to apply the localisation method to field theories on curved backgrounds like S^4, and how some of the results on partition functions outlined in Sect. 2.1 have been obtained.

Chapter "Line Operators in Supersymmetric Gauge Theories and the 2D-4D Relation" by Okuda [V:7] it is discussed how to use localisation techniques for the calculation of expectation values of Wilson and 't Hooft line operators. The results establish direct connections between supersymmetric line operators in gauge theories and the Verlinde line operators known from conformal field theory. Similar results can be used to strongly support connections to the quantum theories obtained from the quantisation of the Hitchin moduli spaces.

Chapter "Surface Operators" by Gukov [V:8] discusses a very interesting class of observables localised on surfaces that attracts steadily growing attention. In the correspondence to conformal field theory some of these observables get related to a class of fields in two dimensions called degenerate fields. These fields satisfy differential equations that can be used to extract a lot of information on the correlation functions. Understanding the origin of these differential equations within gauge theory may help explaining the AGT-correspondence itself.

There are further interesting quantities probing aspect of the non-perturbative physics of theories of class $\mathcal{S}$. Chapter "The Superconformal Index of Theories of Class $\mathcal{S}$" by Rastelli and Razamat [V:9] reviews the superconformal index. It is often simpler to calculate than instanton partition functions, but nevertheless allows one to perform many nontrivial checks of conjectured dualities. It turns out to admit a representation in terms of a new type of topological field theory associated to the Riemann surfaces C parameterising the class $\mathcal{S}$ theories.

The correspondence between four-dimensional supersymmetric gauge theories and two-dimensional conformal field theories discovered in [AGT] has a very interesting relative, a correspondence between three-dimensional gauge theories and three-dimensional Chern-Simons theories with complex gauge group. It is related to the the correspondence of [AGT], but of interest in its own right. In order to see relations with the AGT-correspondences one may consider four-dimensional field theories of class $\mathcal{S}$ on half-spaces separated by three-dimensional defects. The partition functions of the three-dimensional gauge theories on the defect turns out to be calculable by means of localisation, and the results have a deep meaning within conformal field theory or within the quantum theory of Hitchin moduli spaces. How to apply the localisation method to (some of) the three-dimensional gauge theories that appear in this correspondence is explained in the Chapter "A Review on SUSY Gauge Theories on S^3" by Hosomichi [V:10]. The correspondences between three-dimensional gauge theories and three-dimensional Chern-Simons theory with complex gauge group are the subject of Chapter "3D Superconformal Theories from Three-Manifolds" by Dimofte [V:11].

Chapter "Supersymmetric Gauge Theories, Quantization of $\mathcal{M}_{\text{flat}}$, and Liouville Theory" by the author [V:12] describes an approach to understanding the AGT-correspondence by establishing a triangle of relations between the zero mode quantum mechanics obtained by localisation of class S theories, the quantum theory obtained by quantisation of Hitchin moduli spaces, and conformal field theory. This triangle offers an explanation for the relations discovered in [AGT].

Some aspects of the string-theoretical origin of these results are discussed in the final chapters of our volume.

Chapter "Gauge/Vortex Duality and AGT" by Aganagic and Shakirov, [V:13], describes one way to understand an important part of the AGT-correspondence in terms of a triality between four-dimensional gauge theory, the two-dimensional theory of its vortices, and conformal field theory. This triality is related to, and inspired by known large N dualities of the topological string. It leads to a proof of some cases of the AGT-correspondence, and most importantly, of a generalisation of this correspondence to certain five-dimensional gauge theories.

Chapter "B-Model Approaches to Instanton Counting", Krefl and Walcher [V:14] discusses the relation between the instanton partition functions and the partition function of the topological string from the perspective of the B-model. The instanton partition functions provide solutions to the holomorphic anomaly equations characterising the partition functions of the topological string.

5 What Is Missing?

This collection of articles can only review a small part of the exciting recent progress on $N = 2$ supersymmetric field theories. Many important developments in this field could not be covered here even if they are related to the material discussed in our collection of articles in various ways. In the following we want to indicate some of the developments that appear to have particularly close connections to the subjects discussed in this volume.

5.1 BPS Spectrum, Moduli Spaces of Vacua and Hitchin Systems

BPS states are states in the Hilbert space of a supersymmetric field theory which are forming distinguished "small" representations of the supersymmetry algebra, a feature which excludes various ways of "mixing" with generic states of the spectrum that would exist otherwise. Gaiotto, Moore and Neitzke have initiated a vast program aimed at the study of the spectrum of BPS-states in the $\mathcal{N} = 2$ supersymmetric gauge theories $\mathcal{G}_C$ of class $\mathcal{S}$ [GMN1, GMN2, GMN3], see the article [V:3] for a review of some aspects. To this aim it has turned out to be useful to consider at intermediate steps of the analysis a compactification of the theories $\mathcal{G}_C$ to space-times of the form $\mathbb{R}^3 \times S^1$. The moduli space of vacua of the compactified theory is "twice as large" compared to the one of $\mathcal{G}_C$ on $\mathbb{R}^4$, and it can be identified with Hitchin's moduli space of solutions to the self-duality equations on Riemann surfaces [Hi].

The list of beautiful results that has been obtained along these lines includes:

- A new algorithm for computing the spectrum of BPS states which has a nontrivial, but piecewise constant dependence on the point on the Coulomb-branch of the moduli space of vacua of $\mathcal{G}_C$ on $\mathbb{R}^4$. The spectrum of BPS states may change along certain "walls" in the moduli space of vacua. Knowing the spectrum on one side of the wall one may compute how it looks like on the other side using the so-called wall-crossing formulae. Similar formulae were first proposed in the work of Kontsevich and Soibelman on Donaldson-Thomas invariants.
- Considering the gauge theory $\mathcal{G}_C$ compactified on $\mathbb{R}^3 \times S^1$ one may study natural line operators including supersymmetric versions of the Wilson- or 't Hooft loop observables. Such observables can be constructed using either the fields of the UV Lagrangian, or alternatively those of a Wilsonian IR effective action. The vacuum expectation values of such line operators furnish coordinates on the moduli space $\mathcal{M}$ of vacua of $\mathcal{G}_C$ on $\mathbb{R}^3 \times S^1$ which turn out to coincide with natural sets of coordinates for Hitchin's moduli spaces. The coordinates associated to observables defined in the IR reveal the structure of Hitchin's moduli spaces as a cluster algebra, closely related to the phenomenon of wall-crossing in the spectrum of BPS-states. Considering the observables constructed from the fields in the UV-Lagrangian one gets coordinates describing the Hitchin moduli spaces as algebraic varieties. The relation between these sets of coordinates is the renormalisation group (RG) flow, here protected by supersymmetry, and therefore sometimes calculable [GMN3].

Even if the main focus of this direction of research is the spectrum of BPS-states, it turns out to deeply related to the relations discovered in [AGT], as is briefly discussed in [V:12].

5.2 Relations to Integrable Models

It has been observed some time ago that the description of the prepotentials characterising the low-energy physics of $\mathcal{N} = 2$ supersymmetric field theories provided by Seiberg-Witten theory is closely connected to the mathematics of integrable systems [GKMMM, MW, DW]. There are arguments indicating that such relations to integrable models are generic consequences of $\mathcal{N} = 2$ supersymmetry: $\mathcal{N} = 2$ super-symmetry implies that the Coulomb branch of vacua carries a geometric structure called special geometry. Under certain integrality conditions related to the quantisation of electric and magnetic charges of BPS states one may canonically construct an integrable system in action-angle variables from the data characterising the special geometry of the Coulomb branch [Fr].

The connections between four-dimensional field theories with $\mathcal{N} = 2$-supersymmetry and integrable models have been amplified enormously in a recent series of papers starting with [NSc].[11] It was observed that a *partial* Omega-deformation of many $\mathcal{N} = 2$ field theories localised only on one of the half-planes spanning $\mathbb{R}^4$ is related to the quantum integrable model obtained by quantising the classical integrable model related to Seiberg-Witten theory. The Omega-deformation effectively localises the fluctuations to the origin of the half plane. This can be used to argue that the low-energy physics can be effectively represented by a two-dimensional theory with $(2, 2)$-supersymmetry [NSc] living on the half-plane in $\mathbb{R}^4$ orthogonal to the support of the Omega-deformation. The supersymmetric vacua of the four-dimensional theory are determined by the twisted superpotential of the effective two-dimensional theory which can be calculated by taking the relevant limit of the instanton partition functions. This was used in [NSc] to argue that the supersymmetric vacua are in one-to-one correspondence with the eigenstates of the quantum integrable model obtained by quantising the integrable model corresponding to the Seiberg-Witten theory of the four-dimensional gauge theory under consideration.

This line of thought has not only lead to many new exact results on large families of four-dimensional $\mathcal{N} = 2$ gauge theories [NRS, NP, NPS], it has also created a new paradigm for the solution of algebraically integrable models. More specifically

- For gauge theories $\mathcal{G}_C$ of class $\mathcal{S}$ an elegant description for the two-dimensional superpotential characterising the low-energy physics in the presence of a partial Omega-deformation was given in [NRS] in terms of the mathematics of certain flat connections called opers living on the Riemann C specifying the gauge theory $\mathcal{G}_C$.

- In [NP] the instanton calculus was generalised to a large class of $\mathcal{N} = 2$ gauge theories $\mathcal{G}_\Gamma$ parameterised by certain diagrams Γ called quivers. A generalisation of the techniques from [NO03] allowed the authors to determine Seiberg-Witten type descriptions of the low-energy physics for all these theories, and to identify the integrable models whose solution theory allows one to calculate the corresponding prepotentials.

- The subsequent work [NPS] generalised the results of [NP] to the cases where one has a one-parametric Omega-deformation preserving two-dimensional supersymmetry. The results of [NPS] imply a general correspondence between certain supersymmetric observables and the generating functions of conserved quantities in the models obtained by quantising the integrable models describing the generalisations of Seiberg-Witten-theory relevant for the gauge theories $\mathcal{G}_\Gamma$.

---

[11] This paper is part of a program initiated in [NSa, NSb] investigating even more general connections between field theories with $\mathcal{N} = 2$-supersymmetry and integrable models.

The relations between these developments and the relations discovered in [AGT] deserve further studies. One of the existing relations for A_1-theories of class S is briefly discussed in the article [V:12]. These results suggest that the AGT-correspondence and many related developments can ultimately be understood as consequences of the integrable structure in $\mathcal{N} = 2$ supersymmetric field theories. This point of view is also supported by the relations between the quantisation of Hitchin moduli spaces and conformal field theory described in [Te10].

5.3 Other Approaches to the AGT-Correspondence

In this special volume we collect some of the basic results related to the AGT-correspondence. The family of results on this subject is rapidly growing, and many important developments have occurred during the preparation of this volume. The approaches to proving or deriving the AGT-correspondences and some generalisations include

- Representation-theoretic proofs [AFLT, FL, BBFLT] that W-algebra conformal blocks can be represented in terms of instanton partition functions. This boils down to proving existence of a basis for W-algebra modules in which the matrix elements of chiral vertex operators coincide with the so-called bifundamental contributions representing the main building blocks of instanton partition functions.
- Another approach [MMS, MS] establishes relations between the series expansions for the instanton partition functions and the expressions provided by the free field representation for the conformal blocks developed by Feigin and Fuchs, and Dotsenko and Fateev.
- Mathematical proofs [SchV, MaOk, BFN] that the cohomology of instanton moduli spaces naturally carries a structure as a W-algebra module. This leads to a proof of the versions of the AGT-correspondence relevant for pure $\mathcal{N} = 2$ supersymmetric gauge theories for all gauge groups of type A, D or E. The instanton partition functions get related to norms of Whittaker vectors in modules of W-algebras in these cases. For a physical explanation of this fact see [Tan]. Some aspects of this approach are described in [V:4].
- Physical arguments leading to the conclusion that the six-dimensional $(2, 0)$-theory on certain compact four-manifolds or on four-manifolds M^4 with Omega-deformation can effectively be represented in terms of two-dimensional conformal field theory [Y12, CJ14], or as a $(2, 2)$-supersymmetric sigma model with Hitchin target space [NW].
- Considerations of the geometric engineering of supersymmetric gauge theories within string theory have led to the suggestion that the instanton partition functions of the gauge theories from class S should be related to the partition functions of chiral free fermion theories on suitable Riemann surfaces [N], see [ADKMV,

DHSV, DHS] for related developments.[12] It was proposed in [CNO] that the relevant theory of chiral free fermions is defined on the Riemann surface C specifying the gauge theories $\mathcal{G}_C$ of class $\mathcal{S}$. These relations were called BPS-CFT correspondence in [CNO]. A mathematical link between BPS-CFT correspondence and the AGT-correspondence was exhibited in [ILTe].

5.4 Less Supersymmetry

A very interesting direction of possible future research concerns possible generalisations of the results discussed here to theories with less ($\mathcal{N} = 1$) supersymmetry. Recent progress in this direction includes descriptions of the moduli spaces of vacua resembling the one provided by Seiberg-Witten theory for field theories with $\mathcal{N} = 2$ supersymmetry.

The rapid growth of the number of publications on this direction of research makes it difficult to offer a representative yet concise list of references on this subject here.

Acknowledgments The author is grateful to D. Krefl, K. Maruyoshi, E. Pomoni, L. Rastelli, S. Razamat, Y. Tachikawa and J. Walcher for very useful comments and suggestions on a previous draft of this article.

References

[ABBR]   Andersen, M., Beem, C., Bobev, N., Rastelli, L.: Holographic uniformization. Commun. Math. Phys. **318**, 429–471 (2013). arXiv:1109.3724

[AD]   Argyres, P.C., Douglas, M.R.: New phenomena in SU(3) supersymmetric gauge theory. Nucl. Phys. **B448**, 93–126 (1995). arXiv:hep-th/9505062

[ADKMV]   Aganagic, M., Dijkgraaf, R., Klemm, A., Marino, M., Vafa, C.: Topological strings and integrable hierarchies. Commun. Math. Phys. **261**, 451–516 (2006). arXiv:hep-th/0312085

[AFHNZa]   Antoniadis, I., Florakis, I., Hohenegger, S., Narain, K.S., Zein Assi, A.: Worldsheet realization of the refined topological string. Nucl. Phys. **B875**, 101–133 (2013). arXiv:1302.6993

[AFHNZb]   Antoniadis, I., Florakis, I., Hohenegger, S., Narain, K.S., Zein Assi, A.: Nonperturbative Nekrasov partition function from string theory. Nucl. Phys. **B880**, 87–108 (2014). arXiv:1309.6688

[AFLT]   Alba, V.A., Fateev, V.A., Litvinov, A.V., Tarnopolsky, G.M.: On combinatorial expansion of the conformal blocks arising from AGT conjecture. Lett. Math. Phys. **98**, 33–64 (2011). arXiv:1012.1312

[AGMOO]   Aharony, O., Gubser, S.S., Maldacena, J., Ooguri, H., Oz, Y.: Large N field theories, string theory and gravity. Phys. Rep. **323**, 183–386 (2000). arXiv:hep-th/9905111

---

[12]The relations between the topological vertex and free fermion theories discussed in [ADKMV] imply general relations between topological string partition functions of local Calabi-Yau manifolds, integrable models and theories of free fermions on certain Riemann surfaces; possible implications for four-dimensional gauge theories were discussed in [DHSV, DHS].

[AGT]   Alday, L.F., Gaiotto, D., Tachikawa, Y.: Liouville correlation functions from four-dimensional gauge theories. Lett. Math. Phys. **91**, 167–197 (2010). arXiv:0906.3219

[AKMV]   Aganagic, M., Klemm, A., Marino, M., Vafa, C.: The topological vertex. Commun. Math. Phys. **254**, 425–478 (2005). arXiv:hep-th/0305132

[AOSV]   Aganagic, M., Ooguri, H., Saulina, N., Vafa, C.: Black holes, q-deformed 2d Yang-Mills, and non-perturbative topological strings. Nucl. Phys. B **715**, 304 (2005). arXiv:hep-th/0411280

[APSW]   Argyres, P.C., Plesser, M.R., Seiberg, N., Witten, E.: New $N = 2$ superconformal field theories in four-dimensions. Nucl. Phys. **B461**, 71–84 (1996). arXiv:hep-th/9511154

[AS]   Argyres, P.C., Seiberg, N.: S-duality in $N = 2$ supersymmetric gauge theories. JHEP **0712**, 088 (2007). arXiv:0711.0054

[AS12a]   Aganagic, M., Schaeffer, K.: Refined black hole ensembles and topological strings. JHEP **1301**, 060 (2013). arXiv:1210.1865

[AS12b]   Aganagic, M., Shakirov, S.: Refined Chern-Simons theory and topological string. Preprint arXiv:1210.2733 [hep-th]

[BBFLT]   Belavin, A.A., Bershtein, M.A., Feigin, B.L., Litvinov, A.V., Tarnopolsky, G.M.: Instanton moduli spaces and bases in coset conformal field theory. Commun. Math. Phys. **319**, 269–301 (2013). arXiv:1111.2803

[BE]   Braverman, A., Etingof, A.: Instanton Counting Via Affine Lie Algebras. II. From Whittaker Vectors to the Seiberg-Witten Prepotential. Studies in Lie Theory. Progress in Mathematics, vol. 243, pp. 61–78. Birkhäuser, Boston (2006). arXiv:math/0409441

[BEFP]   Bobev, N., Elvang, H., Freedman, D.Z., Pufu, S.S.: Holography for $N = 2$ on S^4. JHEP **1407**, 1 (2014). arXiv:1311.1508

[Bei]   Beisert, N., et al.: Review of AdS/CFT integrability: an overview. Lett. Math. Phys. **99**, 3 (2012). arXiv:1012.3982

[BFMT]   Bruzzo, U., Fucito, F., Morales, J.F., Tanzini, A.: Multiinstanton calculus and equivariant cohomology. JHEP **0305**, 054 (2003). arXiv:hep-th/0211108

[BFN]   Braverman, A., Finkelberg, M., Nakajima, H.: Instanton moduli spaces and W-algebras. Preprint arXiv:1406.2381

[Bi]   Bilal, A.: Duality in $N = 2$ SUSY SU(2) Yang-Mills theory: a pedagogical introduction to the work of Seiberg and Witten. In: G. 't Hooft et al. (eds.) Proceedings, NATO Advanced Study Institute, Quantum Fields and Quantum Space Time, Cargese, France, 22 July–3 August 1996. Plenum, New York (1997). arXiv:hep-th/9601007

[BMPTY]   Bao, L., Mitev, V., Pomoni, E., Taki, M., Yagi, F.: Non-Lagrangian theories from Brane junctions. JHEP **1401**, 175 (2014). arXiv:1310.3841

[BMT]   Bonelli, G., Maruyoshi, K., Tanzini, A.: Wild quiver gauge theories. JHEP **1202**, 031 (2012). arXiv:1112.1691

[BNP]   Baggio, M., Niarchos, V., Papadodimas, K.: Exact correlation functions in SU(2) $N = 2$ superconformal QCD. Preprint arXiv:1409.4217

[BPZ]   Belavin, A.A., Polyakov, A.M., Zamolodchikov, A.B.: Infinite conformal symmetry in two-dimensional quantum field theory. Nucl. Phys. **B241**, 333–380 (1984)

[BRZ]   Buchel, A., Russo, J.G., Zarembo, K.: Rigorous test of non-conformal holography: Wilson loops in $N = 2^*$ theory. JHEP **1303**, 062 (2013). arXiv:1301.1597

[CD]   Chacaltana, O., Distler, J.: Tinkertoys for Gaiotto duality. JHEP **1011**, 099 (2010). arXiv:1008.5203

[CJ13]   Cordova, C., Jafferis, D.L.: Complex Chern-Simons from M5-branes on the squashed three-sphere. Preprint arXiv:1305.2891

[CJ14]   Cordova, C., Jafferis, D.: Talk at Strings. Princeton, June 25 (2014)

[CKK]   Choi, J., Katz, S., Klemm, A.: The refined BPS index from stable pair invariants. Commun. Math. Phys. **328**, 903–954 (2014). arXiv:1210.4403

[CNO]   Carlsson, E., Nekrasov, N., Okounkov, A.: Five dimensional gauge theories and vertex operators. Moscow Math. J. **14**, 39–61 (2014). arXiv:1308.2465

[CT]   Curtright, T.L., Thorn, C.B.: Conformally invariant quantization of the Liouville theory. Phys. Rev. Lett. **48**, 1309 (1982) (Erratum-ibid. **48**, 1768 (1982))

[DHKM]   Dorey, Nick, Hollowood, Timothy J., Khoze, Valentin V., Mattis, Michael P.: The calculus of many instantons. Phys. Rept. **371**, 231–459 (2002). arXiv:hep-th/0206063

[DHS]   Dijkgraaf, R., Hollands, L., Sułkowski, P.: Quantum curves and $\mathcal{D}$-modules. JHEP **0911**, 047 (2009). arXiv:0810.4157

[DHSV]   Dijkgraaf, R., Hollands, L., Sułkowski, P., Vafa, C.: Supersymmetric gauge theories, intersecting branes and free fermions. JHEP **0802**, 106 (2008). arXiv:.0709.4446

[DOt]   Dorn, H., Otto, H.-J.: Two and three-point functions in Liouville theory. Nucl. Phys. B **429**, 375–388 (1994). arXiv:hep-th/9403141

[DOs]   Dolan, F.A., Osborn, H.: Applications of the superconformal index for protected operators and q-hypergeometric identities to $N = 1$ dual theories. Nucl. Phys. **B818**, 137–178 (2009). arXiv:0801.4947

[DPh]   D'Hoker, E., Phong, D.H.: Lectures on supersymmetric Yang-Mills theory and integrable systems. In: Saint-Aubin, Y., Vinet, L. (ed.) Proceedings. 9th CRM Summer School, Theoretical Physics at the End of the Twentieth Century, Banff, Canada, 1999. Springer, New York (2002). arXiv:hep-th/9912271

[DW]   Donagi, R., Witten, E.: Supersymmetric Yang-Mills theory and integrable systems. Nucl. Phys. **B460**, 299–334 (1996). arXiv:hep-th/9510101

[EK]   Eguchi, T., Kanno, H.: Topological strings and Nekrasov's formulas. JHEP **0312**, 006 (2003). arXiv:hep-th/0310235

[FL]   Fateev, A.V., Litvinov, A.V.: Integrable structure, W-symmetry and AGT relation. JHEP **1201**, 051 (2012). arXiv:1109.4042

[FP]   Flume, R., Poghossian, R.: An algorithm for the microscopic evaluation of the coefficients of the Seiberg-Witten prepotential. Int. J. Mod. Phys. **A18**, 2541 (2003). arXiv:hep-th/0208176

[FPS]   Flume, R., Poghossian, R., Storch, H.: The Seiberg-Witten prepotential and the Euler class of the reduced moduli space of instantons. Mod. Phys. Lett. **A17**, 327–340 (2002). arXiv:hep-th/0112211

[Fr]   Freed, D.: Special Kähler manifolds. Commun. Math. Phys. **203**, 31–52 (1999). arXiv:hep-th/9712042

[Ga09]   Gaiotto, D.: $N = 2$ dualities. JHEP **1208**, 034 (2012). arXiv:0904.2715

[GKMMM] Gorsky, A., Krichever, I., Marshakov, A., Mironov, A., Morozov, A.: Integrability and Seiberg-Witten exact solution. Phys. Lett. **B355**, 466–474 (1995). arXiv:hep-th/9505035

[GMN1]   Gaiotto, D., Moore, G., Neitzke, A.: Four-dimensional wall-crossing via three-dimensional field theory. Commun. Math. Phys. **299**, 163–224 (2010). arXiv:0807.4723

[GMN2]   Gaiotto, D., Moore, G., Neitzke, A.: Wall-crossing, Hitchin systems, and the WKB approximation. Adv. Math. **234**, 239–403 (2013). arXiv:0907.3987

[GMN3]   Gaiotto, D., Moore, G., Neitzke, A.: Framed BPS states. Adv. Theor. Math. Phys. **17**, 241–397 (2013). arXiv:1006.0146

[GNO]   Goddard, P., Nuyts, J., Olive, D.I.: Gauge theories and magnetic charge. Nucl. Phys. **B125**, 1 (1977)

[GOP]   Gomis, J., Okuda, T., Pestun, V.: Exact results for 't Hooft loops in gauge theories on S^4, JHEP **1205**, 141 (2012). arXiv:1105.2568

[GPRR]   Gadde, A., Pomoni, E., Rastelli, L., Razamat, S.S.: S-duality and 2d topological QFT. JHEP **1003**, 032 (2010). arXiv:0910.2225

[GRR]   Gaiotto, D., Rastelli, L., Razamat, S.S.: Bootstrapping the superconformal index with surface defects. JHEP **1301**, 022 (2013). arXiv:1207.3577

[GRRY11] Gadde, A., Rastelli, L., Razamat, S.S., Yan, W.: The 4d superconformal index from q-deformed 2d Yang-Mills. Phys. Rev. Lett. **106**, 241602 (2011). arXiv:1104.3850

[GRRY13] Gadde, A., Rastelli, L., Razamat, S.S., Yan, W.: Gauge theories and Macdonald polynomials. Commun. Math. Phys. **319**, 147 (2013). arXiv:1110.3740

[GT]   Gaiotto, D., Teschner, J.: Irregular singularities in Liouville theory and Argyres-Douglas type gauge theories. JHEP **1212**, 050 (2012). arXiv:1203.1052

[HH]   Hama, N., Hosomichi, K.: Seiberg-Witten theories on ellipsoids. JHEP **1209**, 033
          (2012). Addendum-ibid. **1210**, 051 (2012). arXiv:1206.6359

[Hi]   Hitchin, N.: The self-duality equations on a Riemann surface. Proc. Lond. Math. Soc.
          55(3), 59–126 (1987)

[HIV]   Hollowood, T.J., Iqbal, A., Vafa, C.: Matrix models, geometric engineering and elliptic
          genera. JHEP **0803**, 069 (2008). arXiv:hep-th/0310272

[HK]   Huang, M.-X., Klemm, A.: Direct integration for general Ω backgrounds. Adv. Theor.
          Math. Phys. **16**, 805–849 (2012). arXiv:1009.1126

[HKK]   Huang, M.-X., Kashani-Poor, A.-K., Klemm, A.: The Ω deformed B-model for rigid
          $\mathcal{N} = 2$ theories. Annales Henri Poincare **14**, 425–497 (2013). arXiv:1109.5728

[HKN]   Hayashi, H., Kim, H.-C., Nishinaka, T.: Topological strings and 5d TN partition func-
          tions. JHEP **1406**, 014 (2014). arXiv:1310.3854

[Ho1]   Hollowood, T.J.: Calculating the prepotential by localization on the moduli space of
          instantons. JHEP **03**, 038 (2002). arXiv:hep-th/0201075

[Ho2]   Hollowood, T.J.: Testing Seiberg-Witten theory to all orders in the instanton expansion.
          Nucl. Phys. **B639**, 66–94 (2002). arXiv:hep-th/0202197

[IK02]   Iqbal, A., Kashani-Poor, A.-K.: Instanton counting and Chern-Simons theory. Adv.
          Theor. Math. Phys. **7**, 457–497 (2004). arXiv:hep-th/0212279

[IK03]   Iqbal, A., Kashani-Poor, A.-K.: SU(N) geometries and topological string amplitudes.
          Adv. Theor. Math. Phys. bf **10**, 1–32 (2006). arXiv:hep-th/0306032

[IKV]   Iqbal, A., Kozcaz, C., Vafa, C.: The refined topological. JHEP **0910**, 069 (2009).
          arXiv:hep-th/0701156

[ILTe]   Iorgov, N., Lisovyy, O., Teschner, J.: Isomonodromic tau-functions from Liouville con-
          formal blocks. Preprint arXiv:1401.6104

[ILTy]   Its, A., Lisovyy, O., Tykhyy, Yu.: Connection problem for the sine-Gordon/Painlevé III
          tau function and irregular conformal blocks. Preprint arXiv:1403.1235

[KKV]   Katz, S.H., Klemm, A., Vafa, C.: Geometric engineering of quantum field theories.
          Nucl. Phys. **B497**, 173–195 (1997). arXiv:hep-th/9609239

[KLMVW]   Klemm, A., Lerche, W., Mayr, P., Vafa, C., Warner, N.P.: Selfdual strings and $N = 2$
          supersymmetric field theory. Nucl. Phys. **B477**, 746–766 (1996). arXiv:hep-th/9604034

[KMMR]   Kinney, J., Maldacena, J., Minwalla, S., Raju, S.: An Index for 4 dimensional super
          conformal theories. Commun. Math. Phys. **275**, 209–254 (2007). arXiv:hep-th/0510251

[KMST]   Kanno, H., Maruyoshi, K., Shiba, S., Taki, M.: W_3 irregular states and isolated $N = 2$
          superconformal field theories. JHEP **1303**, 147 (2013). arXiv:1301.0721

[KMV]   Katz, S., Mayr, P., Vafa, C.: Mirror symmetry and exact solution of 4D $N = 2$ gauge
          theories. I. Adv. Theor. Math. Phys. **1**, 53–114 (1997). arXiv:hep-th/9706110

[KMZ]   Kanno, S., Matsuo, Y., Zhang, H.: Extended conformal symmetry and recursion for-
          mulae for nekrasov partition function. JHEP **1308**, 028 (2013). arXiv:1306.1523

[KPW]   Kozcaz, C., Pasquetti, S., Wyllard, N.: A & B model approaches to surface operators
          and Toda theories. JHEP **1008**, 042 (2010). arXiv:1004.2025

[KW]   Krefl, D., Walcher, J.: Extended holomorphic anomaly in gauge theory. Lett. Math.
          Phys. **95**, 67–88 (2011). arXiv:1007.0263

[Le]   Lerche, W.: Introduction to Seiberg-Witten theory and its stringy origin. Nucl.
          Phys. Proc. Suppl. **55B**, 83–117 (1997). Fortsch. Phys. **45**, 293–340 (1997).
          arXiv:hep-th/9611190

[LMN]   Losev, A.S., Marshakov, A.V., Nekrasov, N.A.: Small instantons, little strings and free
          fermions. From Fields to Strings: Circumnavigating Theoretical Physics, vol. 1, pp.
          581–621. World Scientific Publishing, Singapore (2005). arXiv:hep-th/0302191

[LNS]   Losev, A.S., Nekrasov, N.A., Shatashvili, S.: Testing Seiberg-Witten solution. Strings,
          Branes and Dualities (Cargèse, 1997). NATO Advanced Science Institutes Series C:
          Mathematical and Physical Sciences, vol. 520, pp. 359–372. Kluwer Academic Pub-
          lishing, Dordrecht (1999). arXiv:hep-th/9801061

[LY]   Lee, S., Yamazaki, M.: 3d Chern-Simons theory from M5-branes. JHEP **1312**, 035
          (2013). arXiv:1305.2429

[Ma]   Maldacena, J.M.: The large N limit of superconformal field theories and supergravity. Adv. Theor. Math. Phys. **2**, 231–252 (1998). arXiv:hep-th/9711200

[MaOk]   Maulik, D., Okounkov, A.: Quantum groups and quantum cohomology. Preprint arXiv:1211.1287

[MMS]   Mironov, A., Morozov, A., Shakirov, Sh.: A direct proof of AGT conjecture at beta = 1. JHEP **1102**, 067 (2011). arXiv:1012.3137

[MNS1]   Moore, G., Nekrasov, N.A., Shatashvili, S.: Integrating over Higgs branches. Commun. Math. Phys. **209**, 97–121 (2000). arXiv:hep-th/9712241

[MNS2]   Moore, G., Nekrasov, N.A., Shatashvili, S.: D-particle bound states and generalized instantons. Commun. Math. Phys. **209**, 77–95 (2000). arXiv:hep-th/9803265

[MoOI]   Montonen, C., Olive, D.: Magnetic monopoles as gauge particles? Phys. Lett. **B72**, 117 (1977)

[MP14a]   Mitev, V., Pomoni, E.: The exact effective couplings of 4D $N = 2$ Gauge theories. Preprint arXiv:1406.3629

[MP14b]   Mitev, V., Pomoni, E.: Toda 3-point functions from topological strings. Preprint arXiv:1409.6313

[MS]   Morozov, A., Smirnov, A.: Towards the proof of AGT relations with the help of the generalized Jack polynomials. Lett. Math. Phys. **104**, 585–612 (2014). arXiv:1307.2576

[MW]   Martinec, E.J., Warner, N.P.: Integrable systems and supersymmetric gauge theory. Nucl. Phys. **B459**, 97–112 (1996). arXiv:hep-th/9509161

[N]   Nekrasov, N.A.: Seiberg-Witten prepotential from instanton counting. Adv. Theor. Math. Phys. **7**, 831–864 (2003). arXiv:hep-th/0206161

[NO03]   Nekrasov, N., Okounkov, A.: Seiberg-Witten Theory and Random Partitions. The Unity of Mathematics. Progress in Mathematics, vol. 244, pp. 525–596. Birkhäuser, Boston (2006) arXiv:hep-th/0306238

[NO14]   Nekrasov, N., Okounkov, A.: Membranes and sheaves. Preprint arXiv:1404.2323

[NPS]   Nekrasov, N., Pestun, V., Shatashvili, S.: Quantum geometry and quiver gauge theories. Preprint arXiv:1312.6689

[NP]   Nekrasov, N., Pestun, V.: Seiberg-Witten geometry of four dimensional $N = 2$ quiver gauge theories. Preprint arXiv:1211.2240

[NRS]   Nekrasov, N., Rosly, A., Shatashvili, S.: Darboux coordinates, Yang-Yang functional, and gauge theory. Nucl. Phys. Proc. Suppl. **216**, 69–93 (2011). arXiv:1103.3919

[NSa]   Nekrasov, N., Shatashvili, S.: Supersymmetric vacua and Bethe Ansatz. Nucl. Phys. Proc. Suppl. **192–193**, 91–112 (2009). arXiv:0901.4744

[NSb]   Nekrasov, N., Shatashvili, S.: Quantum integrability and supersymmetric vacua. Prog. Theor. Phys. Suppl. **177**, 105–119 (2009). arXiv:0901.4748

[NSc]   Nekrasov, N., Shatashvili, S.: Quantization of integrable systems and four dimensional gauge theories. In: Exner, P. (ed.) Proceedings of the 16th International Congress on Mathematical Physics, Prague, August 2009, pp. 265–289. World Scientific, Singapore (2010). arXiv:0908.4052

[NW]   Nekrasov, N., Witten, E.: The omega deformation, branes, integrability, and Liouville theory. JHEP **1009**, 092 (2010). arXiv:1002.0888

[NY]   Nakajima, H., Yoshioka, K.: Instanton counting on blowup. I. 4-dimensional pure gauge theory. Invent. Math. **162**, 313–355 (2005). arXiv:math/0306198

[Os]   Osborn, H.: Topological charges for $N = 4$ supersymmetric gauge theories and monopoles of spin 1. Phys. Lett. **B83**, 321 (1979)

[Pe97]   Peskin, M.E.: Duality in supersymmetric Yang-Mills theory. In: Efthimiou, C., Greene, B. (eds.) Proceedings, Summer School TASI'96, Fields, Strings and Duality. World Scientific, Singapore (1997). arXiv:hep-th/9702094

[Pe07]   Pestun, V.: Localization of gauge theory on a four-sphere and supersymmetric Wilson loops. Commun. Math. Phys. **313**, 71–129 (2012). arXiv:0712.2824

[Ro]   Romelsberger, C.: Counting chiral primaries in $N = 1$, $d = 4$ superconformal field theories. Nucl. Phys. **B747**, 329–353 (2006). arXiv:hep-th/0510060

[RZ13a]	Russo, J.G., Zarembo, K.: Evidence for large-N phase transitions in $N = 2^*$ theory. JHEP **1304**, 065 (2013). arXiv:1302.6968
[RZ13b]	Russo, J.G., Zarembo, K.: Massive $N = 2$ gauge theories at large N. JHEP **1311**, 130 (2013). arXiv:1309.1004
[ST]	Schwarz, A., Tang, X.: Quantization and holomorphic anomaly. JHEP **0703**, 062 (2007). arXiv:hep-th/0611281
[SchV]	Schiffmann, O., Vasserot, E.: Cherednik algebras, W-algebras and the equivariant cohomology of the moduli space of instantons on $\mathbf{A}^2$, Publ. IHÉS **118**, 213–342 (2013). arXiv:1202.2756
[Sei]	Seiberg, N.: Notes on theories with 16 supercharges. Nucl. Phys. Proc. Suppl. **67**, 158–171 (1998). arXiv:hep-th/9705117
[Sen]	Sen, A.: Dyon—monopole bound states, self-dual harmonic forms on the multi-monopole moduli space, and SL(2, Z) invariance in string theory. Phys. Lett. **B329**, 217–221 (1994). arXiv:hep-th/9402032
[SpV]	Spiridonov, V.P., Vartanov, G.S.: Superconformal indices for $N = 1$ theories with multiple duals. Nucl. Phys. **B824**, 192–216 (2010). arXiv:0811.1909
[St]	Strominger, A.: Open p-branes. Phys. Lett. **B383**, 44–47 (1996). arXiv:hep-th/9512059
[SW1]	Seiberg, N., Witten, E.: Monopole condensation, and confinement in $N = 2$ supersymmetric Yang-Mills theory. Nucl. Phys. **B426**, 19–52 (1994). arXiv:hep-th/9407087
[SW2]	Seiberg, N., Witten, E.: Monopoles, duality and chiral symmetry breaking in $N = 2$ supersymmetric QCD. Nucl. Phys. **B431**, 484–550 (1994). arXiv:hep-th/9408099
[Tac]	Tachikawa, Y.: $N = 2$ supersymmetric dynamics for pedestrians. Lect. Notes Phys. **890**, (2014). arXiv:1312.2684
[Tan]	Tan, M.-C.: M-theoretic derivations of 4d-2d dualities: from a geometric langlands duality for surfaces, to the AGT correspondence, to integrable systems. JHEP **1307**, 171 (2013). arXiv:1301.1977
[Te01]	Teschner, J.: Liouville theory revisited. Class. Quantum Gravity **18**, R153–R222 (2001). arXiv:hep-th/0104158
[Te10]	Teschner, J.: Quantization of the Hitchin moduli spaces, Liouville theory, and the geometric Langlands correspondence I. Adv. Theor. Math. Phys. **15**, 471–564 (2011). arXiv:1005.2846
[TV13]	Teschner, J., Vartanov, G.: Supersymmetric gauge theories, quantisation of moduli spaces of flat connections, and conformal field theory. Adv. Theor. Math. Phys. **19**, 1–135 (2015). arXiv:1302.3778
[VW]	Vafa, C., Witten, E.: A strong coupling test of S-duality. Nucl. Phys. **B431**, 3–77 (1994). arXiv:hep-th/9408074
[W88]	Witten, E.: Topological quantum field theory. Commun. Math. Phys. **117**, 353 (1988)
[W93]	Witten, E.: Quantum background independence in string theory. Published in Salamfest, 0257–275 (1993). arXiv:hep-th/9306122
[W95a]	Witten, E.: Some comments on string dynamics. In: Bars, I., Bouwknegt, P., Minahan, J., Nemeschansky, D., Pilch, K., Saleur, H., Warner N. (eds.) Proceedings. Future Perspectives in String Theory (STRINGS'95). World Scientific, River Edge (1996). arXiv:hep-th/9507121
[W95b]	Witten, E.: Five-branes and M-theory on an orbifold. Nucl. Phys. **B463**, 383–397 (1996). arXiv:hep-th/9512219
[W09]	Witten, E.: Geometric langlands from six dimensions, a celebration of the mathematical legacy of Raoul Bott. In: CRM Proceedings of the Lecture Notes, vol. 50, pp. 281–310. American Mathematical Society, Providence (2010). arXiv:0905.2720
[WO]	Witten, E., Olive, D.: Supersymmetry algebras that include topological charges. Phys. Lett. **B78**, 97 (1978)
[Y12]	Yagi, Y.: Compactification on the Ω-background and the AGT correspondence. JHEP **1209**, 101 (2012). arXiv:1205.6820
[Y13]	Yagi, Y.: 3d TQFT from 6d SCFT. JHEP **1308**, 017 (2013). arXiv:1305.0291
[ZZ]	Zamolodchikov, A.B., Zamolodchikov, Al.B.: Structure constants and conformal bootstrap in Liouville field theory. Nucl. Phys. **B477**, 577–605 (1996). arXiv:hep-th/9506136

Families of $\mathcal{N} = 2$ Field Theories

Davide Gaiotto

The main actors of this review are four-dimensional field theories with $\mathcal{N} = 2$ supersymmetry. There are three well-understood ways to build large classes of $\mathcal{N} = 2$ field theories[1]:

- Standard four-dimensional Lagrangian formulation
- Twisted compactification of a six-dimensional $(2, 0)$ SCFT ("class $\mathcal{S}$")
- Field theory limit of string theory on a Calabi-Yau singularity ("geometric engineering")

These three classes of constructions have large overlaps. Most four-dimensional Lagrangians can be engineered in the class $\mathcal{S}$, and all can be engineered through some Calabi Yau geometry. Class $\mathcal{S}$ theories can be further lifted to Calabi Yau compactifications involving a curve of ADE singularities. Conversely, only a minority of $N = 2$ field theories admits a direct four-dimensional Lagrangian description.

Different UV realizations of the same theory may be better suited to answer specific questions. The six-dimensional or string-theoretic descriptions of a theory can be very powerful for computing properties which are somewhat protected by supersymmetry. On the other hand, some properties, symmetries and probes of a four-dimensional field theory may simply not be inherited from a specific UV definition of the theory. Simple four-dimensional field theory constructions may be hard to lift to six dimensional field theory, and even harder to embed in string theory, where every modification of the theory must involve a dynamical configuration of supergravity fields and D-branes which solves the equations of motion. A reader

---

[1] It is also possible to define four-dimensional $\mathcal{N} = 2$ field theories from a circle compactification of a $\mathcal{N} = 1$ 5d SCFT, or a torus compactification of a six-dimensional $(1, 0)$ SCFT. We are not aware of four-dimensional $\mathcal{N} = 2$ field theories which can only be built that way.

D. Gaiotto (✉)
Perimeter Institute for Theoretical Physics, University of Waterloo,
200 University Ave.W., Waterloo, ON N2L3G1, Canada
e-mail: dgaiotto@gmail.com

© Springer International Publishing Switzerland 2016
J. Teschner (ed.), *New Dualities of Supersymmetric Gauge Theories*,
Mathematical Physics Studies, DOI 10.1007/978-3-319-18769-3_2

of this special volume will have several occasions to appreciate the power of these alternative approaches.

This chapter of the review is intended essentially as a reading guide. We refer the reader to the original references and many excellent reviews available to learn the basic properties of $\mathcal{N} = 2$ field theories. We do not feel we can improve significantly on that available material. We will try to present a global overview of more recent developments.

1 Lagrangian Theories

The requirements of $\mathcal{N} = 2$ supersymmetry and renormalizability impose very strong constraints on the possible couplings in a Lagrangian [1]. We will assume the reader has some familiarity with the construction of $\mathcal{N} = 1$ supersymmetric Lagrangians. Already for an $\mathcal{N} = 1$ theory most of the freedom would lies in the choice of superpotential for the theory. Requiring the presence of $\mathcal{N} = 2$ supersymmetry fixes the form of the superpotential. As a result, the possible $\mathcal{N} = 2$ renormalizable Lagrangians are labelled by a choice of gauge group and of the representations the matter fields sit in.

The gauge fields belong to vectormultiplets, which decompose into a $\mathcal{N} = 1$ gauge multiplet and an adjoint chiral multiplet ϕ. The simplest Lagrangian $\mathcal{N} = 2$ theories are pure gauge theories. The chiral multiplet ϕ has no superpotential, and the only free parameter in the Lagrangian is the complexified gauge coupling

$$\tau = \frac{4\pi i}{g_{YM^2}} + \frac{\theta}{2\pi} \tag{1}$$

The beta functions for the gauge couplings are one-loop exact, and non-Abelian theories are asymptotically free. Abelian gauge groups coupled to matter, on the other hand, are IR free and have a Landau pole. The can only appear in effective theories.

The vectormultiplet kinetic terms can be written as an integral over chiral $\mathcal{N} = 2$ superspace: we can assemble the vectormultiplet into a chiral superfield

$$\Phi(\theta) = \phi + \cdots \tag{2}$$

depending on two sets of chiral superspace variables θ^1_α and θ^2_α. Then the standard kinetic terms take the form

$$\mathcal{L}_{\text{kinetic}} = \tau \int d^4\theta \text{Tr}\Phi^2 + \text{c.c.} \tag{3}$$

A more general choice of kinetic term can be described by a local gauge-invariant holomorphic pre-potential $\mathcal{F}(\Phi)$, as

$$\mathcal{L}_{\text{kinetic}} = \tau \int d^4\theta \mathcal{F}(\Phi) + \text{c.c.} \tag{4}$$

This kind of expression can capture, say, the two-derivative part of a low-energy effective action.

Matter fields can be added in the form of hypermultiplets, which in an $\mathcal{N} = 1$ language can be decomposed to a set of chiral multiplets q^a sitting in a pseudo-real representation of the overall symmetry group, which is the product of the gauge group and possible flavor groups. A renormalizable $\mathcal{N} = 2$ Lagrangian can be written in $\mathcal{N} = 1$ superspace in a standard way, with a superpotential

$$W_{\mathcal{N}=1} = \text{Tr}\phi \left(q^a t_{ab} q^b \right) + \text{Tr}M \left(q^a t_{ab}^f q^b \right) \tag{5}$$

Here t_{ab} are the gauge symmetry generators (they are symmetric, as the representation is pseudoreal) and t_{ab}^f are the flavor symmetry generators. We will use the notation "hypermultiplets in representation R" to indicate a set of chiral fields in the representation $R \otimes \bar{R}$. A set of chiral multiplets in a pseudoreal representation R will be denoted as a "half-hypermultiplet in representation R". In some cases, discrete anomalies prevent half-hypermultiplets from appearing alone.

The complex mass parameters M live in the adjoint of the flavor group. They must be normal, $[M, M^\dagger] = 0$, and can be thought as elements of the Cartan sub algebra of the flavor group.

Thus the only parameters of standard UV complete Lagrangian $\mathcal{N} = 2$ theories are the gauge couplings and the complex mass parameters.[2]

The matter representation is limited by the requirement of asymptotic freedom (or conformality). The beta functions for the gauge couplings are one-loop exact, and receive a positive contribution from every matter field. The limitation of asymptotic freedom allows a systematic classification of all possible Lagrangian $\mathcal{N} = 2$ gauge theories. The full classification and a very nice set of references can be found at [2].

The simplest $\mathcal{N} = 2$ Lagrangian theories with matter are $SU(N)$ gauge theories coupled to fundamental hypermultiplets, i.e. $\mathcal{N} = 2$ SQCD, or to a single adjoint hypermultiplet, i.e. $\mathcal{N} = 4$ SYM (which is denoted as $\mathcal{N} = 2^*$ when the complex mass for the adjoint hypermultiplet is turned on). The beta function for the latter theory vanishes, and the gauge coupling is exactly marginal. For the former theory, the beta function vanishes if the number of flavors is twice the number of colors, i.e. $N_f = 2N$.

A larger class of examples are quiver gauge theories, built from $\prod_a SU(N_a)$ gauge theories coupled to fundamental and bi-fundamental hypermultiplets. The constraint on the number of flavors implies that twice the rank of the gauge group

---

[2]If Abelian gauge groups are present, one could turn on an FI parameter, which breaks explicitly $SU(2)_R$. As Abelian groups coupled to matter have Landau poles, and if an FI parameter is absent in the UV it cannot appear in the IR, they will rarely play a role in this review. A notable exception is the theory of BPS vortices, which can only occur in the presence of an FI parameter.

at any given node must be bigger than the sum of the ranks at adjacent nodes. This is only possible if the quiver takes the form of a Dynkin diagram or, in the absence of fundamental matter, an affine Dynkin diagram. These theories will have two sets of flavor symmetries. Each bifundamental hypermultiplet (and symmetric or anti-symmetric hypers) is rotated by a $U(1)$ flavor symmetry. Each group of M_i fundamentals at the i-th node is rotated by an $U(M_i)$ flavor group.

Many more possibilities exist if we add matter in other representations, and look at more general choices of gauge groups. A possibility which will be important later is to consider $SU(2)^n$ gauge theories coupled to fundamental, bi-fundamental and (half-)trifundamental hypermultiplets. Tri-fundamental hypermultiplets are only allowed by renormalizability for three $SU(2)$ gauge groups, and their existence allows one to build intricate $SU(2)^n$ Lagrangian theories labelled by an arbitrary trivalent graph. This is not possible for other gauge groups.

The low-energy dynamics of $\mathcal{N} = 2$ gauge theories is very rich, and mostly hidden in a UV Lagrangian formulation. Many interesting quantities, even protected by supersymmetry and holomorphicity, receive crucial perturbative and non-perturbative corrections. Initially, the low energy dynamics was understood on a case-by-case basis from a careful analysis of the holomorphicity properties of the $\mathcal{N} = 2$ supersymmetric low-energy effective Lagrangian, starting from simple $SU(2)$ gauge theories [3, 4].

More systematically, many interesting results for very general quiver gauge theories can be computed by localization methods, with the help of the so-called Ω-deformation of $\mathcal{N} = 2$ gauge theories. We refer to [5] for a general analysis and references. Ultimately, as the our mathematical understanding of localization and of the geometry of instanton moduli spaces improves, we may hope to extend such calculations to all Lagrangian $\mathcal{N} = 2$ theories.

An important alternative approach is based on string theory dualities. Many $\mathcal{N} = 2$ Lagrangian field theories can be engineered by brane systems and then mapped through dualities to configurations in M-theory [6] (but not for E and $\hat{E}$-type quivers) and IIB geometric engineering (all quivers) [7]. The power of this approach lies in non-renormalization theorems which allow many protected quantities to be computed classically in the M-theory or IIB descriptions.

These constructions can be thought as providing maps which embeds (most) of the $\mathcal{N} = 2$ Lagrangian theories into larger classes of $\mathcal{N} = 2$ quantum field theories, which are constructed through M-theory or IIB setups which equip each theory with a simple geometric description of its low-energy dynamics, but possibly not a straightforward four-dimensional field-theoretic UV description. These are the other two classes of $\mathcal{N} = 2$ theories mentioned in the introduction.

Before exploring these classes of theories, it is useful to discuss some general properties of $\mathcal{N} = 2$ theories, abstracting from the possible existence of a Lagrangian description.

2 General Properties of $\mathcal{N} = 2$ Field Theories

Most of the facts collected in this section are easily demonstrated for Lagrangian field theories, but appear to be true for all $\mathcal{N} = 2$ UV complete quantum field theories. It is likely that they could be established in full generality by an accurate analysis of the $\mathcal{N} = 2$ tensor and conserved current supermultiplets.

The first general property is the existence of an $SU(2)_R$ R-symmetry. The $\mathcal{N} = 2$ SUSY algebra is compatible with an $SU(2)_R \times U(1)_r$ R-symmetry group. Both factors appear as part of the $\mathcal{N} = 2$ superconformal group, and thus are always symmetries of $\mathcal{N} = 2$ SCFTs. The $U(1)_r$ is broken/anomalous for all asymptotically free or mass-deformed theories, as the breaking of the conformal and $U(1)_r$ symmetries are tied together by supersymmetry.

In a Lagrangian theory, the hypermultiplet scalars sit in a doublet of $SU(2)_R$. In appropriate conventions, the top component of the doublets are $\mathcal{N} = 1$ chiral fields, the bottom are $\mathcal{N} = 1$ anti-chiral fields. The vectormultiplet scalars, on the other hand, are charged under $U(1)_r$. These fields are special examples of two important classes of protected operators: Coulomb branch operators and Higgs branch operators. These two classes of operators control both the parameter spaces of deformations and moduli spaces of vacua preserving $\mathcal{N} = 2$ supersymmetry. The geometry of these spaces is rich and plays a central role throughout this volume.

Coulomb branch operators are operators annihilated by all anti-chiral supercharges: they are chiral operators for every $\mathcal{N} = 1$ sub algebra of the theory. They never belong to non-trivial $SU(2)_R$ representations. In a SCFT they carry an $U(1)_r$ charge proportional to their scaling dimension. The Coulomb branch operators in a Lagrangian theory are holomorphic gauge-invariant polynomials of the vector multiplet scalar fields. For example, if we have some $SU(N)$ gauge fields with scalar super partner ϕ, the traces $\mathrm{Tr}\phi^n$ are all Coulomb branch operators.

A general $\mathcal{N} = 2$ may include many more Coulomb branch operators, not associated to weakly-coupled gauge fields. As long as their scaling dimension is smaller or equal to 2, they will be associated to more general deformation parameters c_i of the theory, written in chiral superspace as

$$\delta c_i \int d^4\theta \mathcal{O}^i \qquad (6)$$

involving the appropriate super-partner $Q^4 \mathcal{O}^i$ of the Coulomb branch operators $\mathcal{O}^i$.

The second class of deformations, complex masses, is also tied to vectormultiplets, but rather than being a coupling in a vectormultiplet Lagrangian, they are vevs of a background, non-dynamical vectormultiplet scalar fields. More precisely, if the theory has a continuous flavor symmetry, with Lie algebra $\mathfrak{g}$, we can couple it to a non-dynamical background vectormultiplet valued in $\mathfrak{g}$. A complex mass deformation is introduced by turning on a vev M for the complex scalar in the background vectormultiplet, such that $[M, M^\dagger] = 0$. Up to a flavor symmetry transformation, we can take M to be valued in the complexified Cartan sub-algebra of $\mathfrak{g}$.

At the level of the Lagrangian, the leading order effect of a complex mass is to add a coupling to a super partner μ^a_{++} of a conserved flavor current J^a

$$\delta M_a d^2 \theta^+ \mu^a_{++} + c.c. \tag{7}$$

The $+$ refers to one component of a $SU(2)_R$ doublet index. Indeed, conserved currents sit in a special supermultiplet which includes an $SU(2)_R$ triplet of moment map operators μ^a_{AB}.

Moment map operators are the typical example of a Higgs branch operator: operators which sit in non-trivial $SU(2)_R$ representations of spin $n/2$, $\mathcal{O}_{A_1 A_2 \cdots A_n}$, and satisfy a shortening condition [8].

$$Q^\alpha_{(A_0} \mathcal{O}_{A_1 A_2 \cdots A_n)} = 0 \qquad \bar{Q}^{\dot\alpha}_{(A_0} \mathcal{O}_{A_1 A_2 \cdots A_n)} = 0 \tag{8}$$

They never carry an $U(1)_r$ charge.

2.1 Parameter Spaces of Vacua and S-Dualities

As the beta functions of gauge couplings are one-loop exact, it is easy to construct conformal invariant Lagrangian $\mathcal{N} = 2$ field theories by tuning the total amount of matter appropriately. These Lagrangian theories will thus have a parameter space of exactly marginal deformations parameterized by the complexified gauge couplings. Although many isolated, strongly-coupled $\mathcal{N} = 2$ SCFTs exist, there are also large classes of non-Lagrangian $\mathcal{N} = 2$ SCFTs with spaces of exactly marginal deformations. Many examples can be defined by coupling standard non-Abelian gauge fields to the flavor symmetry currents of non-Lagrangian isolated $\mathcal{N} = 2$ SCFTs in such a way that the gauge coupling beta function vanishes.

The space of exactly marginal deformations of an $\mathcal{N} = 2$ SCFT is a complex manifold, and several protected quantities are locally holomorphic functions on the space of deformations.[3] Thus a side payoff of exact calculations, done by localization or M-theory/IIB engineering, is a characterization of the complex manifold of marginal couplings.

The results, even for Lagrangian theories, are rather counter-intuitive. Naively, the space of couplings for a Lagrangian theory should consist of a product of several copies of the upper half plane, each parameterized by a complexified gauge coupling τ_a. More precisely, as the gauge theory is invariant under $\tau \to \tau + 1$, one can parameterize the space by the instanton factors $q_a = \exp 2\pi i \tau_a$. (This is a good choice in asymptotically free theories as well, where q_a becomes a dimensionful coupling.)

---

[3] We would like to point out that a general analysis of the geometric properties of the space of exactly marginal deformations of $\mathcal{N} = 2$ SCFTs seems to be missing from the literature.

At weak coupling, protected quantities can be expanded in power series in the q_a, typically convergent in the naive physical range $|q_a| < 1$. Surprisingly, with the exception of $\mathcal{N} = 4$ SYM, the geometry of parameter space is strongly modified at strong coupling, and $|q_a| = 1$ is not a boundary anymore: the theory can be analytically continued beyond $|q_a| = 1$ into a complicated moduli space.

In all known examples, as soon as we move far enough from the original weakly coupled region new dual descriptions of the theory emerge, possibly involving radically different degrees of freedom. Typically, the new descriptions involve weakly coupled gauge fields interacting with intrinsically strongly-coupled matter theories described by isolated $\mathcal{N} = 2$ SCFTs. Only in some cases we find again weakly-coupled Lagrangian theories.

The generic moniker for this type of situation, where seemingly different theories are related by analytic continuation in the space of gauge couplings, is S-duality. The canonical example of S-duality occurs in $\mathcal{N} = 4$ SYM [9, 10]: as one approaches the $|q| = 1$ boundary, new dual descriptions emerge involving magnetic monopoles or dyons which reassemble themselves into weakly coupled $\mathcal{N} = 4$ gauge fields with the same gauge group of the original theory, or its Langlands dual group. In that case, every description covers the whole parameter space, and the couplings are related by $SL(2, Z)$ transformations of the form

$$\tau' = \frac{a\tau + b}{c\tau + d} \tag{9}$$

which map the upper half plane back to itself.

The $N_f = 2N$ SQCD already offers a more general situation: the full parameter space can be described by allowing q to reach arbitrary values. At very large q we have a dual $N_f = 2N$ SQCD description, with coupling $q' = 1/q$. At $q \sim 1$ we have a non-Lagrangian dual description, with an $SU(2)$ weakly coupled gauge field of coupling $q'' \sim 1 - q$, coupled to a fundamental hyper and to an isolated SCFT with $SU(2) \times SU(N_f)$ flavor symmetry [11, 12]. Similar statements hold more general Lagrangian theories. The possible S-dual descriptions of Lagrangian theories which can be mapped to class $\mathcal{S}$ are well understood. Other examples, such as the quiver theories in the shape of an E-type Dynkin diagram, do not appear to have been explored systematically.

There is a neat class of examples of theories with the property that all S-dual descriptions are Lagrangian. This will be our introduction into the class $\mathcal{S}$ theories. The starting point is the observation that for $SU(2)$ $N_f = 4$ SQCD all three S-duality frames, around $q = 0$, $q = \infty$, $q = 1$, are described by a $SU(2)$ $N_f = 4$ SQCD Lagrangian. As the $SU(2)$ fundamental representation is pseudoreal, the flavor group is really $SO(8)$, and the three S-dual descriptions are related by a triality operation of the flavor group: if the quarks in the $q \sim 0$ description sit in the vector representation 8_v, the quarks in the $q \sim \infty$ description sit in the chiral spinor representation 8_s and the quarks in the $q \sim 1$ description sit in the anti-chiral spinor representation 8_c.

This beautiful result, originally found in [4], has far reaching consequences. As we mentioned before, the existence of a trifundamental half-hypermultiplet for three

$SU(2)$ groups allows the construction of a large class of $SU(2)^k$ Lagrangian field theories, with $SU(2)$ gauge groups only [12]. Each gauge group can be coupled to at most two trifundamental blocks, and will be conformal if coupled exactly to two. Thus we can associate such a superconformal field theory to each trivalent graph, with an $SU(2)$ gauge group for every internal edge and an $SU(2)$ flavor group for each external edge.

Two trinions coupled to the same $SU(2)$ essentially consist of four fundamental flavors for that group. A group coupled to two legs of the same trinion, instead, looks like an $SU(2)$ $\mathcal{N} = 4$. Thus if we start from a frame where all couplings are weak, and make a single coupling strong, we can apply one of the two basic S-duality operations to that group, and reach a new S-dual description of the whole theory. The new description is again an $SU(2)^k$ theory, but it is associated to a possibly different trivalent graph. Ultimately, all the theories associated to graphs with n external edges and g loops must belong to the same moduli space of exactly marginal deformations, represent distinct S-dual descriptions of the same underlying SCFT labelled by n and g.

This will be the most basic example of class $\mathcal{S}$ theories. Furthermore, these are the only S-duality frames for these theory. The parameter space of gauge couplings of these theories will be identified to the moduli space of complex structures for a Riemann surface of genus g with n punctures. Each Lagrangian description is associated to a pair of pants decomposition of the Riemann surface, with the couplings q_a identified with the sewing parameters for the surface. The basic S-dualities of individual $SU(2)$ gauge groups represent basic moves relating different pair of pants decompositions.

2.2 Moduli Spaces of Vacua

Generically, the moduli space of $\mathcal{N} = 2$ supersymmetric vacua consist of the union of several branches, each factorized into a "Coulomb branch factor" and a "Higgs branch factor". A Coulomb branch factor is parameterized by the vevs u_i of Coulomb branch operators $\mathcal{O}^i$. A Higgs branch factor is an hyper-Kähler cone parameterized by the vevs of Higgs branch operators. See [13] and references therein for more details.

It is useful to observe that the hyper-Kähler geometry of the Higgs factors does not depend on the couplings. It can thus be usefully computed in convenient corner of parameter space, such as a corner where the theory is weakly coupled. Flavour symmetries act as (tri-holomorphic) isometries on the Higgs branch, and the corresponding mass parameters force the theory to live at fixed points of the corresponding isometries. Often, turning on generic complex masses completely suppresses Higgs branch moduli.

Usually an $\mathcal{N} = 2$ theory has a pure Coulomb branch of vacua, where all Higgs branch operators have zero vet and $SU(2)_R$ is unbroken. We will refer to this branch simply as the Coulomb branch $\mathcal{C}$ of the theory. At special complex singular loci $\mathcal{C}_\alpha$ new branches may open up, of the form $\mathcal{C}_\alpha \times \mathcal{H}_\alpha$ for some Higgs factors $\mathcal{H}_\alpha$.

At low energy on the Coulomb branch, the only massless degrees of freedom are scalar fields which parameterize motion along the Coulomb branch, which sit in Abelian vectormultiplets. Thus the low-energy description of physics on the Coulomb branch involves a $U(1)^r$ gauge theory, where the rank r is the complex dimension of the Coulomb branch. Supersymmetry implies a close interplay between the couplings of the low energy gauge theory and the geometry of the Coulomb branch. This is the main subject of the next section.

The supersymmetry algebra in a sector with Abelian (electric, magnetic and flavour) charges γ admits a central charge function Z_γ, linear in γ [14]. Schematically,

$$[Q, Q] = \bar{Z} \qquad [Q, \bar{Q}] = P \qquad [\bar{Q}, \bar{Q}] = \bar{Z} \tag{10}$$

This implies that charged particles are generically massive, with mass above the BPS bound $|Z_\gamma|$ [15, 16], and can be integrated out at sufficiently low energy at least at generic points in the Coulomb branch.[4]

Particles which saturate the BPS bounds are called BPS particles. They will play an important role in understanding the low energy physics of $\mathcal{N} = 2$ quantum field theories.

2.3 Seiberg-Witten Theory

The low energy dynamics in the Coulomb branch is the subject of Seiberg-Witten theory. The study of the Coulomb branch dynamics was initiated in [3, 4]. See also e.g. [17–19] for reviews of the subject. The central charge function for a charge vector γ including an electric charge γ_e, a magnetic charge γ_m and a flavour charge γ_f takes the form

$$Z_\gamma = a \cdot \gamma_e + a^D \cdot \gamma_m + m \cdot \gamma_f \tag{11}$$

The r complex fields a^I are the super partners of gauge fields. They give a special local coordinate system, where the metric coincides with the imaginary part $\mathrm{Im}\,\tau_{IJ}$ of the complexified gauge couplings, which can be packaged locally in an holomorphic prepotential $\mathcal{F}$:

$$\tau_{IJ} = \frac{\partial^2 \mathcal{F}}{\partial a^I \partial a^J} \tag{12}$$

---

[4]Of course, there could be a separate massless sector which carries no gauge charges. This will happen in theories where the Higgs branch is not fully suppressed at generic points in the Coulomb branch.

The dual fields a_I^D are also given in terms of the prepotential

$$a_I^D = \frac{\partial \mathcal{F}}{\partial a^I} \tag{13}$$

and of course $\tau_{IJ} = \frac{\partial a_I^D}{\partial a^J}$.

The prepotential depends generally on the gauge couplings and mass parameters of the theory. The following relation, valid at fixed masses, is often useful to control the dependence on the couplings:

$$da^I \wedge da_I^D = du^i \wedge dc_i \tag{14}$$

Here u^i is the vev of the Coulomb branch operator dual to c_i.

The low energy description is covariant under electric-magnetic dualities. An electric-magnetic duality transformation rotates the gauge charges by an integer-valued linear transformation which preserves the symplectic pairing

$$\langle \gamma, \gamma' \rangle = \gamma_m \cdot \gamma_e' - \gamma_m' \cdot \gamma_e \tag{15}$$

The action on the gauge couplings is simply encoded by an inverse rotation of (a, a^D), so that the central charge remains invariant. It is also useful to add to the duality group redefinitions of the flavour currents by multiples of the gauge currents. These transformations shift γ_f by multiples of γ_e and γ_m and correspondingly shift a and a^D by multiples of the mass parameters. We will often denote the set of (a, a^D) as "periods", for reasons which will become clear soon.

The crucial insight of Seiberg and Witten is to realize that there is no electric-magnetic duality frame which is globally well-defined over the Coulomb branch. Rather, if we continuously vary the Coulomb branch parameters along a closed path which winds around singular loci in the Coulomb branch, we may come back to an electric-magnetic dual description of the original physics. Thus the (a, a_D) are multivalued functions of the Coulomb branch parameters u^i. It is useful to describe the multi-valuedness in terms of the global structure of the charge lattice Γ: the charge lattice forms a local system of lattices over the Coulomb branch, with monodromies which preserve the symplectic pairing and the sublattice Γ_f of pure flavour charges. The central charge is a globally defined linear map from Γ to the complex numbers.

The singularities of the Coulomb branch must be loci where additional light degrees of freedom appear. In particular, they must be loci where the central charge of some BPS particle goes to zero, as only BPS particles can modify the geometry of the Coulomb branch through loop effects. The extra degrees of freedom must assemble themselves into an infrared free or conformal effective description of the low energy theory. If a Higgs branch opens up at the locus, it must be possible to describe it in terms of the low energy degrees of freedom.

A typical example is a codimension one singularity at which a single BPS hyper-multiplet becomes massless. Without loss of generality, we can go in a duality frame where the BPS hypermultiplet is electrically charged. This is an infrared free setup: the BPS hypermultiplet of charge γ makes the IR gauge coupling run at one loop as

$$\tau_{IJ} \sim -\gamma_I \gamma_J \frac{i}{2\pi} \log a \cdot \gamma \tag{16}$$

The behaviour of the magnetic central charges

$$a_I^D \sim -\gamma_I \frac{i}{2\pi} a \cdot \gamma \log a \cdot \gamma \tag{17}$$

shows the monodromy of the central charge, and thus of the charge lattice:

$$a_I^D \to a_I^D + \gamma_I a \cdot \gamma \qquad q_I^e \to q_I^e - q_m^J \gamma_J \gamma_I \tag{18}$$

In a generic duality frame, we can write the monodromy as

$$q \to q - \langle q, \gamma \rangle \gamma \tag{19}$$

In general, singular loci where a collection of light, IR free electrically charged BPS hypermultiplets appear will be associated to parabolic monodromies similar to (19). If a sufficiently large number of light particles are present, a Higgs branch may open up, described by the vev of the corresponding hypermultiplet fields.

Singular loci where the IR description involves a non-trivial superconformal field theory are associated to more general monodromies. We expect several periods to go to zero at a superconformal points, scaling as interesting, possibly fractional powers of the Coulomb branch coordinates u^i. Thus the monodromies will be in general elliptic. We are not aware of any example which involves hyperbolic monodromies. Interesting superconformal fixed points often arise from the collision/intersection of two or more simple singularities. The collision/intersection of singularities where mutually non-local particles such as an electron and a monopole become light usually produces IR superconformal field theories of the Argyres-Douglas type [20, 21].

A typical example is the collision of a point where one monopole of charge 1 is massless, and one point where n_f particles of electric charge 1 become massless. The combined monodromy is

$$a \to a + a_D \to (1 - n_f)a + a_D \qquad a_D \to a_D \to -n_f a + a_D \tag{20}$$

and has trace $2 - n_f$. For $n_f = 1, 2, 3$ the monodromy is elliptic, and we obtain an Argyres-Douglas theory which possesses an $SU(n_f)$ flavor symmetry rotating the electrically charged particles among themselves.

The interplay between the monodromies of the charge lattice and the spectrum of BPS particles is rather interesting, and is made more intricate by the phenomenon of wall-crossing. General BPS particles belong to supermultiplets which can be described as a (half) hypermultiplet tensored with a spin j representation of the Lorentz group. The one-loop contributions from a BPS particle of generic spin j

will be proportional to $\Omega_j = (-1)^{2j}(2j+1)$. The sum of Ω_j over all BPS particles with a given charge γ is a protected index, which may jump only if the single particle states mix with a continuum of multi particle states Generically, the mass $|Z_\gamma|$ is larger than the mass of constituents of different charge $|Z_{\gamma'}| + |Z_{\gamma-\gamma'}|$ and the index is protected. At walls of marginal stability, where the central charges of particles of different charges align, the index can jump. The jumps in the index are controlled by a specific wall-crossing formula due to Kontsevich and Soibelman [22].

2.4 Seiberg-Witten Curves

There is a tension between two properties of the matrix of gauge couplings τ_{IJ}: it is locally holomorphic in the Coulomb branch parameters u^i, and it has a positive-definite imaginary part. In the absence of intricate monodromies, these properties would actually be incompatible with each other. There is a rather different mathematical problem where a matrix with very similar properties appear: the period matrix of a family of Riemann surfaces. Given a Riemann surface, a set of A cycles α^I and dual B cycles β_I, with

$$\alpha^I \cap \alpha^J = 0 \quad \alpha^I \cap \beta_J = \delta^I_J \quad \beta_I \cap \beta_J = 0 \tag{21}$$

the period matrix τ_{IJ} is computed from the contour integrals of holomorphic differentials ω_I on β_J, normalized so that the contour integral on α^J is δ^J_I.

The period matrix has positive definite imaginary part. If we have a holomorphic family of Riemann surfaces, it will depend holomorphically on the parameters, with appropriate monodromies around loci where the Riemann surface degenerates. Furthermore, if we are given a meromorphic form λ on the Riemann surface, such that the variations of λ along the family are holomorphic differentials, the periods of λ along α^I and β_I will behave in the same way as the periods a^I, a^D_I. More generally, the homology lattice of the Riemann surface behave like the lattice of charges in a gauge theory, with the intersection of cycles playing the role of the $\langle , \rangle$ pairing on the charge lattice and the period of λ on a cycle γ playing the role of the central charge Z_γ. The monodromies around simple degeneration points, where a single cycle γ contracts, take exactly the form (19), and the behaviour of the period matrix is precisely (16). If λ has poles on the Riemann surface, the periods of lambda depend on the homology of the Riemann surface punctured at the poles, and the residues of λ behave like mass parameters.

Originally, this analogy was used by Seiberg and Witten as a simple computational tool to describe their solution for the low energy dynamics of $SU(2)$ gauge theories with various choices of matter. These theories have a one-dimensional Coulomb branch, and the solution was described by simple families of elliptic curves. A priori, there was no reason to believe this tool would be useful for theories with a higher-dimensional Coulomb branch: most matrices τ_{IJ} with positive definite imaginary part are not period matrices of a Riemann surface, because the dimension $3g - 3$ of moduli space of Riemann surfaces of genus g is much smaller than the dimension of the space of $2g \times 2g$ symmetric matrices.

Surprisingly, the great majority of known $\mathcal{N} = 2$ field theories do admit a low-energy description in terms of a Seiberg-Witten curve equipped with an appropriate differential. For many Lagrangian field theories, this fact can be verified through hard localization calculations (see [5] for the broadest possible result, and references therein for previous work).[5] For class $\mathcal{S}$ theories, it follows directly from the properties of the six-dimensional SCFTs. For theories defined by Calabi-Yau compactifications, the situation is less clear. Almost by construction, the low energy physics can be described by periods of the holomorphic three-form on the Calabi-Yau. It is not always obvious if this can be recast in terms of periods of a differential on a Riemann surface.

We record a useful relation which allows one to associate the UV couplings of Seiberg-Witten theories to the corresponding Coulomb branch operators

$$\delta u^i \wedge \delta c_i = \delta a^I \wedge \delta a_I^D = \int_\Sigma \delta\lambda \wedge \delta\lambda \qquad (22)$$

This is derived through the Riemann bilinear identity.

2.5 The Coulomb Branch of Lagrangian Gauge Theories

In a pure $\mathcal{N} = 2$ gauge theory, the D-term equations for the non-Abelian scalar fields Φ take the form

$$[\Phi, \Phi^\dagger] = 0 \qquad (23)$$

Classically, the theory has a family of ($\mathcal{N} = 2$) supersymmetric vacua characterized by a generic complex vev of Φ belonging to some Cartan sub algebra of the gauge group. If the vev is generic, it Higgses the gauge group down to an Abelian subgroup $U(1)^r$, where r is the rank of the group. The off-diagonal components of the vectormultiplet become massive and can be integrated out at low energy.

The same analysis typically holds for theories with matter: a generic Coulomb branch vev suppresses the vevs of hypermultiplets.

$$\Phi \cdot t_{ab} q^b + M \cdot t_{ab}^f q^b = 0 \qquad (24)$$

The Coulomb branch survives quantum-mechanically, but the geometry of the Coulomb branch receives important one-loop and instanton corrections. At very weak coupling, the electric periods can be identified with eigenvalues a_I of Φ. The magnetic periods are derived from the perturbative pre potential, which only receives tree level and one-loop contributions from the massive W-bosons and hypermultiplets

---

[5] It may also be justified through considerations based on surface defects [23].

$$\mathcal{F} = \frac{\tau}{2}a^2 + \sum_{e \in \Delta_+} \frac{i}{2\pi}(a \cdot e)^2 \log a \cdot e - \sum_{(w,w^f) \in R} \frac{i}{4\pi}(a \cdot w + m \cdot w^f)^2 \log(a \cdot w + m \cdot w^f)$$

(25)

We sum over the positive roots e and the weights for the gauge and flavor representation.

Seiberg and Witten [3] observed that this cannot be the end of the story: because of asymptotic freedom, the coefficient of the logarithms makes the gauge couplings negative definite near the locus where a W-boson becomes naively massless. The prepotential receives instanton corrections (in the form of a power series in the instanton factors q for the gauge groups) which must turn the behaviour around, and convert the naive W-boson singularity into singularities at which the gauge couplings have physically acceptable behaviour.

The canonical example is pure $SU(2)$ gauge theory, with Seiberg-Witten curve and differential

$$x^2 = z^3 + 2uz^2 + \Lambda^4 z \qquad \lambda = x \frac{dz}{z^2}$$

(26)

At large values of $u \sim \text{Tr}\Phi^2$ the theory is weakly coupled, and the integral of λ on a circle of unit radius in the z plane gives $a = \sqrt{2u} + \cdots$. The contour integral along a dual contour gives the expected $a_D = \frac{2i}{\pi}\sqrt{2u} \log u + \cdots$. At smaller values of u we encounter two singular loci $u = \pm\Lambda^2$ where a magnetic monopole and a dyon (whose charge add to the W-boson charge 2) become respectively massless.

Similarly, the Seiberg-Witten curve for pure $SU(N)$ gauge theory [24, 25] is

$$y^2 + P_N(x)y + \Lambda^{2N} = 0 \qquad \lambda = x \frac{dy}{y}$$

$$P_N(x) = x^N + u_2 x^{N-2} + \cdots + u_N$$

(27)

The naive W-boson singularity at the discriminant of $P(x)$ is replaced by two simple singular loci, at the discriminants of $P \pm \Lambda^N$. The self-intersections of the two loci produce interesting Argyres-Douglas singularities. For example, the maximal AD singularity corresponds to the curve

$$y^2 = x^N + c_2 x^{N-2} + \cdots + u_N \qquad \lambda = x \, dy$$

(28)

3 Theories in the Class $\mathcal{S}$

The basic starting point for the class $\mathcal{S}$ construction are the six-dimensional $(2, 0)$ SCFTs [26–31]. The known $(2, 0)$ SCFTs have an ADE classifications. These are strongly-interacting generalizations of the free Abelian $(2, 0)$ theory, which consists of a self-dual two-form gauge field, five scalar fields and fermions. The Abelian theory is the world volume theory of a single $M5$ brane. The general SCFTs arise in M-theory as the world-volume theory of N $M5$ branes (the A_{N-1} theory) [27], possibly in the presence of an $O5$ plane (the D-type theories). They also arise in IIB string theory at the locus of an ADE singularity [26]. The string theory construction of these theories makes two properties manifest. A SCFT labeled by the Lie algebra $\mathfrak{g}$

- Provides a UV completion to five-dimensional $\mathcal{N} = 2$ SYM theory with gauge algebra $\mathfrak{g}$
- Has a Coulomb branch of vacua where it reduces to an Abelian 6d theory valued in the Cartan of $\mathfrak{g}$, modulo the action of the Weil group.

To be precise, the 6d theory compactified on a circle of radius R should admit an effective description as 5d SYM with gauge coupling $g^2 = R$. The two statements are compatible: the 6d Abelian theory on the Coulomb branch compactified on a circle gives a 5d Abelian gauge theory, which also describes the Coulomb branch of 5d SYM.

Notice that both theories have an $SO(5)_R$ R-symmetry. In the Abelian theories, which are related in the same way, the R-symmetry rotates the five scalar fields. The Coulomb branch of the $(2, 0)$ SCFT is parameterized by the vevs of Coulomb branch operators, which have the same quantum numbers as the Weil-invariant polynomials in the scalar fields x^a of the Abelian low-energy description [32, 33]. The theory on the Coulomb branch has five central charges, which are carried by strings rather than particles. The BPS strings in the theories carry a charge under the Abelian two-form fields, which coincides with a root of $\mathfrak{g}$. The central charges for such a string of charge e are simply $Z^a = e \cdot x^a$.

The construction of four-dimensional field theories in the class $\mathcal{S}$ involves a twisted compactification of the SCFTs on a Riemann surface C [34]. The twisting uses an $SO(2)$ subgroup of $SO(5)$, and preserves a four-dimensional $\mathcal{N} = 2$ super algebra in the four directions orthogonal to the surface. The $SO(2)$ factor becomes $U(1)_r$. The remaining $SO(3)$ becomes $SU(2)_R$. The six-dimensional Coulomb branch operators which only carry $SO(2)$ charges become Coulomb branch operators for the $\mathcal{N} = 2$ super-algebra. Notice that due to the twisting an operator of $SO(2)$ charge k becomes a k-form on C. The construction of a general theory in the class $\mathcal{S}$ may involve several further modifications of the theory, which preserve the four-dimensional $\mathcal{N} = 2$ super algebra. We will review some details in a later section.

These twisted compactifications have a useful property: the Coulomb branch geometry is independent of the area of C, and can be described exactly at large area in terms of vevs of the scalar fields x of $SO(2)$ charge 1 in the low-energy six-dimensional Abelian description. Because of the twisting, the vevs give a locally holomorphic one-form $\lambda = xdz$ on C, valued in the Cartan of $\mathfrak{g}$ modulo the action of the Weil group. On the other hand, if we make the area of C is arbitrarily small while keeping the Coulomb branch data fixed, we will define a four-dimensional theory which, by definition, is the class $\mathcal{S}$ theory.

Thus the Seiberg-Witten low energy description of a class $\mathcal{S}$ is readily available from its definition. For A_{N-1} theories one can treat λ as a single-valued one-form on a

Riemann surface Σ which is a rank N cover of C [6]. Then Σ, λ can be identified with the Seiberg-Witten curve and differential for the class $\mathcal{S}$ theory. Similar approaches work for general $\mathfrak{g}$.

Much more work is required to find a direct four-dimensional UV descriptions of a given class $\mathcal{S}$ theory, or to find a class $\mathcal{S}$ description of a given Lagrangian four-dimensional theory. We will first describe the examples involving the A_1 theory, where the variety of possible ingredients is more limited, and then sketch the general story. We refer to Sect. 3 of [34] and to [12] for a general discussion of the general story unitary theories and [35] and references therein for a more general discussion.

3.1 A_1 *Theories*

The twisted A_1 6d theory has a single Coulomb branch operator $\hat{\phi}_2$ which behaves upon twisting as a quadratic differential on C. The four-dimensional Coulomb branch is thus parameterized by a holomorphic quadratic differential ϕ_2, the vev of $\hat{\phi}_2$. The dimension of the Coulomb branch, for compact C of genus g, is $3g - 3$. The one-form λ satisfies [12, 34]

$$\lambda^2 = \phi_2 \tag{29}$$

This equation defines simultaneously the double-cover Σ of C as a curve in T^*C, and the differential λ.

The complex structure moduli of C are the exactly marginal UV couplings of this class $\mathcal{S}$ theory. There are exactly as many couplings as operators in the Coulomb branch. This can be understood from the observation that the Coulomb branch operators which come from ϕ_2 have the correct $U(1)_r$ charge to be dual to exactly marginal couplings. We can extract a four-dimensional operator $\hat{u}^i$ from ϕ_2 by contracting it a Beltrami differential. It is natural to associate that operator with the corresponding complex structure deformation. This can be verified from the relation (22) and is discussed in detail in [36].

Thus the regions "at infinity" of the parameter space of exactly marginal deformations should correspond to the boundaries of the complex structure moduli space [6], where the Riemann surface C degenerates and one or more handles pinch. The physical properties of the six-dimensional SCFT confirm this picture. Near a degeneration locus we can pick a metric which makes the pinching handle long and thin compared to the rest of the surface. In that region, we should be allowed to use the effective description as 5d SYM on a long segment, and then find at lower energy a weakly-coupled four-dimensional $SU(2)$ gauge group.

The 4d gauge coupling can be computed and is such that the instanton factor q coincides with the canonical complex structure parameter which describe the length and twist of the handle. In particular, it becomes weak when the surface pinches. If we go to a maximal degeneration locus, where the Riemann surface reduces to a network of $2g - 2$ three-punctured spheres connected by $3g - 3$ handles, we will find

$3g - 3$ $SU(2)$ gauge groups. The calculation of the periods in this limit agrees with the gauge theory picture. The magnetic periods have a logarithmic behaviour which is consistent with the presence of a bloc of trifundamental half-hypermultiplets for each three-punctured sphere. As the gauge groups are conformal, we do not expect any other matter fields coupled to the gauge groups.

This analysis allows us to identify a possible four-dimensional UV description of the class $\mathcal{S}$ theory associated with a Riemann surface of genus g near a maximal degeneration locus: the $SU(2)^{3g-3}$ theory associated to a graph with g loops. The six-dimensional construction provides a global picture of how all the S-dual theories are connected through parameter space, and the low-energy Seiberg-Witten description.

In order to improve our understanding of the physics of decoupling, it is useful to introduce the notion of superconformal defects in the six-dimensional SCFT. A superconformal defect is a local modification of the theory along a hyperplane which preserves the subgroup of the conformal group which fixes the hyperplane, and an appropriate subset of the supercharges. We are interested here in codimension two defects, which preserve a subgroup of the 6d $(2, 0)$ superconformal group which is isomorphic to the 4d $\mathcal{N} = 2$ superconformal group.

Although we have a relatively poor understanding of the six-dimensional theory, there is a simple trick which allows us to define a useful class of defects in terms of the facts we know. We can simply use the twisted compactification strategy to put the theory on a funnel geometry, with an asymptotically flat region connected near the origin to a semi-infinite tube. The configuration preserves $\mathcal{N} = 2$ supersymmetry, and we can flow to the infrared to find something interesting. In the tube region, we flow to the infrared free five-dimensional $SU(2)$ SYM. In the asymptotically flat region, we have the standard 6d theory, modified only at the origin, in some what which allows it to couple to the 5d SYM theory. Thus construction produces a canonical superconformal defect equipped with an $SU(2)$ flavor symmetry. We will call it the regular defect.

This construction clarifies what happens in a degeneration limit of C: the handle can be removed, leaving behind two regular defects weakly coupled to the corresponding $SU(2)$ gauge group. In general, we can now enrich our starting point, and consider a Riemann surface C of genus g with n regular defects at points of C. This gives the six-dimensional realization of the $SU(2)$ quivers associated to a general graph with g loops and n external legs. The use of regular punctures allows us to make contact with standard brane constructions of $\mathcal{N} = 2$ field theories, and verify that the individual three-punctured sphere corresponds to a block of trifundamental half-hypermultiplets.

The Seiberg-Witten geometry in the presence of regular defects is still given by (29), but the quadratic differential is now allowed a double pole at the location of the punctures:

$$\phi_2 \sim \left[\frac{m_a^2}{(z - z_a)^2} + \frac{u_a}{z - z_a} + \cdots \right] dz^2 \tag{30}$$

Here m_a is the $SU(2)$ mass parameter at the puncture and u_a an extra Coulomb branch parameter dual to the position z_a of the puncture, which is a new exactly marginal coupling.

Through appropriate decoupling limits, we can go from these four-dimensional SCFTs to more general asymptotically free $SU(2)$ theories, or generalized Argyres-Douglas theories. In the six-dimensional description, these examples involve "irregular" punctures, where the quadratic differential is allowed poles of order higher than 2. Basic examples are the pure $SU(2)$ Seiberg-Witten theory

$$\phi_2 = \left[\frac{\Lambda^2}{z} + \frac{2u_a}{z^2} + \frac{\Lambda^2}{z^3} \right] dz^2 \tag{31}$$

and the basic Argyres-Douglas theories

$$\phi_2 = P_N(z) dz^2 \tag{32}$$

and

$$\phi_2 = \left[P_N(z) + \frac{u}{z} + \frac{m^2}{z^2} \right] dz^2 \tag{33}$$

where $P_N(z)$ is a degree N polynomial.

3.2 General ADE Theories

The generalization of the A_1 results involves several new ingredients. The Coulomb branch is now described by a family of differentials associated to the Casimirs of $\mathfrak{g}$, with degree of the differential equal to the degree of the Casimir. The exactly marginal couplings still coincide with the space of complex structures of C. The decoupling limit still replaces a handle by a gauge group with Lie algebra $\mathfrak{g}$, and can be understood in terms of a codimension 2 defect with flavor symmetry $\mathfrak{g}$, which we will denote as a full regular defect.

The first difference is that the theory associated to a three-punctured sphere has a non-trivial Coulomb branch, and no couplings: it is an otherwise unknown 4d SCFT with three $\mathfrak{g}$ flavor symmetries. In order to make contact with standard Lagrangian field theories and their brane engineering we need a larger choice of regular punctures. A simple way to understand the possible choices is to realize that the full regular puncture has a Higgs branch, parameterized by the vevs of the moment map operators for the $\mathfrak{g}$ flavor symmetry. The Higgs branch should open up at loci in the Coulomb branch where all the Coulomb branch operators have no pole at the puncture. The Higgs branch conjecturally coincides with the maximal complex nilpotent orbit of $\mathfrak{g}^{\mathbb{C}}$.

If we sit at a generic point of the Higgs branch and flow to the IR, we essentially erase the puncture. If we sit at a non-generic nilpotent element and flow to the IR, we will somewhat "simplify" the full regular puncture to a different type of regular puncture, where the singularities of the Coulomb branch operators are constrained in appropriate patterns. A nilpotent element can always be taken to be the raising operator of an $\mathfrak{su}(2)$ subalgebra ρ of $\mathfrak{g}$. Thus these new regular punctures will be labelled by ρ. A further generalization of regular punctures is possible, in which the operators of the 6d theory undergo a monodromy around the defect, under an outer automorphism of $\mathfrak{g}$ [35]. These general regular punctures allow one to make contact with most superconformal Lagrangian quiver gauge theories. As for the A_1 case, one can also define a large variety of irregular punctures.

4 Calabi-Yau Compactifications

The compactification of string theory on non-compact Calabi-Yau manifolds can also give rise to four-dimensional $\mathcal{N} = 2$ field theories [37]. The low-energy dynamics can be derived in a straightforward way from the geometry in a type IIB duality frame: the periods are identified with the periods of the holomorphic three-form on appropriate cycles in the geometry. On the other hand, the identification of an intermediate UV-complete four-dimensional field theory description of the theory is more laborious. Often, the field theory is engineered through a type IIA construction, and then mirror symmetry gives the map to IIB string theory and thus the low energy solution of the theory.

Theories in the class $\mathcal{S}$ can be embedded in Type IIB string theory by engineering the 6d SCFTs as loci of ADE singularities, fibered appropriately over the curve C. For example, an A_1 theory can be realized through the geometry

$$x^2 + u^2 + v^2 = \phi_2(z) \tag{34}$$

The geometric engineering, though, can provide solutions for theories which do not admit a known six-dimensional construction, such as quiver gauge theories in the shape of E-type Dynkin diagrams. Indeed, it provides a unified picture of all the quiver gauge theories of unitary groups, through geometries where an elliptic singularity is fibered over a complex plane [7]. Remarkably, this provides a description of the space of exactly marginal deformations as a moduli space of flat connections on a torus.

A second remarkable example is a large family of Argyres-Douglas theories, labeled by two ADE labels. Remember the A_1 examples, lifted to a Calabi-Yau

$$u^2 + v^2 + x^2 + P_N(z) = 0 \tag{35}$$

The A_{M-1} generalization is

$$u^2 + v^2 + x^M + z^N + \cdots = 0 \tag{36}$$

The main idea is to write that in terms of ADE polynomials for A_{N-1} and A_{M-1} as

$$W_{A_{N-1}}(u, x) + W_{A_{M-1}}(v, z) = 0 \tag{37}$$

and then replace either polynomials with the ones associated to D type, $u^2 x + x^N$, or E type

$$u^3 + x^4 \qquad u^3 + ux^3 \qquad u^3 + x^5 \tag{38}$$

References

[1] Wess, J., Bagger, J.: Supersymmetry and Supergravity. Princeton, University Press (1983)
[2] Bhardwaj, L., Tachikawa, Y.: Classification of 4d N = 2 gauge theories. JHEP **1312**, 100 (2013). arXiv:1309.5160
[3] Seiberg, N., Witten, E.: Electric-magnetic duality, monopole condensation, and confinement in $\mathcal{N} = 2$ supersymmetric Yang-Mills theory. Nucl. Phys. **B426**, 19–52 (1994). arXiv:hep-th/9407087
[4] Seiberg, N., Witten, E.: Monopoles, duality and chiral symmetry breaking in N = 2 supersymmetric QCD. Nucl. Phys. **B431**, 484–550 (1994). arXiv:hep-th/9408099
[5] Nekrasov, N., Pestun, V.: Seiberg-Witten geometry of four dimensional N = 2 quiver gauge theories. arXiv:1211.2240
[6] Witten, E.: Solutions of four-dimensional field theories via M-theory. Nucl. Phys. **B500**, 3–42 (1997). arXiv:hep-th/9703166
[7] Katz, S., Mayr, P., Vafa, C.: Mirror symmetry and exact solution of 4D N = 2 gauge theories. I. Adv. Theor. Math. Phys. **1**, 53–114 (1998). arXiv:hep-th/9706110
[8] Dolan, F., Osborn, H.: On short and semi-short representations for four-dimensional superconformal symmetry. Ann. Phys. **307**, 41–89 (2003). arXiv:hep-th/0209056
[9] Montonen, C., Olive, D.I.: Magnetic monopoles as gauge particles? Phys. Lett. **B72**, 117 (1977)
[10] Sen, A.: Strong—weak coupling duality in four-dimensional string theory. Int. J. Mod. Phys. **A9**, 3707–3750 (1994). arXiv:hep-th/9402002
[11] Argyres, P.C., Seiberg, N.: S-duality in N = 2 supersymmetric gauge theories. JHEP **0712**, 088 (2007). arXiv:0711.0054
[12] Gaiotto, D.: N = 2 dualities. arXiv:0904.2715
[13] Argyres, P.: Supersymmetric effective actions in four dimensions
[14] Witten, E., Olive, D.I.: Supersymmetry algebras that include topological charges. Phys. Lett. **B78**, 97 (1978)
[15] Prasad, M., Sommerfield, C.M.: An Exact classical solution for the 't Hooft monopole and the Julia-Zee Dyon. Phys. Rev. Lett. **35**, 760–762 (1975)
[16] Bogomolny, E.: Stability of classical solutions. Sov. J. Nucl. Phys. **24**, 449 (1976)
[17] Lerche, W.: Introduction to Seiberg-Witten theory and its stringy origin. Prepared for CERN-Santiago de Compostela-La Plata Meeting on Trends in Theoretical Physics, CERN-Santiago de Compostela-La Plata, Argentina, 28 April–6 May 1997
[18] Freed, D.S.: Special Kaehler manifolds. Commun. Math. Phys. **203**, 31–52 (1999). arXiv:hep-th/9712042
[19] Witten, E.: Dynamics of quantum field theory. In: Quantum Fields and Strings: A Course for Mathematicians. Princeton, vols. 1, 2 (1996/1997), pp. 1119–1424. American Mathematical Society, Providence (1999)
[20] Argyres, P.C., Douglas, M.R.: New phenomena in SU(3) supersymmetric gauge theory. Nucl. Phys. **B448**, 93–126 (1995). arXiv:hep-th/9505062

[21]   Argyres, P.C., Ronen Plesser, M., Seiberg, N., Witten, E.: New N = 2 superconformal field theories in four dimensions. Nucl. Phys. **B461**, 71–84 (1996). arXiv:hep-th/9511154

[22]   Kontsevich, M., Soibelman, Y.: Stability structures, motivic Donaldson-Thomas invariants and cluster transformations. arXiv:0811.2435

[23]   Gaiotto, D.: Surface operators in N = 2 4d gauge theories. JHEP **1211**, 090 (2012). arXiv:0911.1316

[24]   Klemm, A., Lerche, W., Yankielowicz, S., Theisen, S.: Simple singularities and N = 2 supersymmetric Yang-Mills theory. Phys. Lett. **B344**, 169–175 (1995). arXiv:hep-th/9411048

[25]   Argyres, P.C., Faraggi, A.E.: The vacuum structure and spectrum of N=2 supersymmetric SU(n) gauge theory. Phys. Rev. Lett. **74**, 3931–3934 (1995). arXiv:hep-th/9411057

[26]   Witten, E.: Some comments on string dynamics. arXiv:hep-th/9507121

[27]   Strominger, A.: Open p-branes. Phys. Lett. **B383**, 44–47 (1996). arXiv:hep-th/9512059

[28]   Witten, E.: Five-branes and M-theory on an orbifold. Nucl. Phys. **B463**, 383–397 (1996). arXiv:hep-th/9512219

[29]   Seiberg, N., Witten, E.: Comments on string dynamics in six dimensions. Nucl. Phys. **B471**, 121–134 (1996). arXiv:hep-th/9603003

[30]   Seiberg, N.: Notes on theories with 16 supercharges. Nucl. Phys. Proc. Suppl. **67**, 158–171 (1998). arXiv:hep-th/9705117

[31]   Seiberg, N.: New theories in six dimensions and matrix description of M-theory on T^5 and $T^5/\mathbb{Z}_2$. Phys. Lett. **B408**, 98–104 (1997). arXiv:hep-th/9705221

[32]   Aharony, O., Berkooz, M., Seiberg, N.: Light-cone description of (2,0) superconformal theories in six dimensions. Adv. Theor. Math. Phys. **2**, 119–153 (1998). arXiv:hep-th/9712117

[33]   Bhattacharya, J., Bhattacharyya, S., Minwalla, S., Raju, S.: Indices for superconformal field theories in 3, 5 and 6 dimensions. JHEP **02**, 064 (2008). arXiv:0801.1435

[34]   Gaiotto, D., Moore, G.W., Neitzke, A.: Wall-crossing, Hitchin systems, and the WKB approximation. arXiv:0907.3987

[35]   Chacaltana, O., Distler, J., Tachikawa, Y.: Nilpotent orbits and codimension-two defects of 6d N = (2, 0) theories. arXiv:1203.2930

[36]   Gaiotto, D., Teschner, J.: Irregular singularities in Liouville theory and Argyres-Douglas type gauge theories, I. JHEP **1212**, 050 (2012). arXiv:1203.1052

[37]   Katz, S.H., Klemm, A., Vafa, C.: Geometric engineering of quantum field theories. Nucl. Phys. **B497**, 173–195 (1997). arXiv:hep-th/9609239

Hitchin Systems in $\mathcal{N} = 2$ Field Theory

Andrew Neitzke

1 Introduction

This note is a short review of the way Hitchin systems appear in four-dimensional $\mathcal{N} = 2$ supersymmetric field theory.

The literature on the Hitchin system and its role in quantum field theory is a vast one. Restricting attention just to the role of Hitchin systems in $\mathcal{N} = 2$ supersymmetric field theory (thus neglecting such fascinating topics as T-duality on the Hitchin fibration and its relation to the geometric Langlands program [27, 37, 38, 43], the use of Hitchin systems in $\mathcal{N} = 4$ super Yang-Mills [2–4], the role of Higgs bundles in F-theory [16], ...) cuts things down somewhat but still leaves an enormous pool of papers and topics from which to choose. In this article I focus on the points with which I am most personally familiar. In particular, although this review is meant for a special volume devoted to the AGT correspondence, I will have very little to say about that. This is not because I think there is nothing to say—on the contrary, works such as [48, 56] have demonstrated that there clearly is—but because I do not know precisely what to say.

In one sentence, the relation between $\mathcal{N} = 2$ theories and Hitchin systems is that the Hitchin system arises as the *moduli space* of the $\mathcal{N} = 2$ theory compactified on a circle. My aim in this note is to explain a dictionary between various aspects of the field theory (its Coulomb branch, its line defects, its surface defects, ...) and their manifestations in the Hitchin system (the Hitchin base, some distinguished holomorphic functions, some distinguished hyperholomorphic bundles, ...), along with a few ways in which this dictionary gives insight into aspects of the Hitchin system.

A. Neitzke (✉)
Department of Mathematics, University of Texas at Austin, Austin, US
e-mail: neitzke@math.utexas.edu

© Springer International Publishing Switzerland 2016
J. Teschner (ed.), *New Dualities of Supersymmetric Gauge Theories*,
Mathematical Physics Studies, DOI 10.1007/978-3-319-18769-3_3

53

My perspective on this subject has been heavily influenced by a long and enjoyable collaboration with Davide Gaiotto and Greg Moore. It is a pleasure to thank them for this collaboration and for the many things that they have taught me. This work is supported by NSF grant 1151693.

In Sect. 2 we review general facts about $\mathcal{N} = 2$ theories, their relation to integrable systems and hyperkähler geometry, and line and surface defects therein. The Hitchin system does not appear explicitly in this section. In Sect. 3 we specialize to the case of theories of class S; this is the class of $\mathcal{N} = 2$ theories most directly related to Hitchin systems. Finally, in Sect. 4 we give some general background on the Hitchin system, divorced from its role in physics; this section could in principle be read on its own, but is mainly intended as a reference for selected facts which we will need in the other sections.

Each subsection of Sects. 2 and 3 is preceded by a brief slogan. It may be worth reading all the slogans first, to get an idea of what is going on here.

2 $\mathcal{N} = 2$ Theories and Their Circle Compactification

In this section we briefly review some facts about $\mathcal{N} = 2$ theories T in 4 dimensions, and their compactification on a circle R to give theories $T[R]$ in 3 dimensions.

We will describe only general features here, without specializing to any particular theory T; in the next section we will explain how all of these general phenomena are realized in the special case of theories of class S.

2.1 $\mathcal{N} = 2$ Theories in the IR and Integrable Systems

Any $\mathcal{N} = 2$ theory gives rise to a complex integrable system.

Consider an $\mathcal{N} = 2$ supersymmetric theory T in 4 dimensions. Let $\mathcal{B}$ denote the Coulomb branch. $\mathcal{B}$ consists of an open "regular locus" $\mathcal{B}_{\text{reg}}$ plus a "discriminant locus" $\mathcal{B}_{\text{sing}}$.

The IR physics in vacua labeled by points $u \in \mathcal{B}_{\text{reg}}$ is governed by pure abelian $\mathcal{N} = 2$ gauge theory, with gauge group $U(1)^r$, where $r = \dim_{\mathbb{C}} \mathcal{B}$. Locally around any point $u \in \mathcal{B}_{\text{reg}}$, this IR theory can be described in terms of classical fields, namely $r\mathcal{N} = 2$ vector multiplets. The bosonic field content is thus r complex scalars and r abelian gauge fields. However, there is generally no single Lagrangian that describes the IR theory globally on $\mathcal{B}_{\text{reg}}$: rather, we must use different Lagrangians in different patches of $\mathcal{B}_{\text{reg}}$, related to one another by electric/magnetic duality transformations. This story was first worked out in [50, 51]. Although we cannot write a single

Lagrangian that describes the theory globally, there is a single geometric object from which all the local Lagrangians can be derived [14, 15, 47]. This object is a *complex integrable system*: a holomorphic symplectic manifold $\mathcal{I}'$, with a projection $\pi : \mathcal{I}' \to \mathcal{B}_{reg}$, such that the fibers $\mathcal{I}'_u = \pi^{-1}(u)$ are compact complex Lagrangian tori, of complex dimension r. One has the following dictionary:

Fiber of $\mathcal{I}$ over $u \in \mathcal{B}_{reg}$	IR physics at $u \in \mathcal{B}_{reg}$
$H_1(\mathcal{I}'_u, \mathbb{Z})$	EM charge lattice Γ_u
Polarization of $\mathcal{I}'_u$	DSZ pairing on Γ_u
Symplectic basis of $H_1(\mathcal{I}'_u, \mathbb{Z})$	Electric-magnetic splitting
Automorphisms of $\mathcal{I}'_u$	EM duality group ($\simeq Sp(2r, \mathbb{Z})$)
Period matrix of $\mathcal{I}'_u$	Matrix of EM gauge couplings
Point of $\mathcal{I}'_u$	EM holonomies around surface defect

So far we have been discussing the IR physics at points u in the regular locus $\mathcal{B}_{reg}$. At $u \in \mathcal{B}_{sing}$ the simple description of the IR physics by pure abelian gauge theory breaks down, and has to be replaced by something more complicated. Correspondingly, the complex integrable system $\mathcal{I}'$ generally gets extended by adding some singular fibers (degenerations of tori) over $u \in \mathcal{B}_{sing}$. Altogether we get a complete holomorphic symplectic manifold $\mathcal{I}$ fibered over the whole $\mathcal{B}$.

2.2 Compactification of $\mathcal{N} = 2$ Theories on S^1

Compactifying on S^1 turns the integrable system into an honest hyperkähler space.

In Sect. 2.1 we have reviewed the complex integrable system $\mathcal{I}$ which governs the IR physics of the four-dimensional field theory T. In that discussion $\mathcal{I}$ appeared in a somewhat indirect way. Now we describe a way to see $\mathcal{I}$ more directly.

Compactify T on S^1 of length $2\pi R$. At energies $E \ll 1/R$ the resulting physics should be described by a three-dimensional field theory $T[R]$. To get a first approximation to the physics of $T[R]$, we can consider the dimensional reduction of the local IR Lagrangians describing T (at least if we stay away from $\mathcal{B}_{sing}$). Then the fields will be as follows: r complex scalars, r abelian gauge fields, and r periodic real scalars (the holonomies of the gauge fields around S^1). We can moreover dualize the abelian gauge fields to get another r periodic scalars, so altogether we have r complex scalars and $2r$ periodic real scalars. The complex scalars parameterize a sigma model into $\mathcal{B}$, and we can think of the $2r$ periodic real scalars as giving a map into a $2r$-torus; so locally we now have a sigma model into a product of $\mathcal{B}$ with a real $2r$-torus.

To find the global structure of this sigma model, one has to keep track of the EM duality transformations needed to glue together the various local IR Lagrangians of T. After so doing, one finds that $T[R]$ is a sigma model whose target is the complex integrable system $\mathcal{I}'$ which we described in Sect. 2.1. Thus, after compactification the integrable system "comes to life."

We should clarify the meaning of the statement that $T[R]$ is a sigma model into $\mathcal{I}'$. In Sect. 2.1 we described $\mathcal{I}'$ only as a holomorphic symplectic manifold. Now we are getting an actual sigma model into a Riemannian manifold $\mathcal{M}'[R]$. Thus we should ask: how is the Riemannian manifold $\mathcal{M}'[R]$ related to the holomorphic symplectic manifold $\mathcal{I}'$?

The answer is as follows. The constraints of $\mathcal{N} = 4$ supersymmetry in 3 dimensions dictate that the metric on $\mathcal{M}'[R]$ must be hyperkähler [5, 52]. Since $\mathcal{M}'[R]$ is hyperkähler, it carries a family of complex structures J_ζ, parameterized by $\zeta \in \mathbb{CP}^1$, as well as corresponding holomorphic symplectic forms ϖ_ζ. One of these complex structures, J_0, is distinguished. When considered as a holomorphic symplectic manifold in the complex structure J_0, $\mathcal{M}'[R]$ is identical to $\mathcal{I}'$.

The exact IR physics (as opposed to the physics obtained by naive dimensional reduction) is also given by a sigma model into $\mathcal{M}'$. However, the exact hyperkähler metric on $\mathcal{M}'[R]$ is *not* the same as the one obtained by naive dimensional reduction: rather they differ by quantum corrections which can be computed in terms of the spectrum of BPS particles of T_4 [28, 36, 52]. The corrections due to a BPS particle of mass M go like e^{-RM} in the limit $R \to \infty$, so in this limit the two metrics converge to one another uniformly, *except* around points where the mass of some BPS particle goes to zero. The locus where this happens is precisely $\mathcal{B}_{\text{sing}}$, so around $\mathcal{B}_{\text{sing}}$ the quantum corrections are not suppressed even in the $R \to \infty$ limit.[1]

Although the quantum corrections change the metric on $\mathcal{M}[R]$, they do not change what the space looks like in complex structure J_0: even after the corrections, it is still identical to the complex integrable system $\mathcal{I}$ from Sect. 2.1 [52].

2.3 Holomorphic Functions and Line Defects

Vevs of line defects are global holomorphic functions on the hyperkähler space.

The family of complex structures J_ζ on $\mathcal{M}[R]$ (parameterized by $\zeta \in \mathbb{CP}^1$) corresponds to a family of 1/2-BPS subalgebras $\mathcal{A}_\zeta$ of the $\mathcal{N} = 4$ supersymmetry algebra. Vevs of 1/2-BPS local operators $\mathcal{O}$ preserving $\mathcal{A}_\zeta$ thus give J_ζ-holomorphic functions on $\mathcal{M}[R]$.

---

[1]The fact that quantum correction are not suppressed around $\mathcal{B}_{\text{sing}}$ is a good thing: exactly at this locus the naive metric becomes singular, and the quantum corrections smooth out these singularities, in such a way that the exact corrected metric extends to a complete space $\mathcal{M}[R]$ which includes fibers over $\mathcal{B}_{\text{sing}}$. This smoothing requires a correction which is of order 1, not suppressed in R.

For example, the complex scalars $\mathcal{O}$ which descend from the vector multiplets in the original theory T preserve $\mathcal{A}_0$. It follows that the vevs of these complex scalars are J_0-holomorphic functions on $\mathcal{M}[R]$. Said otherwise, the projection $\mathcal{M}[R] \to \mathcal{B}$ is a J_0-holomorphic map. Of course this is just what we expect, since we have already said that in complex structure J_0, $\mathcal{M}[R]$ is the complex integrable system $\mathcal{I}$, with base $\mathcal{B}$.

The four-dimensional origin of operators $\mathcal{O}$ preserving the other subalgebras $\mathcal{A}_\zeta$, $\zeta \in \mathbb{C}^\times$, is a bit different: we consider 1/2-BPS *line defects* in the original theory T. Such line defects can preserve various different subalgebras of the 4-dimensional supersymmery. Upon circle compactification, the line defects reduce to point operators, and their preserved subalgebras reduce to the various 1/2-BPS subalgebras $\mathcal{A}_\zeta$. Thus, the vevs of supersymmetric line defects wrapped on S^1 are J_ζ-holomorphic functions on $\mathcal{M}[R]$.

Among the supersymmetric line defects there is a distinguished subset of "simple" defects, characterized by the property that a simple defect is not expressible (in correlation functions) as a nontrivial sum of other defects. We expect that every defect can be uniquely decomposed as a sum of simple defects (though this statement is not entirely trivial—see [11, 30] for more discussion.)

The existence of simple line defects implies in particular that there should be a distinguished vector space basis of the space of J_ζ-holomorphic functions on $\mathcal{M}[R]$. Distinguished bases for coordinate rings of various algebraic spaces (and their quantum deformations) have been studied in Lie theory (following pioneering work of Lusztig, e.g. [46]) and more generally in algebraic geometry (see e.g. [33]). Indeed, the investigation of these "canonical bases" was an important motivation for the theory of cluster algebras [23]. On the other hand, it has turned out independently that cluster algebras are closely related to the algebras of line defects (see e.g. [8, 29, 59, 60] for more on this.) Thus it seems natural to suspect that the canonical bases studied in mathematics can be identified with the ones coming from simple line defects. This point remains to be understood more precisely.

2.4 Hyperholomorphic Bundles and Surface Defects

Surface defects give hyperholomorphic bundles on the hyperkähler space.

Now, as in [1, 26, 30, 34], let us consider 1/2-BPS *surface defects* in the four-dimensional theory T. We focus on defects which are massive in the IR, with finitely many vacua. Let $\mathbb{S}$ be such a defect. Upon compactification of both T and $\mathbb{S}$ on S^1, $\mathbb{S}$ has a finite-dimensional Hilbert space of ground states, which we denote $V(\mathbb{S})$.

To be more precise, the Hilbert space $V(\mathbb{S})$ actually depends on which vacuum of the theory $T[R]$ we are in. Thus we have a family of Hilbert spaces varying over the moduli space $\mathcal{M}[R]$. Said otherwise, $V(\mathbb{S})$ is a Hermitian vector bundle over $\mathcal{M}$. The supersymmetry in the situation implies that this vector bundle is *hyperholomorphic*: in particular, it admits a family of holomorphic structures, one for each $\zeta \in \mathbb{CP}^1$, compatible with the family of underlying complex structures J_ζ on $\mathcal{M}[R]$.

Suppose given two such defects, labeled $\mathbb{S}$ and $\mathbb{S}'$, and a 1/2-BPS *interface* between them. As observed in [35], such an interface can be viewed as a kind of line defect which is restricted to live on the surface defect rather than roaming free in the 4-dimensional bulk. Upon circle compactification, this picture reduces to a pair of line defects separated by a local operator. The local operator preserves $\mathcal{A}_\zeta$ for some ζ (just as in the case with no surface defects). The vev of this local operator is then a J_ζ-holomorphic section of $\mathrm{Hom}(V(\mathbb{S}), V(\mathbb{S}'))$.

2.5 Line Defects in the IR

UV line defects can be expanded in terms of IR ones; the coefficients of this expansion are integers which jump as parameters are varied.

In Sect. 2.3 we considered line defects from the UV perspective. On the other hand, we could also consider the theory in the IR, in the vacuum labeled by some $u \in \mathcal{B}$. As we recalled in Sect. 2.1, the IR physics is governed by *abelian* $\mathcal{N} = 2$ gauge theory. In pure abelian gauge theory, for any $\zeta \in \mathbb{C}^\times$ we can concretely describe the full set of simple ζ-supersymmetric line defects: for every γ in the EM charge lattice Γ_u, there is a ζ-supersymmetric abelian Wilson-'t Hooft operator $L(\gamma)$.

Now, given a ζ-supersymmetric line defect L_{UV}, we can ask how the same defect appears in the IR. It will look like some integer linear combination of the simple defects of the IR theory:

$$L_{UV} \rightsquigarrow \sum_{\gamma \in \Gamma_u} \overline{\underline{\Omega}}(L_{UV}, \gamma) L_{IR}(\gamma) \tag{2.1}$$

The coefficients $\overline{\underline{\Omega}}(L_{UV}, \gamma) \in \mathbb{Z}$ of this expansion can be interpreted as indices counting supersymmetric ground states of the theory with L_{UV} inserted at some fixed spatial point, extended in the time direction. These states were called *framed BPS states* in [29].

Importantly, the $\overline{\underline{\Omega}}(L_{UV}, \gamma)$ can jump as we vary the parameters (u, ζ): this is the phenomenon of (framed) wall-crossing. The jumps occur when a framed BPS bound state decays or forms, by binding or releasing an unframed BPS state; thus the precise way in which the $\overline{\underline{\Omega}}(L_{UV}, \gamma)$ jump is determined by the (unframed) BPS degeneracies of the theory. Indeed, studying the jumps of the $\overline{\underline{\Omega}}(L_{UV}, \gamma)$ gives a lot of information about the unframed BPS degeneracies: in particular, it is one way of establishing that these degeneracies obey the celebrated Kontsevich-Soibelman wall-crossing formula [44].

Now, let us again consider compactifying on S^1 and taking vevs. Then (2.1) becomes an equation relating the vev of L_{UV} to a sum of vevs of defects L_{IR}:

$$\langle L_{UV} \rangle = \sum_{\gamma \in \Gamma_u} \overline{\underline{\Omega}}(L_{UV}, \gamma) \langle L_{IR}(\gamma) \rangle \tag{2.2}$$

However, the quantities $\langle L_{IR}(\gamma) \rangle$ are not as simple as they would be in the pure abelian gauge theory — they are significantly corrected by contributions from higher-dimension operators. Indeed, to get an indication of how subtle these quantities are, note that $\langle L_{UV} \rangle$ should be continuous as a function of the parameters (u, ζ) (since the UV theory T has no phase transition), while we have just said that the coefficients $\overline{\underline{\Omega}}(L_{UV}, \gamma)$ jump at some walls in the (u, ζ) parameter space.[2] Thus, for (2.2) to be consistent, the vevs $\langle L_{IR}(\gamma) \rangle$ must also jump at these $\mathcal{K}$-walls. As with $\overline{\underline{\Omega}}(L_{UV}, \gamma)$, the jumps of $\langle L_{IR}(\gamma) \rangle$ are completely determined by the unframed BPS degeneracies.

We will return to the meaning of (2.2) when we consider theories of class S, below.

2.6 Asymptotics

Vevs of line defects are *asymptotically* related to functions on the Coulomb branch.

For each $\zeta \in \mathbb{C}^\times$, the ζ-supersymmetric IR line defect vev $\langle L_{IR}(\gamma) \rangle$ is a J_ζ-holomorphic function on $\mathcal{M}[R]$. These functions have an important asymptotic property: as $\zeta \to 0$, they behave as [28, 29]

$$\langle L_{IR}(\gamma) \rangle \sim c(\gamma) \exp(\zeta^{-1} \pi R Z_\gamma) \tag{2.3}$$

where $c(\gamma)$ is some ζ-independent constant, and Z_γ is the central charge function, pulled back from the Coulomb branch $\mathcal{B}$.

These asymptotics are realized in a rather nontrivial way. As we have emphasized, the $\langle L_{IR}(\gamma) \rangle$ are not continuous, but have jumps corresponding to BPS states of the theory T. If we fix a point of $\mathcal{M}[R]$ and look only at the ζ dependence, then the loci where the jumps occur are rays in the ζ-plane, all of which run into the origin. These

---

[2]These walls are known by various names: "BPS walls" in [29], "$\mathcal{K}$-walls" in [31], "walls of second kind" in [44], or parts of the "scattering diagram" in [32].

jumps however do not destroy the asymptotics—rather the discontinuity across each ray becomes trivial in the $\zeta \to 0$ limit. This is an example of the Stokes phenomenon.

One concrete consequence is that the expansion of each $\langle L_{IR}(\gamma) \rangle$ around $\zeta = 0$ will be given only by an *asymptotic* series, not a convergent one (a convergent series would necessarily converge to a continuous function, but $\langle L_{IR}(\gamma) \rangle$ is not continuous in any disc around $\zeta = 0$).

3 Theories of Class S and Hitchin Systems

In this section we specialize from general theories T to theories of class S, $T = S[\mathfrak{g}, C]$. These are theories obtained by compactification of the $(2, 0)$ theory from 6 to 4 dimensions; for the definition see [25, 27], or [V:2] in this volume. In these theories we will see the role of the Hitchin system.

3.1 Theories of Class S

For theories of class S, the hyperkähler manifold which appears upon compactification to three dimensions is a Hitchin system.

Now suppose that T is a theory of class S, $T = S[\mathfrak{g}, C]$. The general discussion of Sect. 2.2 applies to this particular theory. Thus compactifying T on S^1 gives a sigma model $T[R]$ into an hyperkähler manifold $\mathcal{M}$. In this case, we can understand concretely what $\mathcal{M}$ is, as follows.

The 3-dimensional theory $T[R]$ has several descriptions summarized in this picture (arrows mean "compactify and take IR limit"):

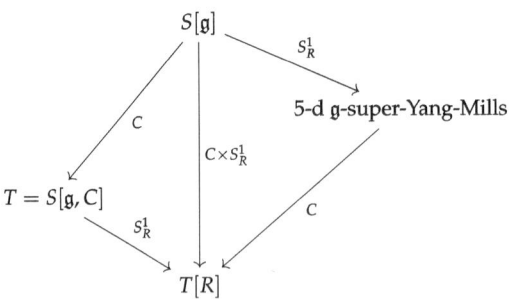

$$(3.1)$$

The left side of the picture is how we have described $T[R]$ up to now: $T[R]$ is the IR limit of the compactification of $S[\mathfrak{g}, C]$ on S_R^1, and $S[\mathfrak{g}, C]$ in turn can be understood as the IR limit of the compactification of the six-dimensional theory $S[\mathfrak{g}]$ on C. Altogether this means that $T[R]$ is simply the IR limit of the compactification of $S[\mathfrak{g}]$ on $C \times S^1$, as indicated by the middle arrow of the picture. Finally we may do this compactification in the opposite order, obtaining the right side of the picture. We first compactify $S[\mathfrak{g}]$ on S_R^1 and take an IR limit to obtain 5-dimensional super Yang-Mills with gauge algebra $\mathfrak{g}$.[3] Then we compactify this 5-dimensional super Yang-Mills theory on C and take an IR limit to get $T[R]$. This leads to the statement that $T[R]$ is a sigma model into the moduli space of vacuum configurations of 5d super Yang-Mills on $C \times \mathbb{R}^{2,1}$ which are translation invariant in the $\mathbb{R}^{2,1}$ directions.

The requirement of translation invariance along $\mathbb{R}^{2,1}$ means that the BPS equations on $C \times \mathbb{R}^{2,1}$ reduce to equations for fields on C. These equations turn out to be some celebrated equations in gauge theory: they are the Hitchin equations (4.1), which we discuss in Sect. 4.1 below. (More precisely, the equations which appear are the Hitchin equations modified by the rescaling $\varphi \to R\varphi$.) This was essentially observed in [6, 37] (in a slightly different context, but the mathematical problem is the same); see also [9, 10] where some important special cases were rediscovered in a context closer to ours.

Thus the target $\mathcal{M}[R]$ of the sigma model $T[R]$ is the moduli space of solutions of Hitchin equations. For the moment we do not need the detailed form of these equations: we will just need a few basic properties of $\mathcal{M}[R]$. In particular,

- $\mathcal{M}[R]$ is a hyperkähler space (Sect. 4.4), as required by $\mathcal{N} = 4$ supersymmetry in three dimensions.
- In its complex structure J_0, $\mathcal{M}[R]$ can be identified with a complex integrable system $\mathcal{I}$ (Sect. 4.6), as expected from Sects. 2.1–2.2.

Let us say a bit more about this integrable system, specializing to the case $\mathfrak{g} = A_{K-1}$ for concreteness.

- The base of the integrable system $\mathcal{I}$ is the "Hitchin base" (Sect. 4.6). On the other hand, from Sect. 2.1 we know that the base should be the Coulomb branch of T. Thus the Coulomb branch $\mathcal{B}$ of T can be identified with the Hitchin base. In particular, the points $u \in \mathcal{B}$ correspond to algebraic curves $\Sigma_u \subset T^*C$ which are K-fold covers of C, better known as "Seiberg-Witten curves."
- The torus fibers $\mathcal{I}_u$ have a concrete algebro-geometric meaning in terms of the Seiberg-Witten curves Σ_u, as follows: a point of $\mathcal{I}_u$ corresponds to a holomorphic line bundle $\mathcal{L}$ over Σ_u, with the extra property that the determinant of the pushforward bundle $\pi_*\mathcal{L}$ is trivial, where $\pi : \Sigma_u \to C$ denotes the covering map (Sect. 4.6).

---

[3] Actually, specifying $\mathfrak{g}$ does not quite determine the 5-dimensional theory; for that we should really specify a particular Lie group G with Lie algebra $\mathfrak{g}$. Which G we get depends on a subtle discrete choice which appears upon compactification, as described e.g. in [24], using some subtleties of the 6-dimensional $S[\mathfrak{g}]$ explained in [58].

3.2 Line Defects

In the theory $S[A_1, C]$, vevs of line defects are holonomies of flat connections along C.

In Sect. 2.3 we explained that for any $\mathcal{N} = 2$ theory the vevs of ζ-supersymmetric line defects compactified on S^1 should give J_ζ-holomorphic functions on $\mathcal{M}[R]$. In theories of class S, these functions turns out to be something quite concrete and understandable in terms of the curve C, as follows.

We will need one more fact about $\mathcal{M}[R]$ (reviewed in Sect. 4.3 below): $\mathcal{M}[R]$ is diffeomorphic to the moduli space $\mathcal{M}_{flat}$ of flat $G_{\mathbb{C}}$-connections, via a map f_ζ which is J_ζ-holomorphic,

$$\mathcal{M} \xrightarrow{f_\zeta} \mathcal{M}_{flat} \tag{3.2}$$

$$x \mapsto \nabla(x, \zeta) \tag{3.3}$$

Thus, if we fix a holomorphic function F on $\mathcal{M}_{flat}$, we can get a J_ζ-holomorphic function F_ζ on $\mathcal{M}$ by pullback:

$$F_\zeta(x) = F(\nabla(x, \zeta)). \tag{3.4}$$

The vevs of ζ-supersymmetric holomorphic line defects arise in this way: each type of simple line defect L corresponds to some holomorphic function $F = F_L$ on $\mathcal{M}_{flat}$.

What are the functions F_L concretely? Let us restrict our attention to the case $\mathfrak{g} = A_1$. In these theories we have a complete understanding of the set of supersymmetric line defects following [19, 29]. The story is especially simple if C has only regular punctures. In that case, for any $\zeta \in \mathbb{C}^\times$, there are simple ζ-supersymmetric line defects corresponding to pairs $\{(\wp, a)\}$, where $\wp$ is a non-self-intersecting closed curve on C, and a a nonnegative integer:

$$L \leftrightarrow (\wp, a). \tag{3.5}$$

The corresponding function F_L on $\mathcal{M}_{flat}$ is

$$F_L(\nabla) = \mathrm{Tr}\,(P_\nabla(\wp, a)), \tag{3.6}$$

where $P_\nabla(\wp, a)$ means the parallel transport of the connection ∇ around the path $\wp$, in the $(a + 1)$-dimensional representation of $SL(2, \mathbb{C})$.[4]

For general $\mathfrak{g}$ it seems very likely that there are line defects whose vevs give holonomies of $SL(K, \mathbb{C})$ connections along closed paths, as well as defects corresponding to more general "spin networks"; however, the story has not yet been completely developed, and in particular it is not yet known how to describe a complete set of simple line defects. Some examples have very recently been worked out in [60]; see also [45, 53] for related mathematical work.

3.3 Interfaces Between Surface Defects

In the theory $S[A_1, C]$, interfaces between surface defects correspond to parallel transport of flat connections along open paths on C.

All the discussion of Sect. 3.2 has a natural extension where we replace line defects by interfaces between surface defects [1, 26, 30], as follows.

In the theory $S[A_1, C]$ there is a natural family of surface defects $\mathbb{S}_z^a$, labeled by an integer $a > 0$ and a point $z \in C$. As we have described in Sect. 2.4, each such defect corresponds to a hyperholomorphic vector bundle $V(\mathbb{S}_z^a)$ over $\mathcal{M}$. In this case, $V(\mathbb{S}_z^a)$ is the a-th symmetric power of the *universal harmonic bundle*, restricted to $z \in C$ (see Sect. 4.5). In particular, when we view it as a holomorphic bundle in complex structure J_ζ, $V(\mathbb{S}_z^a)$ is the a-th symmetric power of the *universal flat bundle*, restricted to $z \in C$.

So much for the surface defects by themselves: how about interfaces between surface defects? Much like (3.5), we have a correspondence

$$L \leftrightarrow (\wp, a) \tag{3.7}$$

where $\wp$ now denotes an *open* path $\wp$ from z to z', and the corresponding L is a ζ-supersymmetric *interface* between $\mathbb{S}_z^a$ and $\mathbb{S}_{z'}^a$. The corresponding vev F_L should be a holomorphic section of $\mathrm{Hom}(V(\mathbb{S}_z^a), V(\mathbb{S}_{z'}^a))$. That section is

$$F_L(\nabla) = P_\nabla(\wp, a). \tag{3.8}$$

Thus: in the theory $S[A_1, C]$, vevs of interfaces between surface defects are parallel transports of $SL(2, \mathbb{C})$ connections along open paths on C.

---

[4]Slightly more generally, there are also simple line defects corresponding to mutually non-intersecting *collections* of closed curves on C, with nonnegative integer weights; the vev of such a defect is simply the product of the traces associated to the individual curves in the collection.

Note that giving the operators $P_\nabla(\wp, a)$ for *all* paths $\wp$ is equivalent to giving the connection ∇ itself. Thus, for any fixed $\zeta \in \mathbb{C}^\times$, studying ζ-supersymmetric interfaces between surface defects in the theory $S[A_1, C]$ is *equivalent* to studying flat $SL(2, \mathbb{C})$-connections on C. (Indeed, this gives an alternative derivation of the fact that $\mathcal{M}[R]$ in complex structure J_ζ is isomorphic to the moduli space of flat $SL(2, \mathbb{C})$-connections).

3.4   Line Defects in the IR

> Vevs of IR line defects give local coordinate systems on the Hitchin moduli space; one can get Fock-Goncharov and Fenchel-Nielsen coordinates in this way.

We have said in (3.4) and (3.6) that the vevs $\langle L_{UV} \rangle$ are the f_ζ-pullback from $\mathcal{M}_{flat}$ of some particular holomorphic functions, namely the trace functions attached to closed paths on C.

Something similar is true for the IR vevs $\langle L_{IR}(\gamma) \rangle$. As we have commented, these functions are not quite globally holomorphic on $\mathcal{M}[R]$: rather they jump at some codimension-1 loci ($\mathcal{K}$-walls). However, suppose that we initially restrict to a small neighborhood of some initial u, and then (if we like) extend $\langle L_{IR}(\gamma) \rangle$ to a larger domain by analytic continuation. In this case we obtain an honest holomorphic function F_γ, defined on some domain in $\mathcal{M}[R]$. These holomorphic functions should be regarded as IR analogues of the $\langle L_{UV} \rangle$ we considered above, and in some respects they are similar: in particular, they are also the f_ζ-pullback of some holomorphic functions F_γ on $\mathcal{M}_{flat}$.

Precisely what functions F_γ we get in this way depends on our choice of an initial u, and also on the parameter ζ. For any fixed choice, considering all F_γ at once gives a local *coordinate system* on $\mathcal{M}_{flat}$. In particular, one can obtain in this way both the Fock-Goncharov and complexified Fenchel-Nielsen coordinate systems on $\mathcal{M}_{flat}$ [27, 41]. The Fock-Goncharov coordinates are obtained for generic choices of (u, ζ) while some special "real" (u, ζ) (related to Strebel differentials on C) give Fenchel-Nielsen.

Incidentally, the coefficients of the expansion (2.1), i.e. the framed BPS indices, have a concrete geometric interpretation: they are counting geometric objects on the curve C, called "millipedes" in [29].

3.5 (Non)abelianization

Vevs of IR line defects can be viewed as giving flat $\mathbb{C}^\times$-connections over spectral curves.

There is another way of viewing the vevs of IR line defects. The charge lattice Γ_u in the theory $S[A_1, C]$ can be described concretely in terms of the Seiberg-Witten curve Σ_u. Indeed, Γ_u sits inside $H_1(\Sigma_u, \mathbb{Z})$.[5] Thus in the IR we have line defects corresponding to paths on Σ. Moreover, there is a $\mathbb{C}^\times$ connection $\nabla^{\mathrm{ab}}(\zeta)$ over Σ, such that the vev of the simple ζ-supersymmetric line defect corresponding to the homology class γ is the holonomy of $\nabla^{\mathrm{ab}}(\zeta)$ around any path in the homology class γ. One can think of $\nabla^{\mathrm{ab}}(\zeta)$ as an "abelianization" of $\nabla(\zeta)$.

This construction can be summarized in a commutative diagram:

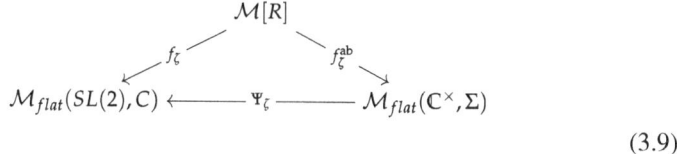

$$(3.9)$$

The left arrow f_ζ is the "UV" map which takes a vacuum of $T[R]$ to its corresponding $SL(2, \mathbb{C})$-connection $\nabla(\zeta)$ over C. The right arrow f_ζ^{ab} is the "IR" map which takes a vacuum of $T[R]$ to its corresponding $\mathbb{C}^\times$-connection $\nabla^{\mathrm{ab}}(\zeta)$ over Σ. The two differ by a third map Ψ_ζ, which we call "nonabelianization" since it takes an abelian connection ∇^{ab} over Σ to a nonabelian one ∇ over C. From the fact that the framed BPS counts $\overline{\Omega}$ are piecewise constant, it follows that Ψ_ζ depends in a piecewise constant way on ζ, and its jumps are controlled by the BPS spectrum of the theory $S[A_1, C]$.

The story is expected to be similar for arbitrary $\mathfrak{g}$; in particular, the nonabelianization map Ψ_ζ was described in detail in [31] for $\mathfrak{g} = gl(K)$.

3.6 Asymptotics

Fock-Goncharov and Fenchel-Nielsen coordinates have nice asymptotic properties, when evaluated along special 1-parameter families of connections coming from points of the Hitchin system.

---

[5]To be precise, consider the projection map $\pi_* : H_1(\Sigma_u, \mathbb{Z}) \to H_1(C, \mathbb{Z})$; the lattice Γ_u is the kernel of this projection.

Fix a point $x \in \mathcal{M}[R]$ and some $\zeta_0 \in \mathbb{C}^\times$. As we have said in Sect. 3.2, there is a corresponding real 1-parameter family of connections $\nabla(t) = f_{t\zeta_0}(x), t > 0$, i.e. a real path in $\mathcal{M}_{flat}$. On the other hand, as we have explained in Sect. 3.4, the choice of (x, ζ_0) also determines a particular local coordinate system on $\mathcal{M}_{flat}$ (by taking vevs of ζ_0-supersymmetric IR line operators, analytically continued from the initial point x.) In particular, these may be Fock-Goncharov or Fenchel-Nielsen coordinates on $\mathcal{M}_{flat}$.

As $t \to 0$, the coordinates $F_\gamma(\nabla(t))$ thus behave according to (2.3). The general statement (2.3) involves the central charges of the theory, but for this particular theory they can be written more concretely:

$$Z_\gamma = \frac{1}{\pi} \int_\gamma \lambda \tag{3.10}$$

for λ the Liouville 1-form on T^*C. Thus (2.3) becomes

$$F_\gamma(\nabla(t)) \sim c(\gamma) \exp\left(t^{-1}\zeta_0^{-1} R \oint_\gamma \lambda\right). \tag{3.11}$$

Let us make a few remarks about (3.11):

- (3.11) is a version of the WKB approximation, applied to the special family of connections $\nabla(t)$; indeed, $\nabla(t)$ has the form (see (4.6))

$$\nabla(t) = t^{-1}\varphi + \cdots \tag{3.12}$$

where $\zeta_0^{-1}\varphi$ is a 2×2 matrix-valued 1-form on C, whose 2 eigenvalues are the values of $R\lambda$ on the 2 sheets of Σ.
- (3.11) provides a link between the Fock-Goncharov or Fenchel-Nielsen coordinates and the periods of the spectral curve. This link plays some role in the AGT correspondence — e.g. for Fenchel-Nielsen coordinates it seems to be used in [56]. The nature of the link is somewhat nontrivial (cf. the comments in Sect. 2.6 about Stokes phenomena); this is in some sense to be expected, since the two objects we are relating are holomorphic in different complex structures on the Hitchin space. I would very much like to know whether these Stokes phenomena have some significance in AGT.
- All of this is expected to generalize to $\mathfrak{g} = gl(K)$, as outlined in [31]. The coordinate systems which appear there seem to be more general than Fock-Goncharov or Fenchel-Nielsen. Presumably it generalizes further to any $\mathfrak{g}$ of ADE type, but this generalization has not yet been worked out.

3.7 Operator Products and their Quantization

Keeping track of spins of framed BPS states leads to a natural quantization of the Hitchin system.

As we have described above, the vevs of supersymmetric line defects give a natural basis for the space of J_ζ-holomorphic functions on $\mathcal{M}[R]$. The algebra structure on this space also has a natural meaning in terms of line defects: it corresponds to the operator product. Indeed, writing LL' for the operator product between ζ-supersymmetric line defects L and L', we have

$$\langle LL' \rangle = \langle L \rangle \langle L' \rangle. \tag{3.13}$$

In particular, this vev does not depend on the direction in which L approaches L'; this is a consequence of the more general fact that moving L or L' changes it only by a term which vanishes in ζ-supersymmetric correlators.

There is an interesting deformation of this story, as follows. Let J_3 be the generator of a spatial $U(1) \subset SO(3)$, and let I_3 be a generator of some $U(1)_R \subset SU(2)_R$. We are going to make a modification of the quantum field theory T, which is most convenient to describe in Hamiltonian language: we insert the operator $(-y)^{2(J_3+I_3)}$ in all correlation functions (so all correlation functions become functions of the auxiliary parameter y, and when $y = -1$ we reduce to the original T). The modified theory is still supersymmetric, but now line defects can be supersymmetric only if they are inserted along the axis $x^1 = x^2 = 0$. As a result, in computing the operator product of supersymmetric line defects we are constrained to consider them approaching one another along this axis. Once again moving the defects along the line does not affect supersymmetric correlators, but there are now two possible orderings of the defects along the line, which have no reason to be equivalent. Thus, at least as far as supersymmetric correlation functions are concerned, we have a noncommutative (but still associative) deformation of the operator product of the original theory. Upon taking vevs, this then induces a corresponding deformation of the algebra of J_ζ-holomorphic functions on $\mathcal{M}[R]$. This deformation has been discussed in various places including [18, 29, 42] (see also [49] in this volume), and essentially also in [21, 22].

For IR line defects we can compute directly in the abelian theory to find the simple deformation

$$\langle L_{IR}(\gamma) L_{IR}(\gamma') \rangle = y^{2\langle \gamma, \gamma' \rangle} \langle L_{IR}(\gamma') L_{IR}(\gamma) \rangle \tag{3.14}$$

(this boils down to working out the angular momentum stored in the crossed electromagnetic fields between two dyons of charges γ and γ'.) On the other hand, as we have described, the $\langle L_{IR}(\gamma) \rangle$ are local coordinates on $\mathcal{M}[R]$; in fact they are even local *Darboux* coordinates, i.e.

$$\{L_{IR}(\gamma), L_{IR}(\gamma')\} = \langle \gamma, \gamma' \rangle L_{IR}(\gamma) L_{IR}(\gamma'). \tag{3.15}$$

Thus (3.14) says that the deformation we are considering is a *quantization* of the Poisson algebra of functions on $\mathcal{M}[R]$.

The precise deformation (3.14) ("quantum torus") had appeared earlier in [13, 44] in the context of the wall-crossing formulas for refined BPS invariants. Here we are encountering the same deformation in our discussion of line defects and their framed BPS states. This is not a coincidence; indeed the refined wall-crossing formula can be understood as a necessary consistency condition for the wall-crossing of framed BPS states [29].

In theories of class $S[A_1]$ the operator product and its quantization are given by "skein relations" like those familiar in Chern-Simons theory (here for the 3-manifold $C \times \mathbb{R}$).

4 Basics on the Hitchin System

In this section we present some background on the Hitchin system, without reference to physics. Fix a compact Riemann surface C, and a compact Lie group G.

4.1 Harmonic Bundles

Hitchin's equations [39] are a system of partial differential equations on C. They concern a triple (E, D, φ) where

- E is a G-bundle on C,
- D is a G-connection in E,
- φ is an element of $\Omega^1(\text{End } E)$.

For example, if $G = SU(K)$, then E can be considered concretely as a Hermitian vector bundle of rank K, with trivial determinant; in a local unitary gauge, D is of the form $D = \partial + A$; and both A and φ are represented by 1-form-valued skew-Hermitian matrices.

The equations are

$$F_D - [\varphi, \varphi] = 0, \tag{4.1a}$$
$$D\varphi = 0, \tag{4.1b}$$
$$D \star \varphi = 0. \tag{4.1c}$$

Call a triple (E, D, φ) obeying these equations a *harmonic bundle*.

When G is abelian, these equations are linear and it is relatively easy to describe the harmonic bundles (it boils down to Hodge theory for 1-forms on the curve C). For G nonabelian, harmonic bundles are harder to describe explicitly. Nevertheless they do exist, as we will discuss below.

Note that from (4.1a) it follows that E must be topologically trivial. There is a generalization to nontrivial bundles, but we will not consider it here.

4.2 Higgs Bundles and Flat Bundles

Given a harmonic bundle, by "forgetting" some of the structure one can obtain either a Higgs bundle or a flat bundle. Remarkably, this "forgetful" map turns out to be invertible, so that we can actually reconstruct the harmonic bundle from a Higgs bundle or a flat bundle. Let us now describe how this works.

Start from a harmonic bundle (E, D, φ). Now suppose we replace E by its complexification, a $G_{\mathbb{C}}$-bundle $E_{\mathbb{C}}$.[6] For example, when $G = SU(K)$, E is a Hermitian vector bundle of rank K, and passing from E to $E_{\mathbb{C}}$ corresponds to forgetting the Hermitian metric and remembering only the underlying complex vector bundle. Let us also decompose D and φ into their $(1, 0)$ and $(0, 1)$ components:

$$(D, \varphi) \rightarrow (D^{(0,1)}, D^{(1,0)}, \varphi^{(1,0)}, \varphi^{(0,1)}). \tag{4.2}$$

4.2.1 Higgs Bundles

Now, suppose that of the four parts (4.2) we remember only the pair

$$(D^{(0,1)}, \varphi^{(1,0)}). \tag{4.3}$$

Then what do we have?

The operator $D^{(0,1)}$ induces the structure of holomorphic $G_{\mathbb{C}}$-bundle on $E_{\mathbb{C}}$ (namely, holomorphic sections are the ones which are annihilated by $D^{(0,1)}$.) Let E_h denote $E_{\mathbb{C}}$ equipped with this holomorphic structure. Equations (4.1b)–(4.1c) together imply that $\varphi^{(1,0)}$ is a holomorphic section of $\mathrm{End}(E_h)$. Let $\phi = \varphi^{(1,0)}$.

Thus, starting from a harmonic bundle, we have produced a pair (E_h, ϕ) where E_h is a holomorphic $G_{\mathbb{C}}$-bundle and ϕ is an $\mathrm{End}(E_h)$-valued holomorphic 1-form. Such a pair is called a *Higgs bundle*.

It looks difficult to recover the original harmonic bundle data (4.2) just from the Higgs bundle data (4.3). If we remembered the underlying G-structure, we could reconstruct $D^{(1,0)}$ from $D^{(0,1)}$, and $\varphi^{(0,1)}$ from $\varphi^{(1,0)}$, just by taking adjoints. However, we have forgotten the G-structure, so we do not have a notion of adjoint. Choosing

---

[6]The notation $E_{\mathbb{C}}$ expresses the fact that the gauge group has been complexified; to avoid confusion we emphasize that the corresponding associated vector bundles do not get complexified.

a random G-structure will not do: this would allow us to construct *some* (D, φ), but there is no reason why Hitchin's equations would be satisfied.

Nevertheless, the remarkable fact [39, 54] is that given a Higgs bundle there is a unique way to find a G-structure such that Hitchin's equations are indeed satisfied! (Strictly speaking this is not quite true for every Higgs bundle, but it is almost true: one only needs to impose an appropriate condition of "stability." This condition holds for a generic Higgs bundle.)

So altogether we have two inverse constructions: one trivial forgetful map from harmonic bundles to Higgs bundles, and one very nontrivial reconstruction map from Higgs bundles to harmonic bundles.

4.2.2 Anti-Higgs Bundles

All of what we have just said has a conjugate version, where instead of (4.3) we remember only the pair

$$(D^{(1,0)}, \varphi^{(0,1)}). \tag{4.4}$$

These give directly the antiholomorphic version of a Higgs bundle, which we might call an *anti-Higgs bundle*. Just as above, we have a forgetful map from harmonic bundles to anti-Higgs bundles, and an inverse reconstruction map from anti-Higgs bundles to harmonic bundles. Complex conjugation exchanges Higgs and anti-Higgs bundles, in a way that commutes with all the above maps.

4.2.3 Flat Bundles

Now suppose instead that we remember a more interesting combination of the data (4.2): fix some $\zeta \in \mathbb{C}^\times$, and remember only the pair

$$(D^{(0,1)} + \zeta\varphi^{(0,1)}, D^{(1,0)} + \zeta^{-1}\varphi^{(1,0)}). \tag{4.5}$$

We may regard these two pieces as the two halves of a complex connection ∇ in $E_\mathbb{C}$:

$$\nabla = \zeta^{-1}\varphi^{(1,0)} + D + \zeta\varphi^{(0,1)}. \tag{4.6}$$

From (4.1) it follows that ∇ is flat. Thus, given a harmonic bundle (E, D, φ) and a parameter $\zeta \in \mathbb{C}^\times$, we have obtained a flat bundle $(E_\mathbb{C}, \nabla)$.

Given only the pair $(E_\mathbb{C}, \nabla)$ it is not obvious how to recover the full harmonic bundle (E, D, φ). Nevertheless this can indeed be done, in a unique way [12, 17] (again under an appropriate "stability" condition, which is generically satisfied). So the story is parallel to what we said above for Higgs bundles: we have a trivial forgetful map from harmonic bundles to flat bundles, and a nontrivial reconstruction map from flat bundles to harmonic bundles. In fact, here we have a *family* of forgetful and reconstruction maps, parameterized by $\zeta \in \mathbb{C}^\times$.

4.2.4   Limits of Parameters

There is a relation between these two constructions, as follows. Evidently, for any $\zeta \in \mathbb{C}^\times$, remembering (4.5) is equivalent to remembering the pair

$$(D^{(0,1)} + \zeta \varphi^{(0,1)}, \zeta D^{(1,0)} + \varphi^{(1,0)}). \tag{4.7}$$

In the limit $\zeta \to 0$ this becomes (4.3). Thus the map between harmonic bundles and Higgs bundles is the $\zeta \to 0$ limit of our family of maps between harmonic bundles and flat connections. Similarly, the map between harmonic bundles and anti-Higgs bundles arises in the $\zeta \to \infty$ limit.

4.2.5   Summing up

Starting from a harmonic bundle, by "forgetting" some information — in a way depeding on a parameter $\zeta \in \mathbb{CP}^1$ — we can produce one of three objects:

1. a Higgs bundle (this arises at $\zeta = 0$),
2. a flat bundle (this arises for any $\zeta \in \mathbb{C}^\times$),
3. an anti-Higgs bundle (this arises at $\zeta = \infty$).

4.3   Moduli Spaces

Now we want to discuss the *moduli space* of harmonic bundles.

It is convenient to fix a single topologically trivial C^∞ bundle E once and for all. Having done so, the remaining equivalences are given by the "gauge group"

$$\mathcal{G} = \{\text{smooth sections of Aut } E\}. \tag{4.8}$$

This $\mathcal{G}$ has an action on (D, φ), under which D transforms as usual for a connection while φ transforms in the adjoint representation. Equations (4.1) are invariant under this action. In particular, $\mathcal{G}$ acts on the space of harmonic bundles.

Similarly, the $G_\mathbb{C}$-bundle appearing in the definition of "Higgs bundles," "flat bundle" or "anti-Higgs bundle" is determined up to equivalence by the same discrete topological invariant. Thus it will be convenient to fix this $G_\mathbb{C}$-bundle to be $E_\mathbb{C}$. Then the remaining equivalences are given by the "complexified gauge group"

$$\mathcal{G}_\mathbb{C} = \{\text{smooth sections of Aut } E_\mathbb{C}\}, \tag{4.9}$$

which thus acts on the space of Higgs bundles, flat bundles or anti-Higgs bundles.

Given two $\mathcal{G}$-equivalent harmonic bundles, the corresponding Higgs bundles are $\mathcal{G}_{\mathbb{C}}$-equivalent, and vice versa; similarly, given two $\mathcal{G}$-equivalent harmonic bundles, the corresponding flat bundles are $\mathcal{G}_{\mathbb{C}}$-equivalent, and vice versa.

Now we can describe the equivalences discussed above, at the level of *moduli spaces* (and once again ignoring stability conditions):

- Let $\mathcal{M} = \mathcal{M}(G, C, E)$ be the moduli space of harmonic bundles modulo $\mathcal{G}$.
- Let $\mathcal{M}_{Higgs} = \mathcal{M}_{Higgs}(G_{\mathbb{C}}, C, E_{\mathbb{C}})$ be the moduli space of Higgs bundles modulo $\mathcal{G}_{\mathbb{C}}$.
- Let $\mathcal{M}_{flat} = \mathcal{M}_{flat}(G_{\mathbb{C}}, E_{\mathbb{C}})$ be the moduli space of $G_{\mathbb{C}}$-flat connections modulo $\mathcal{G}_{\mathbb{C}}$.[7]
- Let $\mathcal{M}_{anti-Higgs} = \mathcal{M}_{anti-Higgs}(G_{\mathbb{C}}, C, E_{\mathbb{C}})$ be the moduli space of anti-Higgs bundles modulo $\mathcal{G}_{\mathbb{C}}$.

If we choose the topology of E appropriately — for example if we take $G = PSU(K)$ and take the Stiefel-Whitney class of E to be a generator of $\mathbb{Z}/K\mathbb{Z}$ — then $\mathcal{M}, \mathcal{M}_{Higgs}, \mathcal{M}_{flat}$ are actually smooth manifolds. For a more general choice of E there will be some singularities to deal with, but I will mostly ignore this issue in what follows.

What we have said above implies that there are diffeomorphisms $f_0 : \mathcal{M} \to \mathcal{M}_{Higgs}$, $f_\infty : \mathcal{M} \to \mathcal{M}_{anti-Higgs}$, and a *family* of diffeomorphisms $f_\zeta : \mathcal{M} \to \mathcal{M}_{flat}$ parameterized by $\zeta \in \mathbb{C}^\times$:

$$
\begin{array}{ccc}
 & \mathcal{M} & \\
\swarrow^{f_0} & \downarrow^{f_\zeta\,f_\zeta\,f_\zeta} & \searrow^{f_\infty} \\
\mathcal{M}_{Higgs} & \mathcal{M}_{flat} & \mathcal{M}_{anti-Higgs}
\end{array}
$$

(4.10)

In particular, this leads to the very nontrivial statement that $\mathcal{M}_{Higgs}$ and $\mathcal{M}_{flat}$ are actually diffeomorphic (via, say, the map $f_1 \circ f_0^{-1}$).

$\mathcal{M}_{Higgs}, \mathcal{M}_{flat}$ and $\mathcal{M}_{anti-Higgs}$ all carry natural complex structures. It follows that $\mathcal{M}$ is also complex, in many different ways: for any $\zeta \in \mathbb{CP}^1$, the diffeomorphism f_ζ endows $\mathcal{M}$ with a complex structure. We write J_ζ for this complex structure on $\mathcal{M}$.

4.4 Hyperkahler Structure

So far we have explained that the moduli space $\mathcal{M}$ of harmonic bundles carries a natural family of complex structures J_ζ, parameterized by $\zeta \in \mathbb{CP}^1$. This might sound exotic at first encounter, but actually there is a natural "explanation" for this family

---

[7]We drop C here to emphasize that $\mathcal{M}_{flat}$ can be defined without using the complex structure on C, e.g. as the space of representations $\pi_1(M) \to G_{\mathbb{C}}$ up to equivalence, although of course it does still depend on the genus of C.

of complex structures: it comes from the fact that $\mathcal{M}$ carries a natural hyperkähler metric, as we now explain.

Let us fix a G-bundle E as we did above. Then let $\mathcal{C}$ denote the space consisting of pairs (D, φ) as in Sect. 4.1, now *without* imposing the Hitchin equations (4.1). $\mathcal{C}$ is an infinite-dimensional affine space, with a natural hyperkähler structure. Moreover $\mathcal{C}$ is naturally acted on by the gauge group $\mathcal{G}$. This action preserves the hyperkähler structure and has a moment map $\vec{\mu}$; the Hitchin equations say that the three components of $\vec{\mu}$ vanish.

Thus $\mathcal{M} = \vec{\mu}^{-1}(0)/\mathcal{G}$. But this is precisely the *hyperkähler quotient* $\mathcal{C} /\!\!/ \mathcal{G}$, as defined in [40]. In particular, this implies $\mathcal{M}$ is hyperkähler [39]. Now, every hyperkähler manifold carries a canonical family of complex structures parameterized by $\zeta \in \mathbb{CP}^1$, and for our $\mathcal{M}$, this family is precisely the family J_ζ we discussed in Sect. 4.3.

4.5 Universal Bundle

A point of $\mathcal{M}$ corresponds to a harmonic bundle on C up to isomorphism. It is thus natural to ask whether there is a *universal bundle*, i.e. a bundle V over $C \times \mathcal{M}$ equipped with some geometric structure, which when restricted to a given $x \in \mathcal{M}$ gives a harmonic bundle over C in the isomorphism class x. Such a bundle need not quite exist, but at least it exists up to some twisting (so more precisely it exists as a section of a certain gerbe over $C \times \mathcal{M}$). Locally on $\mathcal{M}$ we may ignore this twisting, and pretend that we have an honest universal bundle.

For our purposes the most important fact about this universal bundle is that it is *hyperholomorphic* [30]: it carries a single unitary connection D, whose curvature is of type $(1, 1)$ relative to *all* of the complex structures J_ζ on $\mathcal{M}$ (see [20, 57] for some background on this notion).

4.6 Spectral Curves and Hitchin Fibration

The different complex structures on $\mathcal{M}$ expose different features of the space. Let us focus for a moment on the complex structure J_0. In this complex structure, as we have explained, $\mathcal{M}$ is identified with $\mathcal{M}_{Higgs}$. One of the fundamental facts about this space is that it is a *complex integrable system*. In particular, it is a fibration over a complex base space $\mathcal{B}$, where the generic fiber is a compact complex torus.

Let us describe where this fibration structure comes from. To be concrete we will focus on the case where $G = SU(K)$ or $G = PSU(K)$.

Suppose we are given a Higgs bundle (E_h, ϕ). Then the eigenvalues of ϕ in the standard representation of G give a K-sheeted branched cover of C:

$$\Sigma = \{(z \in C, \lambda \in T_z^* C) : \det(\phi(z) - \lambda) = 0\} \subset T^* C. \qquad (4.11)$$

Σ is the *spectral curve* corresponding to the Higgs bundle (E_h, ϕ). The branch points of the covering $\Sigma \to C$ are those points $z \in C$ where $\phi(z)$ has a repeated eigenvalue.

Now, let $\mathcal{B}$ be the space of *all* K-sheeted branched covers $\Sigma \subset T^*C$ of C. Concretely, $\mathcal{B}$ is a finite-dimensional complex vector space. Passing from the Higgs bundle (E_h, ϕ) to its spectral curve gives a projection known as the "Hitchin fibration,"

$$\mathcal{M}_{Higgs} \to \mathcal{B}. \tag{4.12}$$

$\mathcal{B}$ is thus called the "Hitchin base."

We let $\mathcal{B}_{\mathrm{reg}} \subset \mathcal{B}$ be the locus of *smooth* spectral curves, and $\mathcal{B}_{\mathrm{sing}} = \mathcal{B} \setminus \mathcal{B}_{\mathrm{reg}}$. $\mathcal{B}_{\mathrm{sing}}$ is a divisor in $\mathcal{B}$ (discriminant locus).

Now, we have claimed that the fibers of $\mathcal{M}_{Higgs}$ over $\mathcal{B}_{\mathrm{reg}}$ are complex tori: where does that come from? To understand it, note that a smooth spectral curve Σ comes with a tautological holomorphic line bundle $\mathcal{L}$, namely the bundle whose fiber over (z, λ) is the λ-eigenspace of $\phi(z)$. Moreover, by pushforward one can recover the original Higgs bundle (E_h, ϕ) from $(\Sigma, \mathcal{L})$.

Roughly speaking, then, the fiber of $\mathcal{M}_{Higgs}$ over a given $\Sigma \in \mathcal{B}_{\mathrm{reg}}$ is the set of all holomorphic line bundles over Σ, with the correct degree (so that their pushforward has the same degree as E). This set is well known to be a compact complex torus.[8] More precisely, this torus is not quite the one we want, because $\mathcal{L}$ is not an arbitrary bundle—it constrained by the requirement that the pushforward of $\mathcal{L}$ to C should produce a bundle with trivial determinant. Thus the correct statement is that the torus fiber of $\mathcal{M}_{Higgs}$ is parameterizing those holomorphic line bundles $\mathcal{L}$ obeying this constraint. This torus is known as the Prym variety of the covering $\Sigma \to C$.

4.7 Allowing Singularities

For the application to gauge theory, it is useful to slightly extend our discussion: instead of taking (D, φ) to be regular everywhere, we may require them to have singularities at some points of C, of some constrained sort.[9] Broadly speaking there are two classes of singularity which we might consider: either *regular* singularities where the eigenvalues of φ have only a simple pole, or *irregular* ones where the eigenvalues have singularities of higher order.

Essentially all of the mathematical statements we have reviewed in this section have direct extensions to the case with singularities; the main references are [55] for the regular case, and [7] for the irregular case.

---

[8]This is a consequence of Hodge theory for $(0, 1)$-forms on Σ. One concrete way of thinking about it, in the case where $\mathcal{L}$ has degree zero, is that every $\mathcal{L}$ of degree zero admits a metric for which the Chern connection is flat, and this gives an isomorphism between the set of such $\mathcal{L}$ and the set of unitary flat connections over Σ, which is evidently a torus.

[9]These singularities correspond to the *punctures* usually included in the definition of the theories of class S.

One important point to keep in mind is that in the singular case one has some "local parameters" keeping track of the behavior near each singularity: for example, in the case of a regular singularity, the local parameters are the residues of the eigenvalues of φ and the monodromy of D. In defining the moduli space of harmonic bundles one then has to choose whether to hold these parameters fixed or let them vary. If one wants the resulting moduli space to carry a natural hyperkähler metric, one should hold them fixed (morally the reason is that the metric is given by the L^2 inner product of the fluctuations, and variations which change the local parameters around a singularity turn out to be non-normalizable).

References

[1] Alday, L.F., Gaiotto, D., Gukov, S., Tachikawa, Y., Verlinde, H.: Loop and surface operators in N = 2 gauge theory and Liouville modular geometry. JHEP **1001**, 113 (2010). arXiv:0909.0945 [hep-th]

[2] Alday, L.F., Gaiotto, D., Maldacena, J.: Thermodynamic bubble Ansatz (2009). arXiv:0911.4708 [hep-th]

[3] Alday, L.F., Maldacena, J.: Null polygonal Wilson loops and minimal surfaces in Anti-de-Sitter space (2009). arXiv:0904.0663 [hep-th]

[4] Alday, L.F., Maldacena, J., Sever, A., Vieira, P.: Y-system for scattering amplitudes (2010). arXiv:1002.2459 [hep-th]

[5] Alvarez-Gaume, L., Freedman, D.Z.: Geometrical structure and ultraviolet finiteness in the supersymmetric sigma model. Commun. Math. Phys. **80**, 443 (1981)

[6] Bershadsky, M., Johansen, A., Sadov, V., Vafa, C.: Topological reduction of 4-d SYM to 2-d sigma models. Nucl. Phys. B448, 166–186 (1995). eprint:hep-th/9501096

[7] Biquard, O., Boalch, P.: Wild nonabelian Hodge theory on curves (2002). eprint:math/0111098

[8] Cecotti, S., Neitzke, A., Vafa, C.: R-twisting and 4d/2d correspondences (2010). arXiv:1006.3435 [hep-th]

[9] Cherkis, S.A., Kapustin, A.: New hyperkaehler metrics from periodic monopoles. Phys. Rev. D 65, 084015 (2002). eprint:hep-th/0109141

[10] Cherkis, S.A., Kapustin, A.: Periodic monopoles with singularities and N = 2 super-QCD. Commun. Math. Phys. **234**, 1–35 (2003). arXiv:hep-th/0011081

[11] Cordova, C., Neitzke, A.: Line defects, tropicalization, and multi-centered quiver quantum mechanics (2013). arXiv:1308.6829 [hep-th]

[12] Corlette, K.: Flat G-bundles with canonical metrics. J. Differ. Geom. **28**(3), 361–382 (1988)

[13] Dimofte, T., Gukov, S.: Refined, motivic, and quantum. Lett. Math. Phys. **91**, 1 (2010). arXiv:0904.1420 [hep-th]

[14] Donagi, R.Y.: Seiberg-Witten integrable systems (1997). eprint:alg-geom/9705010

[15] Donagi, R.Y., Witten, E.: Supersymmetric Yang-Mills theory and integrable systems. Nucl. Phys. B460, 299–334 (1996). eprint:hep-th/9510101

[16] Donagi, R., Wijnholt, M.: Higgs bundles and UV completion in F-theory (2009). eprint:0904.1218

[17] Donaldson, S.K.: Twisted harmonic maps and the self-duality equations. Proc. Lond. Math. Soc. 3 55(1), 127–131 (1987)

[18] Drukker, N., Gomis, J., Okuda, T., Teschner, J.: Gauge theory loop operators and Liouville theory (2009). arXiv:0909.1105 [hep-th]

[19] Drukker, N., Morrison, D.R., Okuda, T.: Loop operators and S-duality from curves on Riemann surfaces (2009). arXiv:0907.2593 [hep-th]

[20] Feix, B.: Hypercomplex manifolds and hyperholomorphic bundles. Math. Proc. Camb. Philos. Soc. **133**, 443–457 (2002)

[21] Fock, V.V., Goncharov, A.B.: Cluster ensembles, quantization and the dilogarithm. Ann. Sci. Éc. Norm. Supér. (4) 42(6), 865–930 (2009). eprint:math/0311245

[22] Fock, V., Goncharov, A.: Moduli spaces of local systems and higher Teichmüller theory. Publ. Math. Inst. Hautes Études Sci. (103), 1–211 (2006). eprint:math/0311149

[23] Fomin, S., Zelevinsky, A.: Cluster algebras. I. Foundations. J. Am. Math. Soc. 15(2), 497–529 (2002) (electronic)

[24] Freed, D.S., Teleman, C.: Relative quantum field theory (2012). arXiv:1212.1692 [hep-th]

[25] Gaiotto, D.: N = 2 dualities (2009). arXiv:0904.2715 [hep-th]

[26] Gaiotto, D.: Surface operators in N = 2 4d gauge theories (2009). arXiv:0911.1316 [hep-th]

[27] Gaiotto, D., Moore, G.W., Neitzke, A.: Wall-crossing, Hitchin systems, and the WKB approximation (2009). arXiv:0907.3987 [hep-th]

[28] Gaiotto, D., Moore, G.W., Neitzke, A.: Four-dimensional wall-crossing via three-dimensional field theory. Commun. Math. Phys. 299, 163–224 (2010). arXiv:0807.4723 [hep-th]

[29] Gaiotto, D., Moore, G.W., Neitzke, A.: Framed BPS states (2010). arXiv:1006.0146 [hep-th]

[30] Gaiotto, D., Moore, G.W., Neitzke, A.: Wall-crossing in coupled 2d–4d systems. (2011). arXiv:1103.2598 [hep-th]

[31] Gaiotto, D., Moore, G.W., Neitzke, A.: Spectral networks (2012). arXiv:1204.4824 [hep-th]

[32] Gross, M., Siebert, B.: From real affine geometry to complex geometry. Ann. Math. 174, 1301–1428 (2011). eprint:math/0703822

[33] Gross, M., Siebert, B.: Theta functions and mirror symmetry (2012). eprint:1204.1991

[34] Gukov, S.: Surface operators and knot homologies (2007). eprint:0706.2369

[35] Gukov, S., Witten, E.: Gauge theory, ramification, and the geometric Langlands program (2006). eprint:hep-th/0612073

[36] Hanany, A., Pioline, B.: (Anti-)instantons and the Atiyah-Hitchin manifold. JHEP 07, 001 (2000). eprint:hep-th/0005160

[37] Harvey, J. A., Moore, G.W., Strominger, A.: Reducing S duality to T duality. Phys. Rev. D 52, 7161–7167 (1995). eprint:hep-th/9501022

[38] Hausel, T., Thaddeus, M.: Mirror symmetry, Langlands duality, and the Hitchin system (2002). eprint:math.AG/0205236

[39] Hitchin, N.J.: The self-duality equations on a Riemann surface. Proc. Lond. Math. Soc. (3) 55(1), 59–126 (1987)

[40] Hitchin, N.J., Karlhede, A., Lindstrom, U., Roček, M.: Hyperkähler metrics and supersymmetry. Commun. Math. Phys. 108, 535 (1987)

[41] Hollands, L., Neitzke, A.: Spectral networks and Fenchel-Nielsen coordinates (2013). arXiv:1312.2979 [math.GT]

[42] Ito, Y., Okuda, T., Taki, M.: Line operators on $S^1 \times R^3$ and quantization of the Hitchin moduli space (2011). eprint:1111.4221

[43] Kapustin, A., Witten, E.: Electric-magnetic duality and the geometric Langlands program (2006). eprint:hep-th/0604151

[44] Kontsevich, M., Soibelman, Y.: Stability structures, motivic Donaldson-Thomas invariants and cluster transformations (2008). eprint:0811.2435

[45] Le, I.: Higher laminations and affine. Build. eprint:1209.0812 (2012)

[46] Lusztig, G.: Canonical bases arising from quantized enveloping algebras. J. Am. Math. Soc. 3(2), 447–498 (1990)

[47] Martinec, E.J., Warner, N.P.: Integrable systems and supersymmetric gauge theory. Nucl. Phys. B459, 97–112 (1996). eprint:hep-th/9509161

[48] Nekrasov, N., Witten, E.: The omega deformation, branes, integrability, and Liouville theory (2010). arXiv:1002.0888 [hep-th]

[49] Okuda, T.: Line operators in supersymmetric gauge theories and the 2d–4d relation (2014). arXiv:1412.7126 [hep-th]

[50] Seiberg, N., Witten, E.: Electric-magnetic duality, monopole condensation, and confinement in $\mathcal{N} = 2$ supersymmetric Yang-Mills theory. Nucl. Phys. B426, 19–52 (1994). eprint:hep-th/9407087

[51] Seiberg, N., Witten, E.: Monopoles, duality and chiral symmetry breaking in $\mathcal{N} = 2$ super-symmetric QCD. Nucl. Phys. B431, 484–550 (1994). eprint:hep-th/9408099

[52] Seiberg, N., Witten, E.: Gauge dynamics and compactification to three dimensions (1996). eprint:hep-th/9607163

[53] Sikora, A.S.: Generating sets for coordinate rings of character varieties (2011). eprint:1106.4837

[54] Simpson, C.T.: Constructing variations of Hodge structure using Yang-Mills theory and applications to uniformization. J. Am. Math. Soc. 1(4), 867–918 (1988)

[55] Simpson, C.T.: Harmonic bundles on noncompact curves. J. Am. Math. Soc. 3(3), 713–770 (1990)

[56] Teschner, J.: Quantization of the Hitchin moduli spaces, Liouville theory, and the geometric Langlands correspondence I. Adv. Theor. Math. Phys. 15, 471–564 (2011). arXiv:1005.2846 [hep-th]

[57] Verbitsky, M.: Hyperholomorphic bundles over a hyper-Kähler manifold. J. Algebraic Geom. 5(4), 633–669 (1996). eprint:alg-geom/9307008

[58] Witten, E.: Geometric Langlands from six dimensions (2009). arXiv:0905.2720 [hep-th]

[59] Xie, D.: Network, cluster coordinates and $N = 2$ theory I. (2012). eprint:1203.4573

[60] Xie, D.: Higher laminations, webs and $N = 2$ line operators (2013). arXiv:1304.2390 [hep-th]

A Review on Instanton Counting and W-Algebras

Yuji Tachikawa

Abstract Basics of the instanton counting and its relation to W-algebras are reviewed, with an emphasis toward physics ideas. We discuss the case of U(N) gauge group on $\mathbb{R}^4$ to some detail, and indicate how it can be generalized to other gauge groups and to other spaces. This is part of a combined review on the recent developments on exact results on $\mathcal{N} = 2$ supersymmetric gauge theories, edited by J. Teschner.

1 Introduction

1.1 Instanton Partition Function

After the indirect determination of the low-energy prepotential of $\mathcal{N} = 2$ supersymmetric SU(2) gauge theory in [1, 2], countless efforts were spent in obtaining the same prepotential in a much more direct manner, by performing the path integral over instanton contributions. After the first success in the 1-instanton sector [3, 4], people started developing techniques to perform multi-instanton computations. Years of study culminated in the publication of the review [5] carefully describing both the explicit coordinates of and the integrand on the multi-instanton moduli space.

A parallel development was ongoing around the same time, which utilizes a powerful mathematical technique, called equivariant localization, in the instanton calculation. In [6], the authors studied equivariant integrals over various hyperkähler manifolds, including the instanton moduli spaces. From the start, their approach uti-

---

A citation of the form [V:x] refers to article number x in this volume.

Y. Tachikawa (✉)
Faculty of Science, Department of Physics, University of Tokyo, Bunkyo-ku,
Tokyo 133-0022, Japan
e-mail: yuji.tachikawa@ipmu.jp

Y. Tachikawa
Kavli Institute for the Physics and Mathematics of the Universe, University of Tokyo,
Kashiwa, Chiba 277-8583, Japan

© Springer International Publishing Switzerland 2016
J. Teschner (ed.), *New Dualities of Supersymmetric Gauge Theories*,
Mathematical Physics Studies, DOI 10.1007/978-3-319-18769-3_4

lized the equivariant localization, but it was not quite clear at that time exactly which physical quantity they computed. Later in [7–9], the relation between the localization computation and the low-energy Seiberg-Witten theory was explored. Finally, there appeared the seminal paper by Nekrasov [10], where it was pointed out that the equivariant integral in [6], applied to the instanton moduli spaces, is exactly the integral in [5] which can be used to obtain the low-energy prepotential.

In [10], a physical framework was also presented, where the appearance of the equivariant integral can be naturally understood. Namely, one can deform the theory on $\mathbb{R}^4$ by two parameters $\epsilon_{1,2}$, such that a finite partition function $Z(\epsilon_{1,2}; a_i)$ is well-defined, where a_i are the special coordinates on the Coulomb branch of the theory. Then, one has

$$\log Z(\epsilon_{1,2}; a_i) \rightarrow \frac{1}{\epsilon_1 \epsilon_2} F(a_i) + \text{less singular terms} \qquad (1.1)$$

in the $\epsilon_{1,2} \rightarrow 0$ limit. The function $Z(\epsilon_{1,2})$ is called under various names, such as *Nekrasov's partition function*, the *deformed partition function*, or the *instanton partition function*. As the partition function is expressed as a discrete, infinite sum over instanton configurations, the method is dubbed *instanton counting*. In [11–14], it was also noticed that the integral presented in [5] is the integral of an equivariant Euler class, but the crucial idea of using $\epsilon_{1,2}$ is due to [10].

For $SU(N)$ gauge theory with fundamental hypermultiplets, the function Z can be explicitly written down [10, 13, 15, 16]. The equality of the prepotential as defined by (1.1) and the prepotential as determined by the Seiberg-Witten curve is a rigorous mathematical statement which was soon proven by three groups by three distinct methods [17–20]. The calculational methods were soon generalized to quiver gauge theories, other matter contents, and other classical gauge groups [21–27]. It was also extended to calculations on the orbifolds of $\mathbb{R}^4$ in [28]. We now also know a uniform derivation of the Seiberg-Witten curves from the instanton counting for SU quiver gauge theories with arbitrary shape thanks to [29, 30]. Previous summaries and lecture notes on this topic can be found e.g. in [31, 32].

An $\mathcal{N} = 2$ gauge theory can often be engineered by considering type IIA string on an open Calabi-Yau. It turned out [33–36] that the topological A-model partition function as calculated by the topological vertex [37, 38] is then equal to Nekrasov's partition function of the five-dimensional version of the theory, when $\epsilon_1 = -\epsilon_2$ is identified with the string coupling constant in the A-model. This suggested the existence of a refined, i.e. two-parameter version of the topological string, and a refined formula for the topological vertex was formulated in [39–43], so that the refined topological A-model partition function equals Nekrasov's partition function at $\epsilon_1 + \epsilon_2 \neq 0$. The relation between instanton partition functions and refined topological vertex was further studied in e.g. [44, 45]. The same quantity can be computed in the mirror B-model side using the holomorphic anomaly equation [46–49], which also provided an independent insight to the system.

We will derive the instanton partition function of four-dimensional gauge theories by considering a five-dimensional system and then taking the four-dimensional limit.

Therefore the review should prepare the reader so that they can understand systems in either dimensions. In this review, we mostly concentrate on four-dimensional theories, with only a cursory mention of the systems in five dimensions.

1.2 Relation to W-Algebras

Another recent developments concerns the two-dimensional CFT structure on the instanton partition function, which was first observed in [50, 51] in the case of SU(2) gauge theory on $\mathbb{R}^4$, and soon generalized to SU(N) in [52], to other classical groups by [26, 27], and to arbitrary gauge groups by [53].

This observation was motivated from a general construction found in [54] and reviewed in [V:2, V:3] in this volume. Namely, the 6d $\mathcal{N} = (2, 0)$ theory compactified on a Riemann surface C gives rise to 4d $\mathcal{N} = 2$ theories labeled by C. Put the 4d theories thus obtained on S^4. The partition function can be computed as described in [55, 56] and reviewed in [V:6], which is given by an integral of the one-loop part and the instanton part. The one-loop part is given by a product of double-Gamma functions, and the instanton part is the product (one for the north pole and the other for the south pole) of two copies of the instanton partition function as reviewed in this review. As the one-loop part happens to be equal to that of the Liouville-Toda conformal field theory on C as is reviewed in [V:12], the instanton part should necessarily be equal to the conformal blocks of these CFTs. The conformal blocks have a strong connection to matrix models, and therefore the instanton partition functions can also be analyzed from this point of view. This will be further discussed in [V:5] in this volume.

We can also consider instanton partition functions of gauge group U(N) on $\mathbb{R}^4/\mathbb{Z}_n$ where $\mathbb{Z}_n$ is an subgroup of SU(2) acting on $\mathbb{R}^4 \simeq \mathbb{C}^2$. Then the algebra which acts on the moduli space is guessed to be the so-called nth para-W_N algebra [57–63]. For U(2) on $\mathbb{R}^4/\mathbb{Z}_2$, we have definite confirmation that there is the action of a free boson, the affine algebra SU(2)$_2$, together with the $\mathcal{N} = 1$ supersymmetric Virasoro algebra [57, 64].

A further variation of the theme is to consider singularities in the configuration of the gauge field along $\mathbb{C} \subset \mathbb{C}^2$. This is called a surface operator, and more will be discussed in [V:8] in this volume. The simplest of these is characterized by the singular behavior $A_\theta d\theta \to \mu d\theta$ where θ is the angular coordinate transverse to the surface $\mathbb{C}$ and μ is an element of the Lie algebra of the gauge group G. The algebra which acts on the moduli space of instanton with this singularity is believed to be obtained by the Drinfeld-Sokolov reduction of the affine algebra of type G [65–67]. In particular, when μ is a generic semisimple element, the Drinfeld-Sokolov reduction does not do anything in this case, and the algebra is the affine algebra of type G itself when G is simply-laced. This action of the affine algebra was constructed almost ten years ago [19, 20], which was introduced to physics community in [68].

Organization

We begin by recalling why the instantons configurations are important in gauge
theory in Sect. 2. A rough introduction to the structure of the instanton moduli space
is also given there. In Sect. 3, we study the U(N) gauge theory on $\mathbb{R}^4$. We start in
Sect. 3.1 by considering the partition function of generic supersymmetric quantum
mechanics. In Sect. 3.2, we will see how the instanton partition function reduces
to the calculation of a supersymmetric quantum mechanics in general, which is
then specialized to U(N) gauge theory in Sect. 3.3, for which explicit calculation is
possible. The result is given a mathematical reformulation in Sect. 3.4 in terms of the
equivariant cohomology, which is then given a physical interpretation in Sect. 3.5.
The relation to the W-algebra is discussed in Sect. 3.6. Its relation to the topological
vertex is briefly explained in Sect. 3.7; more details will be given in [V:13] in this
volume. In Sects. 4 and 5, we indicate how the analysis can be extended to other
gauge groups and to other spacetime geometries, respectively.

Along the way, we will be able to see the ideas of three distinct mathematical
proofs [17–20] of the agreement of the prepotential as obtained from the instanton
counting and that as obtained from the Seiberg-Witten curve. The proof by Nekrasov
and Okounkov will be indicated in Sect. 3.3, the proof by Braverman and Etingof in
Sect. 5.1, and the proof by Nakajima and Yoshioka in Sect. 5.3.

In this paper we are not going to review standard results in W-algebras, which can
all be found in [69, 70]. The imaginary unit $\sqrt{-1}$ is denoted by i, as we will often
use i for the indices to sum over.

If the reader understands Japanese, an even more introductory account of the
whole story can be found in [71].

2   Gauge Theory and the Instanton Moduli Space

2.1   Instanton Moduli Space

Let us first briefly recall why we care about the instanton moduli space. We are
interested in the Yang-Mills theory with gauge group G, whose partition function is
given by

$$Z = \int [DA_\mu] e^{-S} \quad \text{where} \quad S = \frac{1}{2g^2} \int \operatorname{tr} F_{\mu\nu} F_{\mu\nu}, \tag{2.1}$$

or its supersymmetric generalizations. Configurations with smaller action S contribute more significantly to the partition function. Therefore it is important to find
the action-minimizing configuration:

$$\operatorname{tr} F_{\mu\nu} F_{\mu\nu} = \frac{1}{2} \operatorname{tr}(F_{\mu\nu} \pm \tilde{F}_{\mu\nu})^2 \mp \operatorname{tr} F_{\mu\nu} \tilde{F}_{\mu\nu} \geq \mp \operatorname{tr} F_{\mu\nu} \tilde{F}_{\mu\nu}. \tag{2.2}$$

For a finite-action configuration, it is known that the quantity

$$n := -\frac{1}{16\pi^2} \int d^4x \, \text{tr} \, F_{\mu\nu} \tilde{F}_{\mu\nu} \tag{2.3}$$

is always an integer for the standard choice of the trace tr for SU(N) gauge field. For other gauge groups, we normalize the trace symbol tr so that this property holds true. Then we find

$$\int d^4x \, \text{tr} \, F_{\mu\nu} F_{\mu\nu} \geq 16\pi^2 |n| \tag{2.4}$$

which is saturated only when

$$F_{\mu\nu} + \tilde{F}_{\mu\nu} = 0 \quad \text{or} \quad F_{\mu\nu} - \tilde{F}_{\mu\nu} = 0 \tag{2.5}$$

depending if $n > 0$ or $n < 0$, respectively. This is the instanton equation. As it sets the (anti-)self-dual part of the Yang-Mills field strength to be zero, it is also called the (anti)-self dual equation, or the (A)SD equation for short.

The equation is invariant under the gauge transformation $g(x)$. We identify two solutions which are related by gauge transformations such that $g(x) \to 1$ at infinity. The parameter space of instanton solutions is called the instanton moduli space, and we denote it by $M_{G,n}$ in this paper.

For the simplest case $G = \text{SU}(2)$ and $n = 1$, a solution is parameterized by eight parameters, namely

- four parameters for the center, parameterizing $\mathbb{R}^4$,
- one parameter for the size, parameterizing $\mathbb{R}_{>0}$,
- and three parameters for the global gauge direction $\text{SU}(2)/\mathbb{Z}_2 \sim S^3/\mathbb{Z}_2$.

The last identification by $\mathbb{Z}_2$ is due to the fact that the Yang-Mills field is in the triplet representation and therefore the element $\text{diag}(-1, -1) \in \text{SU}(2)$ doesn't act on it. The instanton moduli space is then

$$M_{\text{SU}(2),1} = \mathbb{R}^4 \times \mathbb{R}^4/\mathbb{Z}_2 \tag{2.6}$$

where we combined $\mathbb{R}_{>0}$ and S^3 to form an $\mathbb{R}^4$.

As the Eq. (2.5) is scale invariant, an instanton can be shrunk to a point. This is called the small instanton singularity, which manifests in (2.6) as the $\mathbb{Z}_2$ orbifold singularity at the origin.

For a general gauge group G and still with $n = 1$, it is known that every instanton solution is given by picking an SU(2) 1-instanton solution and regarding it as an instanton solution of gauge group G by choosing an embedding $\text{SU}(2) \to G$. It is known that such embeddings have $4h^\vee(G) - 5$ parameters, where $h^\vee(G)$ is the dual Coxeter number of G. Together with the position of the center and the size, we have $4h^\vee(G)$ parameters in total. Equivalently, the instanton moduli space $M_{G,1}$ is real $4h^\vee(G)$ dimensional. It is a product of $\mathbb{R}^4$ and the minimal nilpotent orbit of $\mathfrak{g}_\mathbb{C}$: this fact will be useful in Sect. 4.3.

When $n > 0$, one way to construct such a solution is to take n 1-instanton solutions with well-separated centers, superimpose them, and add corrections to satisfy the Eq. (2.5) necessary due to its nonlinearity. It is a remarkable fact that this operation is possible even when the centers are close to each other. The instanton moduli space $M_{G,n}$ then has real $4h^\vee(G)n$ dimensions. There is a subregion of the moduli space where one out of n instantons shrink to zero size, and gives rise to the small instanton singularity. There, the gauge configuration is given by a smooth $(n-1)$-instanton solution with a pointlike instanton put on top of it. Therefore, the small instanton singularity has the form [72]

$$\mathbb{R}^4 \times M_{G,n-1} \subset M_{G,n}. \tag{2.7}$$

2.2 Path Integral Around Instanton Configurations

Now let us come back to the evaluation of the path integral (2.1). We split a general gauge field A_μ of instanton number n into a sum

$$A_\mu = A_\mu^{\text{inst}} + \delta A_\mu \tag{2.8}$$

where A_μ^{inst} is the instanton solution closest to the given configuration A_μ. When δA_μ is small, we have

$$S = \frac{8\pi^2 |n|}{g^2} + \int d^4x [(\text{terms quadratic in } \delta A_\mu) + (\text{higher terms})] \tag{2.9}$$

and the path integral becomes

$$Z = \int [DA_\mu] e^{-S} = \sum_n \int_{M_{G,n}} d^{4h^\vee(G)n} X \int [\delta A_\mu] e^{-\frac{8\pi^2|n|}{g^2} + \cdots} \tag{2.10}$$

where $X \in M_{G,n}$ labels an instanton configuration.

It was 't Hooft who first tried to use this decomposition to study the dynamics of quantum Yang-Mills theory [73]. It turned out that the integral over the fluctuations δA_μ around the instanton configuration makes the computation in the strongly coupled, infrared region very hard in general.

For a supersymmetric model with a weakly coupled region, however, the fermionic fluctuations and the gauge fluctuations cancel, and often the result can be written as an integral over $M_{G,n}$ of a tractable function with explicit expressions; the state of the art at the turn of the century was summarized in the reference [5]. One place the relation between supersymmetry and the instanton equation (2.5) manifests itself is the supersymmetry transformation law of the gaugino, which is roughly of the form

$$\delta\lambda_\alpha = F_{\alpha\beta}\epsilon^\beta, \qquad \delta\bar{\lambda}_{\dot{\alpha}} = F_{\dot{\alpha}\dot{\beta}}\bar{\epsilon}^{\dot{\beta}}. \tag{2.11}$$

Here, $F_{\alpha\beta}$ and $F_{\dot{\alpha}\dot{\beta}}$ are (A)SD components of the field strength written in the spinor notation. Therefore, if the gauge configuration satisfies (2.5), then depending on the sign of n, half of the supersymmetry corresponding to ϵ^α or $\epsilon^{\dot{\alpha}}$ remains unbroken. In general, in the computation of the partition function in a supersymmetric background, only configurations preserving at least some of the supersymmetry gives non-vanishing contributions in the path integral. This is the principle called the supersymmetric localization. In this review we approach this type of computation from a rather geometric point of view.

3 U(N) Gauge Group on $\mathbb{R}^4$

3.1 Toy Models

We will start by considering supersymmetric quantum mechanics, as we are going to reduce the field theory calculations to supersymmetric quantum mechanics on instanton moduli spaces in Sect. 3.2.

3.1.1 Supersymmetric Quantum Mechanics on $\mathbb{C}^2$

Let us first consider the quantum mechanics of a supersymmetric particle on $\mathbb{C}^2$, parameterized by (z, w). Let the supersymmetry be such that z, w are invariant, and $(\bar{z}, \psi_{\bar{z}})$ and $(\bar{w}, \psi_{\bar{w}})$ are paired. This system also has global symmetries J_1 and J_2, such that $(J_1, J_2) = (1, 0)$ for z and $(J_1, J_2) = (0, 1)$ for w.

Let us consider its supersymmetric partition function

$$Z(\beta; \epsilon_1, \epsilon_2) = \text{tr}_{\mathcal{H}}(-1)^F e^{i\beta\epsilon_1 J_1} e^{i\beta\epsilon_2 J_2} \tag{3.1}$$

where $\mathcal{H}$ is the total Hilbert space. As there is a cancellation within the pairs $(\bar{z}, \psi_{\bar{z}})$ and $(\bar{w}, \psi_{\bar{w}})$, we have the equality

$$Z(\beta; \epsilon_1, \epsilon_2) = \text{tr}_{\mathcal{H}_{\text{susy}}} e^{i\beta\epsilon_1 J_1} e^{i\beta\epsilon_2 J_2} \tag{3.2}$$

where $\mathcal{H}_{\text{susy}}$ is the subspace consisting of supersymmetric states, which in this case is

$$\mathcal{H}_{\text{susy}} \simeq \bigoplus_{m,n \geq 0} \mathbb{C}z^m w^n. \tag{3.3}$$

The partition function is then

$$Z(\beta; \epsilon_1, \epsilon_2) = \frac{1}{1 - e^{i\beta\epsilon_1}} \frac{1}{1 - e^{i\beta\epsilon_2}}. \tag{3.4}$$

In the $\beta \to 0$ limit, we have

$$(-i\beta)^2 Z(\beta; \epsilon_1, \epsilon_2) \to \frac{1}{\epsilon_1 \epsilon_2}. \tag{3.5}$$

3.1.2 Supersymmetric Quantum Mechanics on $\mathbb{CP}^1$

Next, consider a charged supersymmetric particle moving on $S^2 \simeq \mathbb{CP}^1$, under the influence of a magnetic flux of charge $j = 0, \frac{1}{2}, 1$, etc. Let us use the complex coordinate z so that $z = 0$ is the north pole and $z = \infty$ is the south pole. The supersymmetric Hilbert space is then

$$\mathcal{H}_{\text{susy}} \simeq \bigoplus_{k=0}^{2j} \mathbb{C}z^k(\partial_z)^{\otimes j}, \tag{3.6}$$

and is the spin j representation of $SU(2)$ acting on $\mathbb{CP}^1$. Let the global symmetry J to rotate z with charge 1. Then we have

$$Z(\beta; \epsilon) = \text{tr}_{\mathcal{H}_{\text{susy}}} e^{i\beta\epsilon J} = e^{ij\beta\epsilon} + e^{i(j-1)\beta\epsilon} + \cdots + e^{-ij\beta\epsilon}. \tag{3.7}$$

This partition function can be re-expressed as

$$Z(\beta; \epsilon) = \frac{e^{+ij\beta\epsilon}}{1 - e^{-i\beta\epsilon}} + \frac{e^{-ij\beta\epsilon}}{1 - e^{+i\beta\epsilon}}. \tag{3.8}$$

Its $\beta \to 0$ limit is finite:

$$Z(\beta; \epsilon) \to 2j + 1. \tag{3.9}$$

3.1.3 Localization Theorem

These two examples illustrate the following *localization theorem*: consider a quantum mechanics of a supersymmetric particle moving on a smooth complex space M of complex dimension d with isometry $U(1)^n$, under the influence of a magnetic flux corresponding to a line bundle L on M. Then the space of the supersymmetric states is the space of holomorphic sections of L. When L is trivial, it is just the space of holomorphic functions on M.

Assume the points fixed by $U(1)^n$ on M are isolated. Denote the generators of $U(1)^n$ by $J_1, \ldots, J_n$. Then the following relation holds:

$$Z(\beta; \epsilon_1, \ldots, \epsilon_n) \equiv \mathrm{tr}_{\mathcal{H}}(-1)^F e^{i\beta \sum_i \epsilon_i J_i} = \sum_p \frac{e^{i\beta \sum_i j(p)_i \epsilon_i}}{\prod_{a=1}^d (1 - e^{i\beta \sum_i k(p)_{i,a} \epsilon_i})}, \quad (3.10)$$

see e.g. [74]. Here, the sum runs over the set of fixed points p on M, and $j(p)_i$ and $k(p)_{i,a}$ are defined so that

$$\mathrm{tr}_{TM|_p} e^{i\beta \sum_i \epsilon_i J_i} = \sum_{a=1}^d e^{i\beta \sum_i k(p)_{i,a} \epsilon_i} \quad (3.11)$$

and

$$\mathrm{tr}_{L|_p} e^{i\beta \sum_i \epsilon_i J_i} = e^{i\beta \sum_i j(p)_i J_i}. \quad (3.12)$$

In the following, it is convenient to abuse the notation and identify a vector space and its character under $U(1)^N$. Then we can just write

$$TM|_p = \sum_{a=1}^d e^{i\beta \sum_i k(p)_{i,a} \epsilon_i}, \qquad L|_p = e^{i\beta \sum_i j(p)_i J_i}. \quad (3.13)$$

We will also use $+$, $\times$, $-$ instead of $\oplus$, $\otimes$ and $\ominus$.

In (3.4), the only fixed point is at $(z, w) = (0, 0)$, and in (3.8), there are two fixed points, one at $z = 0$ and $z = \infty$. It is easy to check that the general theorem reproduces (3.4) and (3.8).

It is also clear that in the $\beta \to 0$ limit, we have

$$(-i\beta)^d Z(\beta; \epsilon_1, \ldots, \epsilon_n) \to \sum_p \frac{1}{\prod_{a=1}^d \sum_i k(p)_{i,a} \epsilon_i}, \quad (3.14)$$

which is zero if M is compact.

3.1.4 Supersymmetric Quantum Mechanics on $\mathbb{C}^2/\mathbb{Z}_2$

Let us make the identification by the $\mathbb{Z}_2$ action $(z, w) \sim (-z, -w)$ in the model of Sect. 3.1.1. Then the supersymmetric Hilbert space (3.3) becomes

$$\mathcal{H}_{\mathrm{susy}} = \bigoplus_{m,n:\ \mathrm{even}} \mathbb{C}z^m w^n \oplus \bigoplus_{m,n:\ \mathrm{odd}} \mathbb{C}z^m w^n \quad (3.15)$$

and the partition function is therefore

$$Z(\beta; \epsilon_1, \epsilon_2) = \frac{1 + e^{i\beta(\epsilon_1 + \epsilon_2)}}{(1 - e^{2i\beta\epsilon_1})(1 - e^{2i\beta\epsilon_2})}. \tag{3.16}$$

The $\beta \to 0$ limit is then

$$(i\beta)^2 Z(\beta; \epsilon_1, \epsilon_2) \to \frac{1}{2\epsilon_1\epsilon_2}. \tag{3.17}$$

The additional factor 2 with respect to (3.5) is due to the $\mathbb{Z}_2$ identification.

The localization theorem is not directly applicable, as the fixed point $(z, w) = (0, 0)$ is singular. Instead, take the blow-up M of $\mathbb{C}^2/\mathbb{Z}_2$, which is the total space of the canonical line bundle of $\mathbb{CP}^1$. The space is now smooth, with two fixed points. At the north pole n,

$$\mathrm{tr}_{TM|_n} e^{i\beta(\epsilon_1 J_1 + \epsilon_2 J_2)} = e^{2i\beta\epsilon_1} + e^{-i\beta(\epsilon_1 - \epsilon_2)}, \tag{3.18}$$

and at the south pole s,

$$\mathrm{tr}_{TM|_s} e^{i\beta(\epsilon_1 J_1 + \epsilon_2 J_2)} = e^{i\beta(\epsilon_1 - \epsilon_2)} + e^{2i\beta\epsilon_2}. \tag{3.19}$$

Then we have

$$Z(\beta; \epsilon_1, \epsilon_2) = \frac{1}{(1 - e^{2i\beta\epsilon_1})(1 - e^{-i\beta(\epsilon_1 - \epsilon_2)})} + \frac{1}{(1 - e^{i\beta(\epsilon_1 - \epsilon_2)})(1 - e^{2i\beta\epsilon_2})} \tag{3.20}$$

from the localization theorem, which agrees with (3.16).

3.2 Instanton Partition Function: Generalities

Let us now come to the real objective of our study, namely the four-dimensional $\mathcal{N} = 2$ supersymmetric gauge theory. The data defining the theory is its gauge group G, the flavor symmetry F, and the hypermultiplet representation $R \oplus \bar{R}$ under $G \times F$. With the same data, we can consider the five-dimensional $\mathcal{N} = 1$ supersymmetric gauge theory, with the same gauge group and the same hypermultiplet representation. We put this five-dimensional theory on a $\mathbb{C}^2$ bundle over S^1 given by taking $\mathbb{C}^2 \times [0, \beta)$ parameterized by (z, w, ξ^5), and making the identification

$$(z, w, 0) \sim (e^{i\beta\epsilon_1} z, e^{i\beta\epsilon_2} w, \beta). \tag{3.21}$$

See Fig. 1 for a picture. This background space-time is often called the Ω background.

Fig. 1 The five-dimensional spacetime. The vertical direction is ξ^5, and the $\mathbb{R}^4$ planes at $\xi^5 = 0$ and $\xi^5 = \beta$ are identified after a rotation

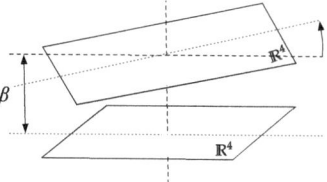

We set the vacuum expectation value of the gauge field at infinity, such that its integral along the ξ^5 direction is given by

$$\text{diag}(e^{i\beta a_1}, e^{i\beta a_2}, \ldots, e^{i\beta a_r}) \in U(1)^r \subset G. \tag{3.22}$$

We also set the background vector field which couples to the flavor symmetry, such that its integral along the ξ^5 direction is given by

$$\text{diag}(e^{i\beta m_1}, e^{i\beta m_2}, \ldots, e^{i\beta m_f}) \in U(1)^f \subset F. \tag{3.23}$$

m_i becomes the mass parameters when we take the four-dimensional limit $\beta \to 0$.

We are interested in the supersymmetric partition function in this background:

$$Z(\beta; \epsilon_{1,2}; a_{1,\ldots,r}; m_{1,\ldots,f}) = \text{tr}_{\mathcal{H}_{\text{QFT}}}(-1)^F e^{i\beta(\epsilon_1 J_1 + \epsilon_2 J_2 + \sum_{s=1}^r a_s Q_s + \sum_{s=1}^f m_s F_s)} \tag{3.24}$$

where $\mathcal{H}_{\text{QFT}}$ is the Hilbert space of the five-dimensional field theory on $\mathbb{R}^4$; $J_{1,2}$, $Q_{1,\ldots,r}$ and $F_{1,\ldots,f}$ are the generators of the spatial, gauge and flavor rotation, respectively.

We are mostly interested in the non-perturbative sector, where one has instanton configurations on $\mathbb{R}^4$ with instanton number n. Here we assume that G is a simple group; the generalization is obvious.

Energetically, five-dimensional configurations which are close to a solution of the instanton equation (2.5) at every constant time slice are favored within the path integral, similarly as discussed in Sect. 2.1. We can visualize such a configuration as one where the parameters describing the n-instanton configuration is slowly changing according to time. Therefore, the system can be approximated by the quantum mechanical particle moving within the instanton moduli space. This approach is often called the moduli space approximation. With supersymmetry, this approximation becomes exact, and we have

$$Z_{\text{inst}}(\beta; \epsilon_{1,2}; a_{1,\ldots,r}) = \sum_{n \geq 0} e^{-\frac{8\pi^2 n\beta}{g^2}} \text{tr}_{\mathcal{H}_n}(-1)^F e^{i\beta(\epsilon_1 J_1 + \epsilon_2 J_2 + \sum_{s=1}^r a_s Q_s + \sum_{s=1}^f m_s F_s)}$$

$$\tag{3.25}$$

where g is the five-dimensional coupling constant, and $\mathcal{H}_n$ is the Hilbert space of the supersymmetric quantum mechanics on the n-instanton moduli space. Its bosonic part $M_{G,n}$ is the moduli space of n-instantons of gauge group G, which we reviewed in Sect. 2.1. It has complex dimension $2h^\vee(G)n$. In addition, the fermionic direction $\mathcal{V}(R)$ has complex dimension $k(R)n$, where $k(R)$ is the quadratic Casimir normalized so that it is $2h^\vee(G)$ for the adjoint representation. This $\mathcal{V}(R)$ is a vector bundle over the instanton moduli space $M_{G,n}$, and is often called the matter bundle.

$M_{G,n}$ has a natural action of $U(1)^2$ which rotates the spacetime $\mathbb{C}^2$, and a natural action of G which performs the spacetime independent gauge rotation. These actions extend equivariantly to the matter bundle $\mathcal{V}(R)$.

Then, if $M_{G,n}$ were smooth and if the fixed points p under $U(1)^{2+r} \subset U(1)^2 \times G$ were isolated, we can apply the localization theorem to compute the instanton partition function:

$$Z_{\text{inst}}(\beta; \epsilon_{1,2}; a_{1,\ldots,r}) = \sum_{n \geq 0} e^{-\frac{8\pi^2 n\beta}{g^2}} \sum_p \frac{\prod_{t=1}^{k(R)n}(1 - e^{i\beta w_t(p)})}{\prod_{t=1}^{2h^\vee(G)n}(1 - e^{i\beta v_t(p)})} \tag{3.26}$$

where $v_t(p)$ and $w_t(p)$ are linear combinations of $\epsilon_{1,2}$, $a_{1,\ldots,r}$ and $m_{1,\ldots,f}$ such that we have

$$TM_{G,n}|_p = \sum_{t=1}^{2h^\vee(G)n} e^{i\beta v_t(p)}, \quad \mathcal{V}(R)|_p = \sum_{t=1}^{k(R)n} e^{i\beta w_t(p)}. \tag{3.27}$$

As was explained in Sect. 2.1, $M_{G,n}$ has small instanton singularities and the formula above is not directly applicable. One of the technical difficulties in the instanton computation is how to deal with this singularity. Currently, the explicit formula is known (or, at least the method to write it down is known) for the following cases: (i) $G = U(N)$ with arbitrary representations, (ii) $G = SO(N)$ with representations appearing in the tensor powers of the vector representation, and (iii) $G = USp(2N)$ with arbitrary representations. We will discuss $U(N)$ with (bi)fundamentals in Sect. 3.3, and $SO(N)$ and $USp(2N)$ with fundamentals in Sect. 4.1. For other representations, see [24, 25].

The 5d gauge theory can have a Chern-Simons coupling, it induces a magnetic flux to the supersymmetric quantum mechanics on the instanton moduli space, which will introduce a factor in the numerator of (3.26) as dictated by the localization theorem (3.10) [75].

The four-dimensional limit $\beta \to 0$ needs to be taken carefully. In principle threre can be multiple interesting choices of the scaling of the variables, resulting in different four dimensional dynamics. Here we only consider the standard one. We would like to take the limit $\beta \to 0$ keeping $\epsilon_{1,2}$ and a_i finite. Note that each term in the sum (3.26) with fixed instanton number n has $(2h^\vee(G) - k(R))n$ more factors in the denominator, producing a factor $\propto \beta^{-(2h^\vee(G)-k(R))n}$. In order to compensate it, we express the classical contribution to the action in (3.26) as

$$e^{-\frac{8\pi^2\beta}{g^2}} = (-i\beta)^{2h^\vee(G)-k(R)}q \tag{3.28}$$

and keep q fixed while taking $\beta \to 0$. The four-dimensional limit of the partition function is then

$$Z_{\text{inst}}(\beta; \epsilon_{1,2}; a_{1,\ldots,r}) = \sum_{n\geq 0} q^n \sum_p \frac{\prod_{t=1}^{k(R)n} w_t(p)}{\prod_{t=1}^{2h^\vee(G)n} v_t(p)}. \tag{3.29}$$

Note that the naive four-dimensional coupling g_{4d} is given by the five-dimensional coupling g_{5d} by the relation

$$\frac{8\pi^2}{g_{4d}^2} = \frac{8\pi^2\beta}{g_{5d}^2}. \tag{3.30}$$

Therefore, the relation (3.28), where q is fixed and β is varied, can be thought of as describing the running of g_{4d} when we change the UV cutoff scale β^{-1}. We see that the relation (3.28) correctly reproduces the logarithmic one-loop running of g_{4d}, controlled by the one-loop beta function coefficient $2h^\vee(G) - k(R)$. The dynamical scale Λ is given by $q = \Lambda^{2h^\vee(G)-k(R)}$. It is somewhat gratifying to see that the logarithmic running arises naturally in this convoluted framework.

This definition of the four-dimensional instanton partition function does not explain why its limit

$$F(a_{1,\ldots,r}) = \lim_{\epsilon_{1,2}\to 0} \epsilon_1\epsilon_2 \log Z_{\text{inst}}(\epsilon_{1,2}; a_{1,\ldots,r}) \tag{3.31}$$

is the prepotential of the four-dimensional gauge theory. For field theoretical explanations, see [10] or the Appendix of [75].

3.3 Instanton Partition Function: Unitary Gauge Groups

The instanton moduli space is always singular as explained in Sect. 2.1. Therefore, we need to do something in order to apply the idea outlined in the previous section. When the gauge group is $U(N)$, there is a standard way to deform the singularities so that the resulting space is smooth [9, 76].

ADHM construction Let the instanton number be n, and introduce the space $M_{G,n,t}$ via

$$M_{G,n,t} := \{\mu_{\mathbb{C}}(x) = t \mid x \in X_{G,n}\}/\text{GL}(n). \tag{3.32}$$

- Here $X_{G,n}$ is a linear space constructed from two vector spaces V, W described below as follows

$$X_{G,n} = (T_1^{\otimes -1} \oplus T_2^{\otimes -1}) \otimes V \otimes V^* \oplus W^* \otimes V \oplus T_1^{\otimes -1} \otimes T_2^{\otimes -1} \otimes V^* \otimes W. \tag{3.33}$$

Here, T_i is a one-dimensional space on which the generator J_i has the eigenvalue $+1$. As it is very cumbersome to write a lot of $\otimes$ and $\oplus$, we abuse the notation as already introduced above, by identifying the vector space and its character:

$$X_{G,n} = (e^{-i\beta\epsilon_1} + e^{-i\beta\epsilon_2})VV^* + W^*V + e^{-i\beta(\epsilon_1+\epsilon_2)}V^*W. \tag{3.34}$$

- $V \simeq \mathbb{C}^n$ is a space with a natural $GL(n)$ action and,
- $W \simeq \mathbb{C}^N$ is a space with a natural $U(N)$ action.
- The $*$ operation is defined naturally by setting $i^* = -i$, $\epsilon_{1,2}{}^* = \epsilon_{1,2}$, and $a_{1,\dots,r}{}^* = a_{1,\dots,r}$,
- and $\mu_{\mathbb{C}}$ is a certain quadratic function on $X_{G,n}$ taking value in the Lie algebra of $GL(n)$,
- and finally t is a deformation parameter taking value in the center of the Lie algebra of $GL(n)$. For generic t the space is smooth, but it becomes singular when $t = 0$.

This is called the ADHM construction, and the space at $t = 0$, $M_{G,n,0}$, is the instanton moduli space $M_{G,n}$.

The trick we use is to replace $M_{G,n}$ by $M_{G,n,t}$ with $t \neq 0$ and apply the localization theorem. The answer does not depend on t as long as it is non-zero. The deformation by t can be physically realized by the introduction of the spacetime noncommutativity [76], but this physical interpretation does not play any role here. Mathematically, this deformation corresponds to considering not just bundles but also torsion free sheaves, see e.g. [9]. Note that it is not known how to perform such deformation in other gauge groups at present.

The fixed points of the $U(1)^{2+N}$ action on $M_{G,n,t}$ was classified in [16], which we will describe below. A fixed point p is labeled by N Young diagrams $\vec{Y} = (Y_1, \dots, Y_N)$ such that the total number of the boxes $|\vec{Y}|$ is n. Let us denote by $(i, j) \in Y$ when there is a box at the position (i, j) in a Young diagram Y. Then, the fixed point labeled by $p = (Y_1, \dots, Y_N)$ corresponds to the action of $U(1)^2$ and $U(1)^r \subset G$ on V and W such that

$$W_p = \sum_{s=1}^{N} e^{i\beta a_s}, \qquad V_p = \sum_{s=1}^{N} \sum_{(i,j)\in Y_s} e^{i\beta(a_s+(1-i)\epsilon_1+(1-j)\epsilon_2)}. \tag{3.35}$$

Then we have

$$TM|_p = W_p^* V_p + e^{i\beta(\epsilon_1+\epsilon_2)} V_p^* W_p - (1 - e^{i\beta\epsilon_1})(1 - e^{i\beta\epsilon_2}) V_p V_p^*, \tag{3.36}$$

from which you can read off $v(p)_t$ in (3.27). As for $w(p)_t$, we have

$$\mathcal{V}(\text{fundamental})_p = e^{-i\beta m} V_p, \qquad \mathcal{V}(\text{adjoint})_p = e^{-i\beta m} TM|_p \tag{3.37}$$

where m is the mass of the hypermultiplets. In the case of a bifundamental of $U(N_1) \times U(N_2)$, the zero modes are determined once the instanton configurations p, q of $U(N_{1,2})$ are specified:

$$\mathcal{V}(\text{bifundamental})_{p,q} = e^{-i\beta m}(W_p^* V_q + e^{i\beta(\epsilon_1+\epsilon_2)} V_p^* W_q - (1 - e^{i\beta\epsilon_1})(1 - e^{i\beta\epsilon_2}) V_p^* V_q).$$
(3.38)

Note that both the adjoint and the fundamental are special cases of the bifundamental, namely, the adjoint is when $p = q$, and the fundamental is when p is empty.

Then it is just a combinatorial exercise to write down the explicit formula for the four-dimensional partition function (3.29) in terms of Young diagrams labeling the fixed points. The explicit formulas are given below. However, before writing them down, the author would like to stress that to implement it in a computer algebra system, it is usually easier and less error-prone to just directly use the formulas (3.35)–(3.38) to compute the characters and then to read off $v(p)_t$ and $w(p)_t$ via (3.27), which can then be plugged in to (3.29).

Explicit formulas Let $Y = (\lambda_1 \geq \lambda_2 \geq \cdots)$ be a Young tableau where λ_i is the height of the ith column. We set $\lambda_i = 0$ when i is larger than the width of the tableau. Let $Y^T = (\lambda_1' \geq \lambda_2' \geq \cdots)$ be its transpose. For a box s at the coordinate (i, j), we let its arm-length $A_Y(s)$ and leg-length $L_Y(s)$ with respect to the tableau Y to be

$$A_Y(s) = \lambda_i - j, \qquad L_Y(s) = \lambda_j' - i,$$
(3.39)

see Fig. 2. Note that they can be negative when the box s is outside the tableau. We then define a function E by

$$E(a, Y_1, Y_2, s) = a - \epsilon_1 L_{Y_2}(s) + \epsilon_2 (A_{Y_1}(s) + 1).$$
(3.40)

We use the vector symbol $\vec{a}$ to stand for N-tuples, e.g. $\vec{Y} = (Y_1, Y_2, \ldots, Y_N)$, etc.

Then, the contribution of an $SU(N)$ vector multiplet from the fixed point p labeled by an N-tuple of Young diagrams $\vec{Y}$ is the denominator of (3.29), where $v_t(p)$ can be read off from the characters of $TM_{G,n}|_p$ once we have the form (3.27). This is done by plugging (3.35) and (3.36). The end result is

$$z_{\text{vect}}(\vec{a}, \vec{Y}) = \frac{1}{\prod_{i,j=1}^{N} \prod_{s \in Y_i} E(a_i - a_j, Y_i, Y_j, s) \prod_{t \in Y_j} (\epsilon_1 + \epsilon_2 - E(a_j - a_i, Y_j, Y_i, t))}.$$
(3.41)

Note that there are $2Nn$ factors in total. This is as it should be, as $TM_{G,n}$ is complex $2Nn$ dimensional, and there are $2Nn$ eigenvalues at each fixed point.

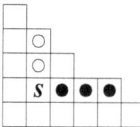

Fig. 2 Definition of the arm-length and the leg-length. For a box s in a Young tableau displayed above, the leg-length is the number of boxes to the *right* of s, marked by *black disks*, and the arm-length is the number of boxes on top of s

The contribution from (anti)fundamental hypermultiplets is given by

$$z_{\text{fund}}(\vec{a}, \vec{Y}, m) = \prod_{i=1}^{N}\prod_{s \in Y_i}(\phi(a_i, s) - m + \epsilon_1 + \epsilon_2), \tag{3.42}$$

$$z_{\text{antifund}}(\vec{a}, \vec{Y}, m) = z_{\text{fund}}(\vec{a}, \vec{Y}, \epsilon_1 + \epsilon_2 - m) \tag{3.43}$$

where $\phi(a, s)$ for the box $s = (i, j)$ is defined as

$$\phi(a, s) = a + \epsilon_1(i - 1) + \epsilon_2(j - 1). \tag{3.44}$$

They directly reflect the characters of V_p in (3.35).

When we have gauge group $SU(N) \times SU(M)$ and a bifundamental charged under both, the contribution from the bifundamental depends on the gauge configuration of both factors of the gauge group. Namely, for the fixed point p of $M_{SU(N),n,t}$ labeled by the Young diagram $\vec{Y}$ and the fixed point q of $M_{SU(N),n}$ labeled by the Young diagram $\vec{W}$, the contribution of a bifundamental is [21, 25]:

$$z_{\text{bifund}}(\vec{a}, \vec{Y}; \vec{b}, \vec{W}; m) =$$

$$\prod_{i}^{N}\prod_{j}^{M}\prod_{s \in Y_i}(E(a_i - b_j, Y_i, W_j, s) - m)\prod_{t \in W_j}(\epsilon_1 + \epsilon_2 - E(b_j - a_i, W_j, Y_i, t) - m) \tag{3.45}$$

where $\vec{a}$ and $\vec{b}$ are the chemical potentials for $SU(N)$ and $SU(M)$ respectively.

The contribution of an adjoint hypermultiplet is a special case where $p = q$ and $\vec{a} = \vec{b}$. It is

$$z_{\text{adj}}(\vec{a}, \vec{Y}, m) = z_{\text{bifund}}(\vec{a}, \vec{Y}, \vec{a}, \vec{Y}, m). \tag{3.46}$$

This satisfies

$$z_{\text{vector}}(\vec{a}, \vec{Y}) = 1/z_{\text{adj}}(\vec{a}, \vec{Y}, 0). \tag{3.47}$$

Note that there are several definitions of the mass parameter m. Another definition with

$$m' = m - \frac{1}{2}(\epsilon_1 + \epsilon_2) \tag{3.48}$$

is also common. For their relative merits, the reader is referred to the thorough discussion in [77].

Let us write down, as an example, the instanton partition function of $\mathcal{N} = 2^*$ $SU(N)$ gauge theory, i.e. an $SU(N)$ theory with a massive adjoint multiplet. We just have to multiply the contributions determined above, and we have

$$Z = \sum_{n \geq 0} q^n \sum_{\vec{Y}, |\vec{Y}| = n} z_{\text{adj}}(\vec{a}, \vec{Y}, m) z_{\text{vector}}(\vec{a}, \vec{Y})$$

$$= \sum_{\vec{Y}} q^{|\vec{Y}|} \prod_{i,j=1}^{N} \frac{\prod_{s \in Y_i} (E(a_i - a_j, Y_i, Y_j, s) + m) \prod_{t \in Y_j} (\epsilon_1 + \epsilon_2 - E(a_j - a_i, Y_j, Y_i, t) - m)}{\prod_{s \in Y_i} E(a_i - a_j, Y_i, Y_j, s) \prod_{t \in Y_j} (\epsilon_1 + \epsilon_2 - E(a_j - a_i, Y_j, Y_i, t))}.$$

(3.49)

Nekrasov-Okounkov For $G = U(N)$, the final result is a summation over N-tuples of Young diagrams $p = (Y_1, \ldots, Y_N)$ of a rational function of $\epsilon_{1,2}, a_{1,\ldots,r}$ and $m_{1,\ldots,f}$. The prepotential can be extracted by taking the limit $\epsilon_{1,2} \to 0$. There, the summation can be replaced by an extremalization procedure over the asymptotic shape of the Young diagrams. Applying the matrix model technique, one finds that the prepotential as obtained from this instanton counting is the same as the prepotential as defined by the Seiberg-Witten curve [18].

Explicit evaluation for $U(2)$ **with 1-instanton** Before proceeding, let us calculate the instanton partition function for the pure $U(2)$ gauge theory at 1-instanton level explicitly. It would be a good exercise, as the machinery used so far has been rather heavy, and the formulas are although concrete rather complicated.

In fact, the calculation is already done in Sect. 3.1, since the moduli space in question is $\mathbb{C}^2 \times \mathbb{C}^2/\mathbb{Z}_2$. Here the first factor $\mathbb{C}^2$ is the position of the center of the instanton, and $\mathbb{C}^2/\mathbb{Z}^2 \sim \mathbb{R}_{>0} \times S^3/\mathbb{Z}_2$ parameterizes the gauge orientation of the instanton via $S^3/\mathbb{Z}_2 \simeq SO(3) \simeq SU(2)/\mathbb{Z}_2$ and the size of the instanton via $\mathbb{R}_{>0}$. Introduce the coordinates (z, w, u, v) with the identification $(u, v) \sim (-u, -v)$. The action of $e^{i\beta(\epsilon_1 J_1 + \epsilon_2 J_2)}$ is given by

$$(z, w, u, v) \to (e^{i\beta\epsilon_1} z, e^{i\beta\epsilon_2} w, e^{i\beta(\epsilon_1 + \epsilon_2)/2} u, e^{i\beta(\epsilon_1 + \epsilon_2)/2} v) \qquad (3.50)$$

and (u, v) form a doublet under the $SU(2)$ gauge group. Then for $\text{diag}(e^{i\beta a}, e^{-i\beta a}) \in SU(2)$, we have

$$(u, v) \to (e^{i\beta a} u, e^{-i\beta a} v). \qquad (3.51)$$

Then the instanton partition function is given by combining (3.5) and (3.17):

$$Z_{\text{inst}}(\epsilon_{1,2}; a) = \frac{1}{\epsilon_1 \epsilon_2} \frac{1}{2} \frac{1}{(\epsilon_1 + \epsilon_2)/2 - a} \frac{1}{(\epsilon_1 + \epsilon_2)/2 + a}. \qquad (3.52)$$

It is an instructive exercise to reproduce this from the general method explained earlier in this section.

3.4 A Mathematical Reformulation

Let us now perform a mathematical reformulation, following the idea of [78]. For $G = U(N)$, consider the vector space

$$\mathbb{V}_{G,\vec{a}} = \bigoplus_{n=0}^{\infty} \mathbb{V}_{G,\vec{a},n} \tag{3.53}$$

where

$$\mathbb{V}_{G,\vec{a},n} = \bigoplus_{p} \mathbb{C}|p\rangle \tag{3.54}$$

where p runs over the fixed points of $U(1)^{2+r}$ action on $M_{G,n,t}$. We define the inner product by taking the denominator of (3.29):

$$\langle p|q \rangle = \delta_{p,q} \frac{1}{\prod_t v(p)_t}. \tag{3.55}$$

Note that the basis vectors are independent of $\vec{a}$, but the inner product does depend on $\vec{a}$. We introduce an operator N such that $\mathbb{V}_{G,\vec{a},n}$ is the eigenspace with eigenvalue n.

Let us introduce a vector

$$|\text{pure}\rangle = \sum_{n=0}^{\infty} \sum_{p} |p\rangle \in \mathbb{V}_{G,\vec{a}}. \tag{3.56}$$

Then the partition function (3.29) of the pure $SU(N)$ gauge theory is just

$$Z(\epsilon_{1,2}; \vec{a}) = \langle \text{pure}|q^{\mathsf{N}}|\text{pure}\rangle. \tag{3.57}$$

A bifundamental charged under $G_1 = SU(N_1)$ and $G_2 = SU(N_2)$ defines a linear map

$$\Phi_{\vec{b},m,\vec{a}} : \mathbb{V}_{G_1,\vec{a}} \to \mathbb{V}_{G_2,\vec{b}} \tag{3.58}$$

such that

$$\langle q|\Phi_m|p\rangle = \prod_t w(p,q)_t \tag{3.59}$$

where the right hand side comes from the decomposition

$$\mathcal{V}(\text{bifundamental})_{p,q} = \sum_t e^{-i\beta w(p,q)_t}. \tag{3.60}$$

Using this linear map $\Phi_{\vec{b},m,\vec{a}}$, we can concisely express the partition function of quiver gauge theories. For example, consider $SU(N)_1 \times SU(N)_2$ gauge theory with bifundamental hypermultiplets charged under $SU(N)_1 \times SU(N)_2$ with mass m, see Fig. 3 (4d). Then the instanton partition function (3.29) is just

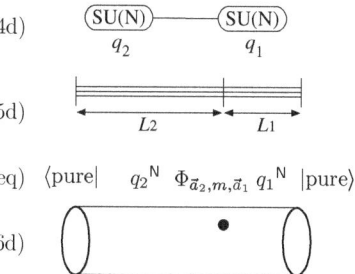

Fig. 3 Higher-dimensional setup of the quiver theory. The horizontal direction is x^5. 4d) the $SU(N) \times SU(N)$ quiver gauge theory in the infrared. 5d) 5d maximally supersymmetric $SU(N)$ theory on a segment. eq) its partition function, regarding the fifth direction as "time". 6d) 6d $\mathcal{N} = (2, 0)$ theory on a cylinder

$$Z(\epsilon_{1,2}; \vec{a}_1, \vec{a}_2; \vec{m}) = \langle \text{pure}|q_2^N \Phi_{\vec{a}_2,m,\vec{a}_1} q_1^N|\text{pure}\rangle. \tag{3.61}$$

In this section, we introduced the vector space $\mathbb{V}_{G,\vec{a},n}$ together with its inner product using fixed points of $M_{G,n,t}$. It is known that this vector space is a natural mathematical object called the equivariant cohomology:

$$\mathbb{V}_{G,\vec{a},n} = H^*_{G \times U(1)^2}(M_{G,n,t}) \otimes \mathcal{S}_G \tag{3.62}$$

where $\mathcal{S}$ is the quotient field of $H^*_{G \times U(1)^2}(pt)$. A vector called the fundamental class $[M_{G,n,t}]$ is naturally defined as an element in $H^*_{G \times U(1)^2}(M_{G,n,t})$. Then the vector $|\text{pure}\rangle$ above is

$$|\text{pure}\rangle = \bigoplus_{n=0}^{\infty} [M_{G,n,t}] \in \bigoplus_{n=0}^{\infty} H^*_{G \times U(1)^2}(M_{G,n,t}). \tag{3.63}$$

For general G, there is only the singular space $M_{G,n}$ and not the smooth version $M_{G,n,t}$. Still, using the equivariant intersection cohomology, one can write the partition function of the pure $\mathcal{N} = 2$ gauge theory with arbitrary gauge group G in the form (3.57), see e.g. [79].

3.5 Physical Interpretation of the Reformulation

The reformulation in the previous section can be naturally understood by considering a five-dimensional setup; it is important to distinguish it from another five-dimensional set-up we already used in Sect. 3.2.

Take the maximally supersymmetric $SU(N)$ gauge theory with coupling constant g. We put the system on $\mathbb{R}^{1,3}$ times a segment in the x^5 direction, which is $[0, L_1] \cup$

$[L_1, L_1+L_2]$. We put boundary conditions at $x^5 = 0, L_1$ and L_1+L_2. This necessarily breaks the supersymmetry to one half of the original, making it to a system with 4d $\mathcal{N} = 2$ supersymmetry. At $x^5 = L_1$, we put $N \times N$ hypermultiplets, to which the $SU(N)$ gauge group on the left and the $SU(N)$ gauge group on the right couple by the left and the right multiplication. At $x^5 = 0$ and $x^5 = L_1 + L_2$, we put a boundary condition which just terminates the spacetime without introducing any hypermultiplet. See Fig. 3 (5d).

In the scale larger than $L_{1,2}$, the theory effectively becomes the quiver gauge theory treated, because the segment $x^5 \in [L_1 + L_2, L_1]$ gives rise to an $SU(N)$ gauge group with 4d gauge inverse square coupling L_2/g_{5d}^2, and the segment $[L_1, 0]$ another $SU(N)$ gauge group with 4d inverse square coupling L_1/g_{5d}^2. Therefore we have, in (3.61),

$$\log q_1 / \log q_2 = L_1/L_2. \tag{3.64}$$

The final idea is to consider the x^5 direction as the time direction. At each fixed value of x^5, one has a state in the Hilbert space of this quantum field theory, which is $\mathbb{V}_{G,\vec{a}}$ introduced in the previous section. Then every factor in the partition function of the quiver theory (3.61) has a natural interpretation, see Fig. 3 (eq):

- $|\text{pure}\rangle$ is the state created by the boundary condition at $x^5 = 0$.
- $q_1^N = e^{(\log q_1)N} = e^{-L_1 E}$ is the Euclidean propagation of the system by the length L_1.
- $\Phi_{\vec{a}_2,m,\vec{a}_1}$ is the operation defined by the bifundamental hypermultiplet at $x^5 = L_1$.
- $q_2^N = e^{(\log q_2)N} = e^{-L_2 E}$ is the Euclidean propagation of the system by the length L_2.
- $\langle\text{pure}|$ is the state representing the boundary condition at $x^5 = L_1 + L_2$.

3.6 W-Algebra Action and the Sixth Direction

For $G = U(N)$, it is a mathematical fact [80, 81] that there is a natural action of the W_N algebra on $\mathbb{V}_{G,\vec{a}}$. The W_N algebra is generated by two-dimensional holomorphic spin-d currents $W_d(z), d = 2, 3, \ldots, N$, and in particular contains the Virasoro subalgebra generated by $T(z) = W_2(z)$. The L_0 of the Virasoro subalgebra is identified with N acting on $\mathbb{V}_{G,\vec{a}}$. In particular, L_{-m} maps $\mathbb{V}_{G,\vec{a},n}$ to $\mathbb{V}_{G,\vec{a},n+m}$. Figuratively speaking, L_{-m} adds m instantons into the system. Furthermore, for generic value of $\vec{a}$, $\mathbb{V}_{G,\vec{a}}$ is the Verma module of the W_N-algebra times a free boson. The central charge of the Virasoro subalgebra of this W_N algebra is given by the formula

$$c = (N - 1) + N(N^2 - 1)\frac{(\epsilon_1 + \epsilon_2)^2}{\epsilon_1 \epsilon_2}. \tag{3.65}$$

Furthermore, it is believed that there is a natural decomposition

$$\mathbb{V}_{G,\vec{a}} = V_{\vec{a}'} \otimes H_m \tag{3.66}$$

into a W_N Verma module $V_{\vec{a}'}$, and a free boson Fock space H_m. Here, we define m and $\vec{a}'$ via

$$m = \sum a_i, \quad \vec{a}' = \vec{a} - m\left(\frac{1}{N}, \ldots, \frac{1}{N}\right). \tag{3.67}$$

Note that $\vec{a}'$ lives in an $N - 1$ dimensional subspace. Then $V_{\vec{a}'}$ is the Verma module of the W_N algebra constructed from $N - 1$ free scalar fields with zero mode eigenvalue by $\vec{a}'$ and the background charge

$$\vec{Q} = \left(b + \frac{1}{b}\right)\left(\frac{N}{2}, \frac{N}{2} - 1, \ldots, 1 - \frac{N}{2}, -\frac{N}{2}\right), \quad b^2 = \frac{\epsilon_1}{\epsilon_2}, \tag{3.68}$$

and H_m is the free boson Fock space with zero mode eigenvalue m. The action of a free boson on H_m was constructed in [78]. The decomposition above was also studied in [82, 83].

When $\epsilon_1 + \epsilon_2 = 0$, we have $b + 1/b = 0$ and the background charges (3.68) vanish. In this case the system becomes particularly simple, and it was already studied in [84–86].

The vector $|\text{pure}\rangle \in \mathbb{V}_{G,\vec{a}}$, from this point of view, is a special vector called a Whittaker vector, which is a kind of a coherent state of the W-algebra [51, 87–89]. Small number of hypermultiplets in the fundamental representation also is a boundary condition which also corresponds to a special state, studied in [90].

The linear map $\Phi_{\vec{a},m,\vec{b}}$ defined by a bifundamental hypermultiplet (3.58) should be a natural map between two representations of W_N algebras. A natural candidate is an intertwiner of the W_N algebra action, or equivalently, it is an insertion of a primary operator of W_N. If that is the case, the partition function of a cyclic quiver with the gauge group $SU(N)_1 \times SU(N)_2 \times SU(N)_3$,

$$\text{tr } q_1^N \Phi_{\vec{a},m_1,\vec{b}} q_2^N \Phi_{\vec{b},m_2,\vec{c}} q_3^N \Phi_{\vec{c},m_3,\vec{a}}, \tag{3.69}$$

for example, is the conformal block of the W_N algebra on the torus $z \sim q_1 q_2 q_3 z$ with three insertions at $z = 1$, q_1, and $q_1 q_2$. This explains the observation first made in [50].

Therefore the mathematically missing piece is to give the proof that $\Phi_{\vec{a},m,\vec{b}}$ is the primary operator insertion. For $N = 2$ when W_N is the Virasoro algebra, this has been proven in [91, 92], but the general case is not yet settled. At least, there are many studies which show the agreement up to low orders in the q-expansion [93, 94]. Also, the decomposition (3.66) predicts the existence of a rather nice basis in the Verma module of W_N algebra times a free boson which was not know before, whose property was studied in [95]. The decomposition was also studied from the point of view of the $W_{1+\infty}$ algebra [96, 97] corresponding to the case $\epsilon_1 + \epsilon_2 = 0$. Its generalization to the case $\epsilon_1 + \epsilon_2 \neq 0$ was done in [98].

When one considers a bifundamental charged under $SU(N_1) \times SU(N_2)$ with $N_1 > N_2$, we have a linear operator

$$\Phi : \mathbb{V}_{SU(N_1)} \to \mathbb{V}_{SU(N_2)}, \qquad (3.70)$$

and we have an action of W_{N_i} on $\mathbb{V}_{SU(N_i)}$. The 6d construction using $\mathcal{N} = (2, 0)$ theory of type $SU(N_1)$ [54] suggests that it can also be represented as a map

$$\Phi : \mathbb{V}_{SU(N_1)} \to \mathbb{V}', \qquad (3.71)$$

where we still have an action of W_{N_1} on $\mathbb{V}'$. Then $\mathbb{V}'$ is no longer a Verma module, even for generic values of parameters. $\mathbb{V}'$ are believed to be the so-called semi-degenerate representations of W_{N_1} algebras determined by N_2, and there are a few checks of this idea [99–101].

3.7 String Theoretical Interpretations

As seen in Sect. 3.5, the operator N is the Hamiltonian generating the translation along x^5. It is therefore most natural to make the identification $\log |z| = x^5$. Although the circle direction $x^6 = \arg z$ was not directly present in the setup of Sect. 3.5, it also has a natural interpretation. Namely, the maximally supersymmetric 5d gauge theory with gauge group $U(N)$ on a space X is in fact the six-dimensional $\mathcal{N} = (2, 0)$ theory of type $U(N)$ on a space $X \times S^1$, such that the Kaluza-Klein momentum along the S^1 direction is the instanton number of the 5d gauge theory. This again nicely fits with the fact that L_n creates n instantons, as the operator L_n has n Kaluza-Klein momenta along S^1. The quiver gauge theory treated at the end of Sect. 3.5 can now be depicted as in Fig. 3 (6d). There, the boundary conditions at both ends correspond to the state $|\text{pure}\rangle$ in $\mathbb{V}$. The operator $\Phi_{\vec{a},m,\vec{b}}$ is now an insertion of a primary field.

If one prefers string theoretical language, it can be further rephrased as follows. We consider N D4-branes on the space X, in a Type IIA set-up. This is equivalent to N M5-branes on the space $X \times S^1$ in an M-theory set-up. The Kaluza-Klein momenta around S^1 are the D0-branes in the Type IIA description, which can be absorbed into the world-volume of the D4-branes as instantons. The insertion of a primary is an intersection with another M5-brane. This reduces in the type IIA limit an intersection with an NS5-brane, which gives the bifundamental hypermultiplet.

In the discussions so far, we introduced two vector spaces associated to the n-instanton moduli space $M_{G,n}$, and saw the appearance of *three* distinct extra space-time directions, ξ^5, x^5 and x^6.

- First, we introduced $\mathcal{H}_n$ in Sect. 3.2. We put $\mathcal{N} = 1$ supersymmetric 5d gauge theory with hypermultiplets on the Ω background $\mathbb{R}^4 \times S^1$ so that $\mathbb{R}^4$ is rotated when we go around S^1. We then considered S^1 as the time direction. We called this direction ξ^5. The supersymmetric, non-perturbative part of the field theory Hilbert space reduces to the Hilbert space of the supersymmetric quantum mechanics on the moduli space of n-instantons plus the hypermultiplet zero modes. We did not use the inner product in this Hilbert space. Mathematically, it is the space of holomorphic functions on the moduli space.
- Second, we introduced $\mathbb{V}_n$ in Sect. 3.4. We put the maximally-supersymmetric 5d gauge theory on $\mathbb{R}^4 \times$ a segment parameterized by x^5, and considered the segment as the time direction. The supersymmetric, non-perturbative part of the field theory Hilbert space reduces to the space $\mathbb{V}_n$. It has an inner product, defined by means of the trace on $\mathcal{H}_n$. Mathematically, $\mathbb{V}_n$ is the equivariant cohomology of the moduli space. In this second setup, another circular direction x^6 automatically appears, so that it combines with x^5 to form a complex direction $\log z = x^5 + ix^6$.

It is important to keep in mind that in this second story with x^5 and x^6 we kept the radius β of ξ^5 direction to be zero. If we keep it to a nonzero value instead, the inner product on $\mathbb{V}_{G,\vec{a}}$ (3.55) is instead modified to

$$\langle p|q \rangle = \delta_{p,q} \frac{1}{\prod_t 1 - e^{i\beta v(p)_t}}. \tag{3.72}$$

Let us distinguish the vector space with this modified inner product from the original one by calling it $\tilde{\mathbb{V}}_{G,\vec{a}}$. The W_N action is no longer there. Instead, we have [80, 81, 102–104] an action of q-deformed W_N algebra on $\tilde{\mathbb{V}}_{G,\vec{a}}$, which does not contain a Virasoro subalgebra. Therefore, we do not generate additional direction x^6 anymore. String theoretically, the set up with ξ^5 and x^5 corresponds to having N D5-branes in Type IIB, and it is hard to add another physical direction to the system.

Relation to the refined topological vertex Now, let us picturize this last Type IIB setup. We depict N D5-branes as N lines as in Fig. 4 (1). The horizontal direction is x^5, the vertical direction is x^9, say. We do not show the spacetime directions $\mathbb{R}^4$ or the compactified direction ξ^5. In the calculation of the instanton partition function, we assign a Young diagram to each D5-brane.

The boundary condition at fixed value of x^5, introducing a bifundamental hyper-multiplet, is realized by an NS5-brane cutting across N D5-branes, which can be depicted as in Fig. 4 (2). When an NS5-brane crosses an D5-brane, they merge to form a $(1, 1)$ 5-brane, which needs to be tilted to preserve supersymmetry; the figure shows this detail.

Therefore, the whole brane set-up describing a five-dimensional quiver gauge theory on a circle can be built from a vertex joining three 5-branes Fig. 4 (3), and a line representing a 5-brane Fig. 4 (4). Any 5-brane is obtained by an application of the $SL(2, \mathbb{Z})$ duality to the 5-brane, so one can associate a Young diagram to any line. The basic quantity is then a function $Z_{\text{vertex}}(\epsilon_1, \epsilon_2; Y_1, Y_2, Y_3)$ which is called the refined topological vertex. The partition function of the system is obtained by

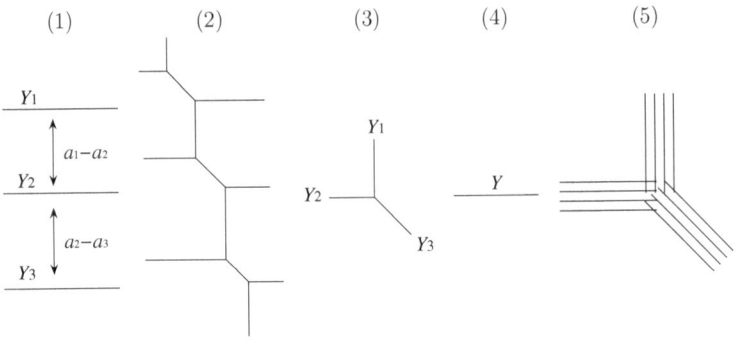

Fig. 4 Type IIB setup, or equivalently the toric diagram.

multiplying the refined topological vertex for all the junctions of three 5-branes, multiplying a propagator factor $\Delta(Y)$ for each of the internal horizontal line, and summing over all the Young diagrams.

The phrase 'refined topological' is used due to the following situation where it was originally discovered. A review of the detail can be found in [V:13] in this volume, so we will be brief here. We apply a further chain of dualities to the setup we have arrived, so that the diagrams in Fig. 4 are now considered as specifying the toric diagram of a non-compact toric Calabi-Yau space on which M-theory is put. The direction ξ^5 is now the M-theory circle. Nekrasov's partition function of this setup when $\epsilon_1 = -\epsilon_2 = g_s$ is given by the partition function of the topological string on the same Calabi-Yau with the topological string coupling constant at g_s. This gives the unrefined version of the topological vertex. The generalized case $\epsilon_1 \neq -\epsilon_2$ should correspond to a refined version of the topological string on the Calabi-Yau, and the function Z_{vertex} for general ϵ_1, ϵ_2 is called the refined topological vertex. The unrefined version was determined in [37, 38] and the refined version was determined in [39, 40, 105].

In this discussion, we implicitly used the fact that the logarithm of the partition function on the Ω background (3.21) is equal to the prepotential in the presence of the graviphoton background, which is further equal to the free energy of the topological string. In the unrefined case this identification goes back to [106, 107]. The refined case is being clarified, see e.g. [108–110].

As an aside, we can also perform a T-duality along the ξ_5 direction in the type IIB configuration above. This gives rise to a type IIA configuration in the fluxtrap solution, which lifts to a configuration of M5-branes with four-form background [111–113]. For a certain class of gauge theories, we can also go to a duality frame where we have D3-branes in an orbifold singularity with a particular RR-background. It has been directly checked that the partition function in this setup reproduces Nekrasov's partition function in the unrefined case $\epsilon_1 + \epsilon_2 = 0$ [114–116].

Let us come back to the discussion of the refined topological vertex itself. The summation over the Young diagrams in the internal lines of Fig. 4 (2) can be carried out explicitly using the properties of Macdonald polynomials, and correctly reproduces the numerator of the partition function (3.26) coming from a bifundamental, given by the weights in (3.38). The denominator basically comes from the propagator factors associated to N horizontal lines [36, 41].

Here, it is natural to consider an infinite dimensional vector space

$$\tilde{\mathbb{V}}_1 = \bigoplus_Y \mathbb{C}|Y\rangle \qquad (3.73)$$

whose basis is labeled by a Young diagram, such that the inner product is given by the propagator factor $\Delta(Y)$ of the topological vertex. Now, the space $\tilde{\mathbb{V}}_1$ is known to have a natural action of an algebra called the Ding-Iohara algebra $\mathcal{DI}$ [117]. It might be helpful to know that this algebra is also called the elliptic Hall algebra, or the quantum toroidal GL(1) algebra; see e.g. [118] for the quantum toroidal algebras. Then the refined topological vertex Z_{vertex} is an intertwiner of this algebra:

$$Z_{\text{vertex}} : \tilde{\mathbb{V}}_1 \otimes \tilde{\mathbb{V}}_1 \rightarrow \tilde{\mathbb{V}}_1. \qquad (3.74)$$

The q-deformed W_N-algebra action on $\tilde{\mathbb{V}}_{\text{U}(N)}$, from this point of view, should be understood from its relation to the action of the Ding-Iohara algebra $\mathcal{DI}$ on

$$\tilde{\mathbb{V}}_1^{\otimes N} \simeq \tilde{\mathbb{V}}_{\text{U}(N)}. \qquad (3.75)$$

The W_N action on $\mathbb{V}_{\text{U}(N)}$ should follow when one takes the four-dimensional limit when the radius β of the ξ^5 direction goes to zero.

This formulation has an advantage that the instanton partition function on S^1 of a 5d non-Lagrangian theory, such as the T_N theory corresponding to Fig. 4 (5), can be computed, by just multiplying the vertex factors and summing over Young diagrams. Indeed this computation was performed in [119, 120], where the E_6 symmetry of the partition function of T_3 was demonstrated.

It should almost be automatic that the resulting partition function Z_{T_N} of T_N is an intertwiner of q-deformed W_N algebra, because the linear map

$$Z_{T_N} : \tilde{\mathbb{V}}_{\text{U}(N)} \otimes \tilde{\mathbb{V}}_{\text{U}(N)} \rightarrow \tilde{\mathbb{V}}_{\text{U}(N)} \qquad (3.76)$$

is obtained by composing $N(N-1)/2$ copies of Z_{vertex} according to Fig. 4 (5). One can at least hope that the intertwining property of Z_{vertex}, together with the naturality of the map (3.75), should translate to the intertwining property of Z_{T_N}.

4 Other Gauge Groups

In this section, we indicate how the instanton calculations can be extended to gauge groups other than (special) unitary groups. We do not discuss the details, and only point to the most relevant results in the literature.

4.1 Classical Gauge Groups

Let us consider classical gauge groups $G = \mathrm{SO}(2n)$, $\mathrm{SO}(2n+1)$ and $\mathrm{USp}(2n)$. Physically, nothing changes from what is stated in Sect. 3.2; we need to perform localization on the n-instanton moduli space $M_{G,n}$ of gauge group G. A technical problem is that there is no known way to resolve and/or deform the singularity of $M_{G,n}$ to make it smooth, when G is not unitary.

To proceed, we first re-think the way we performed the calculation when $G = \mathrm{U}(N)$. For classical G, the instanton moduli space has the ADHM description, just as in the unitary case recalled in (3.32):

$$M_{G,n} = \{\mu_{\mathbb{C}}(x) = 0 \mid x \in X_{G,n}\}/K(G,n) \tag{4.1}$$

Here, $K(G,n)$ is a complexified compact Lie group, and $X_{G,n}$ is a vector space, given as in (3.34) by a tensor product and a direct sum starting from vector spaces V and W which are the fundamental representations of $K(G,n)$ and G respectively. One can formally rewrite the integral which corresponds to the localization on $M_{G,n}$ as an integral over

$$X_{G,n} \oplus \mathfrak{k}_{\mathbb{R}}(G,n), \tag{4.2}$$

where $\mathfrak{k}(G,n)$ is the Lie algebra of $K(G,n)$. The integral along $X_{G,n}$ can be easily performed, and the integration on $\mathfrak{k}_{\mathbb{R}}(G,n)$ can be reduced to an integration on the Cartan subalgebra $\mathfrak{h}_{\mathbb{R}}(G,n)$ of $\mathfrak{k}_{\mathbb{R}}(G,n)$, resulting in a formal expression

$$Z_{\mathrm{inst},n}(\epsilon_{1,2}; a_{1,\ldots,r}; m_{1,\ldots,f}) = \int_{\phi \in \mathfrak{h}_{\mathbb{R}}(G,n)} f(\epsilon_{1,2}; a_{1,\ldots,r}; m_{1,\ldots,f}; \phi) \tag{4.3}$$

where f is a rational function.

The fact that $M_{G,n}$ is singular is reflected in the fact that the poles of the rational function are on the integration locus $\mathfrak{h}_{\mathbb{R}}(G,n)$. When G is unitary, the deformation of the instanton moduli space $M_{G,n}$ to make it smooth corresponds to a systematic deformation of the half-dimensional integration contour $\mathfrak{h}_{\mathbb{R}}(G,n) \subset \mathfrak{h}_{\mathbb{C}}(G,n)$. Furthermore, the poles are in one-to-one correspondence with the fixed points on the smoothed instanton moduli space. A pole is given by a specific value $\phi \in \mathfrak{h}_{\mathbb{C}}(G,n)$ which is a certain linear combinations of $\epsilon_{1,2}$, $a_{1,\ldots,r}$, and $m_{1,\ldots,f}$. In other words, the position of a pole is given by specifying the action of $\mathrm{U}(1)^{2+r} \subset \mathrm{U}(1)^2 \times G$ on the vector space V, which is naturally a representation of $K(G,n)$. Finally, the residues give the summand in the localization formula (3.29).

Although the deformation of the moduli space is not possible when G is not unitary, the systematic deformation of the integration contour $\mathfrak{h}_{\mathbb{R}}(G,n) \subset \mathfrak{h}_{\mathbb{C}}(G,n)$ is still possible. The poles are still specified by the actions

$$\phi_p : \mathrm{U}(1)^{2+r} \curvearrowright V. \tag{4.4}$$

Then the instanton partition function (4.3) can be written down explicitly as

$$Z_{\text{inst},n}(\epsilon_{1,2}; a_{1,\dots,r}; m_{1,\dots,f}) = \sum_p \text{Res}_{\phi=\phi_p} f(\epsilon_{1,2}; a_{1,\dots,r}; m_{1,\dots,f}; \phi). \tag{4.5}$$

This calculation was pioneered in [22, 23], and further elaborated in [26].

4.2 Effect of Finite Renormalization

Let us in particular consider $\mathcal{N} = 2$ SU(2) gauge theory with four fundamental hypermultiplets, with all the masses set to zero for simplicity. Its instanton partition function can be calculated either as the $N = 2$ case of SU(N) theory, or as the $N = 1$ case of USp($2N$) theory, using the ADHM construction either of the SU instantons or of the USp instantons. What was found in [26] is that the n-instanton contribution calculated in this manner, are all different:

$$Z_{\text{inst},n}^{\text{SU}(2)}(\epsilon_{1,2}; a) \neq Z_{\text{inst},n}^{\text{USp}(2)}(\epsilon_{1,2}; a). \tag{4.6}$$

They also found that the total instanton partition functions

$$Z^G(q; \epsilon_{1,2}; a) = q^{a^2/\epsilon_1\epsilon_2} \sum_{n\geq 0} q^n Z_{\text{inst},n}^G(\epsilon_{1,2}; a; m) \tag{4.7}$$

becomes the same,

$$Z^{\text{SU}(2)}(q_{\text{SU}(2)}; \epsilon_{1,2}; a; m) = Z^{\text{USp}(2)}(q_{\text{USp}(2)}; \epsilon_{1,2}; a; m) \tag{4.8}$$

once we set

$$q_{\text{SU}(2)} = q_{\text{USp}(2)}\left(1 + \frac{q_{\text{USp}(2)}}{4}\right)^{-2}. \tag{4.9}$$

The physical coupling $q_{\text{IR}} = e^{2\pi\tau_{\text{IR}}}$ in the infrared is then given in terms of the prepotential:

$$2\pi\tau_{\text{IR}} \lim_{\epsilon_{1,2}\to 0} \epsilon_1\epsilon_2 \log Z^G(q; \epsilon_{1,2}). \tag{4.10}$$

This is given by

$$q_{\text{SU}(2)} = \frac{\theta_2(q_{\text{IR}})^4}{\theta_3(q_{\text{IR}})^4} \tag{4.11}$$

This finite discrepancy between the UV coupling $q_{SU(2)}$ and the IR coupling q_{IR} was first clearly recognized in [121], and the all order form was conjectured by [48]. We see that the UV coupling $q_{USp(2)}$ is different from both.

These subtle difference among $q_{SU(2)}$, $q_{USp(2)}$ and q_{IR} reflects a standard property of any well-defined quantum field theory. The factor weighting the instanton number, q_G, is an ultra-violet dimensionless quantity, and is renormalized, the amount of which depends on the regularization chosen. The choice of the ADHM construction of the SU(2) = USp(2) instanton moduli space and the subsequent deformation of the contours are part of the regularization. The final physical answer should be independent (4.8), once the finite renormalization is correctly performed, as in (4.9).

In this particular case, there is a natural geometric understanding of the relations (4.9) and (4.11) [26]. The SU(2) theory with four flavors can be realized by putting 2 M5-branes on a sphere C with four punctures a_1, a_2, b_1, b_2, whose cross ratio is the UV coupling $q_{SU(2)}$. The Seiberg-Witten curve of the system is the elliptic curve E which is a double-cover of C with four branch points at a_1, a_2, b_1, b_2. The IR gauge coupling is the complex structure of E, and this gives the relation (4.11).

The same system can be also realized by putting 4 M5-branes on top of the M-theory orientifold 5-plane on a sphere C' with four punctures, x, y, a, b, whose cross ratio is the coupling $q_{USp(2)}$. Here we also have the orientifold action around the puncture x, y. There is a natural 2-to-1 map $C \to C'$ with branch points at x and y, so that $a_{1,2}$ and $b_{1,2}$ on C are inverse images of a and b on C', respectively. This gives the relation (4.9).

4.3 Exceptional Gauge Groups

For exceptional gauge groups G, not much was known about the instanton moduli space $M_{G,n}$, except at instanton number $n = 1$, because we do not have ADHM constructions. To perform the instanton calculation in full generality in the presence of matter hypermultiplets, we need to know the properties of various bundles on $M_{G,n}$. For the pure gauge theory, the knowledge of the ring of the holomorphic functions on $M_{G,n}$ would suffice. Any instanton moduli space decomposes as $M_{G,n} = \mathbb{C}^2 \times M_{G,n}^{centered}$, where $\mathbb{C}^2$ parameterize the center of the instanton, and $M_{G,n}^{centered}$ is called the centered instanton moduli space. Therefore the question is to understand the centered instanton moduli space better.

The centered one-instanton moduli space of any gauge group G is the minimal nilpotent orbit of $\mathfrak{g}_{\mathbb{C}}$, i.e. the orbit under $G_{\mathbb{C}}$ of a highest weight vector. The ring of the holomorphic functions on the minimal nilpotent orbit is known [122–124], and thus the instanton partition function of pure exceptional gauge theory can be computed up to instanton number 1 [53].

There are 4d $\mathcal{N} = 2$ quantum field theories "of class S" whose Higgs branch is $M_{E_r,n}^{centered}$ [125]. There is now a conjectured formula which computes the ring of holomorphic functions on the Higgs branch of a large subclass of class S theories [126]. A review can be found in [V:9] in this volume. This method can be used to study $M_{E_r,2}$ explicitly, from which the instanton partition function of E-type gauge theories can be found [127–129].

Moreover, the Higgs branch of any theories of class S is obtained [130] by the hyperkähler modification [131] of the Higgs branch of the so-called T_G theory. The Higgs branch of the T_G theory is announced to be rigorously constructed [132]. Therefore, we now have a finite-dimensional construction of $M_{E_r,n}$. This should allow us to perform any computation on the instanton moduli space, at least in principle.

4.4 Relation to W-Algebras

We can form an infinite-dimensional vector space $\mathbb{V}_G$ as in Sect. 3.4. When $G = SU(N)$, there was an action of the W_N algebra on $\mathbb{V}_G$. There is a general construction of W-algebras starting from arbitrary affine Lie algebras $\hat{G}$ and twisted affine Lie algebras $\hat{G}^{(s)}$ where $s = 2, 3$ specifies the order of the twist; in this general notation, the W_N algebra is $W(\widehat{SU(N)})$ algebra. For a comprehensive account of W-algebras, see the review [69] and the reprint volume [70].

When G is simply-laced, i.e. $G = SU(N)$, $SO(2N)$ or E_N, $\mathbb{V}_G$ has an action of the $W(\hat{G})$ algebra; this can be motivated from the discussion as in Sect. 3.6. We start from the 6d $\mathcal{N} = (2, 0)$ theory of type G, and put it on $\mathbb{R}^4 \times C_2$ where C_2 is a Riemann surface, so that we have $\mathcal{N} = 2$ supersymmetry in four dimensions. Then, we should have some kind of two-dimensional system on C_2. The central charge of this two-dimensional system can be computed [133, 134] starting from the anomaly polynomial of the 6d theory, which results in

$$c = \text{rank } G + h^\vee(G) \dim G \frac{(\epsilon_1 + \epsilon_2)^2}{\epsilon_1 \epsilon_2}. \tag{4.12}$$

This is the standard formula of the central charge of the $W(\hat{G})$ algebra, when G is simply-laced.

When G is not simply-laced, we can use the physical 5d construction in Sect. 3.5, but there is no 6d $\mathcal{N} = (2, 0)$ theory of the corresponding type. Rather, one needs to pick a simply-laced J and a twist σ of order $s = 2, 3$, such that the invariant part of J under σ is Langlands dual to G, see the Table 1. Then, the 5d maximally

Table 1 The type of the 6d theory, the choice of the outer-automorphism twists, and the 5d gauge group

Γ	A_{2n-1}	D_{n+1}	D_4	E_6
s	2	2	3	2
G	B_n	C_n	G_2	F_4

supersymmetric theory with gauge group G lifts to a 6d theory of type J, with the twist by σ around x^6. This strongly suggests that the W-algebra which acts on $\mathbb{V}_G$ is $W(\hat{G}^{(s)})$. This statement was checked to the one-instanton level in [53] by considering pure G gauge theory. A full mathematical proof for simply-laced G is available in [135].

5 Other Spaces

5.1 With a Surface Operator

Generalities Let us consider a gauge theory with a simple gauge group G, with a surface operator supported on $\mathbb{C} \subset \mathbb{C}^2$. A detailed review can be found in [V:8], so we will be brief here. A surface operator is defined in the path integral formalism as in the case of 't Hooft loop operators, by declaring that fields have prescribed singularities there. In our case, we demand that the gauge field has the divergence

$$A_\theta d\theta \rightarrow \mu d\theta \tag{5.1}$$

where θ is the angular coordinate in the plane transverse to the surface operator, μ is an element in $\mathfrak{g}$; the behavior of other fields in the theory is set so that the surface operator preserves a certain amount of supersymmetry.

On the surface operator, the gauge group is broken to a subgroup L of G commuting with μ. Let us say there is a subgroup $U(1)^k \subset L$. Then, the restriction of the gauge field on the surface operator can have nontrivial monopole numbers $n_1, \ldots, n_k$. Together with the instanton number n_0 in the bulk, they comprise a set of numbers classifying the topological class of the gauge field. Thus we are led to consider the moduli space $M_{G,L,\mu,n_0,n_1,\ldots,n_k}$. It is convenient to redefine $n_0, \ldots, n_k$ by an integral linear matrix so that these instanton moduli spaces are nonempty if and only if $n_0, \ldots, n_k \geq 0$. The instanton partition function is schematically given by

$$Z_{\text{inst}}(\epsilon_{1,2}; a_i; q_{0,1,\ldots,k}) = \sum_{n_0,n_1,\ldots,n_k \geq 0} q_0^{n_0} \cdots q_k^{n_k} Z_{\text{inst},n_0,\ldots,n_k}(\epsilon_{1,2}; a_i) \tag{5.2}$$

where $Z_{\text{inst},n_0,\ldots,n_k}(\epsilon_{1,2}; a_i)$ is given by a geometric quantity associated to $M_{G,L,\mu,n_0,n_1,\ldots,n_k}$.

This space is not well understood unless G is unitary. Suppose G is $SU(N)$. Then the singularity is specified by

$$\mu = \text{diag}(\underbrace{\mu_1, \ldots, \mu_1}_{m_1}, \underbrace{\mu_2, \ldots, \mu_2}_{m_2}, \ldots, \underbrace{\mu_{k+1}, \ldots, \mu_{k+1}}_{m_{k+1}}). \tag{5.3}$$

Then the group L is

$$L = S \left[\prod_{i=1}^{k+1} U(m_i) \right] \qquad (5.4)$$

which has a $U(1)^k$ subgroup.

Here, we can use a mathematical result [136, 137] which says that the moduli space $M_{G,L,\mu,n_0,n_1,...,n_k}$ in this case is equivalent as a complex space to the moduli space of instantons on an orbifold $\mathbb{C}/\mathbb{Z}_{k+1} \times \mathbb{C}$. As we will review in the next section, the instanton moduli space on an arbitrary Abelian orbifold of $\mathbb{C}^2$ can be easily obtained from the standard ADHM construction, resulting in the quiver description of the instanton moduli space with a surface operator [138, 139]. The structure of the fixed points can also be obtained starting from that of the fixed points on $\mathbb{C}^2$. Then the instanton partition function can be explicitly computed [68, 140], although the details tend to be rather complicated when $[m_1, ..., m_{k+1}]$ is generic [66, 67, 141].

Corresponding W-algebra An infinite dimensional vector space $\mathbb{V}_{G,L,\vec{a}}$ can be introduced as in Sect. 3.4:

$$\mathbb{V}_{G,L,\vec{a}} = \bigoplus_{n_0,...,n_k \geq 0} \mathbb{V}_{G,L;\vec{a};n_0,...,n_k} \qquad (5.5)$$

where $\mathbb{V}_{G,L;\vec{a};n_0,...,n_k}$ is the equivariant cohomology of $M_{G,L,\mu,n_0,n_1,...,n_k}$ with the equivariant parameter of $SU(N)$ given by $\vec{a}$. As $\mathbb{V}$ does not depend on the continuous deformation of μ with fixed L, we dropped μ from the subscript of $\mathbb{V}$.

The W-algebra which is believed to be acting on $\mathbb{V}_{G,L}$ is obtained as follows, when $G = SU(N)$ and L is given as in (5.4). Introduce an N-dimensional representation of $SU(2)$

$$\rho_{[m_1,...,m_{k+1}]} : SU(2) \to SU(N) \qquad (5.6)$$

such that the fundamental representation of $SU(N)$ decomposes as the direct sum of $SU(2)$ irreducible representations with dimensions $m_1, ..., m_{k+1}$. Let us define a nilpotent element via

$$\nu_{[m_1,...,m_{k+1}]} = \rho_{[m_1,...,m_{k+1}]}(\sigma^+). \qquad (5.7)$$

Then we perform the quantum Drinfeld-Sokolov reduction of $\hat{G}$ algebra via this nilpotent element, which gives the algebra $W(\hat{G}, \nu_{[m_1,...,m_{k+1}]})$ which is what we wanted to have. In particular, when $[m_1, ..., m_{k+1}] = [1, ..., 1]$, the nilpotent element is $\nu = 0$, and the resulting W-algebra is $\hat{G}$. When $[m_1, ..., m_{k+1}] = [N]$, there is no singularity, and the W-algebra is the standard $W(\hat{G})$ algebra. The general W-algebras $W(\hat{G}, \nu)$ were introduced in [142].

Let F be the commutant of $\rho_{[m_1,\ldots,m_{k+1}]}(\mathrm{SU}(2))$ in $\mathrm{SU}(N)$. Explicitly, it is

$$F = S\left[\prod_{s=1}^{t} U(\ell_t)\right] \tag{5.8}$$

where $\ell_{1,\ldots,t}$ is defined by writing

$$[m_1,\ldots,m_k] = [n_1{}^{\ell_1},\ldots,n_t{}^{\ell_t}]. \tag{5.9}$$

Note that the rank of F is k. The W-algebra $W(\hat{G}, \nu_{[m_1,\ldots,m_{k+1}]})$ contains an affine subalgebra $\hat{F}$. Therefore, the dimension of the Cartan subalgebra of $W(\hat{G}, \nu_{[m_1,\ldots,m_{k+1}]})$ is rank $F + 1 = k + 1$, and any representation of the W-algebra is graded by integers $n_0,\ldots,n_k$. This matches with the fact that $\mathbb{V}_{G,L}$ is also graded by the same set of integers (5.5).

Higher-dimenisonal interpretation From the 6d perspective advocated in Sect. 3.5, one considers a codimension-2 operator of the 6d $\mathcal{N} = (2,0)$ theory of type $\mathrm{SU}(N)$, extending along x^5 and x^6. Such a codimension-2 operator is labeled by a set of integers $[m_1,\ldots,m_k]$, and is known to create a singularity of the form (5.1) and (5.3) in the four-dimensional part [54, 143]. Furthermore, the operator is known to have a flavor symmetry F as in (5.8). Therefore, it is as expected that the W-algebra $W(\hat{G}, \nu_{[m_1,\ldots,m_{k+1}]})$ has the $\hat{F}$ affine subalgebra. Its level can be computed by starting from the anomaly polynomial of the codimension-2 operator; a few checks of this line of ideas were performed in [61, 67, 144].

The partition function with surface operator of type $[N-1,1]$ can also be represented as an insertion of a degenerate primary field Φ in the standard W_N algebra [44, 145, 146]. When $N = 2$, we therefore have two interpretations: one is that the surface operator changes the Virasoro algebra to $\widehat{\mathrm{SU}(2)}$, the other is that the surface operator is a degenerate primary field of the Virasoro algebra. These can be related by the Ribault-Teschner relation [147, 148], but the algebraic interpretation is not clear.

For general simply-laced G and L, the W-algebra which acts on $\mathbb{V}_{G,L}$ is thought to be $W(\hat{G}, \nu)$, where ν is a generic nilpotent element in L. But there is not many explicit checks of this general statement, except when L is the Cartan subgroup.

Braverman-Etingof When L is the Cartan subgroup, $\nu = 0$, and the W-algebra is just the $\hat{G}$ affine algebra. Its action on $\mathbb{V}_{G,L}$ was constructed in [19]. The instanton partition function Z of the pure G gauge theory with this surface operator was then analyzed in [20]. The limit

$$F = \lim_{\epsilon_{1,2}\to 0} \epsilon_1\epsilon_2 \log Z \tag{5.10}$$

was shown to be independent of the existence of the surface operator; the surface operator contributes only a term of order $1/\epsilon_1$ to $\log Z$ at most. The structure of the $\hat{G}$ affine Lie algebra was then used to show that F is the prepotential of the Toda system of type G, thus proving that the instanton counting gives the same prepotential as determined by the Seiberg-Witten curve.

Before proceeding, let us consider the contribution from the bifundamental hypermultiplet. Again as in Sect. 3.4, it determines a nice linear map

$$\Phi_{\vec{a},m,\vec{b}} : \mathbb{V}_{G,L,\vec{a}} \to \mathbb{V}_{G,L,\vec{b}} \tag{5.11}$$

where m is the mass of the hypermultiplet. This $\Phi_{\vec{a},m,\vec{b}}$ is expected to be a primary operator insertion of this W-algebra. This is again proven when $v = 0$ and the W-algebra is just the $\hat{G}$ affine algebra [149].

The author does not know how to incorporate hypermultiplet matter fields in this approach.

5.2 On Orbifolds

Let us now consider the moduli space of instantons on an orbifold of $\mathbb{C}^2$ by the $\mathbb{Z}_p$ action

$$g : (z, w) \to (e^{2\pi i s_1/p} z, e^{2\pi i s_2/p} w). \tag{5.12}$$

This was analyzed by various groups, e.g. [150, 151]. We need to specify how this action embeds in $G = U(N)$. This is equivalent to specify how the N-dimensional subspace W in (3.34) transforms under $\mathbb{Z}_p \times G$:

$$W = e^{2\pi i t_1/p} e^{i\beta a_1} + \cdots + e^{2\pi i t_N/p} e^{i\beta a_N}. \tag{5.13}$$

The moduli space $M_{G,n}$ has a natural action of $U(1)^2 \times G$, to which we now have an embedding of $\mathbb{Z}_p$ via (5.12) and (5.13). Then the moduli space of instantons on the orbifold, $M_{G,n}^g$, is just the $\mathbb{Z}_p$ invariant part of $M_{G,n}$.

A fixed point of $M_{G,n}^g$ under $U(1)^{2+r}$ is still a fixed point in $M_{G,n}$. Therefore, it is still specified by W and V as in (3.35). The vector space V now has an action of g, which is fixed to be

$$V_p = \sum_{v=1}^{N} \sum_{(i,j)\in Y_v} e^{2\pi i(t_v+(1-i)s_1+(1-j)s_2)/p} e^{i\beta(a_v+(1-i)\epsilon_1+(1-j)\epsilon_2)}. \tag{5.14}$$

Then, the tangent space at the fixed point and/or the hypermultiplet zero modes can be just obtained by projecting down (3.36)–(3.38) to the part invariant under the $\mathbb{Z}_p$ action.

It is now a combinatorial exercise to write down a general formula for the instanton partition function on the orbifold; as reviewed in the previous section, this includes the case with surface operator. It is again to be said that, however, it is easier to implement the algorithm as written above, than to first write down a combinatorial formula and then implement it in a computer algebra system.

Let us now focus on the case when $(s_1, s_2) = (1, -1)$. Then the orbifold $\mathbb{C}^2/\mathbb{Z}_p$ is hyperkähler. Let us consider $U(N)$ gauge theory on it. We can construct the infinite dimensional space $\mathbb{V}_{G,p}$ as before, by taking the direct sum of the equivariant cohomology of the moduli spaces of $U(N)$ instantons on it. The vector space $\mathbb{V}_{G,p}$ is long known to have an action of the affine algebra $SU(p)_N$ [152, 153], but this affine algebra is not enough to generate all the states in $\mathbb{V}_{G,p}$. It is now believed [83, 154, 155] that $\mathbb{V}_{G,p}$ is a representation of a free boson, $SU(p)_N$, and the pth para-W_N algebra:

$$\frac{\hat{SU}(N)_p \times \hat{SU}(N)_k}{\hat{SU}(N)_{p+k}} \tag{5.15}$$

where k is a parameter determined by the ratio ϵ_1/ϵ_2. For $p = 2$ and $N = 2$, the 2nd para-W_2 algebra is the standard $\mathcal{N} = 1$ super Virasoro algebra, and many checks have been made [57–60, 64]. See also [156] for the analysis of the case $N = 1$ for general p.

5.3 On Non-compact Toric Spaces

There is another way to study $G = U(N)$ instantons on the $\mathbb{Z}_p$ orbifolds (5.12), as they can be resolved to give a smooth non-compact toric spaces X, where instanton counting can be performed [17, 157, 158].

The basic idea is to realize that the fixed points under $U(1)^{2+N}$ of the n-instanton moduli space $M_{G,n}$ on $\mathbb{C}^2$ correspond to point-like n instantons at the origin of $\mathbb{C}^2$, which are put on top of each other. The deformation of the instanton moduli space was done to deal with this singular configuration in a reliable way. The toric space X has an action of $U(1)^2$, whose fixed points $P_1, \ldots, P_k$ are isolated. The action of $U(1)^2$ at each of the fixed points can be different:

$$TX|_{P_i} = e^{i\beta\epsilon_{1;i}} + e^{i\beta\epsilon_{2;i}} \tag{5.16}$$

where $\epsilon_{1,2;i}$ are integral linear combinations of $\epsilon_{1,2}$. Then an $U(N)$ instanton configuration on X fixed under $U(1)^{2+N}$, is basically given by assigning a $U(N)$-instanton configuration on $\mathbb{C}^2$, at each P_i. Another data are the magnetic fluxes $\vec{m}_j = (m_{j,1}, \ldots, m_{j,N})$ through compact 2-cycles C_j of X. Here it is interesting not just to compute the partition function but also correlation functions of certain operators $\mu(C_j)$ which are supported on C_j. Then the correlation function has a schematic form

$$Z_X(\mu(C_1)^{d_1}\mu(C_2)^{d_2}\cdots;\epsilon_1,\epsilon_2) = \sum_{\vec{m}} q^{\vec{m}_i C^{ij}\vec{m}_j} f_{d_1,d_2,\ldots}(\vec{m}_1, \vec{m}_2, \ldots) \prod_{P_i} Z_{\mathbb{C}^2}(\epsilon_{1;i}, \epsilon_{2;i}) \tag{5.17}$$

where C^{ij} is the intersection form of the cycles C_j and $f_{d_1,d_2,\ldots}(\vec{m}_1, \vec{m}_2, \ldots)$ is a prefactor expressible in a closed form. For details, see the papers referred to above.

Nakajima-Yoshioka When X is the blow-up $\hat{\mathbb{C}}^2$ of $\mathbb{C}^2$ at the origin, there are two fixed points P_1 and P_2, with

$$T\hat{\mathbb{C}}^2|_{P_1} = e^{i\beta\epsilon_1} + e^{i\beta(\epsilon_2-\epsilon_1)}, \quad T\hat{\mathbb{C}}^2|_{P_2} = e^{i\beta(\epsilon_1-\epsilon_2)} + e^{i\beta\epsilon_2}. \quad (5.18)$$

We have one compact 2-cycle C. Then we have a schematic relation

$$Z_{\hat{\mathbb{C}}^2}(\mu(C)^d; \epsilon_1, \epsilon_2) = \sum_{\vec{m}} q^{\vec{m}\cdot\vec{m}} f_d(\vec{m}) Z_{\mathbb{C}^2}(\epsilon_1, \epsilon_2 - \epsilon_1) Z_{\mathbb{C}^2}(\epsilon_1 - \epsilon_2, \epsilon_2). \quad (5.19)$$

We can use another knowledge here that the instanton moduli space on $\hat{\mathbb{C}}^2$ and that on $\mathbb{C}^2$ can be related via the map $\hat{\mathbb{C}}^2 \to \mathbb{C}^2$. Let us assume that c_1 of the bundle on $\hat{\mathbb{C}}^2$ is zero. Then we have a relation schematically of the form

$$Z_{\hat{\mathbb{C}}^2}(\mu(C)^d; \epsilon_1, \epsilon_2) = \begin{cases} Z_{\mathbb{C}^2}(\epsilon_1, \epsilon_2) & (d = 0), \\ 0 & (d > 0) \end{cases} \quad (5.20)$$

for $d = 0, 1, \ldots, 2N - 1$. The combination of (5.19) and (5.20) allows us to write down a recursion relation of the form

$$Z_{\mathbb{C}^2}(\epsilon_1, \epsilon_2) = \sum_{\vec{m}} q^{\vec{m}\cdot\vec{m}} c(\vec{m}) Z_{\mathbb{C}^2}(\epsilon_1, \epsilon_2 - \epsilon_1) Z_{\mathbb{C}^2}(\epsilon_1 - \epsilon_2, \epsilon_2). \quad (5.21)$$

This allows one to compute the instanton partition function on $\mathbb{C}^2$ recursively as an expansion in q [17], starting from the trivial fact that the zero-instanton moduli space is just a point. From this, a recursive formula for the prepotential can be found, which was studied and written down in [7, 8]. The recursive formula was proved from the analysis of the Seiberg-Witten curve in [7, 8], while it was derived from the analysis of the instanton moduli space in [17]. This gives one proof that the Seiberg-Witten prepotential as defined by the Seiberg-Witten curve is the same as the one defined via the instanton counting. This method has been extended to the case with matter hypermultiplets in the fundamental representation [159].

The recursive formula, although mathematically rigorously proved only for SU gauge groups, has a form transparently given in terms of the roots of the gauge group involved. This conjectural version of the formula for general gauge groups can then be used to determine the instanton partition function for any gauge group. This was applied to $E_{6,7}$ gauge theories in [129] and the function thus obtained agreed with the one computed via the methods of Sect. 4.3.

The CFT interpretation of this formula was explored in [61]. A similar formula can be formulated for the orbifolds of $\mathbb{C}^2$ and was studied in [150]. It is also found that the instanton counting on $\mathbb{C}^2/\mathbb{Z}_p$ and that on its blowup can have a subtle but controllable difference [160].

Acknowledgments The author thanks his advisor Tohru Eguchi, for suggesting him to review Nekrasov's seminal work [10] as a project for his master's thesis. The interest in instanton counting never left him since then. The Mathematica code to count the instantons which he wrote during that project turned out to be crucial when he started a collaboration five years later with Fernando Alday and Davide Gaiotto, leading to the paper [50]. He thanks Hiraku Nakajima for painstakingly explaining basic mathematical facts to him. He would also like to thank all of his collaborators on this interesting arena of instantons and W-algebras.

It is a pleasure for the author also to thank Miranda Cheng, Hiroaki Kanno, Vasily Pestun, Jaewon Song, Masato Taki, Joerg Teschner and Xinyu Zhang for carefully reading an early version of this review and for giving him constructive and helpful comments.

This work is supported in part by JSPS Grant-in-Aid for Scientific Research No. 25870159 and in part by WPI Initiative, MEXT, Japan at IPMU, the University of Tokyo.

References

[1]   Seiberg, N., Witten, E.: Monopole condensation, and confinement in $\mathcal{N} = 2$ supersymmetric Yang-Mills theory. Nucl. Phys. **B426**, 19–52 (1994). arXiv:hep-th/9407087

[2]   Seiberg, N., Witten, E.: Monopoles, duality and chiral symmetry breaking in $\mathcal{N} = 2$ supersymmetric QCD. Nucl. Phys. **B431**, 484–550 (1994). arXiv:hep-th/9408099

[3]   Finnell, D., Pouliot, P.: Instanton calculations versus exact results in four-dimensional SUSY gauge theories. Nucl. Phys. **B453**, 225–239 (1995). arXiv:hep-th/9503115

[4]   Ito, K., Sasakura, N.: One-instanton calculations in $\mathcal{N} = 2$ supersymmetric $SU(N_C)$ Yang-Mills theory. Phys. Lett. **B382**, 95–103 (1996). arXiv:hep-th/9602073

[5]   Dorey, N., Hollowood, T.J., Khoze, V.V., Mattis, M.P.: The calculus of many instantons. Phys. Rep. **371**, 231–459 (2002). arXiv:hep-th/0206063

[6]   Moore, G.W., Nekrasov, N., Shatashvili, S.: Integrating over Higgs branches. Commun. Math. Phys. **209**, 97–121 (2000). arXiv:hep-th/9712241 [hep-th]

[7]   Losev, A., Nekrasov, N., Shatashvili, S.L.: Issues in topological gauge theory. Nucl. Phys. **B534**, 549–611 (1998). arXiv:hep-th/9711108

[8]   Losev, A., Nekrassov, N., Shatashvili, S.L.: Testing Seiberg-Witten solution. Strings, Branes and Dualities. NATO ASI Series, vol. 520, pp. 359–372. Springer, The Netherlands (1999). arXiv:hep-th/9801061

[9]   Losev, A., Nekrasov, N., Shatashvili, S.L.: The Freckled instantons. arXiv:hep-th/9908204 [hep-th]

[10] Nekrasov, N.A.: Seiberg-Witten prepotential from instanton counting. Adv. Theor. Math. Phys. **7**, 831–864 (2004). arXiv:hep-th/0206161

[11] Flume, R., Poghossian, R., Storch, H.: The Seiberg-Witten prepotential and the Euler class of the reduced moduli space of instantons. Mod. Phys. Lett. **A17**, 327–340 (2002). arXiv:hep-th/0112211

[12] Hollowood, T.J.: Calculating the prepotential by localization on the moduli space of instantons. JHEP **03**, 038 (2002). arXiv:hep-th/0201075

[13] Flume, R., Poghossian, R.: An algorithm for the microscopic evaluation of the coefficients of the Seiberg-Witten prepotential. Int. J. Mod. Phys. **A18**, 2541 (2003). arXiv:hep-th/0208176

[14] Hollowood, T.J.: Testing Seiberg-Witten theory to all orders in the instanton expansion. Nucl. Phys. **B639**, 66–94 (2002). arXiv:hep-th/0202197

[15] Bruzzo, U., Fucito, F., Morales, J.F., Tanzini, A.: Multiinstanton calculus and equivariant cohomology. JHEP **0305**, 054 (2003). arXiv:hep-th/0211108 [hep-th]

[16] Nakajima, H., Yoshioka, K.: Lectures on instanton counting. In: Hurturbise, J., Markman, E. (eds.) Workshop on Algebraic Structures and Moduli Spaces: CRM Workshop. AMS, July 2003. arXiv:math/0311058

[17] Nakajima, H., Yoshioka, K.: Instanton counting on blowup. I. Invent. Math. **162**, 155–313 (2005). arXiv:math/0306198

[18] Nekrasov, N., Okounkov, A.: Seiberg-Witten theory and random partitions. arXiv:hep-th/0306238

[19] Braverman, A.: Instanton counting via affine Lie algebras I: equivariant J-functions of (affine) flag manifolds and Whittaker vectors. In: Hurturbise, J., Markman, E. (eds.) Workshop on Algebraic Structures and Moduli Spaces: CRM Workshop. AMS, July 2003. arXiv:math/0401409

[20] Braverman, A., Etingof, P.: Instanton counting via affine Lie algebras. II: from Whittaker vectors to the Seiberg-Witten prepotential. In: Bernstein, J., Hinich, V., Melnikov, A. (eds.) Studies in Lie Theory: Dedicated to A. Joseph on his 60th Birthday. Birkhäuser, Boston (2006). arXiv:math/0409441

[21] Fucito, F., Morales, J.F., Poghossian, R.: Instantons on quivers and orientifolds. JHEP **10**, 037 (2004). arXiv:hep-th/0408090

[22] Mariño, M., Wyllard, N.: A note on instanton counting for $\mathcal{N} = 2$ gauge theories with classical gauge groups. JHEP **05**, 021 (2004). arXiv:hep-th/0404125

[23] Nekrasov, N., Shadchin, S.: ABCD of instantons. Commun. Math. Phys. **252**, 359–391 (2004). arXiv:hep-th/0404225

[24] Shadchin, S.: Saddle point equations in Seiberg-Witten theory. JHEP **10**, 033 (2004). arXiv:hep-th/0408066

[25] Shadchin, S.: Cubic curves from instanton counting. JHEP **03**, 046 (2006). arXiv:hep-th/0511132

[26] Hollands, L., Keller, C.A., Song, J.: From SO/Sp instantons to W-algebra blocks. JHEP **03**, 053 (2011). arXiv:1012.4468 [hep-th]

[27] Hollands, L., Keller, C.A., Song, J.: Towards a 4D/2D correspondence for Sicilian quivers. JHEP **10**, 100 (2011). arXiv:1107.0973 [hep-th]

[28] Fucito, F., Morales, J.F., Poghossian, R.: Multi instanton calculus on ALE spaces. Nucl. Phys. **B703**, 518–536 (2004). arXiv:hep-th/0406243 [hep-th]

[29] Fucito, F., Morales, J.F., Pacifici, D.R.: Deformed Seiberg-Witten curves for ADE quivers. JHEP **1301**, 091 (2013). arXiv:1210.3580 [hep-th]

[30] Nekrasov, N., Pestun, V.: Seiberg-Witten geometry of four dimensional $\mathcal{N} = 2$ quiver gauge theories. arXiv:1211.2240 [hep-th]

[31] Shadchin, S.: Status report on the instanton counting. SIGMA **2**, 008 (2006). arXiv:hep-th/0601167 [hep-th]

[32] Bianchi, M., Kovacs, S., Rossi, G.: Instantons and supersymmetry. Lect. Notes Phys. **737**, 303–470 (2008). arXiv:hep-th/0703142 [hep-th]

[33] Iqbal, A., Kashani-Poor, A.-K.: Instanton counting and Chern-Simons theory. Adv. Theor. Math. Phys. **7**, 457–497 (2004). arXiv:hep-th/0212279

[34] Iqbal, A., Kashani-Poor, A.-K.: $SU(N)$ geometries and topological string amplitudes. Adv. Theor. Math. Phys. **10**, 1–32 (2006). arXiv:hep-th/0306032 [hep-th]

[35] Eguchi, T., Kanno, H.: Topological strings and Nekrasov's formulas. JHEP **0312**, 006 (2003). arXiv:hep-th/0310235 [hep-th]

[36] Iqbal, A., Kashani-Poor, A.-K.: The vertex on a strip. Adv. Theor. Math. Phys. **10**, 317–343 (2006). arXiv:hep-th/0410174

[37] Iqbal, A.: All genus topological string amplitudes and 5-brane Webs as Feynman diagrams. arXiv:hep-th/0207114

[38] Aganagic, M., Klemm, A., Mariño, M., Vafa, C.: The topological vertex. Commun. Math. Phys. **254**, 425–478 (2005). arXiv:hep-th/0305132

[39] Awata, H., Kanno, H.: Instanton counting, Macdonald functions and the moduli space of D-branes. JHEP **05**, 039 (2005). arXiv:hep-th/0502061

[40] Iqbal, A., Kozçaz, C., Vafa, C.: The refined topological vertex. JHEP **0910**, 069 (2009). arXiv:hep-th/0701156 [hep-th]

[41] Taki, M.: Refined topological vertex and instanton counting. JHEP **03**, 048 (2008). arXiv:0710.1776 [hep-th]

[42]   Awata, H., Kanno, H.: Refined BPS state counting from Nekrasov's formula and Macdonald functions. Int. J. Mod. Phys. **A24**, 2253–2306 (2009). arXiv:0805.0191 [hep-th]

[43]   Awata, H., Feigin, B., Shiraishi, J.: Quantum algebraic approach to refined topological vertex. JHEP **1203**, 041 (2012). arXiv:1112.6074 [hep-th]

[44]   Kozçaz, C., Pasquetti, S., Wyllard, N.: A & B model approaches to surface operators and Toda theories. JHEP **1008**, 042 (2010). arXiv:1004.2025 [hep-th]

[45]   Taki, M.: Surface operator, bubbling Calabi-Yau and AGT relation. JHEP **1107**, 047 (2011). arXiv:1007.2524 [hep-th]

[46]   Grimm, T.W., Klemm, A., Mariño, M., Weiss, M.: Direct integration of the topological string. JHEP **08**, 058 (2007). arXiv:hep-th/0702187

[47]   Huang, M.-x., Klemm, A.: Holomorphicity and modularity in Seiberg-Witten theories with matter. JHEP **1007**, 083 (2010). arXiv:0902.1325 [hep-th]

[48]   Huang, M.-x., Klemm, A.: Direct integration for general Ω backgrounds. Adv. Theor. Math. Phys. **16**(3), 805–849 (2012). arXiv:1009.1126 [hep-th]

[49]   Huang, M.-x., Kashani-Poor, A.-K., Klemm, A.: The Ω deformed B-model for rigid $N = 2$ theories. Annales Henri Poincare **14**, 425–497 (2013). arXiv:1109.5728 [hep-th]

[50]   Alday, L.F., Gaiotto, D., Tachikawa, Y.: Liouville correlation functions from four-dimensional gauge theories. Lett. Math. Phys. **91**, 167–197 (2010). arXiv:0906.3219 [hep-th]

[51]   Gaiotto, D.: Asymptotically free $\mathcal{N} = 2$ theories and irregular conformal blocks. arXiv:0908.0307 [hep-th]

[52]   Wyllard, N.: A_{N-1} conformal Toda field theory correlation functions from conformal $\mathcal{N} = 2$ $SU(N)$ quiver gauge theories. JHEP **11**, 002 (2009). arXiv:0907.2189 [hep-th]

[53]   Keller, C.A., Mekareeya, N., Song, J., Tachikawa, Y.: The ABCDEFG of instantons and W-algebras. JHEP **1203**, 045 (2012). arXiv:1111.5624 [hep-th]

[54]   Gaiotto, D.: $\mathcal{N} = 2$ dualities. JHEP **1208**, 034 (2012). arXiv:0904.2715 [hep-th]

[55]   Pestun, V.: Localization of gauge theory on a four-sphere and supersymmetric Wilson loops. Commun. Math. Phys. **313**, 71–129 (2012). arXiv:0712.2824 [hep-th]

[56]   Hama, N., Hosomichi, K.: Seiberg-Witten theories on ellipsoids. JHEP **1209**, 033 (2012). arXiv:1206.6359 [hep-th]

[57]   Belavin, V., Feigin, B.: Super Liouville conformal blocks from $\mathcal{N} = 2$ $SU(2)$ quiver gauge theories. JHEP **1107**, 079 (2011). arXiv:1105.5800 [hep-th]

[58]   Bonelli, G., Maruyoshi, K., Tanzini, A.: Instantons on ALE spaces and super Liouville conformal field theories. JHEP **1108**, 056 (2011). arXiv:1106.2505 [hep-th]

[59]   Belavin, A., Belavin, V., Bershtein, M.: Instantons and 2D superconformal field theory. JHEP **1109**, 117 (2011). arXiv:1106.4001 [hep-th]

[60]   Bonelli, G., Maruyoshi, K., Tanzini, A.: Gauge theories on ALE space and super Liouville correlation functions. Lett. Math. Phys. **101**, 103–124 (2012). arXiv:1107.4609 [hep-th]

[61]   Wyllard, N.: Coset conformal blocks and $\mathcal{N} = 2$ gauge theories. arXiv:1109.4264 [hep-th]

[62]   Ito, Y.: Ramond sector of super Liouville theory from instantons on an ALE space. Nucl. Phys. **B861**, 387–402 (2012). arXiv:1110.2176 [hep-th]

[63]   Alfimov, M., Tarnopolsky, G.: Parafermionic Liouville field theory and instantons on ALE spaces. JHEP **1202**, 036 (2012). arXiv:1110.5628 [hep-th]

[64]   Belavin, A., Mukhametzhanov, B.: $\mathcal{N}=1$ superconformal blocks with Ramond fields from AGT correspondence. JHEP **1301**, 178 (2013). arXiv:1210.7454 [hep-th]

[65]   Braverman, A., Feigin, B., Finkelberg, M., Rybnikov, L.: A finite analog of the AGT relation I: finite W-algebras and quasimaps' spaces. Commun. Math. Phys. **308**, 457–478 (2011). arXiv:1008.3655 [math.AG]

[66]   Wyllard, N.: Instanton partition functions in $\mathcal{N} = 2$ $SU(N)$ gauge theories with a general surface operator, and their W-algebra duals. JHEP **1102**, 114 (2011). arXiv:1012.1355 [hep-th]

[67]   Kanno, H., Tachikawa, Y.: Instanton counting with a surface operator and the chain-saw quiver. JHEP **06**, 119 (2011). arXiv:1105.0357 [hep-th]

[68] Alday, L.F., Tachikawa, Y.: Affine $SL(2)$ conformal blocks from 4D gauge theories. Lett. Math. Phys. **94**, 87–114 (2010). arXiv:1005.4469 [hep-th]

[69] Bouwknegt, P., Schoutens, K.: W-symmetry in conformal field theory. Phys. Rep. **223**, 183–276 (1993). arXiv:hep-th/9210010

[70] Bouwknegt, P., Schoutens, K. (eds.): W-Symmetry. Advanced Series in Mathematical Physics, vol. 22. World Scientific, Singapore (1995)

[71] Tachikawa, Y.: A strange relationship between 2D CFT and 4D gauge theory. arXiv:1108.5632 [hep-th]

[72] Uhlenbeck, K.K.: Removable singularities in Yang-Mills fields. Commun. Math. Phys. **83**(1), 11–29 (1982)

[73] 't Hooft, G.: Computation of the quantum effects due to a four-dimensional pseudoparticle. Phys. Rev. **D14**, 3432–3450 (1976)

[74] Atiyah, M.F.: Elliptic Operators and Compact Groups. Lecture Notes in Mathematics, vol. 401. Springer, Berlin (1974)

[75] Tachikawa, Y.: Five-dimensional Chern-Simons terms and Nekrasov's instanton counting. JHEP **02**, 050 (2004). arXiv:hep-th/0401184

[76] Nekrasov, N., Schwarz, A.S.: Instantons on noncommutative $\mathbb{R}^4$ and (2,0) superconformal six-dimensional theory. Commun. Math. Phys. **198**, 689–703 (1998). arXiv:hep-th/9802068 [hep-th]

[77] Okuda, T., Pestun, V.: On the instantons and the hypermultiplet mass of N=2* super Yang-Mills on S^4. JHEP **1203**, 017 (2012). arXiv:1004.1222 [hep-th]

[78] Carlsson, E., Okounkov, A.: Ext and vertex operators. Duke Math. J. **161**, 1797–1815 (2012). arXiv:0801.2565 [math.AG]

[79] Braverman, A., Finkelberg, M., Nakajima, H.: Private communication (2014)

[80] Schiffmann, O., Vasserot, E.: Cherednik algebras, W algebras and the equivariant cohomology of the moduli space of instantons on $\mathbb{A}^2$. Publications mathématiques de l'IHÉS **118**(1), 213–342 (2013). arXiv:1202.2756 [math.QA]

[81] Maulik, D., Okounkov, A.: Quantum groups and quantum cohomology. arXiv:1211.1287 [math.QA]

[82] Fateev, V., Litvinov, A.: Integrable structure, W-symmetry and AGT relation. JHEP **1201**, 051 (2012). arXiv:1109.4042 [hep-th]

[83] Belavin, A., Bershtein, M., Feigin, B., Litvinov, A., Tarnopolsky, G.: Instanton moduli spaces and bases in coset conformal field theory. Commun. Math. Phys. **319**, 269–301 (2013). arXiv:1111.2803 [hep-th]

[84] Losev, A.S., Marshakov, A., Nekrasov, N.A.: Small instantons, little strings and free fermions. From Fields to Strings: Circumnavigations Theoretical Physics (Ian Kogan Memorial Collection), pp. 581–621. World Scientific, Singapore (2005). arXiv:hep-th/0302191 [hep-th]

[85] Aganagic, M., Dijkgraaf, R., Klemm, A., Mariño, M., Vafa, C.: Topological strings and integrable hierarchies. Commun. Math. Phys. **261**, 451–516 (2006). arXiv:hep-th/0312085 [hep-th]

[86] Dijkgraaf, R., Hollands, L., Sulkowski, P., Vafa, C.: Supersymmetric gauge theories, intersecting branes and free fermions. JHEP **0802**, 106 (2008). arXiv:0709.4446 [hep-th]

[87] Marshakov, A., Mironov, A., Morozov, A.: On non-conformal limit of the AGT relations. Phys. Lett. **B682**, 125–129 (2009). arXiv:0909.2052 [hep-th]

[88] Taki, M.: On AGT conjecture for pure super Yang-Mills and W-algebra. JHEP **1105**, 038 (2011). arXiv:0912.4789 [hep-th]

[89] Yanagida, S.: Whittaker vectors of the Virasoro algebra in terms of Jack symmetric polynomial. J. Algebra **333**, 278–294 (2011). arXiv:1003.1049 [math.QA]

[90] Kanno, H., Taki, M.: Generalized Whittaker states for instanton counting with fundamental hypermultiplets. JHEP **1205**, 052 (2012). arXiv:1203.1427 [hep-th]

[91] Fateev, V.A., Litvinov, A.V.: On AGT conjecture. JHEP **02**, 014 (2010). arXiv:0912.0504 [hep-th]

[92]   Hadasz, L., Jaskolski, Z., Suchanek, P.: Proving the AGT relation for $N_F = 0, 1, 2$ anti-fundamentals. JHEP **1006**, 046 (2010). arXiv:1004.1841 [hep-th]

[93]   Mironov, A., Mironov, S., Morozov, A., Morozov, A.: CFT exercises for the needs of AGT. Teor. Mat. Fiz. **165**, 503–542 (2010). arXiv:0908.2064 [hep-th]

[94]   Mironov, A., Morozov, A.: On AGT relation in the case of U(3). Nucl. Phys. **B825**, 1–37 (2010). arXiv:0908.2569 [hep-th]

[95]   Alba, V.A., Fateev, V.A., Litvinov, A.V., Tarnopolskiy, G.M.: On combinatorial expansion of the conformal blocks arising from AGT conjecture. Lett. Math. Phys. **98**, 33–64 (2011). arXiv:1012.1312 [hep-th]

[96]   Kanno, S., Matsuo, Y., Shiba, S.: W(1+Infinity) algebra as a symmetry behind AGT relation. Phys. Rev. **D84**, 026007 (2011). arXiv:1105.1667 [hep-th]

[97]   Kanno, S., Matsuo, Y., Zhang, H.: Virasoro constraint for Nekrasov instanton partition function. JHEP **1210**, 097 (2012). arXiv:1207.5658 [hep-th]

[98]   Kanno, S., Matsuo, Y., Zhang, H.: Extended conformal symmetry and recursion formulae for Nekrasov partition function. JHEP **1308**, 028 (2013). arXiv:1306.1523 [hep-th]

[99]   Kanno, S., Matsuo, Y., Shiba, S., Tachikawa, Y.: $\mathcal{N} = 2$ gauge theories and degenerate fields of Toda theory. Phys. Rev. **D81**, 046004 (2010). arXiv:0911.4787 [hep-th]

[100] Drukker, N., Passerini, F.: (De)Tails of Toda CFT. JHEP **1104**, 106 (2011). arXiv:1012.1352 [hep-th]

[101] Kanno, S., Matsuo, Y., Shiba, S.: Analysis of correlation functions in Toda theory and AGT-W relation for $SU(3)$ quiver. Phys. Rev. **D82**, 066009 (2010). arXiv:1007.0601 [hep-th]

[102] Awata, H., Yamada, Y.: Five-dimensional AGT conjecture and the deformed Virasoro algebra. JHEP **1001**, 125 (2010). arXiv:0910.4431 [hep-th]

[103] Awata, H., Yamada, Y.: Five-dimensional AGT relation and the deformed beta-ensemble. Prog. Theor. Phys. **124**, 227–262 (2010). arXiv:1004.5122 [hep-th]

[104] Carlsson, E., Nekrasov, N., Okounkov, A.: Five dimensional gauge theories and vertex operators. arXiv:1308.2465 [math.RT]

[105] Iqbal, A., Kozçaz, C.: Refined topological strings and toric Calabi-Yau threefolds. arXiv:1210.3016 [hep-th]

[106] Antoniadis, I., Gava, E., Narain, K., Taylor, T.: Topological amplitudes in string theory. Nucl. Phys. **B413**, 162–184 (1994). arXiv:hep-th/9307158 [hep-th]

[107] Bershadsky, M., Cecotti, S., Ooguri, H., Vafa, C.: Kodaira-Spencer theory of gravity and exact results for quantum string amplitudes. Commun. Math. Phys. **165**, 311–428 (1994). arXiv:hep-th/9309140 [hep-th]

[108] Antoniadis, I., Hohenegger, S., Narain, K., Taylor, T.: Deformed topological partition function and Nekrasov backgrounds. Nucl. Phys. **B838**, 253–265 (2010). arXiv:1003.2832 [hep-th]

[109] Nakayama, Y.: Refined cigar and omega-deformed conifold. JHEP **1007**, 054 (2010). arXiv:1004.2986 [hep-th]

[110] Nakayama, Y., Ooguri, H.: Comments on worldsheet description of the omega background. Nucl. Phys. **B856**, 342–359 (2012). arXiv:1106.5503 [hep-th]

[111] Hellerman, S., Orlando, D., Reffert, S.: String theory of the omega deformation. JHEP **1201**, 148 (2012). arXiv:1106.0279 [hep-th]

[112] Reffert, S.: General omega deformations from closed string backgrounds. JHEP **1204**, 059 (2012). arXiv:1108.0644 [hep-th]

[113] Hellerman, S., Orlando, D., Reffert, S.: The omega deformation from string and M-theory. JHEP **1207**, 061 (2012). arXiv:1204.4192 [hep-th]

[114] Billó, M., Frau, M., Fucito, F., Lerda, A.: Instanton calculus in R-R background and the topological string. JHEP **0611**, 012 (2006). arXiv:hep-th/0606013 [hep-th]

[115] Ito, K., Nakajima, H., Sasaki, S.: Instanton calculus in R-R 3-form background and deformed $\mathcal{N} = 2$ super Yang-Mills theory. JHEP **0812**, 113 (2008). arXiv:0811.3322 [hep-th]

[116] Ito, K., Nakajima, H., Saka, T., Sasaki, S.: $\mathcal{N} = 2$ instanton effective action in Ω-background and D3/D(-1)-brane system in R-R background. JHEP **1011**, 093 (2010). arXiv:1009.1212 [hep-th]

[117] Awata, H., Feigin, B., Hoshino, A., Kanai, M., Shiraishi, J. et al.: Notes on Ding-Iohara algebra and AGT conjecture. arXiv:1106.4088 [math-ph]
[118] Ginzburg, V., Kapranov, M., Vasserot, É.: Langlands reciprocity for algebraic surfaces. Math. Res. Lett. 2(2), 147–160 (1995)
[119] Bao, L., Mitev, V., Pomoni, E., Taki, M., Yagi, F.: Non-Lagrangian theories from brane junctions. arXiv:1310.3841 [hep-th]
[120] Hayashi, H., Kim, H.-C., Nishinaka, T.: Topological strings and 5D T_N partition functions. arXiv:1310.3854 [hep-th]
[121] Dorey, N., Khoze, V.V., Mattis, M.P.: On mass deformed $\mathcal{N} = 4$ supersymmetric Yang-Mills theory. Phys. Lett. $\mathbf{B396}$, 141–149 (1997). arXiv:hep-th/9612231 [hep-th]
[122] Vinberg, E.B., Popov, V.L.: On a class of quasihomogeneous affine varieties. Math. USSR-Izv. $\mathbf{6}$, 743 (1972)
[123] Garfinkle, D.: A new construction of the Joseph ideal (1982). http://hdl.handle.net/1721.1/15620 (quoted in Chap. III)
[124] Benvenuti, S., Hanany, A., Mekareeya, N.: The Hilbert series of the one instanton moduli space. JHEP $\mathbf{06}$, 100 (2010). arXiv:1005.3026 [hep-th]
[125] Benini, F., Benvenuti, S., Tachikawa, Y.: Webs of five-branes and $\mathcal{N} = 2$ superconformal field theories. JHEP $\mathbf{09}$, 052 (2009). arXiv:0906.0359 [hep-th]
[126] Gadde, A., Rastelli, L., Razamat, S.S., Yan, W.: Gauge theories and Macdonald polynomials. Commun. Math. Phys. $\mathbf{319}$, 147–193 (2013). arXiv:1110.3740 [hep-th]
[127] Gaiotto, D., Razamat, S.S.: Exceptional indices. JHEP $\mathbf{1205}$, 145 (2012). arXiv:1203.5517 [hep-th]
[128] Hanany, A., Mekareeya, N., Razamat, S.S.: Hilbert series for moduli spaces of two instantons. JHEP $\mathbf{1301}$, 070 (2013). arXiv:1205.4741 [hep-th]
[129] Keller, C.A., Song, J.: Counting exceptional instantons. JHEP $\mathbf{1207}$, 085 (2012). arXiv:1205.4722 [hep-th]
[130] Moore, G.W., Tachikawa, Y.: On 2D TQFTs whose values are holomorphic symplectic varieties. String-Math 2011, vol. 85, pp. 191–207. AMS, Providence (2012). arXiv:1106.5698 [hep-th]
[131] Bielawski, R.: Hyperkähler structures and group actions. J. Lond. Math. Soc. $\mathbf{55}$, 400 (1997)
[132] Ginzburg, V., Kazhdan, D.: Private communication (2011)
[133] Bonelli, G., Tanzini, A.: Hitchin systems, $\mathcal{N}=2$ gauge theories and W-gravity. Phys. Lett. $\mathbf{B691}$, 111–115 (2010). arXiv:0909.4031 [hep-th]
[134] Alday, L.F., Benini, F., Tachikawa, Y.: Liouville/Toda central charges from M5-branes. Phys. Rev. Lett. $\mathbf{105}$, 141601 (2010). arXiv:0909.4776 [hep-th]
[135] Braverman, A., Finkelberg, M., Nakajima, H.: Instanton moduli spaces and W-algebras. arXiv:1406.2381 [math.QA]
[136] Mehta, V.B., Seshadri, C.S.: Moduli of vector bundles on curves with parabolic structures. Mathematische Annalen $\mathbf{248}$, 205–239 (1980)
[137] Biswas, I.: Parabolic bundles as orbifold bundles. Duke Math. J. $\mathbf{88}$, 305–325 (1997)
[138] Finkelberg, M., Rybnikov, L.: Quantization of Drinfeld Zastava in type A. J. Eur. Math. Soc. $\mathbf{16}$, 235–271 (2014). arXiv:1009.0676 [math.AG]
[139] Feigin, B., Finkelberg, M., Negut, A., Rybnikov, L.: Yangians and cohomology rings of Laumon spaces. Selecta Mathematica $\mathbf{17}$, 1–35 (2008). arXiv:0812.4656 [math.AG]
[140] Kozçaz, C., Pasquetti, S., Passerini, F., Wyllard, N.: Affine $SL(N)$ conformal blocks from $\mathcal{N} = 2$ $SU(N)$ gauge theories. JHEP $\mathbf{01}$, 045 (2011). arXiv:1008.1412 [hep-th]
[141] Wyllard, N.: W-algebras and surface operators in $\mathcal{N} = 2$ gauge theories. J. Phys. $\mathbf{A44}$, 155401 (2011). arXiv:1011.0289 [hep-th]
[142] de Boer, J., Tjin, T.: The relation between quantum W algebras and Lie algebras. Commun. Math. Phys. $\mathbf{160}$, 317–332 (1994). arXiv:hep-th/9302006
[143] Chacaltana, O., Distler, J.: Tinkertoys for Gaiotto duality. JHEP $\mathbf{1011}$, 099 (2010). arXiv:1008.5203 [hep-th]
[144] Tachikawa, Y.: On W-algebras and the symmetries of defects of 6D $\mathcal{N}=(2, 0)$ theory. JHEP $\mathbf{03}$, 043 (2011). arXiv:1102.0076 [hep-th]

[145]   Drukker, N., Gomis, J., Okuda, T., Teschner, J.: Gauge theory loop operators and Liouville theory. JHEP **02**, 057 (2010). arXiv:0909.1105 [hep-th]

[146]   Alday, L.F., Gaiotto, D., Gukov, S., Tachikawa, Y., Verlinde, H.: Loop and surface operators in $\mathcal{N} = 2$ gauge theory and Liouville modular geometry. JHEP **01**, 113 (2010). arXiv:0909.0945 [hep-th]

[147]   Ribault, S., Teschner, J.: H_3^+ WZNW correlators from Liouville theory. JHEP **06**, 014 (2005). arXiv:hep-th/0502048

[148]   Hikida, Y., Schomerus, V.: H_3^+ WZNW model from Liouville field theory. JHEP **10**, 064 (2007). arXiv:0706.1030 [hep-th]

[149]   Negut, A.: Affine Laumon spaces and the Calogero-Moser integrable system. arXiv:1112.1756 [math.AG]

[150]   Bonelli, G., Maruyoshi, K., Tanzini, A., Yagi, F.: $\mathcal{N} = 2$ gauge theories on toric singularities, blow-up formulae and W-algebrae. JHEP **1301**, 014 (2013). arXiv:1208.0790 [hep-th]

[151]   Bruzzo, U., Pedrini, M., Sala, F., Szabo, R.J.: Framed sheaves on root stacks and supersymmetric gauge theories on ALE spaces. arXiv:1312.5554 [math.AG]

[152]   Nakajima, H.: Instantons on ALE spaces, quiver varieties and Kac-Moody algebras. Duke Math. J. **76**, 365 (1994)

[153]   Nakajima, H.: Heisenberg algebra and Hilbert schemes of points on projective surfaces. Ann. Math. **145**, 379 (1997). arXiv:alg-geom/9507012

[154]   Nishioka, T., Tachikawa, Y.: Central charges of Para-Liouville and Toda theories from M5-branes. Phys. Rev. **D84**, 046009 (2011). arXiv:1106.1172 [hep-th]

[155]   Belavin, A., Bershtein, M., Tarnopolsky, G.: Bases in coset conformal field theory from AGT correspondence and Macdonald polynomials at the roots of unity. JHEP **1303**, 019 (2013). arXiv:1211.2788 [hep-th]

[156]   Pedrini, M., Sala, F., Szabo, R.J.: AGT relations for Abelian quiver gauge theories on ALE spaces. arXiv:1405.6992 [math.RT]

[157]   Nekrasov, N.: Localizing gauge theories. In: Zambrini, J.-C. (ed.) XIVth International Congress on Mathematical Physics, pp. 645–654 (March, 2006). http://www.ihes.fr/~nikita/IMAGES/Lisbon.ps

[158]   Gasparim, E., Liu, C.-C.M.: The Nekrasov conjecture for toric surfaces. Commun. Math. Phys. **293**, 661–700 (2010). arXiv:0808.0884 [math.AG]

[159]   Gottsche, L., Nakajima, H., Yoshioka, K.: Donaldson = Seiberg-Witten from Mochizuki's formula and instanton counting. Publ. Res. Inst. Math. Sci. Kyoto **47**, 307–359 (2011). arXiv:1001.5024 [math.DG]

[160]   Ito, Y., Maruyoshi, K., Okuda, T.: Scheme dependence of instanton counting in ALE spaces. JHEP **1305**, 045 (2013). arXiv:1303.5765 [hep-th]

β-Deformed Matrix Models and 2d/4d Correspondence

Kazunobu Maruyoshi

Abstract We review the β-deformed matrix model approach to the correspondence between four-dimensional $\mathcal{N} = 2$ gauge theories and two-dimensional conformal field theories. The β-deformed matrix model equipped with the log-type potential is obtained as a free field (Dotsenko-Fateev) representation of the conformal block of chiral conformal algebra in two dimensions, with the precise choice of integration contours. After reviewing various matrix models related to the conformal field theories in two-dimensions, we study the large N limit corresponding to turning off the Omega-background $\epsilon_1, \epsilon_2 \to 0$. We show that the large N analysis produces the purely gauge theory results. Furthermore we discuss the Nekrasov-Shatashvili limit ($\epsilon_2 \to 0$) by which we see the connection with the quantum integrable system. We then perform the explicit integration of the matrix model. With the precise choice of the contours we see that this reproduces the expansion of the conformal block and also the Nekrasov partition function. This is a contribution to the special volume on the 2d/4d correspondence, edited by J. Teschner.

1 Introduction

Matrix models have played a crucial role in the studies of theoretical physics. It has turned out that these models compute quantum observables or the partition function of quantum field theory [1] and two-dimensional gravity [2, 3] (see references therein). Rather recent examples are a one-matrix model which describes the low energy effective superpotential of four-dimensional $\mathcal{N} = 1$ supersymmetric gauge theory [4], and the exact partition functions of supersymmetric gauge theories in various dimensions by localization [5, 6] which are itself written as matrix models (Reviews can be found in [V:5, V:6] in this volume). These have already shown the usefulness of the matrix model in theoretical physics.

A citation of the form [V:x] refers to article number x in this volume.

K. Maruyoshi (✉)
The Blackett Laboratory, Imperial Collage London, Prince Concert Rd, London SW7 2AZ, UK
e-mail: k.maruyoshi@imperial.ac.uk

© Springer International Publishing Switzerland 2016
J. Teschner (ed.), *New Dualities of Supersymmetric Gauge Theories*,
Mathematical Physics Studies, DOI 10.1007/978-3-319-18769-3_5

121

This paper reviews the matrix model introduced by Dijkgraaf and Vafa [7] which was proposed to capture the non-perturbative dynamics of four-dimensional $\mathcal{N} = 2$ supersymmetric gauge theory and two-dimensional conformal field theory (CFT). This proposal is strongly related with the remarkable relation between the Nekrasov partition function [8] of four-dimensional $\mathcal{N} = 2$ supersymmetric gauge theory and the conformal block of two-dimensional Liouville/Toda field theory found by [9]. (We refer to this relation as AGT relation.) The four-dimensional gauge theory is obtained by a partially twisted compactification of the six-dimensional $(2, 0)$ theory on a Riemann surface ([10], [V:2]), and the associated conformal block is defined on the same Riemann surface where vertex operators are inserted at the punctures [V:4].

The conformal block has several different representations. The one we focus here on is the Dotsenko-Fateev integral representation [11, 12], which will be interpreted as β-deformed matrix model. This integral representation has long been known, but regarded as describing degenerate conformal blocks where the degenerate field insertion restricts the internal momenta to fixed values depending on the external momenta. However the recent proposal by [7] is that it *does* describe the full conformal block. The point is the prescription of the contours of the integrations which divides integrals into sets of integral contours whose numbers are N_i (with $\sum N_i = N$ where N is the size of the matrix.) In other words, in the large N perspective, we fix the filling fractions when evaluating the matrix model. This gives additional degrees of freedom corresponding to the internal momenta.

This matrix model plays an interesting role to bridge a gap between four-dimensional $\mathcal{N} = 2$ gauge theory on the Ω background and two-dimensional CFT. In addition to the correspondence with the CFT mentioned above, this is because the matrix model has a standard expansion in $1/N$. The large N limit in the matrix model corresponds to the $\epsilon_{1,2} \to 0$ limit on the gauge theory side. Therefore, the matrix model approach is suited for the ϵ expansion of the Nekrasov partition function.

In Sect. 2, we derive the β-deformed matrix model with the logarithmic potential starting from the free scalar field correlator in the presence of background charge. The case of the Lie algebra-valued scalar field is described by the β-deformation of the quiver matrix model [13–15]. We further see that the similar integral representation can be obtained for the correlator on a higher genus Riemann surface. These matrix models are proposed to be identified with the Nekrasov partition functions of four-dimensional $\mathcal{N} = 2$ (UV) superconformal gauge theories and the conformal blocks.

In Sect. 3 we analyze these matrix models, by taking the size of the matrix N large. The leading part of the large N expansion is studied by utilizing the so-called loop equation. We identify the spectral curve of the matrix model with the Seiberg-Witten curve of the corresponding four-dimensional gauge theory in the form of [10, 16]. We then see evidence of the proposal by checking that the free energy at leading order reproduces the prepotential of the gauge theory.

In Sect. 4 another interesting limit which keeps one of the Ω deformation parameter ϵ_1 finite while $\epsilon_2 \to 0$ in the four-dimensional side will be analyzed. This limit was considered in [17–19] to relate the four-dimensional gauge theory on the Ω background with the quantization of the integrable system. We will see that the β-deformation is crucial for the analysis, and that the matrix model indeed captures the quantum integrable system.

In Sect. 5, we will perform a direct calculation of the partition function of the matrix model keeping all the parameters finite. We compare the explicit result of the direct integration with the Virasoro conformal block and with the Nekrasov partition function.

We conclude in Sect. 6 with a couple of discussions. In Appendix, we present the Selberg integral formula and its generalization which will be used in the analysis in Sect. 5.

2 Integral Representation of Conformal Block

In this section, we introduce the β-deformed matrix model as a free field representation of the conformal block, and the proposal [7] that the matrix model is related to the four-dimensional gauge theory. In Sect. 2.1, we see the simplest version of this proposal: the β-deformed one-matrix model with the logarithmic-type potential[1] obtained from the correlator of the single-scalar field theory on a sphere corresponds to the four-dimensional $\mathcal{N} = 2$ $SU(2)$ linear quiver gauge theory. In Sect. 2.2, we will introduce the quiver matrix model corresponding to the gauge theory with higher rank gauge group. We will then generalize this to the one associated with a generic Riemann surface in Sect. 2.3.

2.1 β-Deformed Matrix Model

In [9], it was found that the conformal block on a sphere with n punctures can be identified with the Nekrasov partition function of $\mathcal{N} = 2$ $SU(2)^{n-3}$ superconformal linear quiver gauge theory. We will first review the integral representation of the conformal block, first introduced by Dotsenko and Fateev [11, 12], and interpret it as a β-deformed matrix model [21, 22]. (See [23] for a review of the relation between the matrix model and the CFT.) We then state the conjecture among the matrix model, the Nekrasov partition function, and the conformal block.

We start with the free scalar field $\phi(z)$

$$\phi(z) = q + p \log z + \sum_{n \neq 0} \frac{\alpha_n}{n} z^{-n}, \qquad (2.1)$$

with the following commutation relations

---

[1]The matrix model with a logarithmic potential was first studied by Penner [20] related to the Eular characteristic of a Riemann surface.

$$[\alpha_m, \alpha_n] = -m\delta_{m+n,0}, \quad [p, q] = -1. \tag{2.2}$$

Thus, the OPE of $\phi(z)$ is

$$\phi(z)\phi(w) \sim -\log(z - w). \tag{2.3}$$

The energy-momentum tensor is given by $T(z) = -\frac{1}{2} : \partial\phi(z)\partial\phi(z) :$ with the central charge 1.

Let us introduce a background charge $Q = b + 1/b$ at the point at infinity by changing the energy-momentum tensor

$$T(z) = -\frac{1}{2} : \partial\phi(z)\partial\phi(z) : +\frac{Q}{\sqrt{2}} : \partial^2\phi(z) := \sum_{n\in\mathbb{Z}} \frac{L_n}{z^{n+2}}. \tag{2.4}$$

The central charge with this background is $c = 1 + 6Q^2$.

The Fock vacuum is defined by

$$\alpha_n|0\rangle = 0, \quad \langle 0|\alpha_{-n} = 0, \quad \text{for } n \geq -1. \tag{2.5}$$

The energy-momentum tensor satisfies the Virasoro constraints

$$\langle L_n \rangle = 0, \quad \text{for } n \geq -1. \tag{2.6}$$

Now we consider the correlator $\langle \prod_{k=0}^{n-1} V_{\alpha_k}(w_k) \rangle$, where the vertex operator is defined by $V_\alpha(z) =: e^{\sqrt{2}\alpha\phi(z)} :$ with conformal dimension $\Delta_\alpha = \alpha(Q - \alpha)$. This is nonzero only if the momenta satisfy the condition $\sum_{k=1}^{n} \alpha_k = Q$. To relax the condition, let us consider the following operators

$$Q_+ = \int d\lambda : e^{\sqrt{2}b\phi(\lambda)} :, \quad Q_- = \int d\lambda : e^{\sqrt{2}b^{-1}\phi(\lambda)} : . \tag{2.7}$$

Since the integrand of each operator has conformal dimension 1, the screening operators are dimensionless. Therefore we can insert these operators into the correlator without changing the conformal property. The insertion however changes the momentum conservation condition, thus we refer these as screening operators. By inserting N screening operators Q_+ in the correlator we define

$$\hat{Z} = \left\langle Q_+^N \prod_{k=0}^{n-1} V_{\alpha_k}(w_k) \right\rangle, \tag{2.8}$$

The momentum conservation condition now relates the external momenta and the number of integrals as $\sum_{k=0}^{n-1} \alpha_k + bN = Q$. This adds one more degree of freedom, bN, to the model. Nevertheless, it is important to note that the momenta m_k (or α_k)

cannot be completely arbitrary because N is an integer. This point will be discussed in Sect. 5.

By evaluating the OPEs, it is easy to obtain

$$\hat{Z} = C(m_k, w_k)Z \tag{2.9}$$

where Z is of the matrix model like form

$$Z = \int \prod_{I=1}^{N} d\lambda_I \prod_{I<J} (\lambda_I - \lambda_J)^{-2b^2} e^{-\frac{b}{g_s} \sum_I W(\lambda_I)} \equiv e^{F_m/g_s^2}, \tag{2.10}$$

with the following potential

$$W(z) = \sum_{k=0}^{n-2} 2m_k \log(z - w_k), \quad C(m_k, w_k) = \prod_{k<\ell \leq n-2} (w_k - w_\ell)^{-\frac{2m_k m_\ell}{g_s^2}}. \tag{2.11}$$

We have introduced the parameter g_s by defining $\alpha_k = \frac{m_k}{g_s}$. (We will use parameters α_k and m_k interchangeably below.) We also have taken $w_{n-1} \to \infty$ by which the corresponding term in $W(z)$ disappeared. While the dependence on m_{n-1} cannot be seen in the potential, this is recovered by the momentum conservation condition

$$\sum_{k=0}^{n-1} m_k + bg_s N = g_s Q. \tag{2.12}$$

Note that the hermitian matrix model corresponds to the $b = i$ case because the first factor in the integrand is the familiar vandermonde determinant. Also the cases with $b = i/2$ and $2i$ correspond to an orthogonal matrix and a symplectic matrix respectively. However for generic choice of b, there is no such expression in terms of a matrix. This integral expression is known as β ensemble or β-deformed matrix model with $\beta = -b^2$.

It is useful to rewrite the β deformed matrix model (2.10) as

$$Z = \langle N | \exp \left(\frac{1}{2\pi i \sqrt{2} g_s} \oint dw \, W(w) \partial \phi(w) \right) Q_+^N |0\rangle, \tag{2.13}$$

where we defined $\langle N | := \langle 0 | e^{-\sqrt{2} bNq}$. Thus the insertion of (the derivative of) the scalar field ϕ in the correlator (2.13) is written as

$$\partial \phi(z) = -\frac{W'(z)}{\sqrt{2} g_s} - b\sqrt{2} \sum_I \frac{1}{z - \lambda_I}, \quad \phi(z) = -\frac{W(z)}{\sqrt{2} g_s} - b\sqrt{2} \log \prod_I (z - \lambda_I), \tag{2.14}$$

in the matrix model average $\langle \ldots \rangle$ defined by

$$\langle \mathcal{O} \rangle = \frac{1}{Z} \int \prod_{I=1}^{N} d\lambda_I \prod_{I<J} (\lambda_I - \lambda_J)^{-2b^2} \mathcal{O} \, e^{-\frac{b}{g_s} \sum_I W(\lambda_I)}. \tag{2.15}$$

Note that a similar expression as (2.13) in terms of free fermions was presented in [8] to express the instanton partition function of $\mathcal{N} = 2$ gauge theory.

Relation to conformal block The proposal [7] is that the partition function of this β-deformed matrix model can be identified with the Virasoro conformal block, and the Nekrasov partition function of four-dimensional $\mathcal{N} = 2$ $SU(2)^{n-3}$ linear quiver gauge theory. The relation to the former is

$$Z_0^{-1} \hat{Z}(\alpha_k, N_i, b, w_k) = \mathcal{B}(\alpha_k, \alpha_p^{int}, b, w_k), \tag{2.16}$$

where Z_0 is defined such that the $\hat{Z}$ is expanded in w_k as $\hat{Z} = Z_0(1 + \mathcal{O}(w_k))$. Here $\mathcal{B}$ is the Virasoro n-point conformal block on the sphere and defined such that $\mathcal{B} = 1 + \mathcal{O}(w_k)$. We will review this in Sect. 5.1. The momenta α_k are identified with the external momenta of the conformal block, as it should be. The parameters b and w_k are defined in the conformal block side in the same way as the free field theory. Thus, the only nontrivial point is the identification of the internal momenta α_p^{int} ($p = 1, \ldots, n - 3$).

At the first sight there is no parameter corresponding to the internal momenta in the matrix model. However the prescription to identify them was established by [24–27]: as we will see in Sect. 5.1, the conformal block can be computed from the three-point functions, denoted by the trivalent vertices, and the propagators, denoted by the lines connecting the vertices, as in Fig. 1. The idea is that there are N_i screening operators inserted at each vertex, with $\sum_{i=1}^{n-2} N_i = N$, where the momentum conservation is satisfied as

$$\alpha_1^{int} = \alpha_0 + \alpha_{n-2} + bN_1, \quad \alpha_2^{int} = \alpha_1^{int} + \alpha_{n-3} + bN_2, \ldots, \tag{2.17}$$
$$\alpha_{n-3}^{int} = \alpha_{n-4}^{int} + \alpha_2 + bN_{n-3} = -\alpha_1 - \alpha_{n-1} - bN_{n-2} + Q,$$

In the last equality we used the momentum conservation (2.12). This means that in the integral representation we have $n - 2$ sets of integrals, each number of the integrals is N_i.

Fig. 1 The n-point conformal block. The screening operators are inserted at each vertex to maintain the momentum conservation

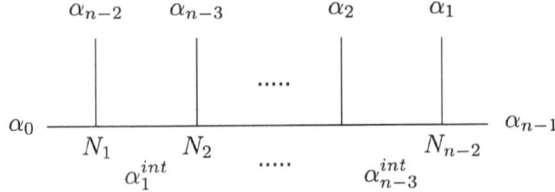

The precise choice of the integration contours will be seen in Sect. 5. Here let us see a rationale of this identification by considering the large N limit shortly. The critical points of the eigenvalues λ_I are obtained from the equations of motion

$$\sum_{k=0}^{n-2} \frac{m_k}{\lambda_I - w_k} + b g_s \sum_{J(\neq I)} \frac{1}{\lambda_I - \lambda_J} = 0. \tag{2.18}$$

Focusing on the first term, when the parameters are generic enough there are $n - 2$ critical points. Let N_i be the number of the matrix eigenvalues which are at the i-th critical point. These critical points are diffused to form line segments by the second term. The integrals are defined such that they include these segments. Now we introduce the filling fractions $\nu_i = b g_s N_i$, and consider the matrix model by fixing these values in the large N limit. Because of the momentum conservation, we have $n - 3$ independent degrees of freedom.

Relation to Nekrasov partition function The relation to the Nekrasov partition function is as follows:

$$Z_{U(1)} Z_0^{-1} \hat{Z}(\alpha_k, N_i, b, w_k) = Z_{\text{Nek}}(m_k, a_p, \epsilon_1, \epsilon_2, q_p), \tag{2.19}$$

under the following identification of the parameters. We choose three insertion points as $w_0 = 0$, $w_1 = 1$ and $w_{n-1} = \infty$. The remaining parameters are identified with the gauge theory coupling constants $q_p = e^{2\pi i \tau_p}$ ($p = 1, \ldots, n - 3$) as follows:

$$w_2 = q_1, \quad w_3 = q_1 q_2, \ldots, \quad w_{n-2} = q_1 q_2 \cdots q_{n-3}. \tag{2.20}$$

We denote the gauge group whose gauge coupling constant is q_p as $SU(2)_p$. Let μ_a^L, μ_b^L and μ_a^R, μ_b^R be the mass parameters of hypermultiplets in the fundamental representation of the $SU(2)_1$ and those of the $SU(2)_{n-3}$ respectively. Let also μ_i ($i = 1, \ldots, n-4$) be the mass parameter of the hypermultiplet in the $(\mathbf{2}, \bar{\mathbf{2}})$ representation of $SU(2)_i \times SU(2)_{i+1}$. Then the mass parameters and the external momenta are identified as

$$m_0 = \frac{\mu_a^L - \mu_b^L}{2} + \frac{g_s Q}{2}, \quad m_{n-2} = \frac{\mu_a^L + \mu_b^L}{2},$$

$$m_{n-1} = \frac{\mu_a^R - \mu_b^R}{2} + \frac{g_s Q}{2}, \quad m_1 = \frac{\mu_a^R + \mu_b^R}{2}, \quad m_{n-2-i} = \mu_i \tag{2.21}$$

The identification of the parameter b with the Ω-deformation parameters is given by

$$\epsilon_1 = b g_s, \quad \epsilon_2 = \frac{g_s}{b}. \tag{2.22}$$

Note that the case $b = i$ corresponds to the self-dual background $\epsilon_1 = -\epsilon_2$. Finally, the vacuum expectation values a_i of the scalar fields in the $SU(2)_i$ vector multiplets are identified as

$$a_p - \mu_a^L - \sum_{q=1}^{p-1} \mu_q = \sum_{q=1}^{p} bN_q, \tag{2.23}$$

for $p = 1, \ldots, n-3$. By using the momentum conservation, a_{n-3} can also be written as $a_{n-3} + \mu_a^R = -bN_{n-2}$.

The first factor in the right hand side of (2.19) is the so-called $U(1)$ factor corresponding to the $U(1)$ part of the gauge theory, which is, e.g., given by

$$Z_{U(1)} = (1-q)^{2\alpha_1\alpha_2}, \tag{2.24}$$

for the $n = 4$ case.[2]

2.2 Quiver Matrix Model and Higher Rank Gauge Theory

In this section, we briefly review the β-deformation of the ADE quiver matrix model [13–15, 28]. We then see that the matrix model can be obtained from the CFT of a free chiral boson valued in Lie algebra. A review of the undeformed quiver matrix model can be found in [29].

Let $\mathfrak{g}$ be a finite dimensional Lie algebra of ADE type with rank r, $\mathfrak{h}$ the Cartan subalgebra of $\mathfrak{g}$, and $\mathfrak{h}^*$ its dual. We denote the natural pairings between $\mathfrak{h}$ and $\mathfrak{h}^*$ by $\langle \cdot, \cdot \rangle$:

$$\alpha(h) = \langle \alpha, h \rangle, \qquad \alpha \in \mathfrak{h}^*, \quad h \in \mathfrak{h}. \tag{2.25}$$

Let $\alpha_a \in \mathfrak{h}^*$ ($a = 1, 2, \ldots, r$) be simple roots of $\mathfrak{g}$ and $(\cdot, \cdot)$ is the inner product on $\mathfrak{h}^*$. Our normalization is chosen as $(\alpha_a, \alpha_a) = 2$. The fundamental weights are denoted by Λ^a ($a = 1, 2, \ldots, r$)

$$(\Lambda^a, \alpha_b^\vee) = \delta_b^a, \qquad \alpha_a^\vee = \frac{2\alpha_a}{(\alpha_a, \alpha_a)}. \tag{2.26}$$

In the Dynkin diagram of $\mathfrak{g}$ we associate $N_a \times N_a$ Hermitian matrices M_a with vertices a for simple roots α_a, and complex $N_a \times N_b$ matrices Q_{ab} and their Hermitian conjugate $Q_{ba} = Q_{ab}^\dagger$ with links connecting vertices a and b. We label links of the Dynkin diagram by pairs of nodes (a, b) with an ordering $a < b$. Let $\mathcal{E}$ and $\mathcal{A}$ be the set of "edges" (a, b) (with $a < b$) and the set of "arrows" (a, b) respectively:

$$\mathcal{E} = \{(a, b) \mid 1 \leq a < b \leq r, \ (\alpha_a, \alpha_b) = -1\},$$
$$\mathcal{A} = \{(a, b) \mid 1 \leq a, b \leq r, \ (\alpha_a, \alpha_b) = -1\}. \tag{2.27}$$

---

[2]This is slightly different from the one in [9]. This is because we consider the Nekrasov partition function where the hypermultiplets are in the fundamental representation of the gauge group. Changing the representation to the anti-fundamental one leads to $\alpha_3 \to Q - \alpha_3$ in this case, then we recover the factor in [9].

The partition function of the quiver matrix model associated with $\mathfrak{g}$ is given by

$$Z = \int \prod_{a=1}^{r} [\mathrm{d}M_a] \prod_{(a,b)\in\mathcal{A}} [\mathrm{d}Q_{ab}] \exp\left(\frac{1}{g_s} W(M, Q)\right), \tag{2.28}$$

where

$$W(M, Q) = i \sum_{(a,b)\in\mathcal{A}} s_{ab} \mathrm{Tr}\, Q_{ba} M_a Q_{ab} - i \sum_{a=1}^{r} \mathrm{Tr}\, W_a(M_a), \tag{2.29}$$

with real constants s_{ab} obeying the conditions $s_{ab} = -s_{ba}$. Note that

$$\prod_{(a,b)\in\mathcal{A}} [\mathrm{d}Q_{ab}] = \prod_{(a,b)\in\mathcal{E}} [\mathrm{d}Q_{ba}\mathrm{d}Q_{ab}], \tag{2.30}$$

$$\sum_{(a,b)\in\mathcal{A}} s_{ab} \mathrm{Tr}\, Q_{ba} M_a Q_{ab} = \sum_{(a,b)\in\mathcal{E}} s_{ab}\big(\mathrm{Tr}\, Q_{ba} M_a Q_{ab} - \mathrm{Tr}\, Q_{ab} M_b Q_{ba}\big). \tag{2.31}$$

The integration measures $[\mathrm{d}M_a]$ and $[\mathrm{d}Q_{ba}\mathrm{d}Q_{ab}]$ are defined by using the metrics $\mathrm{Tr}(\mathrm{d}M_a)^2$ and $\mathrm{Tr}(\mathrm{d}Q_{ba}\mathrm{d}Q_{ab})$ respectively.

Integrations over Q_{ab} are easily performed:

$$\int [\mathrm{d}Q_{ba}\mathrm{d}Q_{ab}] \exp\left(\frac{is_{ab}}{g_s}\big(\mathrm{Tr}\, Q_{ba} M_a Q_{ab} - \mathrm{Tr}\, Q_{ab} M_b Q_{ba}\big)\right) = \det\big(M_a \otimes 1_{N_b} - 1_{N_a} \otimes M_b^T\big)^{-1}, \tag{2.32}$$

where 1_n is the $n \times n$ identity matrix and T denotes transposition. For simplicity we have chosen the normalization of the measure $[\mathrm{d}Q_{ba}\mathrm{d}Q_{ab}]$ to set the proportional constant in the right hand side of (2.32) to be unity. Now the integrand depends only on the eigenvalues of r Hermitian matrices M_a. Let us denote them by $\lambda_I^{(a)}$ ($a = 1, 2, \ldots, r$ and $I = 1, 2, \ldots, N_a$). The partition function of the quiver matrix model reduces to the form of integrations over the eigenvalues of M_a

$$Z = \int \prod_{a=1}^{r} \left\{ \prod_{I=1}^{N_a} \mathrm{d}\lambda_I^{(a)} \right\} \Delta_{\mathfrak{g}}(\lambda) \exp\left(-\frac{i}{g_s} \sum_{a=1}^{r} \sum_{I=1}^{N_a} W_a(\lambda_I^{(a)})\right), \tag{2.33}$$

where W_a is a potential and

$$\Delta_{\mathfrak{g}}(\lambda) = \prod_{a=1}^{r} \prod_{1\le I<J\le N_a} (\lambda_I^{(a)} - \lambda_J^{(a)})^2 \prod_{1\le a<b\le r} \prod_{I=1}^{N_a}\prod_{J=1}^{N_b} (\lambda_I^{(a)} - \lambda_J^{(b)})^{(\alpha_a,\alpha_b)}. \tag{2.34}$$

We then define the β deformation of the above quiver matrix model (with $\beta = -b^2$) by

$$Z = \int \prod_{a=1}^{r} \left\{ \prod_{I=1}^{N_a} d\lambda_I^{(a)} \right\} \left(\Delta_{\mathfrak{g}}(\lambda) \right)^{-b^2} \exp\left(-\frac{b}{g_s} \sum_{a=1}^{r} \sum_{I=1}^{N_a} W_a(\lambda_I^{(a)}) \right). \quad (2.35)$$

At $b = i$, it reduces to the original quiver matrix model (2.33).

The partition function (2.35) can be rewritten in terms of CFT operators. Let $\phi(z)$ be $\mathfrak{h}$-valued massless chiral field and $\phi_a(z) := \langle \alpha_a, \phi(z) \rangle$. Their correlators are given by

$$\phi_a(z)\phi_b(w) \sim -(\alpha_a, \alpha_b) \log(z - w), \qquad a, b = 1, 2, \ldots, r. \quad (2.36)$$

The modes

$$\phi(z) = q + p \log z + \sum_{n \neq 0} \frac{a_n}{n} z^{-n} \in \mathfrak{h} \quad (2.37)$$

obey the commutation relations

$$[\langle \alpha, a_n \rangle, \langle \beta, a_m \rangle] = -n \delta_{n+m,0}(\alpha, \beta), \qquad [\langle \alpha, p \rangle, \langle \beta, q \rangle] = -i(\alpha, \beta), \qquad \alpha, \beta \in \mathfrak{h}^*. \quad (2.38)$$

The Fock vacuum is given by

$$\alpha(a_n)|0\rangle = 0, \qquad \langle 0|\alpha(a_{-n}) = 0, \qquad n \geq 0, \qquad \alpha \in \mathfrak{h}^*. \quad (2.39)$$

Let

$$\langle \{N_a\}| := \langle 0| \exp\left(-b \sum_{a=1}^{r} N_a \alpha_a(\phi_0) \right). \quad (2.40)$$

It is convenient to introduce the $\mathfrak{h}^*$-valued potential $W(z)$ by

$$W(z) := \sum_{a=1}^{r} W_a(z)\Lambda^a \in \mathfrak{h}^*. \quad (2.41)$$

Note that $W_a(z) = (\alpha_a^\vee, W(z))$.

As in the previous subsection, we put the background charge $Q = b + 1/b$ which leads to the energy-momentum tensor

$$T(z) = -\frac{1}{2} : \mathcal{K}\big(\partial\phi(z), \partial\phi(z)\big) : +Q\langle \rho, \partial^2\phi(z) \rangle, \quad (2.42)$$

where $\mathcal{K}$ is the Killing form and ρ is the Weyl vector of $\mathfrak{g}$, half the sum of the positive roots. Let H^i ($i = 1, 2, \ldots, r$) be an orthonormal basis of the Cartan subalgebra $\mathfrak{h}$ with respect to the Killing form: $\mathcal{K}(H^i, H^j) = \delta^{ij}$. In this basis, the components of the $\mathfrak{h}$-valued chiral boson are just r independent free chiral bosons:

$$\phi(z) = \sum_{i=1}^{r} H^i \phi_i(z), \qquad \phi_i(z)\phi_j(w) \sim -\delta_{ij} \log(z-w), \tag{2.43}$$

and the energy-momentum tensor in this basis is given by

$$T(z) = -\frac{1}{2} \sum_{i=1}^{r} : \left(\partial \phi_i(z)\right)^2 : + Q \sum_{i=1}^{r} \rho^i \partial^2 \phi_i(z). \tag{2.44}$$

The central charge is given by

$$c = r + 12 Q^2(\rho, \rho) = r\left\{1 + h(h+1)Q^2\right\}. \tag{2.45}$$

Here h is the Coxeter number of the simply-laced Lie algebra $\mathfrak{g}$ whose rank is r. Explicitly, $h_{A_{n-1}} = n$ (with $r = n-1$), $h_{D_r} = 2r - 2$, $h_{E_6} = 12$, $h_{E_7} = 18$ and $h_{E_8} = 30$.

Note that for a root α, $[H^i, E_\alpha] = \alpha^i E_\alpha$ with $\alpha^i = \alpha(H^i) = \langle \alpha, H^i \rangle$. Then, the bosons $\phi_a(z)$ associated with the simple roots α_a are expressed in this basis as follows:

$$\phi_a(z) = \langle \alpha_a, \phi(z) \rangle = \sum_{i=1}^{r} \alpha_a^i \phi_i(z) \equiv \alpha_a \cdot \phi(z), \qquad a = 1, 2, \ldots, r. \tag{2.46}$$

For roots α and β, the inner product on the root space is expressed in their components as $(\alpha, \beta) = \sum_{i=1}^{r} \alpha^i \beta^i$. Here $\alpha^i = \alpha(H^i)$ and $\beta^i = \beta(H^i)$.

Let us now consider the four-point correlator of this theory. The vertex operator is defined by

$$V_{\hat\mu}(z) =: e^{\langle \hat\mu, \phi(z) \rangle} :, \tag{2.47}$$

where $\hat\mu \in \mathfrak{h}^*$. As in the one-matrix case, we introduce the screening operators associated with the simple roots are defined by

$$Q_a := \int dz : e^{b\phi_a(z)} :, \qquad a = 1, 2, \ldots, r. \tag{2.48}$$

We define the chiral four-point correlation function

$$\hat{Z} = \left\langle : \prod_{k=0}^{3} e^{\langle \hat\mu_k, \phi(w_k) \rangle} : Q_1^{N_1} Q_2^{N_2} \cdots Q_r^{N_r} \right\rangle. \tag{2.49}$$

For later convenience, we set $m_k := g_s \hat{\mu}_k$ $(k = 0, 1, 2, 3)$. The momentum conservation condition is required

$$\sum_{k=0}^{3} m_k + \sum_{a=1}^{r} b g_s N_a \alpha_a = 0. \tag{2.50}$$

Using this four-point function, we define the partition function of the β deformed quiver matrix model by sending $w_3 \to \infty$ (2.35) with the potential $W_a(z)$:

$$W_a(z) = \sum_{k=0}^{2} (m_k, \alpha_a) \log(w_k - z). \tag{2.51}$$

We will set $w_0 = 0$, $w_1 = 1$ and $w_2 = q$. Using these definitions, the partition function (2.33) can be written as follows

$$Z = \langle \{N_a\}| \exp\left(\frac{1}{2\pi i g_s} \oint_\infty dz \langle W(z), \partial \phi(z) \rangle\right) (Q_1)^{N_1} \cdots (Q_r)^{N_r} |0\rangle. \tag{2.52}$$

2.3 Higher Genus Case

A generalization of the matrix model to a higher genus Riemann surface has also been considered in [7]. The integral representation is basically obtained by changing the two-point function of the free field on a sphere to the one on a Riemann surface, which can be written in terms of the prime form, and by adding a term to the action which is the integral of the holomorphic differentials on the Riemann surface. For the conformal block on a torus with n punctures, for instance, the two-point function is proportional to the theta function and the integral representation is given by [7, 30]

$$Z = \int \prod_{I=1}^{N} d\lambda_I \prod_{1 \le I < J \le N} \theta_1 (\lambda_I - \lambda_J)^{-2b^2} \exp\left(-\frac{b}{g_s} \sum_{I=1}^{N} W(\lambda_I)\right), \tag{2.53}$$

where $\theta_1(z) = 2q^{1/8} \sin z \prod_{n=1}^{\infty} (1 - q^n)(1 - 2q^n \cos 2z + q^{2n})$, $q = \exp(2\pi i \tau)$, and

$$W(z) = \sum_{k=1}^{n} 2m_k \log \theta_1 (z - w_k) + 4\pi i a z. \tag{2.54}$$

The last term in $W(z)$ is the integral of the holomorphic differential on the torus, dz, as mentioned above. Since the factor $\prod_{1 \le I < J \le N} \theta_1 (\lambda_I - \lambda_J)^{-2b^2}$ can be regarded as the generalization of the Vandermonde determinant, we refer to the integral (2.53) as "generalized matrix model". In [7], the potential (2.54) of the generalized matrix model was expected from the geometrical argument of topological string theory.

In the following, we explain how the generalized matrix model is obtained from the full Liouville correlation function [30] for the torus case and [31] for the generic Riemann surface, based on the perturbative argument of [32]. This method is different from the one seen in the previous subsection, although the both use the free field formalism.

The n-point function of the Liouville theory on a genus g Riemann surface C_g is formally given by the following path integral

$$A \equiv \left\langle \prod_{k=1}^{n} e^{2\alpha_k \phi(w_k, \bar{w}_k)} \right\rangle_{\text{Liouville on } C_g} = \int \mathcal{D}\phi(z, \bar{z}) e^{-S[\phi]} \prod_{k=1}^{n} e^{2\alpha_k \phi(w_k, \bar{w}_k)}, \quad (2.55)$$

where the Liouville action is given by

$$S[\phi] = \frac{1}{4\pi} \int d^2 z \sqrt{g} (\partial_a \phi \partial^a \phi + Q R \phi + 4\pi \mu e^{2b\phi}). \quad (2.56)$$

Here R is Ricci scalar and μ is a constant. We divide the Liouville field into the zero mode and the fluctuation $\phi(z, \bar{z}) = \phi_0 + \tilde{\phi}(z, \bar{z})$. By integrating over ϕ_0, we obtain

$$A = \frac{\mu^N \Gamma(-N)}{2b} \int \mathcal{D}\tilde{\phi}(z, \bar{z}) e^{-S_0[\tilde{\phi}]} e^{-\frac{Q}{4\pi} \int d^2 z R \tilde{\phi}} \left(\int d^2 z \, e^{2b\tilde{\phi}(z, \bar{z})} \right)^N \prod_{k=1}^{n} e^{2\alpha_k \tilde{\phi}(w_k, \bar{w}_k)}, \quad (2.57)$$

where

$$N = -\sum_{k=1}^{n} \frac{\alpha_k}{b} + \frac{Q}{b}(1 - g), \quad (2.58)$$

and S_0 is the free scalar field action. When $N \in \mathbb{Z}_{\geq 0}$, the correlator diverges due to the factor $\Gamma(-N)$. The residues at these poles A_N are evaluated in the perturbation theory in b around the free field action:

$$A_N = \frac{(-\mu)^N}{2bN!} \int \prod_{I=1}^{N} d^2 z_I \left\langle e^{-\frac{Q}{4\pi} \int d^2 z R \tilde{\phi}} \prod_{I=1}^{N} : e^{2b\phi(z_I, \bar{z}_I)} : \prod_{k=1}^{n} : e^{2\alpha_k \phi(w_k, \bar{w}_k)} : \right\rangle_{\text{free on } C_g}. \quad (2.59)$$

That N is integer ensures the momentum conservation in the free theory.

Now let us focus on the torus case which simplifies the expression. The ℓ-point function of the free theory on a torus is written in terms of the factorized expression by introducing an additional integral as [33–35]

$$
\left\langle \prod_{i=1}^{\ell} : e^{ik_i\phi(z_i,\bar{z}_i)} : \right\rangle_{\text{free on } T^2}
$$

$$
= 2i|\eta(\tau)|^{-2}\delta\left(\sum_i k_i\right)\int_{-\infty}^{\infty} da \left| \left(\prod_{i<j}\left(\frac{\theta_1\,(z_{ij}|\tau)}{\eta(\tau)^3}\right)^{\frac{k_ik_j}{2}}\right) q^{a^2} \exp\left(-2\pi i \sum_{j=1}^{\ell} k_j z_j a\right) \right|^2,
$$

$$
\tag{2.60}
$$

where $z_{ij} \equiv z_i - z_j$, τ is the moduli of the torus and $q = \exp(2\pi i\tau)$. By using the explicit expression (2.60), we find that the n-point function A_N of the Liouville theory reduces to the following integral

$$
A_N = C(\tau, m_k, b) \prod_{1\leq k<l\leq n} |\theta_1(w_{kl})|^{-4m_km_l} \int_{-i\infty}^{i\infty} da|q|^{-2a^2} \int_{T^2} \prod_{I=1}^{N} d^2 z_I \tag{2.61}
$$

$$
\times \left| \exp\left[-2b\sum_{I=1}^{N}\sum_{k=1}^{n} m_k \log\theta_1\,(z_I - w_k) - 2b^2 \sum_{I<J} \log\theta_1(z_{IJ}) - 4\pi iba \sum_{I=1}^{N} z_I\right] \right|^2,
$$

where $w_{kl} \equiv w_k - w_l$, and we have chosen the insertion points w_k such that they satisfy $\sum_k m_k w_k = 0$. The factor $C(\tau, m_k, b)$ in front of the z integral is irrelevant for the analysis below.

The discussion above is valid even for finite N. However, it is not straightforward to divide the integral over the torus into the product of the holomorphic and the anti-holomorphic pieces for generic N. In order to proceed, we evaluate the integral (2.61) in the large N limit. We see that all the three terms in the exponent in (2.61) are $\mathcal{O}(N^2)$. Thus, the integral (2.61) is evaluated at the critical points of the exponent of the integrand. The conditions for the criticality of the exponent are factorized into holomorphic equations and anti-holomorphic equations, which indicates that the integral over the torus in (2.61) can be replaced by the product of the holomorphic and the anti-holomorphic integrals in the large N limit. Thus we define the holomorphic part of the correlation function as in (2.53) after introduction of g_s by $\alpha_k = m_k/g_s$.

Relation to conformal block and gauge theory We propose that this generalized matrix model (2.53) reproduces the full conformal block on the punctured torus (not only in the large N limit), and also the Nekrasov partition function of the $\mathcal{N} = 2$ elliptic $SU(2)^n$ quiver gauge theory which is obtained from two M5-branes on the same torus.

Let us shortly see the relation of the parameters in the conformal block and the generalized matrix model. In the toric conformal block with n punctures, we have n external and n internal momenta, giving $2n$ parameters in total. The parameters m_k are directly identified with the external momenta. Then, the potential (2.54) has n critical points for each variable z_I, assuming that the parameters m_k are generic. Similar to the case in Sect. 2.1, we expect that the n critical points are "diffused" to form line segments due to the "determinant" factor. Then, the partition function is labelled by the filling fractions $\nu_i = bg_s N_i$, in which N_i out of N variables z_I take the value on the i-th line segment. Due to the momentum conservation condition the sum of all ν_i is not independent degree of freedom. Thus we have $n - 1$ independent filling fractions. These and the parameter a in the potential are mapped to the internal momenta. (See [36] for the precise identification in the $n = 1$ case.)

The relation to the gauge theory is stated as follows: the gauge theory coupling constants $q_p = e^{2\pi i \tau_p}$ $(p = 1, \ldots, n)$ are identified with the moduli of the torus as

$$e^{2\pi i w_k} = \prod_{p=k}^{n-1} q_p, \quad q \equiv e^{2\pi i \tau} = \prod_{p=1}^{n} q_p. \tag{2.62}$$

The parameters m_k are directly identified with the mass parameters of the bifunda-mentals. The filling fractions and the parameters in the potential are mapped to the vevs a_p of the scalars in the vector multiplets.

$g > 1$ **case** Finally, let us quickly consider the case of the genus g Riemann surface with n puncture. As stated above the two-point function is written in terms of the prime form, and the generalized matrix model is the one in (2.54) where the theta function is replaced by the prime form and the last term in the potential is the integral of the holomorphic differential, with some additional terms. The precise form is presented in [31]. The parameters are identified as follows [31, 36]: the conformal block is parameterized by $n + (2g - 2 + n)$ parameters, where the first factor is from the external momenta and the second from the internal ones. In general the generalized matrix model corresponding to this Riemann surface has $n\, m_k$ parameters and g parameters including in the term involving the integrals of the holomorphic differentials. Since critical points of the potential lead to $(2g - 2 + n) - 1$ filling fraction (where -1 comes from the momentum conservation), we have the same number of the parameters as the conformal block.

3 Large N Limit

Let us start an analysis of the matrix models introduced in Sect. 2, focusing on the relation with four-dimensional gauge theory. One way to study a hermitian matrix model is to make use of the loop equation [37–39], and take the limit where the size of matrix, N, is large. By this we can calculate the partition function of the matrix model in the iterative way as in [40, 41] (see e.g., [42] for a review). The systematic study of this method, so-called topological recursion has been performed in [43–45], and in [22, 46, 47] for the β-deformed case. An advantage of considering the large N limit (while $g_s N$ kept fixed) of the matrix model introduced above is that the limit nicely corresponds to the one where ϵ_1 and ϵ_2 go to zero in the four-dimensional side, as can be seen from (2.22). Thus, this section is devoted to study this limit and see the correspondence with the four-dimensional gauge theory.

In Sect. 3.1, we derive the loop equation of the β-deformed matrix model. We see that this equation can be interpreted as the Virasoro constraints in the conformal field theory. Then we show in Sect. 3.2 that in the large N limit the spectral curve obtained from the loop equation can be identified with the Seiberg-Witten curve of the corresponding gauge theory. The free energy of the matrix model can also be computed and agrees with the prepotential of the gauge theory. In Sect. 3.3, we turn to the generalized matrix model on torus, and consider the large N limit.

3.1 Loop Equation

Let us define the generator of the multi-trace operators as

$$R(z_1, \ldots, z_k) = (bg_s)^k \sum_{I_1} \frac{1}{z_1 - \lambda_{I_1}} \cdots \sum_{I_k} \frac{1}{z_k - \lambda_{I_k}}. \tag{3.1}$$

When $k = 1$ this is simply the generator of the single trace operators. First of all, we consider the Schwinger-Dyson equation associated to the transformation $\delta\lambda_I = \frac{1}{z - \lambda_I}$, keeping the potential arbitrary

$$\begin{aligned}
0 &= \frac{1}{Z} \int \prod_{I=1}^{N} d\lambda_I \sum_{K} \frac{\partial}{\partial \lambda_K} \left[\frac{1}{z - \lambda_K} \prod_{I<J} (\lambda_I - \lambda_J)^{-2b^2} e^{-\frac{b}{g_s} \sum_I W(\lambda_I)} \right] \\
&= -\frac{1}{g_s^2} \langle R(z, z) \rangle - \frac{b + \frac{1}{b}}{g_s} \langle R(z)' \rangle - \frac{1}{g_s^2} W'(z) \langle R(z) \rangle + \frac{f(z)}{4g_s^2}, \tag{3.2}
\end{aligned}$$

where R' is the z-derivative of the resolvent and we have defined

$$f(z) = 4bg_s \left\langle \sum_I \frac{W'(z) - W'(\lambda_I)}{z - \lambda_I} \right\rangle. \tag{3.3}$$

The expectation value is defined as the matrix model average (2.15). By multiplying (3.2) by $-g_s^2$, we obtain

$$0 = \langle R(z, z) \rangle + (\epsilon_1 + \epsilon_2) \langle R(z)' \rangle + W'(z) \langle R(z) \rangle - \frac{f(z)}{4}. \tag{3.4}$$

In the case of the hermitian matrix model $b = i$, the second term vanishes and the equation reduces to the well-known one.

We now see that this loop equation is interpreted as the Virasoro constraints in the CFT language. To see this, let us write the energy-momentum tensor by using the expression (2.14)

$$g_s^2 T(z) = -\left(\frac{1}{4} W'(z)^2 + \frac{Q}{2} W''(z) + \frac{f(z)}{4} + (\text{r.h.s. of (3.4)}) \right). \tag{3.5}$$

The singular part in z only comes from the last term. (We here assume that the potential is a polynomial.) Therefore the Virasoro constraint $0 = g_s^2 \langle T(z)|_{\text{sing}} \rangle$ is equivalent to the loop equation. The expectation value of $g_s^2 T(z)$ is simply the first three terms in (3.5).

We now define the "quantum" spectral curve as

$$0 = \hat{x}^2 + g_s^2 \langle T(z) \rangle = \langle (\hat{x} + \frac{g_s}{\sqrt{2}} \partial \phi)(\hat{x} - \frac{g_s}{\sqrt{2}} \partial \phi) \rangle, \qquad (3.6)$$

where we introduce the commutation relation $[\hat{x}, z] = -Q g_s$.

3.2 Large N Limit and Seiberg-Witten Theory

We now take the large N limit while the filling fractions $v_i \equiv b g_s N_i$ ($i = 1, \ldots, n-2$) are fixed. As we saw in Sect. 2.1, there are $n-3$ independent filling fractions because of the momentum conservation. Since both $b g_s$ and g_s/b send to zero, this limit corresponds to $\epsilon_{1,2} \to 0$ in the four-dimensional side.

In this limit the resolvent $\langle R(z, z) \rangle$ is factorized to $\langle R(z) \rangle^2$ in the large N. Therefore the loop equation is written as

$$0 = \langle R(z) \rangle^2 + \langle R(z) \rangle W'(z) - \frac{f(z)}{4}, \qquad (3.7)$$

which is solved as

$$\langle R(z) \rangle = -\frac{1}{2} \left(W'(z) - \sqrt{W'(z)^2 + f(z)} \right). \qquad (3.8)$$

The sign has been chosen such that the large z asymptotics agrees with the definition of $R(z)$. The spectral curve (3.6) now becomes "classical" because $[z, x] = 0$:

$$x^2 = \frac{1}{4}(W'(z)^2 + f(z)). \qquad (3.9)$$

It is easy to see that $x = \pm(W'/2 + \langle R \rangle)$ from (3.8) and (3.9), which is indeed the classical value of $\hat{x}$ by using (2.14). Note that the b-dependence has disappeared by defining the resolvent as in (3.1). Thus, in the large N limit we get the same spectral curve for arbitrary b.

Let us then analyze $f(z)$ by specifying the potential to (2.11). In this case,

$$f(z) = \sum_{k=0}^{n-2} \frac{c_k}{z - w_k}, \qquad (3.10)$$

where for $k \geq 2$

$$c_k = -4bg_s \left\langle \sum_I \frac{2m_k}{\lambda_I - w_k} \right\rangle = -4g_s^2 \frac{\partial \log Z}{\partial w_k} = -4 \frac{\partial F_m}{\partial w_k}. \qquad (3.11)$$

The remaining c_0 and c_1 can be written in terms of c_k with $k \geq 2$ as follows. First of all, due to the equations of motion: $\langle \sum_I W'(\lambda_I) \rangle = 0$, the sum of c_k is constrained to vanish $\sum_{k=0}^{n-2} c_k = 0$. In order to find another constraint, we consider the asymptotic at large z of the loop equation. The asymptotic of the resolvent is $\langle R(z) \rangle \sim \frac{bg_s N}{z}$, so that the leading terms at large z in the loop equations satisfy

$$(bg_s N)^2 - (\epsilon_1 + \epsilon_2) bg_s N + bg_s N \sum_{k=0}^{n-2} 2m_k - \sum_{k=0}^{n-2} \frac{w_k c_k}{4} = 0. \qquad (3.12)$$

The leading term of order $1/z$ in $f(z)$ vanishes via the first constraint. Thus, we obtain

$$\sum_{k=0}^{n-2} w_k c_k = -4 \left(\sum_{k=0}^{n-2} m_k + m_{n-1} - g_s Q \right) \left(\sum_{k=0}^{n-2} m_k - m_{n-1} \right) =: M^2, \qquad (3.13)$$

where we have used the momentum conservation (2.12). Therefore, c_0 and c_1 can be written in terms of c_k (3.11). This means that we have $n-3$ undetermined parameters in the matrix model.

By substituting the potential the curve (3.9) is of the form

$$x^2 = \sum_{k=0}^{n-2} \frac{m_k^2}{(z - w_k)^2} + f(z)/4 = \frac{P_{2n-4}(z)}{\prod_{k=0}^{n-2}(z - w_k)^2}, \qquad (3.14)$$

where P_{2n-4} is a polynomial of degree $2n - 4$, and the residues of $f(z)$ at $z = w_k$ (3.10) are nontrivial functions of the vacuum values of single trace operators. The zeros of P_{2n-4} are the branch points on the z-plane, and there are $n-2$ branch cuts. Let us define the meromorphic differential $\lambda_m = \frac{x dz}{2\pi i}$. This has simple poles at $z = w_k, \infty$ with the residues m_k, m_{n-1}, by observing $\langle R \rangle \sim \frac{g_s N}{z}$ and $W'(z) \sim \sum_{k=0}^{n-2} 2m_k/z$ at large z and by using the momentum conservation. By definition, the filling fractions are obtained by the contour integrals of this differential

$$v_i = \oint_{C_i} dz \lambda_m, \qquad (3.15)$$

where C_i ($i = 1, \ldots, n-2$) are the contours around the branch cuts. These equations relate the vevs of the single trace operators included in $f(z)$ with the filling fraction v_i.

This is exactly the form of the Seiberg-Witten curve of the $SU(2)$ linear quiver gauge theory, $x^2 = \phi_2$ where ϕ_2 is a quadratic meromorphic differential on a sphere. Moreover the differential defined above is identified with the Seiberg-Witten differential $\lambda_{SW} = \frac{\sqrt{\phi_2}dz}{2\pi i}$. Indeed, as proposed in Sect. 2, the filling fractions are mapped to the vacuum expectation values of the vector multiplet scalars, since in the Seiberg-Witten theory these are given by contour integrals of the Seiberg-Witten differential exactly in the same way as (3.15). For the case with $n = 4$ associated with the $SU(2)$ gauge theory with four fundamental hypermultiplets, the precise identification between the vevs of single trace operators and the Coulomb moduli parameter has been worked out in [48]. In [49], the standard saddle point analysis developed in [40] has been applied to determine the spectral curve, in particular the positions of branch cuts.

This is in agreement with the argument in [9] that the ϕ_2 appearing in the Seiberg-Witten curve can be identified with the vacuum expectation value of the energy-momentum tensor of the Virasoro CFT

$$\phi_2(z) = g_s^2 \langle T(z) \rangle|_{\epsilon_{1,2} \to 0}, \tag{3.16}$$

by recalling our definition of the spectral curve (3.6).

Free energy So far we have seen the identification of the spectral curve of the matrix model and the Seiberg-Witten curve of the gauge theory. However, it is still not straightforward to see the equivalence of the free energy of the former with the prepotential of the latter, because the special geometry relation of the Seiberg-Witten theory: $a = \oint_A \lambda_{SW}$ and $\frac{\partial \mathcal{F}}{\partial a} = \oint_B \lambda_{SW}$, where $\mathcal{F}$ is the prepotential, is not manifest in the matrix model. The saddle point analysis of the matrix model can be used to obtain the free energy and the equation like the (second) spacial geometry relation, as in [48]. However here let us shortly see a more direct approach to the free energy for the $n = 4$ case considered in [50].

Recall the relation (3.11). In the $n = 4$ case with $w_2 = q$, this is

$$\frac{\partial F_m}{\partial q} = -\frac{c_2}{4}. \tag{3.17}$$

Therefore, what we need to do is to calculate c_2. (Actually we can only derive the q dependent part of the free energy by this method.) As we discussed in the previous subsection, the parameters c_0, c_1 and c_2 in $f(z)$ are related by $\sum c_i = 0$ and (3.13) $c_1 + qc_2 = 4m_3^2 - 4(\sum_{i=0}^2 m_i)^2$ (when $\epsilon_{1,2} = 0$). Thus, we have $(1 - q)c_2 = 4(\sum_{i=0}^2 m_i)^2 - 4m_3^2 - c_0$. Below we will compute c_0 by writing the spectral curve in terms of it.

In what follows, we consider the simple case where all the hypermultiplet masses are equal to m: i.e., $m_0 = m_3 = 0$ and $m_1 = m_2 = m$. In this case, the polynomial P_4 in the spectral curve is reduced to degree 3: $P_3(z) = Cz(z - z_+)(z - z_-)$, where we have introduced $C = c_0 q/4$ and

$$z_{\pm} = \frac{1}{2}\left(1 + q - (1-q)^2\frac{m^2}{C} \pm (1-q)\sqrt{1 - 2(1+q)\frac{m^2}{C} + (1-q)^2\frac{m^4}{C^2}}\right).$$
(3.18)

By taking the C derivative of xdz, we get the holomorphic differential with

$$\frac{\partial}{\partial C}xdz = \frac{1}{2\sqrt{Cz_+}}\frac{dz}{\sqrt{z(1-z)(1-k^2z)}}, \quad k^2 = \frac{z_-^2}{q}.$$
(3.19)

Since the contour integral of this differential gives the C derivative of the filling fraction ν_1 which has been identified with the vevs a by $a = bg_sN_1$. Thus by expanding in $\frac{m^2}{C}$ and integrating over C, we obtain

$$a = \sqrt{C}\left(h_0(q) - h_1(q)\frac{m^2}{C} - \frac{h_2(q)}{3}\frac{m^4}{C^2} + \mathcal{O}\left(\frac{m^6}{C^3}\right)\right),$$
(3.20)

where $h_i(q)$ depend only on q and are given in [50]. By solving for C, substituting it into (3.17), and integrating over q, we finally obtain the free energy

$$F_m = (a^2 - m^2)\log q + \frac{a^4 + 6a^2m^2 + m^4}{2a^2}q$$
$$+ \frac{13a^8 + 100m^2a^6 + 22m^4a^4 - 12m^6a^2 + 5m^8}{64a^6}q^2 + \mathcal{O}(q^3). \quad (3.21)$$

This agrees with the prepotential of the $SU(2)$ gauge theory with four fundamental hypermultiplets. The latter can be obtained from the Nekrasov partition function of $U(2)$ gauge theory by subtracting the terms coming from the $U(1)$ factor.

Subleading order of large N expansion It is interesting to check the subleading order in the large N (small ϵ_1, ϵ_2) expansion. On the four-dimensional side, the Nekrasov partition function is expanded as

$$\mathcal{F} := \epsilon_1\epsilon_2 \ln Z_{\text{Nek}} = \mathcal{F}_0 + (\epsilon_1 + \epsilon_2)H + \epsilon_1\epsilon_2 F_1 + (\epsilon_1 + \epsilon_2)^2 G + \cdots \quad (3.22)$$

Subleading terms H, F_1 and G can be obtained from the geometric data of the Seiberg-Witten theory. (See [51] for detail.) The matrix model analysis for the subleading orders can also be done. In particular, it was shown that the corresponding parts of the free energy agrees with F_1 in [49] and with H and G in [52, 53]. For generic b, this expansion of the matrix model was compared [54] with the finite N calculation which will be explained in Sect. 5.

The method using the topological recursion [43–45] would be useful. In particular, the calculation of the partition function of the β-deformed matrix model with the logarithmic potential was considered in [47, 55] in this context.

3.3 Higher Genus Case

Let us turn to the generalized matrix model corresponding to the torus (2.53), and derive the loop equation. We then see the equivalence of the spectral curve obtained by taking the large N limit and the Seiberg-Witten curve [56].

We now define the toric version of the resolvent

$$R(z_1, \ldots, z_k) = (bg_s)^k \sum_{I_1} \frac{\theta'_1(z_1 - \lambda_{I_1})}{\theta_1(z_1 - \lambda_{I_1})} \cdots \sum_{I_k} \frac{\theta'_1(z_k - \lambda_{I_k})}{\theta_1(z_k - \lambda_{I_k})}. \qquad (3.23)$$

From the Schwinger-Dyson equation for an arbitrary transformation $\delta\lambda_K = \frac{\theta'_1(z-\lambda_K)}{\theta_1(z-\lambda_K)}$, we derive

$$0 = g_s^2 \left\langle \sum_I \left(\frac{\theta'_1(z - \lambda_I)}{\theta_1(z - \lambda_I)} \right)^2 \right\rangle - g_s^2 \left\langle \sum_I \frac{\theta''_1(z - \lambda_I)}{\theta_1(z - \lambda_I)} \right\rangle - bg_s W'(z) \left\langle \sum_I \frac{\theta'_1(z - \lambda_I)}{\theta_1(z - \lambda_I)} \right\rangle$$

$$+ t(z) - 2b^2 g_s^2 \left\langle \sum_{I<J} \frac{\theta'_1(\lambda_I - \lambda_J)}{\theta_1(\lambda_I - \lambda_J)} \left(\frac{\theta'_1(z - \lambda_I)}{\theta_1(z - \lambda_I)} - \frac{\theta'_1(z - \lambda_J)}{\theta_1(z - \lambda_J)} \right) \right\rangle, \qquad (3.24)$$

where we have multiplied by g_s^2 and defined

$$t(z) = bg_s \left\langle \sum_I \frac{\theta'_1(z - \lambda_I)}{\theta_1(z - \lambda_I)} (W'(z) - W'(\lambda_I)) \right\rangle. \qquad (3.25)$$

By using the formula of the theta function and, after some algebra, we obtain [56]

$$0 = -\langle R(z, z) \rangle - (\epsilon_1 + \epsilon_2) \langle R'(z) \rangle - W'(z) \langle R(z) \rangle + b^2 g_s^2 N \left\langle \sum_I \frac{\theta''_1(z - \lambda_I)}{\theta_1(z - \lambda_I)} \right\rangle$$

$$+ t(z) + b^2 g_s^2 \left\langle \sum_{I<J} \frac{\theta''_1(\lambda_I - \lambda_J)}{\theta_1(\lambda_I - \lambda_J)} \right\rangle + 3b^2 g_s^2 \eta_1 N(N - 1), \qquad (3.26)$$

where $\eta_1 = 4 \frac{\partial \ln \eta}{\partial \ln q}$. This equation is valid for an arbitrary potential.

Let us now focus on the potential (2.54). By rewriting $t(z)$ we finally obtain

$$0 = -\langle R(z, z) \rangle - (\epsilon_1 + \epsilon_2) \langle R'(z) \rangle + W'(z) \langle R(z) \rangle - 3bg_s (N + 1) \eta_1 \sum_k m_k \qquad (3.27)$$

$$-2bg_s \sum_{k=1}^n m_k \frac{\theta'_1(z - w_k)}{\theta_1(z - w_k)} \left\langle \sum_I \frac{\theta'_1(\lambda_I - w_k)}{\theta_1(\lambda_I - w_k)} \right\rangle + bg_s N \sum_k m_k \frac{\theta''_1(z - w_k)}{\theta_1(z - w_k)} + 4g_s^2 \frac{\partial \ln Z}{\partial \ln q}.$$

Let us now see the spectral curve in the large N limit. For simplicity we consider the $n = 1$ case, and take $w_1 = 0$. In this case, it is easy to see that the loop equation reduces to

$$0 = -x^2 + m_1^2 \mathcal{P}(z) - 4u, \tag{3.28}$$

where $\mathcal{P}$ is the Weierstrass function, $x = \pm(\langle R \rangle + W'/2)$, and

$$u = -\pi^2 a^2 + \frac{\partial}{\partial \ln q} \left(F_0 - m_1^2 \ln \eta \right). \tag{3.29}$$

We defined the free energy as $F_0 = \lim_{\epsilon_{1,2} \to 0} (\epsilon_1 \epsilon_2) \ln Z$.

This is indeed the Seiberg-Witten curve of the $SU(2)$ $\mathcal{N} = 2^*$ theory.

4 Nekrasov-Shatashvili Limit

It has long been known that the Seiberg-Witten theory is related with the classical integrable system [57–62]. The connection has been considered in [63] from the recent perspective of the 6d (2,0) theory compactification. A review can be found in [V:3]. Quite remarkably it was proposed in [19] that the gauge theory on the Ω background with $\epsilon_2 \to 0$ while ϵ_1 kept fixed is related with the quantization of the integrable system. In this section, we consider this limit from the matrix model side. The limit is translated to $b \to \infty$ and $g_s \to 0$ with bg_s, $g_s \alpha_k$ and $g_s N_i$ kept finite, and corresponds to the semiclassical limit in the CFT.

In this limit, the leading order part of the free energy is obtained from the value of the critical points which solve the equations of motion (2.18), as in the large N limit. We note that two terms in (2.18) are of the same order in the limit because N and ϵ_1 are kept finite. Let us then consider the loop equation (3.4). Again, in this limit, the connected part of (3.1) can be ignored: $\langle R(z, z) \rangle \to \langle R(z) \rangle^2$. Taking this into account, (3.4) becomes

$$0 = \langle \tilde{R}(z) \rangle^2 + \epsilon_1 \langle \tilde{R}(z)' \rangle + \langle \tilde{R}(z) \rangle W'(z) - \frac{\tilde{f}(z)}{4}, \tag{4.1}$$

where $\tilde{R}$ and $\tilde{f}$ are $R|_{\epsilon_2 \to 0}$ and $f|_{\epsilon_2 \to 0}$ respectively. In the following, we will omit the tildes of R and f. Then, in terms of $x = \langle R(z) \rangle + W'(z)/2$, the equation becomes [22, 46, 64]

$$0 = -x^2 - \epsilon_1 x' + U(z), \tag{4.2}$$

where

$$U(z) = \frac{1}{4} \left(W'(z)^2 + 2\epsilon_1 W''(z) + f(z) \right). \tag{4.3}$$

This is a Ricatti type equation. It is then easy to see that this can be written as the Schrödinger-type equation:

$$0 = -\epsilon_1^2 \frac{\partial^2}{\partial z^2} \Psi(z) + U(z)\Psi(z), \qquad (4.4)$$

where the "wave function" $\Psi(z)$ is defined by

$$\Psi(z) = \exp\left(\frac{1}{\epsilon_1}\int^z x(z')dz'\right). \qquad (4.5)$$

This indicates the relation between the β-deformed matrix model and quantum integrable system.

Note that the quantum spectral curve indeed leads to the same conclusion. Equation (3.6) becomes in this limit

$$0 = \hat{x}^2 - U(z). \qquad (4.6)$$

These variables are not commutative $[\hat{x}, z] = -\epsilon_1$. Thus $\hat{x} = -\epsilon_1 \frac{\partial}{\partial z}$ which leads to (4.4).

In [65, 66], it was shown that the conformal block on a sphere with the additional insertion of the degenerate fields $V_{-\frac{1}{2b}}(z) = e^{-\frac{\phi(z)}{\sqrt{2}b}}$ captures the quantization of the integrable systems. The details can be found in [V:10]. (The similar relation between the affine $SL(2)$ conformal block and integrable system has been found in [67].) This has an interpretation in the 4d gauge theory as an insertion of a surface operator [68] (see [V:9] for a review on this part.) In the following, we will show that under the identification of the β-deformed matrix model Z with the n-point conformal block, the integral representation of the conformal block with degenerate field insertions can be written in terms of the resolvent of the original matrix model [56, 69], in the $\epsilon_2 \to 0$ limit. We note that the similar analysis was done in [70] from the topological string viewpoint.

Let us consider the integral representation of the $(n + \ell)$-point conformal block where ℓ degenerate fields are inserted

$$Z_\ell = \left\langle \prod_{i=1}^{\ell} V_{\frac{1}{2b}}(z_i) \left(\int d\lambda e^{\sqrt{2}b\phi(\lambda)}\right)^N \prod_{k=0}^{n-1} V_{\frac{m_k}{g_s}}(w_k) \right\rangle$$

$$= \prod_{i<j}(z_i - z_j)^{-\frac{1}{2b^2}} \prod_{0\leq k<\ell\leq n-2}(w_k - w_\ell)^{-\frac{2m_k m_\ell}{g_s^2}} \prod_{i=1}^{\ell}\prod_{k=0}^{n-2}(z_i - w_k)^{\frac{m_k}{bg_s}}$$

$$\times \int \prod_{I=1}^{N} d\lambda_I \prod_{I<J}(\lambda_I - \lambda_J)^{-2b^2} \prod_{I}\prod_{k=0}^{n-2}(\lambda_I - w_k)^{-\frac{2bm_k}{g_s}} \prod_{i=1}^{\ell}(z_i - \lambda_I), \quad (4.7)$$

where we have taken w_{n-1} to infinity and omitted the factor including this, as we have done above. The momentum conservation is however modified by the degenerate field insertion as

$$\sum_{k=0}^{n-1} m_k - \frac{\ell g_s}{2b} + b g_s N = g_s Q. \tag{4.8}$$

By dividing by Z and taking a log, we obtain

$$\log \frac{Z_\ell}{Z} = -\frac{1}{2b^2} \sum_{i<j} \log(z_i - z_j) + \sum_i \frac{W(z_i)}{2bg_s} + \log \left\langle \prod_{i,I} (z_i - \lambda_I) \right\rangle, \tag{4.9}$$

where the potential $W(z)$ is the same as (2.11). Notice that the expectation value is defined with the modified momentum conservation (4.8). By defining $e^L = \prod_{i,I} (z_i - \lambda_I)$, we notice that $L = \sum_{i,I} \log(z_i - \lambda_I) = \sum_{i,I} \int^{z_i} \frac{dz_i'}{z_i' - \lambda_I}$, where we have ignored irrelevant terms due to the end points of the integrations. Then, we use that the expectation value of e^L can be written as $\log \langle e^L \rangle = \sum_{k=1}^{\infty} \frac{1}{k!} \langle L^k \rangle_{conn}$ [69], where $\langle \ldots \rangle_{conn}$ means the connected part of the correlator, $\langle L^2 \rangle_{conn} = \langle L^2 \rangle - \langle L \rangle^2$, etc., while $\langle L \rangle_{conn} = \langle L \rangle$. Thus, the last term in the right hand side of (4.9) can be expressed as $\sum_{k=1}^{\infty} \frac{1}{k!} \left\langle \left(\sum_{i,I} \int^{z_i} \frac{dz'}{z' - \lambda_I} \right)^k \right\rangle_{conn}$.

In the limit where $\epsilon_2 \to 0$, the terms with $k > 1$ of the previous expression are subleading contributions compared with the $k = 1$ terms since the connected part of the expectation value can be ignored. Also the first term in the right hand side of (4.9) can be neglected in this limit. Thus, we obtain

$$\frac{Z_\ell}{Z} \to \prod_{i=1}^{\ell} \Psi_i(z_i), \quad \Psi_i(z_i) = \exp \left(\frac{1}{\epsilon_1} \int^{z_i} x(z') dz' \right). \tag{4.10}$$

This indicates that the properties of the conformal block with degenerate field insertions are build in the resolvent of the matrix model in the $\epsilon_2 \to 0$ limit. This property of "separation of variables" agrees with the corresponding result of the Virasoro conformal block as in [65, 71]. Furthermore, this Ψ with $\ell = 1$ is exactly the one which satisfied the Schrödinger equation (4.4).

In summary, we have seen that the integral representation corresponding to the insertion of the degenerate fields into the Virasoro conformal block satisfies the Schrödinger equation, whose potential can be obtained from the loop equation.

Relation with Gaudin model The above argument is applicable for an arbitrary potential $W(z)$. Here we return to the logarithmic one (2.11) and see the relation [22, 56, 72, 73] with the Gaudin Hamiltonian. In this case (4.3) becomes [56]

$$U(z) = \sum_{k=0}^{n-2} \frac{m_k(m_k + \epsilon_1)}{(z - w_k)^2} + \sum_k \frac{H_k}{z - w_k} - \sum_{k=0}^{n-2} \frac{c_k/4}{z - w_k}, \tag{4.11}$$

where

$$H_k = \sum_{\ell(\neq k)} \frac{2m_k m_\ell}{w_k - w_\ell}. \tag{4.12}$$

$U(z)$ is indeed the vacuum expectation value of Gaudin Hamiltonian. In particular, $H_k - c_k/4$ are the vacuum energies of the quantum Hamiltonians.

So far, we discussed the case corresponding to the CFT on the sphere. For the toric case, it has been shown that the loop equation of the generalized matrix model in Sect. 3.3 gives the Hamiltonian of the Hitchin system on the torus in [56]. In particular the $n = 1$ case leads to the elliptic Calogero-Moser model.

5 Finite N Analysis

In the previous section we considered the large N limit and the Nekrasov-Shatashvili limit of the $β$-deformed matrix model. Here we will see a different expansion of the matrix model partition function in the complex structures of the Riemann surface. We calculate each order of the expansion by performing the direct integration. Indeed, this expansion is more useful to compare with the Virasoro conformal block and the Nekrasov partition function. We first review the conformal block of the Virasoro algebra in Sect. 5.1. Then we analyze the integral representation in Sect. 5.2.

5.1 Virasoro Conformal Block

Let us review the Virasoro algebra and the conformal block [74]. (See e.g., [75, 76] for detailed computations.) We consider the conformal symmetry generated by the holomorphic energy-momentum tensor $T(z)$ with

$$T(z) = \sum_{n=-\infty}^{\infty} \frac{L_n}{z^{n+2}}. \tag{5.1}$$

The Virasoro algebra is

$$[L_m, L_n] = (m - n)L_{m+n} + \frac{c}{12}(m^3 - m)\delta_{m+n,0}, \tag{5.2}$$

and we consider the case with Liouville like central charge $c = 1 + 6Q^2$ where $Q = b + 1/b$.

The primary field $V_\alpha(x)$ corresponds to the highest weight vector satisfying

$$L_n V_\alpha = 0, \quad L_0 V_\alpha = \Delta_\alpha V_\alpha, \tag{5.3}$$

where $n > 0$. The conformal dimension of the primary is $\Delta_\alpha = \alpha(Q - \alpha)$. By state-operator correspondence we denote the primary state with Δ_α by $|\Delta\rangle$. Then the Verma module $\mathcal{V}$ is formed by the descendants $V_{Y,\alpha} =: L_{-Y}V_\alpha$: which are obtained by acting with the raising operators $L_{-Y} = (L_{-y_1})^{n_1}(L_{-y_2})^{n_2}(L_{-y_3})^{n_3} \ldots$, where $\{y_i\}$ are positive integers with $y_1 < y_2 < \cdots$. Below we use the shorthand notation to denote Y: $Y = [\cdots y_3^{n_3} y_2^{n_2} y_1^{n_1}]$, e.g., for $y_1 = 1$ and $n_1 = 2$, $Y = [1^2]$. Let us denote the sum of $n_i y_i$ as $|Y|$. The dimension of the descendant is $\Delta_\alpha + |Y|$, and we call $|Y|$ as level.

The OPE of these operators is given by

$$V_{Y_1,\alpha_1}(q) V_{Y_2,\alpha_2}(0) = \sum q^{\Delta - \Delta_1 - \Delta_2 - |Y_1| - |Y_2|} C^\alpha_{\alpha_1,\alpha_2} \sum_Y q^{|Y|} \beta^{\Delta,Y}_{\Delta_1,Y_1;\Delta_2,Y_2} V_{Y,\alpha}(0), \tag{5.4}$$

where Δ is the conformal dimension of V_α. $C^\alpha_{\alpha_1,\alpha_2}$ depends on the dynamics of a two-dimensional theory while $\beta^{\Delta,Y}_{\Delta_1,Y_1;\Delta_2,Y_2}$ is determined from the Virasoro algebra only and depends on the conformal dimensions and central charge. We will focus on the latter and ignore the factors $C^{\alpha_3}_{\alpha_1,\alpha_2}$ coming from the dynamics.

Let us now define the two-point function

$$Q_\Delta(Y_1, Y_2) = \langle \Delta | L_{Y_1} L_{-Y_2} | \Delta \rangle. \tag{5.5}$$

This is symmetric under the exchange of Y_1 and Y_2, and vanishes unless $|Y_1| = |Y_2|$. By using this, $\beta^{\Delta,Y}_{\Delta_1,Y_1;\Delta_2,Y_2}$ can be written in terms of the three-point function γ:

$$\gamma_{\Delta_1,\Delta_2,\Delta_3}(Y_1, Y_2, Y_3) = \langle V_{Y_1,\alpha_1}(\infty) V_{Y_2,\alpha_2}(1) V_{Y_3,\alpha_3}(0) \rangle$$
$$= \sum_{Y'} \beta^{\Delta_3,Y}_{\Delta_1,Y_1;\Delta_2,Y_2} Q_{\Delta_3}(Y', Y_3). \tag{5.6}$$

When $|Y_1| = |Y_2| = \emptyset$, the expressions for the β and γ can be simplified. Thus we define in particular

$$\gamma_{\Delta_1,\Delta_2,\Delta_3}(Y) = \gamma_{\Delta_1,\Delta_2,\Delta_3}(\emptyset, \emptyset, Y),$$
$$\beta^{\Delta_3}_{\Delta_1,\Delta_2}(Y) = \beta^{\Delta_3,Y}_{\Delta_1,\emptyset;\Delta_2,\emptyset}. \tag{5.7}$$

These Q and γ can be computed order by order in the level. Let us give a few results of the computation for later convenience:

$$Q_\Delta([1], [1]) = 2\Delta,$$

$$Q_\Delta([2], [2]) = 4\Delta + c/2, \quad Q_\Delta([2], [1^2]) = 6\Delta, \quad Q_\Delta([1^2], [1^2]) = 4\Delta(1 + 2\Delta),$$

$$\vdots$$

$$\gamma_{\Delta_1, \Delta_2, \Delta_3}([1]) = \Delta_1 + \Delta_3 - \Delta_2,$$

$$\gamma_{\Delta_1, \Delta_2, \Delta_3}([2]) = 2\Delta_1 + \Delta_3 - \Delta_2, \quad \gamma_{\Delta_1, \Delta_2, \Delta_3}([1^2]) = (\Delta_1 + \Delta_3 - \Delta_2)(\Delta_1 + \Delta_3 - \Delta_2 + 1),$$

$$\vdots \tag{5.8}$$

We used $[L_n, V_\alpha(z)] = z^n(z\partial_z + (n+1)\Delta)V_\alpha(z)$ following from the conformal Ward identities when computing the three-point function. From this we can calculate β as

$$\beta_{\Delta_1, \Delta_2}^{\Delta_3}([1]) = \frac{\Delta_1 + \Delta_3 - \Delta_2}{2\Delta_3}. \tag{5.9}$$

Now we can write down the conformal block in terms of these functions. Let us focus on the four-point conformal block which we refer to as $\mathcal{B}$. By translation symmetry we put three points at 0, 1 and ∞. Thus the conformal block is written in terms of the cross ratio q which is the position of the remaining vertex operator. Then the conformal block has the following structure:

$$\mathcal{B} = \sum_{k=0}^{\infty} \mathcal{B}_k q^k, \quad \mathcal{B}_k = \sum_{|Y|=|Y'|=k} \gamma_{\Delta_0, \Delta_2, \Delta}(Y) Q_\Delta^{-1}(Y, Y') \gamma_{\Delta_1, \Delta_3, \Delta}(Y'), \tag{5.10}$$

and $\mathcal{B}_0 = 1$. The conformal block is computed by order by order. E.g., the first order coefficient $\mathcal{B}_1$ is computed as

$$\mathcal{B}_1 = \frac{(\Delta + \Delta_0 - \Delta_2)(\Delta + \Delta_1 - \Delta_3)}{2\Delta}. \tag{5.11}$$

5.2 Finite N Matrix Model

Now we consider the integral representation. Let us first see that the prescription for the momentum conservation at the vertex (2.17) is indeed the correct one by checking the equivalence of the three-point functions. To see this, we consider the following OPE in the free scalar theory

$$: L_{-Y_1} V_{\alpha_1}(q) :: L_{-Y_2} V_{\alpha_2}(0) : \prod_{I=1}^{N} \int_0^q d\lambda_I : e^{\sqrt{2}b\phi(\lambda_I)} :$$

$$= C \sum_Y q^{|Y|} \beta_{\Delta_1, Y_1; \Delta_2, Y_2}^{\Delta_{\alpha_1 + \alpha_2 + bN}, Y} \bigg|_{\text{free}} : L_{-Y} V_{\alpha_1 + \alpha_2 + bN}(0) :, \tag{5.12}$$

where C is an irrelevant factor normalizing $\beta^{\Delta_{\alpha_1}+\alpha_2+bN,\emptyset}_{\Delta_1,\emptyset;\Delta_2\emptyset}|_{\text{free}} = 1$. The coefficient $\beta|_{\text{free}}$ corresponds to the three-point function. Thus, it is natural to propose that [26]

$$\beta^{\Delta,Y}_{\Delta_1,Y_1;\Delta_2,Y_2} = \beta^{\Delta_{\alpha_1}+\alpha_2+bN,Y}_{\Delta_1,Y_1;\Delta_2,Y_2}\bigg|_{\text{free}}, \tag{5.13}$$

under the identification of the internal momenta $\alpha = \alpha_1 + \alpha_2 + bN$, where the left hand side is the one obtained in the previous subsection.

Let us focus on the case with $Y_1 = Y_2 = \emptyset$ and analyze the right hand side of (5.12) further. By calculating the OPE in the free field theory, we obtain

$$V_{\alpha_1}(q) V_{\alpha_2}(0) \prod_{I=1}^{N} \int_0^q d\lambda_I : e^{\sqrt{2}b\phi(\lambda_I)} :$$

$$= q^{-2\alpha_1\alpha_2} \prod_{I=1}^{N} \int_0^q d\lambda_I \prod_{I<J} (\lambda_I - \lambda_J)^{-2b^2} \prod_{I=1}^{N} \lambda_I^{-2b\alpha_2} (q - \lambda_I)^{-2b\alpha_1} : e^{\sqrt{2}(\alpha_1\phi(q)+\alpha_2\phi(0)+b\sum_I \phi(\lambda_I))} :$$

$$= q^{\sigma} \prod_{I=1}^{N} \int_0^1 dx_I \prod_{I<J} (x_I - x_J)^{-2b^2} \prod_{I=1}^{N} x_I^{-2b\alpha_2}(1 - x_I)^{-2b\alpha_1} \sum_{Y,Y'} q^{|Y|-|Y'|} H_{Y,Y'} x^{Y'} : V_{Y,\alpha}(0) : . \tag{5.14}$$

In the last equality we have changed the variables $\lambda_I = q x_I$ and defined $H_{Y,Y'}$ such that

$$: e^{\sqrt{2}(\alpha_1\phi(q)+\alpha_2\phi(0)+b\sum_I \phi(\lambda_I))} := \sum_{Y,Y'} q^{|Y|-|Y'|} H_{Y,Y'} \lambda^{Y'} : L_{-Y} e^{\sqrt{2}(\alpha_1+\alpha_2+bN)\phi(0)} :,$$

$$\tag{5.15}$$

where $\lambda^Y = \prod_I \lambda_I^{y_I}$ for the partition $Y = [y_N, \ldots, y_1]$ with $y_1 \leq y_2 \leq \ldots$. We sum over all the possible Y and Y' with $|Y| \geq |Y'|$. By defining the following multiple integral

$$\langle\langle x^Y \rangle\rangle_N = \prod_{I=1}^{N} \int_0^1 dx_I \prod_{I<J} (x_I - x_J)^{-2b^2} x^Y \prod_{I=1}^{N} x_I^{-2b\alpha_2}(1 - x_I)^{-2b\alpha_1}, \tag{5.16}$$

the three-point function β from the free scalar field theory is thus

$$\beta^{\Delta_{\alpha_1}+\alpha_2+bN}_{\Delta_1,\Delta_2}(Y)\bigg|_{\text{free}} = \sum_{Y',|Y'|\leq|Y|} H_{Y,Y'} \frac{\langle\langle x^{Y'} \rangle\rangle_N}{\langle\langle 1 \rangle\rangle_N}. \tag{5.17}$$

The multiple integral (5.16) is of the Selberg type. We will give results of the integration in Appendix. Thus, the right hand side is in principle calculable.

Let us check the equivalence of the first order. In this case $Y = [1]$, $H_{[1],\emptyset} = \frac{\alpha_1}{\alpha_1+\alpha_2+bN}$ and $H_{[1],[1]} = \frac{bN}{\alpha_1+\alpha_2+bN}$. Combining the formula for $\langle\langle x^{Y=[1]} \rangle\rangle_N$ (6.2) we obtain

$$\beta^{\alpha_1+\alpha_2+bN}_{\Delta_1,\Delta_2}([1])\Big|_{\text{free}} = \frac{\Delta+\Delta_1-\Delta_2}{2\Delta}\Big|_{\alpha=\alpha_1+\alpha_2+bN}. \tag{5.18}$$

This agrees with (5.9) with $\Delta_3 \to \Delta$. The strategy to compute the higher order terms is the following: rewrite x^Y in terms of the Jack polynomial $P_W(x)$ which is specified again by the partition W. (See Appendix for detail.) By writing $x^Y = \sum_W P_W(x)C_{Y,W}$, we have

$$\beta^{\Delta}_{\Delta_1,\Delta_2}(Y)\Big|_{\text{free}} = \sum_{Y',W,|Y'|\le|Y|} H_{Y,Y'}C_{Y',W}\frac{\langle\!\langle P_W(x)\rangle\!\rangle_N}{\langle\!\langle 1\rangle\!\rangle_N}. \tag{5.19}$$

The right hand side can be calculated by performing the integration $\langle\!\langle P_W(x)\rangle\!\rangle_{\alpha_1,\alpha_2,b}$ (6.4). The equivalence with the Virasoro three-point function was checked in lower levels in [26].

Note that this equivalence is only valid for an integer N. However, the result is a rational function of N. Therefore we analytically continue N to an arbitrary complex number.

Four-point conformal block Now let us compute the partition function. We will below focus on the matrix model with $n = 4$ which corresponds to a sphere with four punctures. In this case we define

$$\hat{Z} = C(q)\left(\prod_{I=1}^{N_1}\int_0^q d\lambda_I\right)\left(\prod_{I=N_1+1}^N\int_1^\infty d\lambda_I\right)\prod_{I<J}(\lambda_I-\lambda_J)^{-2b^2}e^{-\frac{b}{g_s}\sum_I W(\lambda_I)}, \tag{5.20}$$

where $C(q) = q^{-2\alpha_0\alpha_2}(1-q)^{-2\alpha_1\alpha_2}$. As proposed in (2.17), the internal momentum α is given by

$$\alpha = \alpha_0 + \alpha_2 + bN_1 = -\alpha_1 - \alpha_3 - bN_2 + Q. \tag{5.21}$$

The above prescription of the contour and the relation between N_1, N_2 and the external momenta was first given in [24] (see [64]) and elaborated in [25, 26]. (We are following the choice of the integration contours in [25].) This integral can be expanded in q

$$\hat{Z} = Z_0 J, \quad J = \sum_{k=0}^\infty J_k q^k, \tag{5.22}$$

where J_k are normalized such that $J_0 = 1$, $Z_0 = cq^\delta$, δ is a function of the conformal dimensions, and c is an irrelevant factor. The proposal of the equivalence between the integral representation and the conformal block is thus

$$J_k = \mathcal{B}_k. \tag{5.23}$$

Let us check this below.

For convenience, we change the variables as

$$
\lambda_I = \begin{cases} qx_I & I = 1, \ldots, N_1 \\ 1/y_{I-N_1} & I = N_1 + 1, \ldots, N_1 + N_2 \end{cases} \tag{5.24}
$$

by which the partition function becomes

$$
\hat{Z} = C'(q) \prod_{I=1}^{N_1} \int_0^1 dx_I \prod_{I=1}^{N_1} x_I^{-2b\alpha_0} (1 - x_I)^{-2b\alpha_2} (1 - qx_I)^{-2b\alpha_1} \prod_{1 \leq I < J \leq N_1} (x_I - x_J)^{-2b^2}
$$
$$
\prod_{I=1}^{N_2} \int_0^1 dy_I \prod_{I=1}^{N_2} y_I^{-2b\alpha_3} (1 - y_I)^{-2b\alpha_1} (1 - qy_I)^{-2b\alpha_2} \prod_{1 \leq I < J \leq N_2} (y_I - y_J)^{-2b^2}
$$
$$
\prod_{I=1}^{N_1} \prod_{J=1}^{N_2} (1 - qx_I y_J)^{-2b^2}, \tag{5.25}
$$

where $C'(q) = q^{\Delta - \Delta_0 - \Delta_2} (1 - q)^{-2\alpha_1 \alpha_2}$. This can be thought of as the double Selberg-type integral. By defining $\langle\!\langle \ldots \rangle\!\rangle_{N_1, N_2}$ as the average of the double Selberg integral, the partition function is written as

$$
\hat{Z} = C'(q) \langle\!\langle 1 \rangle\!\rangle_{N_1, N_2} \langle\!\langle \prod_{I=1}^{N_1} \prod_{J=1}^{N_2} (1 - qx_I)^{-2b\alpha_1} (1 - qy_J)^{-2b\alpha_2} (1 - qx_I y_J)^{-2b^2} \rangle\!\rangle_{N_1, N_2}. \tag{5.26}
$$

Therefore, we obtained $c = \langle\!\langle 1 \rangle\!\rangle_{N_1, N_2}$, $\delta = \Delta - \Delta_0 - \Delta_2$ and [25]

$$
J = (1 - q)^{-2\alpha_1 \alpha_2} \langle\!\langle \prod_{I=1}^{N_1} \prod_{J=1}^{N_2} (1 - qx_I)^{-2b\alpha_1} (1 - qy_J)^{-2b\alpha_2} (1 - qx_I y_J)^{-2b^2} \rangle\!\rangle_{N_1, N_2}
$$
$$
= \langle\!\langle \exp \left(2 \sum_{k=1}^{\infty} \frac{q^k}{k} \left(b \sum_I x_I^k + \alpha_2 \right) \left(b \sum_J y_J^k + \alpha_1 \right) \right) \rangle\!\rangle_{N_1, N_2}. \tag{5.27}
$$

For example, the first order term in q is given by

$$
2 \langle\!\langle \left(b \sum_I x_I^k + \alpha_2 \right) \left(b \sum_J y_J^k + \alpha_1 \right) \rangle\!\rangle_{N_1, N_2} = \frac{(\Delta + \Delta_0 - \Delta_2)(\Delta + \Delta_1 - \Delta_3)}{2\Delta}, \tag{5.28}
$$

by using the formulas of the Selberg integral, with the identification (5.21). This is indeed the conformal block at the level 1, $\mathcal{B}_1$. In principle it is possible to compute higher order terms in q by using the Selberg integral formula and its generalization.

Relation to Nekrasov partition function So far we focused on the relation between the integral representation and the conformal block. At the same time, one can argue the relation to the Nekrasov partition function as considered in [77] following [25, 78]. To do that we use the following expression of J instead of (5.27):

$$J = (1-q)^{-2\alpha_1\alpha_2}$$

$$\left\langle\!\!\left\langle \exp\left(b^2 \sum_{k=1}^{\infty} \frac{q^k}{k}\left(\left(\sum_I x_I^k + \frac{2\alpha_2}{b}\right)\sum_J y_J^k + \sum_I x_I^k\left(\sum_J y_J^k + \frac{2\alpha_1}{b}\right)\right)\right)\right\rangle\!\!\right\rangle_{N_1,N_2}.$$

$$(5.29)$$

At this stage we note that the pre-factor $(1-q)^{-2\alpha_1\alpha_2}$ is the inverse of the $U(1)$ factor introduced in Sect. 2.1. Therefore from (2.19) the second line is conjectured to be identified with the Nekrasov partition function. Indeed the second line has a form of summing over two Young diagrams, μ and ν:

$$\sum_{\mu,\nu} q^{|\mu|+|\nu|} Z_{\mu,\nu},$$

$$(5.30)$$

where $Z_{\mu,\nu}$ is a double Selberg average of polynomials specified by μ and ν. In [77], it was found that by using a particular generalization of the Jack polynomial which depends on a pair of Young diagram, $Z_{\mu,\nu}$ is identified with the corresponding Nekrasov partition function $Z_{\text{Nek}\mu,\nu}$ for given μ and ν. While the Selberg average of the generalization of the Jack polynomial is not completely understood, this is a profound way towards showing the AGT correspondence.

A similar calculation has been done in [79, 80] for the A-type quiver matrix model in Sect. 2.2 by making use of the Selberg integral to see the relation with four-dimensional $SU(N)$ gauge theory. This method has also been performed in the generalized matrix model for the one-punctured torus presented in Sect. 2.3 in [36], and the partition function has been checked to agree with the Virasoro conformal block on the torus in the expansion in the complex structure.

6 Conclusion and Discussion

We have reviewed the β-deformed matrix model associated to the conformal block of two-dimensional CFT and instanton partition function of four-dimensional $\mathcal{N} = 2$ gauge theory, introduced in [7]. This matrix model is originally motivated from the topological string theory, and this interesting part will be seen in the accompanying review [V:13] in this volume.

It would be interesting to consider the β-deformed matrix model corresponding to asymptotically free $\mathcal{N} = 2$ gauge theory. Such models were found first in [48] for $SU(2)$ theory with $N_f = 2, 3$ hypermultiplets and in [81–83] for $SU(2)$ theory coupled to superconformal field theory of Argyres-Douglas type, which are related with irregular conformal blocks in the CFT [84–87]. The former model was elaborated in [88] by calculating directly the integral as in Sect. 5 and in [89, 90] by using the loop equation to see the agreement with the subleading expansion in $\epsilon_{1,2}$. It was also found in [91] the matrix model corresponding to the $SU(2)$ super Yang-Mills theory.

Another interesting generalization is the q-deformed matrix model related to the Nekrasov partition function of the five-dimensional gauge theory proposed in [49, 92, 93]. It would be interesting to elaborate this model further in the context of topological string theory.

In [94–97] a different matrix model which describes the Nekrasov partition function has been found. While the form of the potential in particular is quite different, it would be interesting to see the relation with the model in this review.

Acknowledgments The author would like to thank Giulio Bonelli, Tohru Eguchi, Hiroshi Itoyama, Takeshi Oota, Alessandro Tanzini, and Futoshi Yagi for stimulating collaborations on the β deformed matrix model. The author would like to thank Kazuo Hosomichi and Pavel Putrov for helpful discussions and useful comments. This work of the author is supported by the EPSRC programme grant "New Geometric Structures from String Theory", EP/K034456/1.

Appendix: Integral Formulas

Let us define the following multiple integral

$$
\langle\!\langle x^Y \rangle\!\rangle_N = \prod_{I=1}^{N} \int_0^1 dx_I \prod_{I=1}^{N} x_I^\alpha (1 - x_I)^\beta \prod_{1 \leq I < J \leq N} (x_I - x_J)^{2\gamma} x^Y \tag{6.1}
$$

where supposing that $\Re\beta > 0, \ldots$ for convergence of the integrals. When $Y = \emptyset$ and $Y = [1^k]$, this is the Selberg integral [98] and Aomoto integral [99]

$$
\langle\!\langle 1 \rangle\!\rangle_N = \prod_{j=0}^{N-1} \frac{\Gamma(\alpha + 1 + j\gamma)\Gamma(\beta + 1 + j\gamma)\Gamma(1 + (j+1)\gamma)}{\Gamma(\alpha + \beta + 2 + (N + j - 1)\gamma)\Gamma(1 + \gamma)},
$$

$$
\langle\!\langle x^{Y=[1^k]} \rangle\!\rangle_N = \langle\!\langle 1 \rangle\!\rangle_N \prod_{j=1}^{k} \frac{\alpha + 1 + (N - j)\gamma}{\alpha + \beta + 2 + (2N - j - 1)\gamma}. \tag{6.2}
$$

Another multiple integral which appeared in the main text is involving the Jack polynomial $P_Y(x)$. This is a polynomial of $(x_1, x_2, \ldots, x_N)$ and written as

$$
P_Y(x) = m_Y(x) + \sum_{Y' < Y} a_{Y,Y'} m_{Y'}(x), \tag{6.3}
$$

where $m_Y(x)$ is the monomial symmetric polynomial. Then the following integral is given by [100, 101]

$$\langle\!\langle P_Y(x) \rangle\!\rangle_N = \prod_{I=1}^{N} \int_0^1 dx_I \prod_{I=1}^{N} x_I^{\alpha}(1-x_I)^{\beta} \prod_{1 \le I < J \le N}(x_I - x_J)^{2\gamma} P_Y(x) \tag{6.4}$$

$$= \prod_{i \ge 1} \prod_{j=0}^{y_i-1} \frac{\alpha + 1 + j + (N-i)\gamma}{\alpha + \beta + 2 + j + (2N-i-1)\gamma} \frac{\prod_{i \ge 1} \prod_{j=0}^{y_i-1}(N+1-i)\gamma + j}{\prod_{(i,j) \in Y}(y_i - j + (\tilde{y}_j - i + 1)\gamma)},$$

where $Y = [y_1, y_2, \ldots]$ with $y_1 \ge y_2 \ge \ldots$ and $\tilde{Y} = [\tilde{y}_1 \ge \tilde{y}_2 \ge \ldots]$ is the transpose of Y.

References

[1]   Brezin, E., Itzykson, C., Parisi, G., Zuber, J.: Planar diagrams. Commun. Math. Phys. **59**, 35 (1978)

[2]   Ginsparg, P.H., Moore, G.W.: Lectures on 2-D gravity and 2-D string theory. arXiv:hep-th/9304011 [hep-th]

[3]   Di Francesco, P., Ginsparg, P.H., Zinn-Justin, J.: 2-D Gravity and random matrices. Phys. Rep. **254**, 1–133 (1995). arXiv:hep-th/9306153 [hep-th]

[4]   Dijkgraaf, R., Vafa, C.: A perturbative window into nonperturbative physics. arXiv:hep-th/0208048 [hep-th]

[5]   Pestun, V.: Localization of gauge theory on a four-sphere and supersymmetric Wilson loops. Commun. Math. Phys. **313**, 71–129 (2012). arXiv:0712.2824 [hep-th]

[6]   Kapustin, A., Willett, B., Yaakov, I.: Exact results for Wilson loops in superconformal Chern-Simons theories with matter. JHEP **1003**, 089 (2010). arXiv:0909.4559 [hep-th]

[7]   Dijkgraaf, R., Vafa, C.: Toda theories, matrix models, topological strings, and N=2 gauge systems. arXiv:0909.2453 [hep-th]

[8]   Nekrasov, N.A.: Seiberg-Witten prepotential from instanton counting. Adv. Theor. Math. Phys. **7**, 831–864 (2004). arXiv:hep-th/0206161 [hep-th]

[9]   Alday, L.F., Gaiotto, D., Tachikawa, Y.: Liouville correlation functions from four-dimensional gauge theories. Lett. Math. Phys. **91**, 167–197 (2010). arXiv:0906.3219 [hep-th]

[10]   Gaiotto, D.: $\mathcal{N}$=2 dualities. JHEP **1208**, 034 (2012). arXiv:0904.2715 [hep-th]

[11]   Dotsenko, V., Fateev, V.: Conformal algebra and multipoint correlation functions in two-dimensional statistical models. Nucl. Phys. **B240**, 312 (1984)

[12]   Dotsenko, V., Fateev, V.: Four point correlation functions and the operator algebra in the two-dimensional conformal invariant theories with the central charge $c < 1$. Nucl. Phys. **B251**, 691 (1985)

[13]   Marshakov, A., Mironov, A., Morozov, A.: Generalized matrix models as conformal field theories: discrete case. Phys. Lett. **B265**, 99–107 (1991)

[14]   Kharchev, S., Marshakov, A., Mironov, A., Morozov, A., Pakuliak, S.: Conformal matrix models as an alternative to conventional multimatrix models. Nucl. Phys. **B404**, 717–750 (1993). arXiv:hep-th/9208044 [hep-th]

[15]   Kostov, I.: Gauge invariant matrix model for the A-D-E closed strings. Phys. Lett. **B297**, 74–81 (1992). arXiv:hep-th/9208053 [hep-th]

[16]   Witten, E.: Solutions of four-dimensional field theories via M theory. Nucl. Phys. **B500**, 3–42 (1997) arXiv:hep-th/9703166 [hep-th]

[17]   Nekrasov, N.A., Shatashvili, S.L.: Supersymmetric vacua and Bethe Ansatz. Nucl. Phys. Proc. Suppl. **192–193**, 91–112 (2009). arXiv:0901.4744 [hep-th]

[18]   Nekrasov, N.A., Shatashvili, S.L.: Quantum integrability and supersymmetric vacua. Prog. Theor. Phys. Suppl. **177**, 105–119 (2009). arXiv:0901.4748 [hep-th]

[19] Nekrasov, N.A., Shatashvili, S.L.: Quantization of integrable systems and four dimensional gauge theories. arXiv:0908.4052 [hep-th]

[20] Penner, R.: Perturbative series and the moduli space of Riemann surfaces. J. Differ. Geom. **27**, 35 (1988)

[21] Mehta, M.L.: Random Matrices. Pure and Applied Mathematics Series, vol. 142, 3rd edn. Elsevier, Amsterdam (2004)

[22] Eynard, B., Marchal, O.: Topological expansion of the Bethe Ansatz, and non-commutative algebraic geometry. JHEP **0903**, 094 (2009). arXiv:0809.3367 [math-ph]

[23] Kostov, I.K.: Conformal field theory techniques in random matrix models. arXiv:hep-th/9907060 [hep-th]

[24] Mironov, A., Morozov, A., Shakirov, S.: Conformal blocks as Dotsenko-Fateev integral discriminants. Int. J. Mod. Phys. **A25**, 3173–3207 (2010). arXiv:1001.0563 [hep-th]

[25] Itoyama, H., Oota, T.: Method of generating q-expansion coefficients for conformal block and N=2 Nekrasov function by beta-deformed matrix model. Nucl. Phys. **B838**, 298–330 (2010). arXiv:1003.2929 [hep-th]

[26] Mironov, A., Morozov, A., Morozov, A.: Conformal blocks and generalized Selberg integrals. Nucl. Phys. **B843**, 534–557 (2011). arXiv:1003.5752 [hep-th]

[27] Cheng, M.C., Dijkgraaf, R., Vafa, C.: Non-perturbative topological strings and conformal blocks. JHEP **1109**, 022 (2011). arXiv:1010.4573 [hep-th]

[28] Itoyama, H., Maruyoshi, K., Oota, T.: The quiver matrix model and 2d-4d conformal connection. Prog. Theor. Phys. **123**, 957–987 (2010). arXiv:0911.4244 [hep-th]

[29] Chiantese, S., Klemm, A., Runkel, I.: Higher order loop equations for A(r) and D(r) quiver matrix models. JHEP **0403**, 033 (2004). arXiv:hep-th/0311258 [hep-th]

[30] Maruyoshi, K., Yagi, F.: Seiberg-Witten curve via generalized matrix model. JHEP **1101**, 042 (2011). arXiv:1009.5553 [hep-th]

[31] Bonelli, G., Maruyoshi, K., Tanzini, A., Yagi, F.: Generalized matrix models and AGT correspondence at all genera. JHEP **1107**, 055 (2011). arXiv:1011.5417 [hep-th]

[32] Goulian, M., Li, M.: Correlation functions in Liouville theory. Phys. Rev. Lett. **66**, 2051–2055 (1991)

[33] Verlinde, E.P., Verlinde, H.L.: Multiloop calculations in covariant superstring theory. Phys. Lett. **B192**, 95 (1987)

[34] Dijkgraaf, R., Verlinde, E.P., Verlinde, H.L.: C = 1 conformal field theories on Riemann surfaces. Commun. Math. Phys. **115**, 649–690 (1988)

[35] D'Hoker, E., Phong, D.: The geometry of string perturbation theory. Rev. Mod. Phys. **60**, 917 (1988)

[36] Mironov, A., Morozov, A., Shakirov, S.: On 'Dotsenko-Fateev' representation of the toric conformal blocks. J. Phys. **A44**, 085401 (2011). arXiv:1010.1734 [hep-th]

[37] David, F.: Loop equations and nonperturbative effects in two-dimensional quantum gravity. Mod. Phys. Lett. **A5**, 1019–1030 (1990)

[38] Ambjorn, J., Jurkiewicz, J., Makeenko, Y.: Multiloop correlators for two-dimensional quantum gravity. Phys. Lett. **B251**, 517–524 (1990)

[39] Itoyama, H., Matsuo, Y.: Noncritical Virasoro algebra of d < 1 matrix model and quantized string field. Phys. Lett. **B255**, 202–208 (1991)

[40] Ambjorn, J., Chekhov, L., Kristjansen, C., Makeenko, Y.: Matrix model calculations beyond the spherical limit. Nucl. Phys. **B404**, 127–172 (1993). arXiv:hep-th/9302014 [hep-th]

[41] Akemann, G.: Higher genus correlators for the Hermitian matrix model with multiple cuts. Nucl. Phys. **B482**, 403–430 (1996). arXiv:hep-th/9606004 [hep-th]

[42] Marino, M.: Les Houches lectures on matrix models and topological strings. arXiv:hep-th/0410165 [hep-th]

[43] Alexandrov, A., Mironov, A., Morozov, A.: Partition functions of matrix models as the first special functions of string theory. 1. Finite size Hermitian one matrix model. Int. J. Mod. Phys. **A19**, 4127–4165 (2004). arXiv:hep-th/0310113 [hep-th]

[44] Eynard, B.: Topological expansion for the 1-Hermitian matrix model correlation functions. JHEP **0411**, 031 (2004). arXiv:hep-th/0407261 [hep-th]

[45] Eynard, B., Orantin, N.: Invariants of algebraic curves and topological expansion. Commun. Number Theory Phys. **1**, 347–452 (2007). arXiv:math-ph/0702045

[46] Chekhov, L., Eynard, B., Marchal, O.: Topological expansion of the Bethe Ansatz, and quantum algebraic geometry. arXiv:0911.1664 [math-ph]

[47] Chekhov, L.: Logarithmic potential β-ensembles and Feynman graphs. arXiv:1009.5940 [math-ph]

[48] Eguchi, T., Maruyoshi, K.: Penner type matrix model and Seiberg-Witten theory. JHEP **1002**, 022 (2010). arXiv:0911.4797 [hep-th]

[49] Schiappa, R., Wyllard, N.: An A(r) threesome: matrix models, 2d CFTs and 4d N=2 gauge theories. J. Math. Phys. **51**, 082304 (2010). arXiv:0911.5337 [hep-th]

[50] Eguchi, T., Maruyoshi, K.: Seiberg-Witten theory, matrix model and AGT relation. JHEP **1007**, 081 (2010). arXiv:1006.0828 [hep-th]

[51] Nakajima, H., Yoshioka, K.: Lectures on instanton counting. arXiv:math/0311058 [math-ag]

[52] Itoyama, H., Yonezawa, N.: ϵ-corrected Seiberg-Witten prepotential obtained from half genus expansion in beta-deformed matrix model. Int. J. Mod. Phys. **A26**, 3439–3467 (2011). arXiv:1104.2738 [hep-th]

[53] Nishinaka, T., Rim, C.: β-deformed matrix model and Nekrasov partition function. JHEP **1202**, 114 (2012). arXiv:1112.3545 [hep-th]

[54] Morozov, A., Shakirov, S.: The matrix model version of AGT conjecture and CIV-DV prepotential. JHEP **1008**, 066 (2010). arXiv:1004.2917 [hep-th]

[55] Brini, A., Marino, M., Stevan, S.: The uses of the refined matrix model recursion. J. Math. Phys. **52**, 052305 (2011). arXiv:1010.1210 [hep-th]

[56] Bonelli, G., Maruyoshi, K., Tanzini, A.: Quantum Hitchin systems via beta-deformed matrix models. arXiv:1104.4016 [hep-th]

[57] Gorsky, A., Krichever, I., Marshakov, A., Mironov, A., Morozov, A.: Integrability and Seiberg-Witten exact solution. Phys. Lett. **B355**, 466–474 (1995). arXiv:hep-th/9505035 [hep-th]

[58] Martinec, E.J., Warner, N.P.: Integrable systems and supersymmetric gauge theory. Nucl. Phys. **B459**, 97–112 (1996). arXiv:hep-th/9509161 [hep-th]

[59] Nakatsu, T., Takasaki, K.: Whitham-Toda hierarchy and N = 2 supersymmetric Yang-Mills theory. Mod. Phys. Lett. **A11**, 157–168 (1996). arXiv:hep-th/9509162 [hep-th]

[60] Donagi, R., Witten, E.: Supersymmetric Yang-Mills theory and integrable systems. Nucl. Phys. **B460**, 299–334 (1996). arXiv:hep-th/9510101 [hep-th]

[61] Itoyama, H., Morozov, A.: Integrability and Seiberg-Witten theory: curves and periods. Nucl. Phys. **B477**, 855–877 (1996). arXiv:hep-th/9511126 [hep-th]

[62] Itoyama, H., Morozov, A.: Prepotential and the Seiberg-Witten theory. Nucl. Phys. **B491**, 529–573 (1997). arXiv:hep-th/9512161 [hep-th]

[63] Gaiotto, D., Moore, G.W., Neitzke, A.: Wall-crossing, Hitchin systems, and the WKB approximation. arXiv:0907.3987 [hep-th]

[64] Mironov, A., Morozov, A., Shakirov, S.: Matrix model conjecture for exact BS periods and nekrasov functions. JHEP **1002**, 030 (2010). arXiv:0911.5721 [hep-th]

[65] Teschner, J.: Quantization of the Hitchin moduli spaces, Liouville theory, and the geometric Langlands correspondence I. Adv. Theor. Math. Phys. **15**, 471–564 (2011). arXiv:1005.2846 [hep-th]

[66] Maruyoshi, K., Taki, M.: Deformed prepotential, quantum integrable system and Liouville field theory. Nucl. Phys. **B841**, 388–425 (2010). arXiv:1006.4505 [hep-th]

[67] Alday, L.F., Tachikawa, Y.: Affine SL(2) conformal blocks from 4d gauge theories. Lett. Math. Phys. **94**, 87–114 (2010). arXiv:1005.4469 [hep-th]

[68] Alday, L.F., Gaiotto, D., Gukov, S., Tachikawa, Y., Verlinde, H.: Loop and surface operators in N=2 gauge theory and Liouville modular geometry. JHEP **1001**, 113 (2010). arXiv:0909.0945 [hep-th]

[69] Marshakov, A., Mironov, A., Morozov, A.: On AGT relations with surface operator insertion and stationary limit of beta-ensembles. J. Geom. Phys. **61**, 1203–1222 (2011). arXiv:1011.4491 [hep-th]

[70] Aganagic, M., Cheng, M.C., Dijkgraaf, R., Krefl, D., Vafa, C.: Quantum geometry of refined topological strings. JHEP **1211**, 019 (2012). arXiv:1105.0630 [hep-th]

[71] Kozcaz, C., Pasquetti, S., Wyllard, N.: A and B model approaches to surface operators and Toda theories. JHEP **1008**, 042 (2010). arXiv:1004.2025 [hep-th]

[72] Bourgine, J.-E.: Large N limit of beta-ensembles and deformed Seiberg-Witten relations. JHEP **1208**, 046 (2012). arXiv:1206.1696 [hep-th]

[73] Bourgine, J.-E.: Large N techniques for Nekrasov partition functions and AGT conjecture. JHEP **1305**, 047 (2013). arXiv:1212.4972 [hep-th]

[74] Belavin, A., Polyakov, A.M., Zamolodchikov, A.: Infinite conformal symmetry in two-dimensional quantum field theory. Nucl. Phys. **B241**, 333–380 (1984)

[75] Marshakov, A., Mironov, A., Morozov, A.: Combinatorial expansions of conformal blocks. Theor. Math. Phys. **164**, 831–852 (2010). arXiv:0907.3946 [hep-th]

[76] Mironov, A., Mironov, S., Morozov, A., Morozov, A.: CFT exercises for the needs of AGT. arXiv:0908.2064 [hep-th]

[77] Morozov, A., Smirnov, A.: Towards the proof of AGT relations with the help of the generalized Jack polynomials. Lett. Math. Phys. **104**(5), 585–612 (2014). arXiv:1307.2576 [hep-th]

[78] Mironov, A., Morozov, A., Shakirov, S.: A direct proof of AGT conjecture at beta = 1. JHEP **1102**, 067 (2011). arXiv:1012.3137 [hep-th]

[79] Zhang, H., Matsuo, Y.: Selberg integral and SU(N) AGT conjecture. JHEP **1112**, 106 (2011). arXiv:1110.5255 [hep-th]

[80] Mironov, S., Morozov, A., Zenkevich, Y.: Generalized Jack polynomials and the AGT relations for the $SU(3)$ group. arXiv:1312.5732 [hep-th]

[81] Nishinaka, T., Rim, C.: Matrix models for irregular conformal blocks and Argyres-Douglas theories. JHEP **1210**, 138 (2012). arXiv:1207.4480 [hep-th]

[82] Rim, C.: Irregular conformal block and its matrix model. arXiv:1210.7925 [hep-th]

[83] Choi, S.-K., Rim, C.: Parametric dependence of irregular conformal block. arXiv:1312.5535 [hep-th]

[84] Gaiotto, D.: Asymptotically free N=2 theories and irregular conformal blocks. arXiv:0908.0307 [hep-th]

[85] Marshakov, A., Mironov, A., Morozov, A.: On non-conformal limit of the AGT relations. Phys. Lett. **B682**, 125–129 (2009). arXiv:0909.2052 [hep-th]

[86] Bonelli, G., Maruyoshi, K., Tanzini, A.: Wild quiver gauge theories. JHEP **1202**, 031 (2012). arXiv:1112.1691 [hep-th]

[87] Gaiotto, D., Teschner, J.: Irregular singularities in Liouville theory and Argyres-Douglas type gauge theories, I. JHEP **1212**, 050 (2012). arXiv:1203.1052 [hep-th]

[88] Itoyama, H., Oota, T., Yonezawa, N.: Massive scaling limit of beta-deformed matrix model of Selberg type. Phys. Rev. **D82**, 085031 (2010). arXiv:1008.1861 [hep-th]

[89] Fujita, M., Hatsuda, Y., Tai, T.-S.: Genus-one correction to asymptotically free Seiberg-Witten prepotential from Dijkgraaf-Vafa matrix model. JHEP **1003**, 046 (2010). arXiv:0912.2988 [hep-th]

[90] Krefl, D.: Penner type ensemble for gauge theories revisited. Phys. Rev. **D87**, 045027 (2013). arXiv:1209.6009 [hep-th]

[91] Mironov, A., Morozov, A., Shakirov, S.: Brezin-Gross-Witten model as 'pure gauge' limit of Selberg integrals. JHEP **1103**, 102 (2011). arXiv:1011.3481 [hep-th]

[92] Awata, H., Yamada, Y.: Five-dimensional AGT relation and the deformed beta-ensemble. Prog. Theor. Phys. **124**, 227–262 (2010). arXiv:1004.5122 [hep-th]

[93] Mironov, A., Morozov, A., Shakirov, S., Smirnov, A.: Proving AGT conjecture as HS duality: extension to five dimensions. Nucl. Phys. **B855**, 128–151 (2012). arXiv:1105.0948 [hep-th]

[94] Klemm, A., Sulkowski, P.: Seiberg-Witten theory and matrix models. Nucl. Phys. **B819**, 400–430 (2009). arXiv:0810.4944 [hep-th]

[95] Sulkowski, P.: Matrix models for beta-ensembles from Nekrasov partition functions. JHEP **1004**, 063 (2010). arXiv:0912.5476 [hep-th]

[96]   Marshakov, A.: On gauge theories as matrix models. Teor. Mat. Fiz. **169**, 391–412 (2011).
        arXiv:1101.0676 [hep-th]
[97]   Kimura, T.: Matrix model from N = 2 orbifold partition function. JHEP **1109**, 015 (2011).
        arXiv:1105.6091 [hep-th]
[98]   Selberg, A.: Norsk. Mat. Tisdskr. **24**, 71 (1944)
[99]   Aomoto, K.: On the complex Selberg integral. Q. J. Math. **38**, 385–399 (1987)
[100]   Kaneko, J.: Selberg integrals and hypergeometric functions associated with Jack polyno-
        mials. SIAM J. Math. Anal. **24**, 1086–1110 (1993)
[101]   Kadell, K.: The Selberg-Jack symmetric functions. Adv. Math. **130**, 33–102 (1997)

Localization for $\mathcal{N} = 2$ Supersymmetric Gauge Theories in Four Dimensions

Vasily Pestun

Abstract We review the supersymmetric localization of $\mathcal{N} = 2$ theories on curved backgrounds in four dimensions using $\mathcal{N} = 2$ supergravity and generalized conformal Killing spinors. We review some known backgrounds and give examples of new geometries such as local T^2-bundle fibrations. We discuss in detail a topological four-sphere with generic T^2-invariant metric. This review is a contribution to the special volume on recent developments in $\mathcal{N} = 2$ supersymmetric gauge theory and the 2d-4d relation.

1 Introduction

Non-perturbative exact results in interacting quantum field theories (QFTs) are rare and precious and usually we explain them using non-trivial symmetries of QFT, such as quantum groups in the theory of quantum integrable systems [1–3]. Another instrumental symmetry for exact results in QFTs is supersymmetry. For example, Seiberg-Witten solution [4] of four-dimensional $\mathcal{N} = 2$ supersymmetric field theories is explained by rigid constraints imposed by $\mathcal{N} = 2$ supersymmetry on the low-energy effective Lagrangian and certain assumptions on electric-magnetic duality. For a review of $\mathcal{N} = 2$ four-dimensional theory from the modern angle of view see contributions [V:2, V:3] in this volume.

Exact non-perturbative results in supersymmetric QFTs are suited for strong tests of non-perturbative dualities between QFTs that have different microscopic description, they give a practical approximation to interesting physical phenomena in non-supersymmetric QFTs, and they open new perspectives on fascinating geometrical spaces such moduli spaces of instantons, monopoles, complex structures, flat connections, and others.

A citation of the form [V:x] refers to article number x in this volume.

V. Pestun (✉)
IHES, Bures-sur-yvette, France
e-mail: pestun@ihes.fr

© Springer International Publishing Switzerland 2016
J. Teschner (ed.), *New Dualities of Supersymmetric Gauge Theories*,
Mathematical Physics Studies, DOI 10.1007/978-3-319-18769-3_6

A fruitful non-perturbative approach is supersymmetric localization. In finite dimensional geometry localization appeared in the Lefchetz fixed-point formula, Duistermaat-Heckman and Atiyah-Bott formula for integration of equivariantly closed differential forms [5]. In [6] Witten generalized the localization formula for the infinite-dimensional geometry of the path integral of supersymmetric quantum mechanics. Similar approach was proposed to two-dimensional sigma models [7], four-dimensional gauge theories [8] and others. The similarity of these constructions is the topological twist of a given supersymmetric QFT. The topological twist introduces a certain background connection for the local internal R-symmetry of the theory. Usually, this connection is such that there exists a scalar fermionic supersymmetry generator Q for a QFT coupled to a generically curved background metric. In topologically twisted theories the stress-energy tensor is Q-exact, and, consequently, the theory is metric independent. A further twist to the supersymmetric localization of gauge theories, called ϵ-equivariant deformation or Ω-background $\mathbb{R}^4_{\epsilon_1,\epsilon_2}$, has been added by Nekrasov [9] based on the considerations of [10–13]. The construction of the gauge theory instanton partition function is reviewed in [V:4] of this volume. The ϵ-equivariant partition function $Z_{\epsilon_1,\epsilon_2}$, referred as Nekrasov's function, turned out to be a fascinating object of mathematical physics, with profound connections to other branches of research such as topological strings (see review [V:13, V:14] in this volume), matrix models (see review [V:5] in this volume), quantum groups [2, 3] and integrable systems [14]. For a recent study of the instanton partition function $Z_{\epsilon_1,\epsilon_2}$ for a large class of quiver theories see [15, 16]. A profound connection between four-dimensional gauge theory supersymmetric objects (BPS) and two-dimensional conformal field theories (CFT), called BPS/CFT correspondence in [17], was a subject of long research [9, 13, 18–22].

Another version of localization was used in [23] for $\mathcal{N} = 2$ supersymmetric gauge theory on a four-sphere S^4 with an insertion of a Wilson operator [23] or 't Hooft operators [24]. The topological twist is not necessary because of rich $OSp(2|4)$ symmetry that $\mathcal{N} = 2$ QFT on S^4 has. A similar localization was later performed for gauge theories on S^3 [25], on S^2 [26, 27], on S^5 [28, 29], on squashed S^3_b [30, 31], on squashed $S^4_{\epsilon_1,\epsilon_2}$ [32] and other geometries. For a review of 3d localization in this volume see [V:10], for a review of line operators (such as Wilson and 't Hooft operators) in 4d gauge theory see review [V:7] in this volume, and for review of surface operators see [V:8]. The four-sphere partition function of the $\mathcal{N} = 2$ gauge theory of class $\mathcal{S}_\mathfrak{g}$ (see [V:2] in this volume) turned out to be equal to the correlation function of the 2d conformal $\mathfrak{g}$-Toda theory, the statement known as AGT conjecture [33], which provided explicit beautiful realization of the 4d/2d BPS/CFT correspondence. For a review of AGT conjecture (4d/2d BPS/CFT correspondence) in this volume see [V:12], for review of the superconformal index see [V:9] and for a review of the 3d/3d version of the BPS/CFT correspondence see [V:11].

A general procedure to construct a QFT on a curved manifold with some amount of supersymmetry is to couple QFT with supergravity, choose the supegravity background fields in such a way that there exists a non-trivial supersymmetric variation under which these background fields are invariant and then freeze the supegravity fields. This construction was explored for $\mathcal{N} = 1$ supersymmetric four-dimensional theories in [34].

In this note we will partially analyze off-shell $\mathcal{N} = 2$ supersymmetry backgrounds suitable for localization, and review the case of the four-sphere [23] with a generic T^2-invariant deformation of the metric. We employ the formalism of $\mathcal{N} = 2$ supergravity known as superconformal tensor calculus, see [35–38] and reviews [39, 40]. For previous analysis of $\mathcal{N} = 2$ supegravity localization backgrounds see [41–43].

Acknowledgements The author is grateful to Kazuo Hosomichi for the correspondence while preparing this review and discussion of the results of [32], and Takuya Okuda for the elaborate comments on the manuscript.

2 $\mathcal{N} = 2$ Supergravity

2.1 Gravity Multiplet

A way to construct $\mathcal{N} = 2$ Poincare supergravity is to promote the $\mathcal{N} = 2$ superconformal symmetry to local gauge symmery and introduce associated gauge fields. The gauge fields are combined with auxiliary fields to form *Weyl* multiplet. The notations are collected in the Table 1.

The gauge fields $(\omega_\mu^{\hat\mu\hat\nu}, f_\mu^{\hat\mu}, \phi_\mu^i)$ associated to the rotation, the special conformal symmetry and the special conformal supersymmetry are expressed in terms of the other fields from the constraints on superconformal covariant curvatures $\hat{R}$ for the fields $e_\mu^{\hat\mu}, \omega_\mu^{\hat\mu\hat\nu}, \psi_\mu^i$.

Table 1 The *Weyl* multiplet

Symmetry	Gauge field	Constraint	Parameter
Translation	$e_\mu^{\hat\mu}$		
Rotation	$\omega_\mu^{\hat\mu\hat\nu}$	$\hat{R}(e_\mu^{\hat\mu})$	
Special conformal	$f_\mu^{\hat\mu}$	$\hat{R}(\omega_\mu^{\hat\mu\hat\nu})$	
Dilatation	b_μ		
Translational supersymmetry Q	ψ_μ^i		ε^i
Conformal supersymmetry S	ϕ_μ^i	$\hat{R}(\psi_\mu^i)$	η^i
$SU(2)_R$-symmetry	$V_{\mu\,j}^{\ \ i}$		
$U(1)_{\tilde{R}}$-axial symmetry	$\tilde{A}_\mu$		
Auxiliary fields			
Tensor	$T_{\mu\nu a}$		
Spinor	χ^i		
Scalar	M		

Table 2 6d chirality

Weyl multiplet		
Field	Variation	6d chirality
ψ^i_μ	ε^i	$+1$
ϕ^i_μ	η^i	-1
χ^i		$+1$
$T_{\mu\nu a}$		-1
Vector multiplet		
λ^i		$+1$

It is convenient to use 6d spinorial notations for the spinors of 4d $\mathcal{N} = 2$ theories under dimensional reduction. We use index conventions from Appendix 1 and chirality conventions as in Table 2.

To find the action of the $\mathcal{N} = 2$ Poincare supergravity interacting with n_V vector multiplets and n_H hypermutiplets one considers Weyl multiplet coupled with $n_V + 1$ vector multiplets and $n_H + 1$ hypermultiplets and then uses one vector multiplet and one hypermultiplet as auxiliary fields to gauge fix the non-Poincare superconformal gauge symmetries and to integrate out non-Poincare supergravity fields. Finally, one gets the on-shell physical fields of the Poincare $\mathcal{N} = 2$ supergravity: the frame $e^{\hat{\mu}}_\mu$, the gravitino doublet ψ^i_μ and the graviphoton A_μ together with n_v vector multiplets and n_h hypermultiplets (See more details in the diagram [44], p. 81).

To construct gauge theories on fixed curved backgrounds with partially preserved off-shell supersymmetry the full machinery described above is not necessary. It is sufficient to consider the off-shell action and the supersymmetry transformation for the vector multiplets and hypermultiplets coupled to the Weyl gravity multiplet and then freeze the fields of the Weyl multiplet to a supersymmetry invariant background [32, 34, 45].

The supersymmetry transformation is a linear superposition of the Poincare supersymmetry variation ε^i and the conformal supersymmetry variation η^i. Since variation of bosonic fields in Weyl multiplet is proportional to fermionic fields of Weyl multiplet and they are set to zero in the background, the supersymmetric equation is the vanishing variation of the independent fermions ψ^i_μ, χ^i. The field ϕ^i_μ is expressed in terms of ψ^i_μ and χ_i through the curvature constraints, and the vanishing variation of ψ^i_μ and χ^i_μ automatically implies vanishing variation of ϕ^i_μ.

We quote the variation of gravitino ψ^i_μ and auxiliary field χ^i under the Poincare and conformal supersymmetries ε^i and η^i from [40], p. 429. The equations are[1]:

---

[1]In some $\mathcal{N} = 2$ supergravity literature the auxiliary scalar field M in Weyl multiplet is denoted D. For the conventions on the Clifford algebra see Appendix section "Clifford algebra"; the slash symbol on tensors denotes Clifford contraction as in Eq. (4.34).

$$\delta_{\varepsilon,\eta}\psi_\mu^i = D_\mu\varepsilon^i - \tfrac{1}{16}T\gamma_\mu\varepsilon^i - \gamma_\mu\eta^i = 0$$

$$\delta_{\varepsilon,\eta}\chi^i = -\tfrac{1}{24}[D_\mu T]\gamma^\mu\varepsilon^i + \tfrac{1}{6}\left(F_V^R\right)_j^i \varepsilon^j + \tfrac{1}{12}T\eta^i + \tfrac{1}{2}M\varepsilon^i = 0 \qquad (2.1)$$

From the first equation the conformal supersymmetry parameter η can be expressed in terms of the Poincare supersymmetry parameter ε as

$$\eta^i = \tfrac{1}{4}\slashed{D}\varepsilon^i \qquad (2.2)$$

where we used (4.57). Later we use this relation to substitute η^i with $\tfrac{1}{4}\slashed{D}\varepsilon^i$ and vice versa. The second equation, called the *auxiliary* equation, can be transformed using the Lichnerowicz formula for $\slashed{D}^2$ (4.63) and the divergence of the first equation, see Appendix (4.65):

$$D_\mu\varepsilon^i - \tfrac{1}{16}T\gamma_\mu\varepsilon^i - \tfrac{1}{4}\gamma_\mu\slashed{D}\varepsilon^i = 0$$

$$\slashed{D}\eta = -\tfrac{1}{2}\left(\tfrac{1}{6}R + M\right)\varepsilon + \tfrac{1}{16}[D^\mu T]\gamma_\mu\varepsilon \qquad \left(= \tfrac{1}{4}\slashed{D}^2\varepsilon\right) \qquad (2.3)$$

Here R denotes the scalar curvature (4.62) of the background metric.

The Equation (2.3) are called *generalized conformal Killing spinor equations*, and the spinor ε is called *generalized conformal Killing spinor*.

The generalized conformal Killing spinor equations transform covariantly with respect to local Weyl transformation

$$g_{\mu\nu} \mapsto e^{2\Omega}g_{\mu\nu} \qquad (2.4)$$

with the weights

$$\varepsilon \mapsto e^{\frac{1}{2}\Omega}\varepsilon, \qquad M \mapsto e^{-2\Omega}M, \qquad T \mapsto e^{-\Omega}T \qquad (2.5)$$

Therefore the solutions can be classified by their conformal class.

The generalized conformal Killing spinor equations, similarly to the conformal Killing equations, can be rewritten as the generalized parallel transport equations on the section of doubled spinorial bundle

$$\mathcal{D}_\mu \begin{pmatrix} \varepsilon^i \\ \eta^i \end{pmatrix} = 0 \qquad (2.6)$$

for certain $\mathcal{D}_\mu$. This representation could be useful to classify the solutions.

The solution to the generalized conformal Killing equations is particularly simple for conformally flat background with vanishing auxiliary field $T_{\mu\nu a}$, flat R-symmetry gauge connection and vanishing auxiliary scalar M. In the flat $\mathbb{R}^4$ coordinates x^μ, the solution is simply

$$\varepsilon^i(x) = \hat{\varepsilon}^i + \slashed{x}\hat{\eta}^i \qquad (2.7)$$

where ε^i and η^i are arbitrary constant spinor parameters associated with translational and special conformal supersymmetry respectively. The maximal dimension of the space of solutions to the parallel transport equation in a bundle is the rank of this bundle. We see that the conformally flat background with flat R-symmetry connection and vanishing $T_{\mu\nu a}$ is maximally supersymmetric. The 16 sections are generated by 8 components of $\hat{\varepsilon}$ and 8 components of $\hat{\eta}$.

It would be interesting to find the complete classification of the solutions to the generalized conformal Killing equation with various amounts of supersymmetry. In this note we will focus on particular backgrounds interesting for the localization of gauge theories.

2.2 Vector Multiplet

The 4d $\mathcal{N} = 2$ vector multiplet (A_m, λ^i, Y^{ij}) includes the gauge field A_μ and two real scalar fields Φ_a combined into the reduction of 6d gauge field $(A_m) = (A_\mu, \Phi_a)$, the $SU(2)_R$-doublet of gaugino fermions λ^i, and the $SU(2)_R$-triplet of auxiliary fields represented by the matrix Y^{ij} symmetric in (ij). The gaugino λ^i is the reduction of the $SU(2)_R$-doublet of 6d Weyl spinors of chirality $+1$ for γ_*^{6d}. The spinor fields from the $SU(2)_R$-doublet enter into the Lagrangian and supersymmetry variation holomorphically, their complex conjugates never appear in the Euclidean formulation of the theory.

The supersymmetry variation for the vector multiplet is

$$\delta A_m = \tfrac{1}{2}\lambda^i \gamma_m \varepsilon_i$$
$$\delta \lambda^i = -\tfrac{1}{4} F_{mn} \gamma^{mn} \varepsilon^i + Y^i{}_j \varepsilon^j + \Phi_a \gamma^a \eta + \tfrac{1}{8} T_{\mu\nu a} \Phi^a \gamma^{\mu\nu} \varepsilon \qquad (2.8)$$
$$\delta Y^{ij} = -\tfrac{1}{2}\left(\varepsilon^{(i} \slashed{D} \lambda^{j)} \right)$$

where there are two extra two terms for $\delta\lambda^i$ compared to the standard translational supersymmetry. In our conventions the supersymmetry parameters ε^i, η^i are bosonic and $\delta_{\varepsilon,\eta}$ is fermionic, for a field ϕ the field $\delta_{\varepsilon,\eta}\phi$ has opposite statistics of ϕ.

If ε and $\eta = \tfrac{1}{4}\slashed{D}\varepsilon$ solve the generalized conformal Killing spinor equation (2.3), the supersymmetry transformations (2.8) closes off-shell:

$$\delta_{\varepsilon,\eta}^2 A_\mu = \tfrac{1}{4}(\varepsilon\gamma^\nu \varepsilon) F_{\nu\mu} + \tfrac{1}{4}[(\varepsilon\gamma^a \varepsilon)\Phi_a, D_\mu]$$
$$\delta_{\varepsilon,\eta}^2 \Phi_a = \tfrac{1}{4}[(\varepsilon\gamma^m \varepsilon) D_m, \Phi_a] - \tfrac{1}{2}(\eta\gamma_{ab}\varepsilon)\Phi^b + \tfrac{1}{2}(\eta\varepsilon)\Phi_a$$
$$\delta_{\varepsilon,\eta}^2 \lambda^i = \tfrac{1}{4}\left((\varepsilon\gamma^m \varepsilon) D_m \lambda^i + \tfrac{1}{4} D_\mu(\varepsilon\gamma_\nu \varepsilon)\gamma^{\mu\nu}\lambda^i \right) - \tfrac{1}{8}(\eta\gamma_{ab}\varepsilon)\gamma^{ab}\lambda^i \qquad (2.9)$$
$$\qquad\qquad + \tfrac{3}{4}(\eta\varepsilon)\lambda^i + (\eta^{(i}\varepsilon^{j)})\lambda_j$$
$$\delta_{\varepsilon,\eta}^2 Y^{ij} = \tfrac{1}{4}[(\varepsilon\gamma^m \varepsilon) D_m Y^{ij}] + (\eta\varepsilon)Y^{ij} + (\eta^{(k}\varepsilon^{i)})Y^j{}_k + (\eta^{(k}\varepsilon^{j)})Y^i{}_k$$

Table 3 The symmetry
action of $\delta^2_{\varepsilon,\eta}$

$\delta^2_{\varepsilon,\eta}$ Acts by	Parameter
$\mathcal{L}_v$	$v^m = \frac{1}{4}(\varepsilon\gamma^m\varepsilon)$
$SO(2)_{\tilde{R}}$	$\tilde{R}^{ab} = -\frac{1}{2}(\eta\gamma^{ab}\varepsilon)$
$SU(2)_R$	$R^{ij} = (\eta^{(i}\varepsilon^{j)})\lambda_j$
Dilatation	$(\eta\varepsilon)$

The variation $\delta^2_{\varepsilon,\eta}$ contains the Lie derivative action by the 6d reduced vector field

$$v^m = \tfrac{1}{4}(\varepsilon\gamma^m\varepsilon) \tag{2.10}$$

The scalar components $m \equiv a$ generate the gauge transformation by $v^a\Phi_a$ (Table 3).

The Lagrangian of 4d $\mathcal{N} = 2$ vector multiplet coupled to the Weyl gravity multiplet can be found in [37–39], or [40], p. 433

$$S = -\frac{1}{g_{YM}^2}\int\sqrt{g}d^4x\,\mathrm{tr}\left(\tfrac{1}{2}F_{mn}F^{mn} + \lambda^i\gamma^m D_m\lambda_i + \left(\tfrac{1}{6}R + M\right)\Phi_a\Phi^a - 2Y_{ij}Y^{ij} + \right.$$
$$\left. - F^{\mu\nu}T_{\mu\nu a}\Phi^a + \tfrac{1}{4}T_{\mu\nu a}T^{\mu\nu b}\Phi^a\Phi_b\right) \tag{2.11}$$

Provided ε and $\eta = \tfrac{1}{4}\not{D}\varepsilon$ satisfy (2.1) the action S is invariant under $\delta_{\varepsilon,\eta}$

$$\delta_{\varepsilon,\eta}S = 0. \tag{2.12}$$

3 Generalized Conformal Killing Spinor

Presently the complete classification of the solutions to the generalized conformal Killing spinor Equation (2.1) is not available. We list some known examples. In all these examples the $U(1)_{\tilde{R}}$ connection is set to zero, the square of the supersymmetry transformation δ^2_ε generates isometry transformation and possibly $SU(2)_R$ transformation but without dilatation and $U(1)_{\tilde{R}}$ transformation.

3.1 Topologically Twisted Theories

One simple class of solutions which exists on any smooth 4-manifold is the Donaldson-Witten topological twist [8]. One sets R-symmetry $SU(2)_R$ connection to compensate right component of the $Spin(4) = SU(2)_L \times SU(2)_R$ spin-connection. In the twisted theory the 8 components of the 4d $\mathcal{N} = 2$ spinor generators transform as a one-form, self-dual two-form and scalar. The scalar component yields the scalar

supersymmetry charge defined on any smooth 4-manifold. The theory localizes to
the instanton configurations $F_A^+ = 0$.

3.2 Omega Background

Another example is the equivariant twist of the topologically twisted theory on any
manifold with $U(1)$ isometry. To construct such theory, one uses a combination of the
scalar supersymmetry of the topologically twisted theory and the one-form supercharge contracted with the vector field that generates $U(1)$ isometry. Localization
of such theory on $\mathbb{R}^4$ counts equivariant instantons and gives Nekrasov partition
function [9–12, 46], with two equivariant parameters ϵ_1, ϵ_2, each associated to the
rotation of the $\mathbb{R}^2$ planes in the decomposition $\mathbb{R}^4_{\epsilon_1, \epsilon_2} = \mathbb{R}^2_{\epsilon_1} \oplus \mathbb{R}^2_{\epsilon_2}$. For a review of
instanton counting see contribution [V:12] of this volume.

3.3 Conformal Killing Spinor

Another example is conformally flat and $SU(2)_R$-flat metric with $T_{\mu\nu a} = 0$ and
conformal Killing spinor. A spinor of this type has been used to localize the physical
$\mathcal{N} = 2$ gauge theory on S^4 [23]. The isometry vector field has two fixed points: the
north and the south poles of S^4. In the neighborhood of the north pole the theory
is locally isomorphic to the theory in the Omega-background with parameters $\varepsilon_1 =
\varepsilon_2 = r^{-1}$ where r is the radius of S^4, counting equivariant instantons $F_A^+ = 0$. In
the neighborhood of the south pole the theory is conjugate to the theory in Omega-
background, and it counts equivariant anti-instantons $F_A^- = 0$. The complete partition
function on S^4 is the fusion of the Nekrasov partition function and its conjugate:

$$Z_{S^4} = \int [d\mathbf{a}] \, |Z_{\mathbb{R}^4; r^{-1}, r^{-1}}(\mathrm{i}\mathbf{a})|^2 \tag{3.1}$$

where $Z_{\mathbb{R}^4; \epsilon_1, \epsilon_2}(\mathbf{a})$ is the complete partition function in Omega background including
the classical and perturbative factors, and $\mathbf{a}$ is the gauge Lie algebra equivariant argument of $Z_{\mathbb{R}^4; \epsilon_1, \epsilon_2}$ that physically is interpreted as the electrical type special coordinate
on the Coulomb moduli space of the $\mathcal{N} = 2$ theory or boundary conditions at infinity
of $\mathbb{R}^4_{\epsilon_1; \epsilon_2}$ for the scalar field Φ_a of the vector multiplet. In the formula (3.1) we omit
from the arguments of the partition function the parameters of the Lagrangian.

More generally, the cases of Ω-background and conformal Killing spinor could
be viewed as the specialization of local T^2-bundle geometry.

3.4 Local T^2-Bundles

Consider a manifold X_4 endowed with the metric structure of the warped product $X_4 = T^2_{w_1, w_2} \tilde{\times} \Sigma_2$ where Σ_2 is a Riemann surface, possibly with boundaries, and $T^2_{w_1, w_2}$ is a flat 2-torus with basis cycles of length $(2\pi w_1, 2\pi w_2)$. Here (w_1, w_2) are locally arbitrary functions on Σ_2. This geometry generalizes the Omega background on $\mathbb{R}^4$ and the standard conformal Killing spinor geometry on S^4. The ellipsoid solution [32] is a special example of X_4. Another case was studied in [47].

We denote the coordinates along the two circles on T^2 by (ϕ_1, ϕ_2). We pick two real parameters (ϵ_1, ϵ_2) with the aim to get $\delta^2_{\epsilon, \eta}$ action by the vector field

$$v = \epsilon_1 \partial_{\phi_1} + \epsilon_2 \partial_{\phi_2} \tag{3.2}$$

We assume that ϵ_i are such that $(\epsilon_1 w_1)^2 + (\epsilon_2 w_2)^2 \leq 1$ everywhere on Σ. For any generic functions (w_1, w_2) on Σ we can always find local coordinates (θ, ρ) such that

$$\begin{aligned} \cot \theta &:= \frac{\epsilon_1 w_1}{\epsilon_2 w_2} \\ \sin^2 \rho &:= (\epsilon_1 w_1)^2 + (\epsilon_2 w_2)^2 \end{aligned} \quad \Leftrightarrow \quad \begin{aligned} w_1 &= \epsilon_1^{-1} \sin \rho \cos \theta \\ w_2 &= \epsilon_2^{-1} \sin \rho \sin \theta \end{aligned} \tag{3.3}$$

After we have fixed special coordinates (θ, ρ) on Σ, the metric components $g_{\mu\nu}(\rho, \sigma)$ are parametrized by three arbitrary functions $g_{\theta\theta}(\theta, \rho)$, $g_{\theta\rho}(\theta, \rho)$ and $g_{\rho\rho}(\theta, \rho)$.

Next we choose the frame on X^4 of the form

$$\begin{aligned} e^1 &= w_1(\theta, \rho) d\phi_1 & e^3 &= e^3_\theta(\theta, \rho) d\theta + e^3_\rho(\theta, \rho) d\rho \\ e^2 &= w_2(\theta, \rho) d\phi_2 & e^4 &= e^4_\rho(\theta, \rho) d\rho \end{aligned} \tag{3.4}$$

Three functions $e^3_\theta(\theta, \rho)$, $e^3_\rho(\theta, \rho)$, $e^4_\rho(\theta, \rho)$ generically parametrize 2d metric by the relations $g_{\theta\theta} = e^3_\theta e^3_\theta$, $g_{\theta\rho} = e^3_\theta e^3_\rho$ and $g_{\rho\rho} = e^3_\rho e^3_\rho + e^4_\rho e^4_\rho$. It is convenient to denote

$$e^3_\theta(\theta, \rho) \equiv \sin \rho f_1(\theta, \rho) \quad e^3_\rho(\theta, \rho) \equiv f_3(\theta, \rho) \quad e^4_\rho(\theta, \rho) \equiv f_2(\theta, \rho) \tag{3.5}$$

and present solution for the background fields $T_{\mu\nu}$ and $V^i_{\mu j}$ in terms of f_1, f_2, f_3.[2]

---

[2]In these notations the solution can be easily specialized to the Hama-Hosomichi ellipsoid [32] metrically defined by the equation in $\mathbb{R}^5$ with the standard metric

$$r_1^{-2}(X_1^2 + X_2^2) + r_2^{-2}(X_3^2 + X_4^2) + r^{-2}X_5^2 = 1$$

by taking

$$X_1 + \iota X_2 = r_1 \sin \rho \cos \theta e^{\iota \phi_1}, \quad X_3 + \iota X_4 = r_2 \sin \rho \sin \theta e^{\iota \phi_2}, \quad X_5 = r \cos \rho$$

In the γ matrix basis (4.69) we choose the $SU(2)_R$-doublet spinor $(\varepsilon^1, \varepsilon^2)$ in the frame (3.4) to be given by

$$
\varepsilon^1 = e^{\frac{1}{2}(i\phi_1 + i\phi_2)} \left(e^{-i\frac{\theta}{2}} \sin \tfrac{\rho}{2}, -e^{i\frac{\theta}{2}} \sin \tfrac{\rho}{2}, ie^{-i\frac{\theta}{2}} \cos \tfrac{\rho}{2}, -ie^{i\frac{\theta}{2}} \cos \tfrac{\rho}{2} \right)
$$
$$
\varepsilon^2 = e^{-\frac{1}{2}(i\phi_1 + i\phi_2)} \left(e^{-i\frac{\theta}{2}} \sin \tfrac{\rho}{2}, e^{i\frac{\theta}{2}} \sin \tfrac{\rho}{2}, -ie^{-i\frac{\theta}{2}} \cos \tfrac{\rho}{2}, -ie^{i\frac{\theta}{2}} \cos \tfrac{\rho}{2} \right)
$$

(3.6)

Notice that this spinor satisfies the standard reality condition $\varepsilon^2 = c\bar{\varepsilon}^1$ where bar is the complex conjugation and c is the Majorana bilinear matrix (4.69), and that this spinor is the transformation to the $(\phi_1, \phi_2, \rho, \theta)$ frame of the standard conformal Killing spinor $\varepsilon(x) = \varepsilon_s + x^\mu \gamma_\mu \varepsilon_c$ that was used in [23] for S^4, where x^μ are stereographic projection coordinates on S^4.

For the spinor (3.6) we find the bilinear vector field[3]

$$
v^m = \tfrac{1}{4}\varepsilon^i \gamma^m \varepsilon_i = \tfrac{1}{2}\varepsilon^1 \gamma^m \varepsilon^2 : \qquad
\begin{aligned}
v^\mu|_{\mu \in (\phi_1, \phi_2, \theta, \rho)} &= (\epsilon_1, \epsilon_2, 0, 0) \\
v^a|_{a \in (5,6)} &= (-\cos \rho, -i)
\end{aligned}
$$

(3.7)

This vector field is the natural isometry of the T^2-bundle X_4.

Under the ansatz (3.6), the equations on the background fields $V_\mu{}^i{}_j$, $T_{\mu\nu}$, M are inhomogeneous ordinary linear equations, which can be directly solved. Though the system is overdetermined, as there are $32 + 8$ linear equations from $\delta\psi^i_\mu$ and from $\delta\chi^i$ on $12 + 6 + 1 = 19$ components for V, T, M, we find that solution always exists for any T^2-bundle. Moreover, the solution is not unique; the space of solutions forms a vector bundle of rank three. This is completely analogous to the case of the Hama-Hosomichi ellipsoid [32].

Below $T_{\hat{\mu}\hat{\nu}}$ denote the components of T in the frame (3.4) $e^{\hat{\mu}}_\mu$, so that $T_{\mu\nu} = T_{\hat{\mu}\hat{\nu}} e^{\hat{\mu}}_\mu e^{\hat{\nu}}_\nu$. The V denotes the connection one-form of the $SU(2)_R$ gauge field $D = d + V$. The components of the T do not depend on (ϕ_1, ϕ_2), and the components of

---

(Footnote 2 continued)
and

$$
f_1(\theta, \rho) = f_{\mathrm{HH}}(\theta) = \sqrt{r_1^2 \sin^2 \theta + r_2^2 \cos^2 \theta}
$$
$$
f_2(\theta, \rho) = g_{\mathrm{HH}}(\theta, \rho) = \sqrt{r^2 \sin^2 \rho + r_1^2 r_2^2 f_1(\theta)^{-2} \cos^2 \rho}
$$
$$
f_3(\theta, \rho) = h_{\mathrm{HH}}(\theta, \rho) = (-r_1^2 + r_2^2) f_1(\theta)^{-1} \cos \theta \sin \theta \cos \rho
$$

In the case of round sphere S^4 we set

$$
f_1(\theta, \rho) = r \qquad f_2(\theta, \rho) = r \qquad f_3(\theta, \rho) = 0.
$$

[3]In the Eq. (3.7) ε^i denotes the $+1$ chiral $6d$ spinors and γ^m for the $6d$ gamma-matrices, while in the Eq. (3.6) the components of the spinor ε^i are presented with respect to the 4d Clifford algebra representation (4.69).

$V = i\sigma_I V^I$, where σ_I are the standard Pauli matrices, depend on (ϕ_1, ϕ_2) as

$$\begin{pmatrix} V^1 \\ V^2 \end{pmatrix} = \begin{pmatrix} \cos(\phi_1 + \phi_2) & \sin(\phi_1 + \phi_2) \\ -\sin(\phi_1 + \phi_2) & \cos(\phi_1 + \phi_2) \end{pmatrix} \begin{pmatrix} \hat{V}^1 \\ \hat{V}^2 \end{pmatrix}, \qquad V^3 = \hat{V}^3 \qquad (3.8)$$

where $\hat{V}$ is constant in (ϕ_1, ϕ_2).

The particular solution is

$$T_{12} = 0 \qquad\qquad T_{34} = 0$$

$$T_{13} = 2\sin\theta \left(\frac{1}{f_1} - \frac{1}{f_2}\right) \qquad T_{23} = -2\cos\theta \left(\frac{1}{f_1} - \frac{1}{f_2}\right) \qquad (3.9)$$

$$T_{14} = -\frac{2\sin\theta f_3}{f_1 f_2} \qquad\qquad T_{24} = \frac{2\cos\theta f_3}{f_1 f_2}$$

for T components and

$$\hat{V} = \left(-\frac{1}{4\epsilon_1}\sin 2\theta \cos\rho \left(\frac{1}{f_1} - \frac{1}{f_2}\right) + \frac{\sin^2\theta f_3}{2\epsilon_1 f_1 f_2}\right) i\sigma_2 \, d\phi_1 +$$

$$\left(\frac{1}{4\epsilon_2}\sin 2\theta \cos\rho \left(\frac{1}{f_1} - \frac{1}{f_2}\right) + \frac{\cos^2\theta f_3}{2\epsilon_2 f_1 f_2}\right) i\sigma_2 \, d\phi_2 +$$

$$\left(-\frac{1}{2} + \frac{\sin^2\theta}{2\epsilon_1 f_1} + \frac{\cos^2\theta r_1}{2\epsilon_1 f_2} + \frac{\sin 2\theta \cos\rho f_3}{4\epsilon_1 f_1 f_2}\right) i\sigma_3 \, d\phi_1 +$$

$$\left(-\frac{1}{2} + \frac{\cos^2\theta}{2\epsilon_2 f_1} + \frac{\sin^2\theta}{2\epsilon_2 f_2} - \frac{\sin 2\theta \cos\rho f_3}{4\epsilon_2 f_1 f_2}\right) i\sigma_3 \, d\phi_2 +$$

$$\left(\frac{(f_1 - f_2)\cos\rho + \sin\rho \partial_\rho f_1 - \partial_\theta f_3}{2 f_2}\right) i\sigma_1 \, d\theta$$

$$+ \left(\frac{f_3 \left(\sin\rho \, \partial_\rho f_1 + f_1 \cos\rho - \partial_\theta f_3\right) - f_2 \partial_\theta f_2}{2 f_1 f_2 \sin\rho}\right) i\sigma_1 \, d\rho$$

$$(3.10)$$

are the V components. This particular solution can be deformed by three-parametric family

$$\delta T_{12} = c_3 \qquad\qquad\qquad \delta T_{34} = c_3 \cos\rho$$

$$\delta T_{13} = -c_1 \cos\rho \sin\theta - c_2 \cos\theta \qquad \delta T_{23} = c_1 \cos\rho \cos\theta - c_2 \sin\theta \qquad (3.11)$$

$$\delta T_{14} = c_1 \cos\theta - c_2 \cos\rho \sin\theta \qquad \delta T_{24} = c_1 \sin\theta + c_2 \cos\theta \cos\rho$$

together with

$$\delta\hat{V} = \left(-\frac{1}{4}c_1 \sin^2 \rho f_1 d\theta - \frac{1}{4}\sin \rho c_1 f_3 d\rho - \frac{1}{4}c_2 \sin \rho f_2 d\rho\right) i\sigma_1 +$$
$$\left(-\frac{1}{8\epsilon_1}c_1 \sin 2\theta \sin^2 \rho d\phi_1 + \frac{1}{8\epsilon_2}c_2 \sin 2\theta \sin^2 \rho d\phi_2 - \frac{1}{4}c_3 \sin \rho f_2 d\rho\right) i\sigma_2 +$$
$$\left(-\frac{1}{8\epsilon_1}c_2 \sin 2\theta \sin^2 \rho d\phi_1 + \frac{1}{8\epsilon_2}c_2 \sin 2\theta \sin^2 \rho d\phi_2 + \frac{1}{4}c_3 \sin^2 \rho f_1 d\theta\right.$$
$$\left. + \frac{1}{4}c_3 \sin \rho f_3 d\rho\right) i\sigma_3 \tag{3.12}$$

where c_1, c_2, c_3 are arbitrary functions on Σ. The background auxiliary scalar M is

$$-\frac{1}{2}\left(\frac{1}{6}R + M\right) = \frac{1}{4f_1^2} - \frac{1}{4f_2^2} - \frac{1}{f_1 f_2} + \frac{f_3^3}{4f_1^2 f_2^2}$$
$$+ c_1\left(-\frac{\cos\rho}{4f_1} + \frac{3\cos\rho}{4f_2} - \frac{\cot 2\theta f_3}{2f_1 f_2} + \frac{\sin\rho\partial_\rho f_1}{4f_1 f_2} - \frac{\partial_\theta f_3}{4f_1 f_2}\right) + \frac{\sin\rho}{4f_2}\partial_\rho c_1 - \frac{f_3}{4f_1 f_2}\partial_\theta c_1$$
$$+ c_2\left(-\frac{\cot 2\theta}{2f_1} + \frac{\cos\rho f_3}{4f_1 f_2} - \frac{\partial_\theta f_2}{4f_1 f_2}\right) - \frac{1}{4f_1}\partial_\theta c_2 - \frac{1}{16}\sin^2\rho\left(c_1^2 + c_2^2 + c_3^2\right)$$
$$\tag{3.13}$$

3.5 Four-Sphere

A topological four-sphere $X_4 = S^4_{\epsilon_1, \epsilon_2}$ with T^2 invariant metric can be presented as a local $T^2_{w_1, w_2}$ bundle fibered over a two-dimensional digon Σ_2. One of the cycles of $T^2_{w_1, w_2}$ collapses at one edge of the digon, and the other cycle collapses at the other edge:

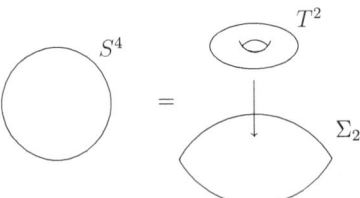

The coordinates (θ, ρ) on the base Σ_2 are in the range $(\theta, \rho) \in [0, \frac{\pi}{2}] \times [0, \pi]$. The w_1 cycle collapses at $\theta = \frac{\pi}{2}$ and the w_2 cycle collapses at $\theta = 0$. Both circles collapse in the corners of the digon. The corner $\rho = 0$ will be called the north pole, and the corner $\rho = \pi$ will be called the south pole. The metric on $S^4_{\epsilon_1, \epsilon_2}$ is smooth at the cusps and the edges of the digon Σ_2 if the functions $f_i(\theta, \rho)$ satisfy asymptotically

$$f_1(\theta, \rho)|_{\theta=0} = \epsilon_2^{-1}, \qquad f_1(\theta, \rho)|_{\theta=\frac{\pi}{2}} = \epsilon_1^{-1}$$

$$f_1(\theta, \rho)|_{\rho=0,\pi} = (\epsilon_1^{-2} \sin^2 \theta + \epsilon_2^{-2} \cos^2 \theta)^{\frac{1}{2}}$$

$$f_3(\theta, \rho)|_{\rho=0,\pi} = \pm(\epsilon_2^{-2} - \epsilon_1^{-2}) f_1(\theta, \rho)^{-1} \cos \theta \sin \theta \qquad (3.14)$$

$$f_2(\theta, \rho)|_{\rho=0,\pi} = (\epsilon_1 \epsilon_2 f_1(\theta, \rho))^{-1}$$

The metric along Σ_2 is arbitrary in the interior. In particular, taking $f_2(\theta, \rho)$ very large in the interior, it is possible to stretch $S^4_{\epsilon_1, \epsilon_2}$ to a very long cylinder with two hemispherical caps attached at the ends. Localization on this geometry presumably is related to the convolution of the ground state topological wave functions with its conjugate by cutting the $S^4_{\epsilon_1, \epsilon_2}$ in the middle at $\rho = \frac{\pi}{2}$, as in the AGT correspondence [33] with Liouville theory and quantum Teichmuller theory [48], [V:2] and Nekrasov-Witten construction [49].

The background fields (3.9), (3.10) and the spinor (3.6) with generic smooth functional parameters c_1, c_2, c_3 (3.11), (3.12) is a supersymmetric background only in the interior of Σ_2. The coordinates are singular at the north and the south poles $\rho = 0$ and $\rho = \pi$. We need to ensure that the spinor ϵ and the background fields V and T are smooth in a proper coordinate system around the poles. At the north pole $\rho = 0$ we choose approximately Cartesian coordinates

$$x_1 = 2\epsilon_1^{-1} \tan \frac{\rho}{2} \cos \theta \cos \phi_1 \qquad\qquad x_2 = 2\epsilon_1^{-1} \tan \frac{\rho}{2} \cos \theta \sin \phi_1$$

$$x_3 = 2\epsilon_2^{-1} \tan \frac{\rho}{2} \sin \theta \cos \phi_2 \qquad\qquad x_4 = 2\epsilon_2^{-1} \tan \frac{\rho}{2} \sin \theta \sin \phi_2 \qquad (3.15)$$

with the standard frame $e_\mu^{\hat{\mu}} = \delta_\mu^{\hat{\mu}}$. In the x-frame the spinor (3.6) becomes

$$^x\varepsilon = e^{-\frac{\pi}{4}\gamma_{12}} e^{-\frac{\phi_1}{2}\gamma_{12} - \frac{\phi_2}{2}\gamma_{34}} e^{-\frac{\pi}{4}\gamma_{24}} e^{\frac{\beta}{2}\gamma_{34}} \varepsilon, \quad \text{with} \quad \sin \beta = \frac{\epsilon_1^{-1} \sin \theta}{\epsilon_1^{-2} \sin^2 \theta + \epsilon_2^{-2} \cos^2 \theta} \qquad (3.16)$$

The spinor $^x\varepsilon$ is not smooth in the x-frame but is $SU(2)_R$ gauge equivalent to the conformal class of the standard smooth spinor in the Omega-background [9] (up to the Weyl transformation $^x\varepsilon_\Omega \to {^x}\varepsilon_\Omega \cos \frac{\rho}{2}$):

$$^x\varepsilon_\Omega := (\hat{e}_s - \tfrac{1}{2}\Omega_{\mu\nu} x^\mu \gamma^\nu \hat{e}_s) \cos \frac{\rho}{2} \qquad (3.17)$$

where non-zero components of Ω are $\Omega_{12} = -\Omega_{12} = \epsilon_1$ and $\Omega_{34} = -\Omega_{43} = \epsilon_2$. Namely, for $\hat{e}_s = (1 + i)(0, 0, 1, 0)$ we find

$$^x\varepsilon_\Omega^1 = (1 + i)(\cos \frac{\rho}{2})(-\tan\frac{\rho}{2} \sin \theta e^{i\phi_2}, -\tan\frac{\rho}{2} \cos \theta e^{i\phi_1}, 1, 0)$$

$$^x\varepsilon^1 = (1 + i)(\cos \frac{\rho}{2})(-\tan\frac{\rho}{2} \sin \frac{\beta+\theta}{2} e^{i\phi_2}, -\tan\frac{\rho}{2} \cos \frac{\beta+\theta}{2} e^{i\phi_1}, \cos \frac{\beta-\theta}{2},$$
$$-\sin\frac{\beta-\theta}{2} e^{i(\phi_1+\phi_2)})$$

with ε^2 found by Majorana conjugation $\varepsilon^2 = c\bar{\varepsilon}^1$.

The $SU(2)_R$ gauge transformation relating spinor $^x\varepsilon$ (3.6) and the Omega-background spinor ε_Ω near the north pole is

$$\varepsilon_\Omega^i = U_j^i \varepsilon^j, \qquad U = \begin{pmatrix} \cos\frac{\theta-\beta}{2} & ie^{i(\phi_1+\phi_2)}\sin\frac{\beta-\theta}{2} \\ ie^{-i(\phi_1+\phi_2)}\sin\frac{\beta-\theta}{2} & \cos\frac{\theta-\beta}{2} \end{pmatrix} \tag{3.18}$$

Requiring that $SU(2)_R$ gauge field $^U V = U dU^{-1} + U V U^{-1}$ (3.8), (3.10), (3.12) is smooth at the origin, and that components $T_{\mu\nu}$ are well defined in the x-frame, we find the parameters

$$c_1 = \left(\frac{1}{f_1} - \frac{1}{f_2}\right)\varphi(\rho), \quad c_2 = -\frac{f_3}{f_1 f_2}\varphi(\rho), \quad c_3 = 0 \tag{3.19}$$

where $\varphi(\rho)$ is any smooth function such that $\varphi(\rho)_{\rho=0} = 1 + O(\rho^2)$ and $\varphi(\rho)_{\rho=\pi} = -1 + O(\rho^2)$. Then the gauge field $^U V$ is smooth everywhere and $T_{\rho=0} = T^-$, $T_{\rho=\pi} = T^+$.

In our conventions the spinor ε is of positive chirality at the north pole (transforms under self-dual spacial rotations) and negative-chirality at the south pole (transforms under anti-self-dual rotation). In the zeroth order approximation the theory around the north pole is topological Donaldson-Witten theory that localizes to configurations $F^+ = 0$, and the theory around the south pole is conjugated and localizes to configurations $F^- = 0$. In the first order approximation the theory around poles is equivalent to the theory in the Omega-background, and localizes respectively to the equivariant instantons $F^+ = 0$ at the north pole and equivariant anti-instantons $F^- = 0$ around the south pole.

With the choice (3.19) at $\rho = 0$ we find that non-zero components of $T = T^-$ are

$$T_{12} = -T_{34} = \epsilon_1 - \epsilon_2 \tag{3.20}$$

If the geometry in the neighborhood of the north pole is approximated by the embedded ellipsoid in $\mathbb{R}^5$ with radian (r_1, r_1, r_2, r_2, r) for $r_1 = \epsilon_1^{-1}, r_2 = \epsilon_2^{-1}$ as in [32] then the curvature of the $SU(2)_R$ background field at $\rho = 0$ is particularly simple

$$\begin{aligned}
F_V = {}& \left(\frac{-2r^2 + r_1^2 + r_2^2}{4r_1^2 r_2^2}(dx_1 \wedge dx_3 - dx_2 \wedge dx_4)\right) i\sigma_1 \\
& + \left(-\frac{-2r^2 + r_1^2 + r_2^2}{4r_1^2 r_2^2}(dx_1 \wedge dx_4 + dx_2 \wedge dx_3)\right) i\sigma_2 \\
& + \left(\frac{r_1^2 - r^2}{2r_1^4}dx_1 \wedge dx_2 + \frac{r_2^2 - r^2}{2r_2^4}dx_3 \wedge dx_4\right) i\sigma_3
\end{aligned} \tag{3.21}$$

One can compare F_V with the metric curvature at the north pole and notice the difference: the $SU(2)_R$ background field differs from the usual topologically twisted theory. The non-zero metric curvature components in the x-frame at the north pole are

$$R^{12}{}_{12} = \frac{r^2}{r_1^2}, \qquad R^{34}{}_{34} = \frac{r^2}{r_2^2}, \qquad R^{24}{}_{24} = R^{13}{}_{13} = R^{14}{}_{14} = R^{23}{}_{23} = \frac{r^2}{r_1^2 r_2^2}$$

$$(3.22)$$

3.6 Superconformal Index

For this geometry the base is the product of an interval and the circle $\Sigma_2 = I_{(\theta)} \times S^1_{(\rho)}$. At the ends of the interval I the two circles of T collapse. The slice of $X^4_{\epsilon_1,\epsilon_2}$ at fixed ρ is topologically an $S^3_{\epsilon_1,\epsilon_2}$, and then $X_4 = S^3_{\epsilon_1,\epsilon_2} \times S^1$. A suitable $SU(2)_R$ background field ensures existence of unbroken supercharge. The partition function on $S^3_{\epsilon_1,\epsilon_2} \times S^1$ computes the superconformal index [50–54] and [V:10], Sect. 4.1 and [V:9] in this volume.

3.7 Other Geometries

It would be interesting to study more general four-manifolds with the structure of local T^2 bundle such as $S^2 \times S^2$ or $T^2 \times \Sigma_2$ where Σ_2 is a Riemann surface.

4 Localization

Often a supersymmetric quantum field theory with a particular choice of the super-charge can be interpreted as infinite-dimensional version of the Cartan model for $\mathcal{G}$-equivariant cohomology on the space of fields of the theory, see e.g. [8, 55]. The supercharge Q plays the role of the equivariant differential. The path integral is interpreted as the infinite-dimensional version of Mathai-Quillen form for the Thom class of the BPS equations bundle over the space of fields [56, 57]. For example, in the Donaldson-Witten topological gauge theory [8], the space of fields is the infinite-dimensional affine space of connections $\mathcal{A}$ in a given principal G-bundle on a four-manifold X_4 for a compact Lie group G, the group $\mathcal{G}$ of the equivariant action is the infinite-dimensional group of gauge transformations, and the fibers of the equation bundle over $\mathcal{A}$ is the space of self-dual adjoint valued two-forms. The Mathai-Quillen form for the Thom class, with a choice of section F_A^+, localizes to the zeroes of the section: instanton configurations. The construction is equivariant with respect to the $\mathcal{G}$ action on $\mathcal{A}$. The path integral over $\mathcal{A}/\mathcal{G}$ reduces to the integration over the instanton moduli space $\mathcal{M}_{\text{inst}} = \{A | F_A^+ = 0\}/\mathcal{G}$.

4.1 Omega Background

See [9–12, 46] and [V:4] in this volume. A conventional 4d $\mathcal{N} = 2$ theory with
Lagrangian formulation is specified by the choice of a compact semi-simple Lie group
G for the gauge group and a representation R of G for the hypermultiplet matter. The
automorphism group of the representation R is the flavor group F. The path integral
of $\mathcal{N} = 2$ theory in Omega background $\mathbb{R}^4_{\epsilon_1,\epsilon_2}$ localizes to the equivariant form on
moduli space of instantons, further integration over moduli space is localized to the
fixed points the equivariant group action. The equivariant group $\mathsf{G} = L \times G \times F$ is
the product of the isometry of the space-time $L = SO(4)$, the gauge group G that acts
on the framing at infinity, and the flavour group F. Let T be the maximal torus of the
equivariant group $\mathsf{T} = T_L \times T_G \times T_F$. The coordinates on the complexified Lie algebra
of T are $(\epsilon, \mathbf{a}, \mathbf{m})$. Physically, the parameters $\mathbf{a}$ are the asymptotics at the space-time
infinity of the scalar field Φ in the gauge vector multiplet, the parameters $\mathbf{m}$ are the
matter fields masses, and the parameters $\epsilon = (\epsilon_1, \epsilon_2)$ are the equivariant space-time
rotation angular momenta, the Ω-background parameteres. In our conventions the
subscript T denotes the dependence on $(\epsilon, \mathbf{a}, \mathbf{m})$.

The partition function Z in the Omega background can be represented as a product
of the classical, perturbative and non-perturbative contributions:

$$Z_\mathsf{T}(\mathbf{q}) = Z_\mathsf{T}^{\text{tree}} Z_\mathsf{T}^{\text{1-loop}} Z_\mathsf{T}^{\text{inst}}(\mathbf{q}) \tag{4.1}$$

Formally,

$$Z_\mathsf{T}(\mathbf{q}) = \sum_\mathbf{k} \mathbf{q}^k \int_{\mathcal{A}/\mathcal{G}_{\text{gauge}}} \mathrm{eu}_\mathsf{T}(\Omega^{2+} \otimes \mathfrak{g}) \; \mathrm{eu}_\mathsf{T}((S^- \ominus S^+) \otimes R) \tag{4.2}$$

where $\mathcal{A}$ is the infinite-dimensional space of G-connections on a principal G-bundle
$E \to M$ with fixed trivialization at infinity, $\mathcal{G}_{\text{gauge}} = \text{Aut}(E)$ is the group of gauge
transformations equal to identity at the space-time infinity, $\Omega^{2+} \otimes \mathfrak{g}$ is the infinite-
dimensional vector bundle over $\mathcal{A}$ with the fiber being the space of self-dual $\mathfrak{g}$-valued
two-forms, $S^\pm \otimes R$ is the infinite-dimensional vector bundle over $\mathcal{A}$ with the fiber
being the space of positive/negative chirality R-valued spinors.

Mathematically, the instanton partition function is

$$Z_\mathsf{T}^{\text{inst}}(\mathbf{q}) = \sum_{\mathbf{k}=0}^{\infty} \mathbf{q}^k \int_{\mathcal{M}_\mathbf{k}} \mathrm{eu}_\mathsf{T}(\mathcal{E}_R). \tag{4.3}$$

Here we are assuming that $G = \times_{i \in I} G_i$ where G_i are simple factors and I denotes
the set of labels for the simple gauge group factors, $\mathbf{q} = \{q_i | i \in I\}$ is the $|I|$-tuple
of the exponentiated complexified gauge coupling constants $q_i = \exp(2\pi i \tau_i)$, $\mathbf{k} =$
$\{k_i | i \in I\}$ is an n-tuple of non-negative integers, k_i is the instanton charge (second

Chern class[4] of G_i-bundle on the space-time $M = \mathbb{R}^4 = \mathbf{S}^4 \backslash \infty = \mathbb{CP}^2 \backslash \mathbb{CP}^1_\infty$ with fixed framing at infinity, the $\mathcal{M}_\mathbf{k} = \times_{i \in I} \mathcal{M}_{G_i, k_i}$ is the instanton moduli space: $\mathcal{M}_{G_i, k_i}$ is moduli space of the anti-self-dual G_i-connections on $\mathbb{R}^4$ with the second Chern class k_i. The integration measure $\mathrm{eu}_\mathsf{T}(\mathcal{E}_R)$ is the T-equivariant Euler class of the matter bundle $\mathcal{E}_R \to \mathcal{M}_\mathbf{k}$ where a fiber of $\mathcal{E}_R$ is the space of the virtual zero modes for the Dirac operator: $\Gamma(S^- \otimes R) \to \Gamma(S^+ \otimes R)$ associated to the hypermultiplet.

Here the classical contribution is

$$Z_\mathsf{T}^{\text{tree}}(\mathbf{q}) = \exp\left(-\frac{1}{2\epsilon_1\epsilon_2} \sum_{i \in I} 2\pi \mathrm{i}\tau_i \langle a_i, a_i \rangle \right) \qquad (4.4)$$

where $\langle \rangle$ is the standard bilinear form on the Lie algebra of G_i normalized such that the long root length squared is 2, and τ is the complexified coupling constant

$$\tau = \frac{4\pi\mathrm{i}}{g_{\text{YM}}^2} + \frac{\theta}{2\pi} \qquad (4.5)$$

Let

$$R = \bigoplus_\ell R_\ell \otimes M_\ell \qquad (4.6)$$

be the decomposition of the matter representation onto the irreducible, with respect to G, components, with the multiplicity spaces $M_\ell \simeq \mathbb{C}^{N_\ell^{\mathrm{f}}}$, on which the masses have the value $m_{\ell,1}, \ldots, m_{\ell, N_\ell^{\mathrm{f}}}$.

The one-loop contribution is expressed in terms of the special function related to Barnes double gamma function

$$G_{\epsilon_1, \epsilon_2}(x) = \mathrm{Reg}[\prod_{n_1, n_2 \geq 0} (x + n_1\epsilon_1 + n_2\epsilon_2)] \qquad (4.7)$$

where $\mathrm{Reg}[]$ denotes regularization of the infinite product with Weierstrass multipliers.

We find the one-loop factors for the theory in the Omega background for vector multiplet and hypermultiplet to be given by

---

[4]For a generic compact simple Lie group G the integer k classifies the topology of G-bundle on S^4 by $\pi_3(G) = \mathbb{Z}$. The instanton number k can be computed as

$$k = \frac{1}{8\pi^2} \int_M \langle F, \wedge F \rangle = -\frac{1}{16\pi^2 h^\vee} \int_M \mathrm{Tr}_{\text{adj}} F \wedge F$$

in the conventions where F is $\mathfrak{g}$-valued two-form, $\langle, \rangle$ is the invariant positive definite bilinear form on $\mathfrak{g}$ induced from the standard bilinear form on $\mathfrak{h}^*$ in which long roots have length squared 2, the Tr_{adj} is the trace in adjoint representation, and $h^\vee$ is the dual Coxeter number for $\mathfrak{g}$. For $G = SU(n)$ the instanton charge k is the second Chern class $k = c_2$.

$$Z_{\mathsf{T}}^{\text{1-loop; vec}}(a; m) = \prod_i \prod_{\alpha \in \Delta_i^+} G_{\epsilon_1, \epsilon_2}(\alpha \cdot a_i) G_{\epsilon_1, \epsilon_2}(\epsilon_1 + \epsilon_2 - \alpha \cdot a_i)$$

$$Z_{\mathsf{T}}^{\text{1-loop; hyper}}(a; m) = \prod_\ell \prod_{f=1}^{N_\ell^f} \prod_{w \in P(R_\ell)} G_{\epsilon_1, \epsilon_2}\left(w \cdot a_i + m_{lf} + \tfrac{1}{2}(\epsilon_1 + \epsilon_2)\right)^{-1} \tag{4.8}$$

Here Δ_i^+ denotes the set of positive roots for the i-th gauge group factor G_i and $P(R_\ell)$ denotes the set of weights for the irreducible representation R_ℓ. These expressions follow from the equivariant index computation by Atiyah-Singer formula for the self-dual complex and the Dirac complex respectively. The Aityah-Singer formula for the equivariant index of complex C evaluated at the group element $g \in \mathsf{G}$

$$\text{ind}(C; g) = \sum_{f \in \text{fixed points}} \frac{\text{tr}_{C_f}(g)}{\det_{T_f}(1 - g)} \tag{4.9}$$

For the self-dual complex

$$\Omega^0 \xrightarrow{d} \Omega^1 \xrightarrow{d} \Omega^{2+} \tag{4.10}$$

on $\mathbb{R}^4 \simeq \mathbb{C}^2_{(z_1, z_2)}$ under the equivariant action $z_1 \to t_1 z_1$, $z_2 \to t_2 z_2$ we find that each two-dimensional weight space of root α and its conjugate $-\alpha$ contributes as

$$\frac{w + \bar{w} + t_1 t_2 w + \bar{t}_1 \bar{t}_2 \bar{w} - (t_1 w + \bar{t}_1 \bar{w} + t_2 w + \bar{t}_2 \bar{w})}{(1 - t_1)(1 - \bar{t}_1)(1 - t_2)(1 - \bar{t}_2)}$$

$$= \bar{w} \frac{1}{(1 - \bar{t}_1)(1 - \bar{t}_2)} + w \frac{1}{(1 - t_1)(1 - t_2)} \tag{4.11}$$

where

$$w = e^{i\alpha \cdot a} \quad t_1 = e^{i\epsilon_1} \quad t_2 = e^{i\epsilon_2} \tag{4.12}$$

Taking $\bar{t}_1 = t_1^{-1}, \bar{t}_2 = t_2^{-1}$ and expanding the index in positive powers of (t_1, t_2) one finds the Chern character of the complex. Converting the Chern character (the sum) into the Euler character (the product) we find (4.8).

For the Dirac complex associated with the hypermultiplet of mass m, the weight space w contributes as

$$\frac{t_1^{\frac{1}{2}} t_2^{\frac{1}{2}}}{(1 - t_1)(1 - t_2)} w\mu \tag{4.13}$$

where $\omega = e^{iw \cdot \alpha}, \mu = e^{im}$. This can be seen from Atiyah-Singer formula for the Dirac complex $S^+ \xrightarrow{D} S^-$ with numerator $t_1^{\frac{1}{2}} t_2^{\frac{1}{2}} + \bar{t}_1^{\frac{1}{2}} \bar{t}_2^{\frac{1}{2}} - \bar{t}_1^{\frac{1}{2}} t_2^{\frac{1}{2}} - t_1^{\frac{1}{2}} \bar{t}_2^{\frac{1}{2}}$ or from the fact that Dirac complex is the twist of Dolbeault complexs twisted by the square root of the canonical bundle. Again, expanding in positive powers of t_1, t_2 and converting the sum to the product we find the equivariant Euler character, or the one-loop determinant (4.8) for the hypermultiplet.

The explicit expression for $Z_T^{inst}(\mathbf{q})$ can be found for example in [22, 46] and in Y. Tachikawa's review [V:4] in this volume.

4.2 Supersymmetric Configurations on $S^4_{\epsilon_1, \epsilon_2}$

The path integral for the partition function of QFT with an action S invariant under a fermionic symmetry $\delta S = 0$ localizes near the *supersymmetric configurations*, which are the field configurations invariant under δ. In other words the supersymmetric configurations are the zeroes of the odd vector field δ in the space of all field configurations [8]. The localization theorem is the infinite-dimensional generalization of the Atiyah-Bott formula [5] for the integration of the equivariantly closed differential forms over a manifold on which a compact Lie group G acts

$$\int_M \alpha = \int_F \frac{i_F \alpha}{e(N_F)} \qquad (4.14)$$

where $F \subset M$ is the fixed point locus of G action on M and $e(N_F)$ is the equivariant Euler class of the normal bundle to F.

From the analysis of the Eq. (2.8), similar to the S^4 [23] and the ellipsoid case [32] we expect that the only smooth field configurations that satisfy $\delta\lambda = 0$ for the topologically trivial gauge bundle is the trivial gauge field, vanishing scalar Φ^5, constant scalar $\Phi^6 = const = \mathring{\Phi}_6$ and a suitable auxiliary field Y^i_j proportional to $\mathring{\Phi}^6$. It is easy to see that such a solution exists. Under the ansatz $F_{mn} = 0$, $\Phi_5 = 0$ the equations turn into an overdetermined algebraic linear system of equations on Φ_6 and Y^i_j, and this system has one-dimensional kernel corresponding to the zero mode of Φ_6. What is more difficult to show is the absence of other solutions, and presumably this can be shown similarly to the analysis in [58].

With the ansatz $F_{mn} = 0$, $\Phi_5 = 0$ and all fermions set to zero, we find the explicit supersymmetric configuration invariant under the $\delta_{\varepsilon, \eta}$ (2.8)

$$Y^i{}_j = \hat{Y}^i{}_j \Phi_6, \quad \hat{Y}^i{}_j = \left(\left(\frac{1}{2f_1} - \frac{\varphi(\rho)\cos\rho}{4f_1} + \frac{\varphi(\rho)\cos\rho}{4f_2}\right)(\sigma_3)^i{}_j\right.$$
$$+ \left(\frac{f_3}{2f_1 f_2} - \frac{\varphi(\rho)\cos\rho f_3}{4f_1 f_2}\right)(e^{\frac{1}{2}(i\phi_1 + i\phi_2)}\sigma_2 e^{-\frac{1}{2}(i\phi_1 + i\phi_2)})^i{}_j\right)$$

$$(4.15)$$

It is straightforward to evaluate the classical action on the supersymmetric configuration (4.15) and find

$$S|_{\text{susy conf}} = -\frac{1}{g_{\text{YM}}^2} \operatorname{tr} \mathring{\Phi}_6^2 \int_{S^4_{\epsilon_1,\epsilon_2}} \sqrt{g} d^4 x \left((\tfrac{1}{6} R + M) + 2 \hat{Y}^i{}_j \hat{Y}^j{}_i - \tfrac{1}{16} T_{\mu\nu} T^{\mu\nu} \right) \quad (4.16)$$

From the explicit solution for the $T_{\mu\nu}$ (3.9), (3.11), (3.19), the $\hat{Y}^i{}_j$ (4.15) and M (3.13) we find that most terms in the action combine into total derivative

$$\sqrt{g}((\tfrac{1}{6} R + M) + 2 \hat{Y}^i{}_j \hat{Y}^j{}_i - \tfrac{1}{16} T_{\mu\nu} T^{\mu\nu}) =$$
$$= f_1 f_2 r_1 r_2 \sin^3 \rho \sin\theta \cos\theta \left(\frac{\varphi f_3 \partial_\theta f_2}{2 f_1 f_2^3} - \frac{\varphi \partial_\theta f_3}{2 f_1 f_2^2} + \frac{\varphi \partial_\rho f_1 \sin\rho}{2 f_1 f_2^2} - \frac{\varphi \partial_\rho f_2 \sin\rho}{2 f_2^3} \right.$$
$$\left. + \frac{\varphi f_3 \tan\theta}{2 f_1 f_2^2} - \frac{\varphi f_3 \cot\theta}{2 f_1 f_2^2} - \frac{\varphi \partial_\rho \sin\rho}{2 f_1 f_2} - \frac{2\varphi \cos\rho}{f_1 f_2} + \frac{3}{f_1 f_2} + \frac{\partial_\rho \varphi \sin\rho}{2 f_2^2} + \frac{2\varphi \cos\rho}{f_2^2} \right) =$$
$$= -\partial_\theta \left(r_1 r_2 \varphi \sin 2\theta \sin^3 \rho \frac{f_3}{4 f_2} \right) + \partial_\rho \left(r_1 r_2 \varphi \sin 2\theta \sin^4 \rho \frac{f_1 - f_2}{4 f_2} \right) + \frac{3}{2} r_1 r_2 \sin 2\theta \sin^3 \rho$$
$$\quad (4.17)$$

The last term is the only term non-vanishing after integration over $S^4_{\epsilon_1,\epsilon_2}$. It gives

$$S|_{\text{susy conf}} = -\frac{8\pi^2 r_1 r_2}{g_{\text{YM}}^2} \operatorname{tr} \mathring{\Phi}_6^2 \quad (4.18)$$

Therefore, the contribution from the smooth configuration of the localization locus for the partition function is[5]

$$Z^{\text{pert}}_{S^4_{\epsilon_1,\epsilon_2}} = \int d\mathring{\Phi}_6 e^{-S|_{\text{susy conf}}} Z_{\text{1-loop}}(\mathring{\Phi}_6) = \int d\mathring{\Phi}_6 e^{-\frac{1}{\epsilon_1 \epsilon_2} \frac{8\pi^2}{g_{\text{YM}}^2} \langle \mathring{\Phi}_6, \mathring{\Phi}_6 \rangle} Z_{\text{1-loop}}(\mathring{\Phi}_6) \quad (4.19)$$

where $Z_{\text{1-loop}}(\Phi_6)$ needs to be computed from the fluctuations of the quantum fields around the supersymmetric background. Since mathematically such determinant is the same as a certain infinite-dimensional equivariant Euler class as in the Eq. (4.14), it can be computed [23] using equivariant Atiyah-Singer index theorem for the transversally elliptic operators [59]. The Atiyah-Singer index theorem computes the index as the sum of the contributions from the fixed points: the north and the south pole of the $S^4_{\epsilon_1,\epsilon_2}$. The result is that the $Z_{\text{1-loop}}$ factorizes into the product of two factors, each related to the one-loop factor $Z^{\text{1-loop}}_T$ (4.8) of the gauge theory partition function in the Omega background coming from the north or the south pole of the $S^4_{\epsilon_1,\epsilon_2}$. Careful application of Atiyah-Singer index theorem for the transversally elliptic operator shows that the north pole contributes the factor $Z^{\text{1-loop}}_T$ obtained from the expansion of the index in the positive powers of the equivariant parameters $t_1 = e^{i\epsilon_1}$, $t_2 = e^{i\epsilon_2}$ (4.8). The contribution of the south pole is obtained from the expansion of the index in the negative powers of the equivariant parameters.

---

[5]In our conventions Φ_6 is an element of the Lie algebra of the gauge group. For $U(N)$ gauge group Φ_6 is represented by anti-Hermitian matrices. The bilinear form $\langle , \rangle$ is the positive definite invariant metric on the Lie algebra normalized such that the length squared of the long root is 2. For $U(N)$ group $\operatorname{tr}_f \Phi^2 = -\langle \Phi, \Phi \rangle$.

The argument of $Z_\mathsf{T}^{\text{1-loop}}$, the equivariant parameter $\mathbf{a}$ of the gauge theory in the Omega background, relates to the scalar fields on $S_{\epsilon_1,\epsilon_2}^4$ in the way

$$\mathbf{a} = v^a \Phi_a \tag{4.20}$$

where v^a is the vector field (3.7). At the north pole for the supersymmetric configuration we find

$$\mathbf{a} = -i\Phi_6 - \Phi_5 = -i\overset{\circ}{\Phi}_6 \tag{4.21}$$

On X^4 it is natural to assume the mass parameter pure imaginary, since the mass can be thought as the fixed background value of the scalar field Φ_6 in the vector multiplet of gauged flavour symmetry, so for convenience we set that mass parameters on X^4 are $i\mathbf{m}$ where $\mathbf{m}$ is real. Then, up to an overall phase, and assuming that the arguments a_i and m_{lf} in (4.8) are pure imaginary we find

$$Z_{\text{1-loop}; S_{\epsilon_1,\epsilon_2}^4} = Z_{\text{1-loop},\mathsf{T}}(i\mathbf{a}, i\mathbf{m}) \tag{4.22}$$

The classical contribution also factorizes, and using (4.4), (4.5) we find the partition function (4.19) can be rewritten as

$$Z_{S_{\epsilon_1,\epsilon_2}^4}^{\text{pert}} = \int [d\mathbf{a}] |Z_\mathsf{T}^{\text{pert}}(\epsilon_1, \epsilon_2; i\mathbf{a}, i\mathbf{m})|^2 \tag{4.23}$$

The above formula for the partition function takes into account only the perturbative contribution in the localization computation around the smooth solution of the supersymmetric equations. However, the complete partition function on $S_{\epsilon_1,\epsilon_2}^4$ is also contributed by the point like instanton/anti-instanton configurations, with point instantons supported at the north pole and the point anti-instantons supported at the south pole [23]. This follows from the analysis of the asymptotics of the localization equations near the north and south poles: the supersymmetric theory on $S_{\epsilon_1,\epsilon_2}^4$ near the north pole is approximated by the gauge theory in the Omega-background, and the supersymmetric theory on $S_{\epsilon_1,\epsilon_2}^4$ near the south pole is approximated by the conjugated version of the gauge theory in the Omega-background. This argument leads to the complete formula

$$Z_{S_{\epsilon_1,\epsilon_2}^4} = \int [d\mathbf{a}] |Z_\mathsf{T}(\epsilon_1, \epsilon_2; i\mathbf{a}; i\mathbf{m}; \mathbf{q})|^2 \tag{4.24}$$

4.3 Hypermultiplets

The treatment of conformal massless hypermultiplets is straightforward and is similar to [32]. The mass-term are added by gauging the flavour symmetry, introducing the vector-multiplet for the flavour-symmetry group and then freezing all the fields of this flavour-symmetry vector field to zero except the constant scalar field $(\overset{\circ}{\Phi}_0)_{\text{flavour}}$ which then plays the role of the mass parameter.

4.4 Open Problem

It should be possible to classify all possible T^2-bundle solutions to generalized conformal Killing equations, construct the supersymmetric theories on such backgrounds and localize the partition function generalizing the result (4.24).

Appendix 1: Conventions and Useful Identities

Indices

For the 4d theories with 8 supercharges ($\mathcal{N} = 2$ supersymmetry in 4d) we use the notations of the (0, 1) 6d supersymmetric theories under the dimensional reduction. The Table 4 summarizes the index notations.

The symmetrization and anti-symmetrization of tensors

$$t_{(m_1...m_r)} = \frac{1}{r!} \sum_{\sigma \in \mathrm{Perm}(r)} t_{m_{\sigma(1)},...m_{\sigma(r)}}$$

$$t_{[m_1...m_r]} = \frac{1}{r!} \sum_{\sigma \in \mathrm{Perm}(r)} (-1)^\sigma t_{m_{\sigma(1)},...m_{\sigma(r)}}$$

(4.25)

Spinors

The spinors λ and ε in the (0, 1) Euclidean supersymmetric 6d theory are the holomorphic $SU(2)_R \simeq Sp(1)_R$ doublets of Weyl four-component spinors, of Weyl chirality $+1$, for the 6d Clifford algebra over complex numbers $\mathbb{C}$. We take $\lambda \equiv (\lambda^i)_{i=1,2}$ where each λ^1 and λ^2 is 6d Weyl fermion. In total the spinor λ has 8 complex components.[6]

Table 4 Indices

Type of indices	Symbol	Range	Fields
4d space-time vectors	μ, ν, ρ, σ	$[1, \ldots, 4]$	Gauge field A_μ
$U(1)_R = SO(2)_R$ vectors	a, b	$[5, 6]$	Scalar field ϕ_a
6d vectors (the sum of above)	m, n, p, q	$[1, \ldots, 6]$	$A_m = (A_\mu, \phi_a)$
$SU(2)_R$	i, j	$1, 2$	Gaugino doublet λ^i; auxiliary triplet $Y^i{}_j$

---

[6]We construct the Lagrangian and supersymmetry algebra using only holomorphic/algebraic dependence on the spinorial components. In other words, the complex conjugate of gaugino (λ^i) never

Clifford Algebra

The 8×8 complex matrices γ_m represent the 6d Clifford algebra

$$\{\gamma_m, \gamma_n\} = 2g_{mn} \tag{4.26}$$

The chirality operator γ_*^{6d} anticommuting with all γ_m is

$$\gamma_* = i\gamma_1 \ldots \gamma_6; \qquad \{\gamma_*, \gamma_m\} = 0; \qquad \gamma_*^2 = 1. \tag{4.27}$$

The chirality of the spinors is the eigenvalue of γ_*. The projection operators that split $\mathcal{S} = \mathcal{S}^+ \oplus \mathcal{S}^-$ are

$$\gamma_\pm = \tfrac{1}{2}(1 \pm \gamma_*), \qquad \varepsilon_\pm = \gamma_\pm \varepsilon_\pm = \pm \gamma_* \varepsilon_\pm \tag{4.28}$$

Explicit form of γ_m matrices is not needed, but for concreteness one can recursively define the $\gamma_m^{(d)}$ matrices of size $2^{d/2} \times 2^{d/2}$ in even dimension d in terms of $\gamma_m^{(d-2)}$ as follows (see e.g. [60])

$$\begin{aligned}
\gamma_m^{(d)} &= \sigma_3 \otimes \gamma_m^{(d-2)}, \quad m \in [1, \ldots, d-2] \\
\gamma_{d-1}^{(d)} &= \sigma_1 \otimes 1, \quad \gamma_d^{(d)} = \sigma_2 \otimes 1 \\
\gamma_*^{(d)} &= \sigma_3 \otimes \gamma_*^{(d-2)}
\end{aligned} \tag{4.29}$$

where $(\sigma_0, \sigma_1, \sigma_2, \sigma_3)$ are the 2×2 Pauli matrices

$$(\sigma_0, \sigma_1, \sigma_2, \sigma_3) = \left(\begin{pmatrix} 1 & 0 \\ 0 & 1 \end{pmatrix}, \begin{pmatrix} 0 & 1 \\ 1 & 0 \end{pmatrix}, \begin{pmatrix} 0 & -i \\ i & 0 \end{pmatrix}, \begin{pmatrix} 1 & 0 \\ 0 & -1 \end{pmatrix} \right). \tag{4.30}$$

We use antisymmetric multi-index notations

$$\gamma_{m_1 \ldots m_r} = \gamma_{[m_1} \ldots \gamma_{m_r]}. \tag{4.31}$$

and we use underline notation for the multi-index

$$\gamma_{\underline{r}} \quad \text{is one of} \quad \gamma_{m_1 \ldots m_r} \tag{4.32}$$

In the contraction of multi-index we use non-repetitive summation

---

(Footnote 6 continued)
appears neither in the Lagrangian, nor in the measure of the path integral, nor in the supersymmetry transformations. The fermionic analogue of the *contour of integration* in the path integral or the *reality condition* is not necessary since evaluation the Pfaffian or top degree form is an algebraic operation.

$$A^{\underline{p}}B_{\underline{p}} \equiv \sum_{m_1 < \cdots < m_p} A^{m_1 \ldots m_p} B_{m_1 \ldots m_p} = \frac{1}{p!} A^{m_1 \ldots m_p} B_{m_1 \ldots m_p} \qquad (4.33)$$

For the forms we use component and slashed notation

$$\omega \equiv \frac{1}{r!} \omega_{\mu_1 \ldots \mu_r} dx^{\mu_1} \wedge \cdots \wedge dx^{\mu_r} \qquad \qquad \not\psi = \omega_{\mu_1 \ldots \mu_r} \gamma^{\mu_1 \ldots \mu_r} \qquad (4.34)$$

Contraction identity

$$\gamma^m \gamma_{\underline{r}} \gamma_m = (-1)^r (d - 2r) \gamma_{\underline{r}} \qquad (4.35)$$

Multi-index contraction identity

$$\gamma^{\underline{p}} \gamma_{\underline{r}} \gamma_{\underline{p}} = \Delta(d, r, p) \gamma_{\underline{r}} \qquad (4.36)$$

with

$$\Delta(d, r, p) = (-1)^{p(p-1)/2} (-1)^{rp} \sum_{q=\max(p+r-d,0)}^{\min(r,p)} (-1)^q \binom{r}{q} \binom{d-r}{p-q} \qquad (4.37)$$

The contraction formula and the completeness of $(\gamma_p)_{p \in [0,\ldots,d]}$ for $d \in 2\mathbb{Z}$ in the matrix algebra of $2^{d/2} \times 2^{d/2}$ matrices implies the Fierz identity

$$(\gamma^{\underline{r}})^{\alpha_1}_{\alpha_2} (\gamma_{\underline{r}})^{\alpha_3}_{\alpha_4} = \sum_{k=0}^{d} \tilde{\Delta}(d, r, k) (\gamma^{\underline{k}})^{\alpha_1}_{\alpha_4} (\gamma_{\underline{k}})^{\alpha_3}_{\alpha_2} \qquad (4.38)$$

where

$$\tilde{\Delta}(d, r, k) = (-1)^{\frac{k(k-1)}{2}} 2^{-\frac{d}{2}} \Delta(d, r, k) \qquad (4.39)$$

The terms with $k > \frac{d}{2}$ in the Fierz identity are conveniently represented as

$$\gamma_{\underline{k}} \gamma_* = (-1)^{r(r-1)/2} i^{-n/2} \gamma_{\underline{k}^\vee} \qquad (4.40)$$

where $\gamma_{\underline{k}^\vee}$ is complementary in indices of $\gamma_{\underline{k}}$ with a proper permutation sign. The Fierz identity is

$$(\gamma^{\underline{l}})^{\alpha_1}_{\alpha_2} (\gamma_{\underline{l}})^{\alpha_3}_{\alpha_4} = \sum_{k=0}^{d/2} \tilde{\Delta}(d, l, k) (\gamma^{\underline{k}})^{\alpha_1}_{\alpha_4} (\gamma_{\underline{k}})^{\alpha_3}_{\alpha_2}$$

$$+ (-1)^{d/2} \sum_{k=0}^{d/2-1} \tilde{\Delta}(d, l, d-k) (\gamma^{\underline{k}} \gamma_*)^{\alpha_1}_{\alpha_4} (\gamma_{\underline{k}} \gamma_*)^{\alpha_3}_{\alpha_2}. \qquad (4.41)$$

This form is useful when applied to the chiral spinors.

Table 5 Symmetries of $C\gamma_{\underline{r}}$

d mod 8	2	4	6	8
C_1	$++--$	$-++-$	$\boxed{--++}$	$+--+$
C_2	$-++-$	$\boxed{--++}$	$+--+$	$++--$

Spinor Bilinears

The spinor representation space $\mathcal{S}$ can be equipped with an invariant complex bilinear form $(,) : \mathcal{S} \otimes \mathcal{S} \to \mathbb{C}$. In components we write[7]

$$(\eta\varepsilon) := \eta^\alpha C_{\alpha\beta}\varepsilon^\beta \tag{4.42}$$

where C is a matrix representing the bilinear form.

All operators $\gamma_{\underline{r}}$ are symmetric or antisymmetric with respect to C. The symmetry of $C\gamma_{\underline{r}}$ depends on the dimension d and is summarized in Table 5.

The entries $s_0 s_1 s_2 s_3$ with $s_r = \pm 1$ denote the transposition symmetry of $C\gamma_{\underline{r}}$ for r mod 4. There are two choices of C denoted by C_1 and C_2 in the table. In representation (4.29) one can take

$$\begin{aligned} C_1 &= \cdots \otimes \sigma_1 \otimes \sigma_2 \otimes \sigma_1 \\ C_2 &= \cdots \otimes \sigma_2 \otimes \sigma_1 \otimes \sigma_2. \end{aligned} \tag{4.43}$$

The bilinear form C_2 for spinors in even dimension d can be also used as the bilinear form for spinors in $d + 1$-dimensions. For the theories with 8 supercharges in $d = 4, 5, 6$ dimensions we are using C highlighted in the Table 5.

The matrices $C\gamma_{\underline{r}}$ represent bilinear forms on $\mathcal{S}$ valued in r-forms, in other words, for spinors η and ε the

$$\omega_{\underline{r}} = (\eta\gamma_{\underline{r}}\varepsilon) \tag{4.44}$$

transform covariantly as the rank r form. Since

$$\gamma_* C = (-1)^{\frac{d}{2}} C\gamma_* \tag{4.45}$$

it follows that[8]

---

[7] Often in the physics literature the dual spinor $\eta_\beta = \eta^\alpha C_{\alpha\beta}$ (an element of the dual space $\mathcal{S}^\vee$) is denoted $\bar\varepsilon$ and is called *Majorana* conjugate to ε. We have chosen here to avoid the bar notation to avoid confusion with complex conjugation.

[8] Consistent with the fact that for $\frac{d}{2} \in 2\mathbb{Z}$ the tensor product $\mathcal{S}_+ \otimes \mathcal{S}_-$ contains odd rank forms; and $\mathcal{S}_+ \otimes \mathcal{S}_+, \mathcal{S}_- \otimes \mathcal{S}_-$ contains even rank forms; in particular for $\frac{d}{2} \in 2\mathbb{Z}$ the representation $\mathcal{S}^\pm$ is dual to $\mathcal{S}^\pm$; while for $\frac{d}{2} \in 2\mathbb{Z} + 1$ the representation $\mathcal{S}^\pm$ is dual to $\mathcal{S}^\mp$.

$$(\eta\gamma_{\underline{r}}\epsilon) = (\eta_+\gamma_{\underline{r}}\epsilon_+) + (\eta_-\gamma_{\underline{r}}\epsilon_-), \qquad \tfrac{d}{2} + r \in 2\mathbb{Z}$$
$$(\eta\gamma_{\underline{r}}\epsilon) = (\eta_-\gamma_{\underline{r}}\epsilon_+) + (\eta_+\gamma_{\underline{r}}\epsilon_-), \qquad \tfrac{d}{2} + r \in 2\mathbb{Z} + 1 \tag{4.46}$$

In $d = 6$ the bilinears in the spinors of the same chirality transform as forms of odd rank; while the bilinears in the spinors of opposite chirality transform as forms of even rank.

$$d = 6: \quad \begin{cases} (\varepsilon_+\gamma_{\underline{r}}\varepsilon'_+) \neq 0 & \text{only for } r \in \{1, 3, 5\} \\ (\eta_-\gamma_{\underline{r}}\varepsilon_+) \neq 0 & \text{only for } r \in \{0, 2, 4, 6\} \end{cases} \tag{4.47}$$

The bilinear form valued in 1-forms is antisymmetric in $d = 6$ for either choice of C. To construct the standard fermionic action $(\lambda\gamma^m D_m\lambda)$ we need the symmetric 1-form valued bilinear form. For the minimal 6d $(0, 1)$ supersymmetry we introduce a $SU(2)_R$-doublet of Weyl fermions $(\lambda^i)_{i=1,2}$ and then use $C \otimes \epsilon$, where $\epsilon = \epsilon_{ij}$ is the standard 2×2 antisymmetric symbol, as the symmetric bilinear form on the $S^+ \otimes \mathbb{C}^2$. The resulting 1-form valued bilinear is symmetric and there is a proper fermionic kinetic action

$$(\lambda^i \not{D}\lambda_i) \equiv (\lambda^i \not{D}\lambda^j)\epsilon_{ji} \tag{4.48}$$

We use the standard antisymmetric 2×2 tensor ϵ_{ij} to raise and lower the $SU(2)_R$ indices i, j in the pattern ${}^i{}_i$:

$$\lambda^i := \epsilon^{ij}\lambda_j, \qquad \lambda_j := \lambda^i \epsilon_{ij}$$
$$\epsilon^{ij}\epsilon_{ik} = \delta^j_k, \qquad (\varepsilon^{[j}\eta^{i]}) = \tfrac{1}{2}\epsilon^{ij}(\varepsilon^k\eta_k) \tag{4.49}$$

When the $SU(2)_R$ indices are omitted, the contraction ${}^i{}_i$ is assumed

$$(\varepsilon\gamma_{\underline{r}}\varepsilon') \equiv (\varepsilon^i\gamma_{\underline{r}}\varepsilon'_i) \tag{4.50}$$

$d = 6$ Fierz Identities

For $d = 6$ and $l = 1$ we find

k	0	1	2	3	4	5	6
$\tilde{\Delta}(6, 1, k)$	$\tfrac{3}{4}$	$-\tfrac{1}{2}$	$-\tfrac{1}{4}$	0	$-\tfrac{1}{4}$	$\tfrac{1}{2}$	$\tfrac{3}{4}$

Notice that $\tilde{\Delta}(6, 1, k) = (-1)^k \tilde{\Delta}(6, 1, 6 - k)$. Therefore if we project Fierz identity (4.41) with γ_+ applied to the α_2 and α_4 indices, we find that terms with even k vanish. In addition the middle term $k = 3$ vanishes too. Finally

$$(\gamma^{\underline{1}})^{\alpha_1}_{\alpha_2}(\gamma_{\underline{1}})_{\alpha_3\alpha_4} = -(\gamma^{\underline{1}})^{\alpha_1}_{\alpha_4}(\gamma_{\underline{1}})_{\alpha_3\alpha_2} \qquad \text{projected by } (\gamma_\pm)^{\alpha_2}_{\alpha'_2}(\gamma_\pm)^{\alpha_4}_{\alpha'_4} \tag{4.51}$$

A frequently used form of the above identity involves cylic permutation of three $+$-chiral spinor doublets ε^i, κ^i, λ^i. Taking the sum of

$$(\varepsilon^j \gamma_m \kappa_j)\gamma^m \lambda^i = -(\varepsilon^j \gamma_m \lambda^i)\gamma^m \kappa_j$$
$$(\lambda^j \gamma_m \kappa_j)\gamma^m \varepsilon^i = -(\lambda^j \gamma_m \varepsilon^i)\gamma^m \kappa_j \tag{4.52}$$

we find

$$\boxed{(\varepsilon\gamma_m \kappa)\gamma^m \lambda + (\lambda\gamma_m \kappa)\gamma^m \varepsilon + (\varepsilon\gamma_m \lambda)\gamma^m \kappa = 0} \tag{4.53}$$

Now we consider projection of 6d Fierz identity at $p = 1$ on spinors of opposite chirality. Take $+$-chiral doublet ε^i and $--$chiral doublet η^i. We find

$$(\varepsilon^j \gamma^1 \varepsilon_j)\gamma_1 \eta^i = \tfrac{3}{2}(\varepsilon^j \eta^i)\varepsilon_j - \tfrac{1}{2}(\varepsilon^j \gamma^2 \eta^i)\gamma_2 \varepsilon_j$$
$$(\varepsilon^j \gamma^2 \eta^i)\gamma_2 \varepsilon_j = -\tfrac{5}{4}(\varepsilon^j \gamma^1 \varepsilon_j)\gamma_1 \eta^i \tag{4.54}$$

where at $l = 2$ the explicit coefficients in (4.41) are given as follows:

k	0	1	2	3	4	5	6
$\tilde{\Delta}(6, 2, k)$	$-\frac{15}{8}$	$-\frac{5}{8}$	$-\frac{1}{8}$	$-\frac{3}{8}$	$+\frac{1}{8}$	$-\frac{5}{8}$	$+\frac{15}{8}$

Hence, from the Eq. (4.54) we find another useful 6d Fierz identity

$$\boxed{(\varepsilon^j \gamma^m \varepsilon_j)\gamma_m \eta^i = 4(\varepsilon^j \eta^i)\varepsilon_j, \qquad (\varepsilon = \gamma_+ \varepsilon, \quad \eta = \gamma_- \eta)} \tag{4.55}$$

6d $(0, 1)$ Theory Conventions

The spinor ε is $+$-chiral, the spinor η is $--$chiral

$$\varepsilon = \gamma_* \varepsilon = \gamma_+ \varepsilon, \qquad\qquad \eta = -\gamma_* \eta = \gamma_- \eta \tag{4.56}$$

The tensor field $T_{\mu\nu a}$ is 6d anti-self-dual, $*_{6d} T = -T$. Useful contraction identities

$$\gamma^\mu \not{F} = \not{F}\gamma^\mu - 4F_{m\mu}\gamma^m, \qquad \not{T} = T_{\mu\nu a}\gamma^{\mu\nu a} \tag{4.57}$$
$$\gamma^\mu \not{T}\gamma_\mu = \gamma^a \not{T}\gamma_a = \gamma^m \not{T}\gamma_m = 0 \tag{4.58}$$
$$T_{\mu\nu a}\gamma^{\mu\nu} = \tfrac{1}{2}\{\not{T}, \gamma_a\}, \qquad T_{\mu\nu b}\gamma^{\nu b} = \tfrac{1}{4}\{\not{T}, \gamma_\mu\} \tag{4.59}$$

The Bianchi identity on the field strength

$$D_m F_{pq} + D_q F_{mp} + D_p F_{qm} = 0, \qquad \gamma^{mpq} D_m F_{pq} = 0 \tag{4.60}$$

Positive chirality of $\varepsilon \equiv \varepsilon^+$ and negative chirality of $T \equiv T^-$ implies

$$T\gamma_{2r}\varepsilon = 0$$
$$(\varepsilon\gamma_r\varepsilon) = 0, \quad r \bmod 4 \in \{2,3\}; \qquad (\varepsilon^{(i}\gamma_r\varepsilon^{j)}) = 0, \quad r \bmod 4 \in \{0,1\}$$
$$\gamma_\rho\{T, \gamma_a\}\varepsilon = \{T, \gamma_\rho\}\gamma_a\varepsilon, \qquad \tfrac{1}{2}T_{\mu\nu a}\gamma_\rho\gamma^{\mu\nu}\varepsilon = T_{\rho\nu b}\gamma^{\nu b}\gamma_a \tag{4.61}$$
$$\gamma^\rho T_{\mu\nu a}\gamma^{\nu a}\varepsilon = T_{\rho\nu a}\gamma^{\nu a}\gamma_\mu\varepsilon, \qquad \gamma_a T_{\mu\nu b}\gamma^{\mu\nu}\varepsilon = T_{\mu\nu a}\gamma^{\mu\nu}\gamma_b\varepsilon$$
$$T_{\mu\nu a}T_{\rho\sigma b}\gamma^\rho\gamma^{\mu\nu}\gamma^{\sigma b}\varepsilon = 4T_{\rho\nu a}\gamma^\nu T_{\rho\sigma b}\gamma^{\sigma b}\varepsilon = 4T_{\rho\nu a}T_{\rho\nu b}\gamma^b\varepsilon$$

The spin-connection and the metric curvatures

$$D_\mu v^{\hat\rho} = \partial_\mu v^{\hat\rho} + \omega^{\hat\rho}{}_{\hat\sigma\mu}v^{\hat\sigma}$$
$$R^{\hat\rho}{}_{\hat\sigma\mu\nu} = [D_\mu, D_\nu]^{\hat\rho}_{\hat\sigma}, \qquad R_{\sigma\nu} = R^\mu{}_{\sigma\mu\nu}, \qquad R = R^\mu{}_\mu \tag{4.62}$$

The covariant derivative on spinors, the curvature and the Lichnerowicz formula

$$D_\mu\varepsilon^i = \partial_\mu\varepsilon^i + \tfrac{1}{4}\omega^{\rho\sigma}{}_\mu\gamma_{\rho\sigma}\varepsilon^i + (V^{\mathrm R}_\mu)^i{}_j\varepsilon^j$$
$$\slashed{D}^2 = D^\mu D_\mu - \tfrac{1}{4}R + \tfrac{1}{2}\slashed{F}^{\mathrm R}_V \tag{4.63}$$

where $(V^{\mathrm R}_\mu)^i{}_j$ is the $SU(2)_{\mathrm R}$-connection.

Supersymmetry Equations

The divergence of the first equation in the system (2.1) implies

$$D^\mu D_\mu\varepsilon - \tfrac{1}{16}[D^\mu T]\gamma_\mu\varepsilon - \tfrac{1}{4}T\eta = \tfrac{1}{4}\slashed{D}^2\varepsilon \tag{4.64}$$

which together with Lichnerowicz formula (4.63) produces

$$\tfrac{1}{4}\slashed{D}^2\varepsilon + \tfrac{1}{3}(\tfrac{1}{4}R\varepsilon - \tfrac{1}{2}\slashed{F}_{\mathrm R}\varepsilon - \tfrac{1}{16}[D^\mu T]\gamma_\mu\varepsilon - \tfrac{1}{4}T\eta) = 0 \tag{4.65}$$

and the linear combination with the second equation in (2.1) produces

$$\tfrac{1}{4}\slashed{D}^2\varepsilon = -\tfrac{1}{2}\left(\tfrac{1}{6}R + M\right)\varepsilon + \tfrac{1}{16}[D^\mu T]\gamma_\mu\varepsilon \tag{4.66}$$

The 6d and 4d Spinor Conventions

As in (4.29) we take

$$\gamma^{(6)}_\mu = \begin{pmatrix} \gamma^{(4)}_\mu & 0 \\ 0 & -\gamma^{(4)}_\mu \end{pmatrix}, \quad \gamma^{(6)}_5 = \begin{pmatrix} 0 & 1 \\ 1 & 0 \end{pmatrix}, \quad \gamma^{(6)}_6 = \begin{pmatrix} 0 & -i \\ i & 0 \end{pmatrix}$$
$$C^{(6)} = \begin{pmatrix} 0 & -c^{(4)}\gamma^{(4)}_* \\ -c^{(4)}\gamma^{(4)}_* & 0 \end{pmatrix}, \quad \varepsilon^{(6)}_+ = \begin{pmatrix} \varepsilon^{(4)}_+ \\ \varepsilon^{(4)}_- \end{pmatrix}, \quad \eta^{(6)}_- = \begin{pmatrix} \eta^{(4)}_- \\ -\eta^{(4)}_+ \end{pmatrix} \tag{4.67}$$

where $\varepsilon_{\pm}^{(4)}$ denote the $\pm$-chiral spinors of the 4d Clifford algebra with respect to $\gamma_*^{(4)}$, the $C^{(6)}$ is the bilinear form for the 6d Clifford algebra of type $(--++)$, and $c^{(4)}$ is the bilinear form of the 4d Clifford algebra of type $(--++)$. In these conventions the bilinears computed in 4d and 6d notations agree:

$$\varepsilon_+^{(6)} C^{(6)} \eta_-^{(6)} = \varepsilon_+^{(4)} c^{(4)} \eta_+^{(4)} + \varepsilon_-^{(4)} c^{(4)} \eta_-^{(4)}$$
$$\varepsilon_+^{(6)} C^{(6)} \gamma_\mu^{(6)} \tilde{\varepsilon}_+^{(6)} = \varepsilon_+^{(4)} c^{(4)} \gamma_\mu^{(4)} \tilde{\varepsilon}_-^{(4)} + \varepsilon_-^{(4)} c^{(4)} \gamma_\mu^{(4)} \tilde{\varepsilon}_+^{(4)} \tag{4.68}$$

For the explicit form of spinors we use 4d gamma-matrices, the 4d chirality operator and 4d bilinear form in terms of (4.30)

$$(\gamma_i, \gamma_4) = (\sigma_2 \otimes \sigma_i, \sigma_1 \otimes \sigma_0),$$
$$\gamma_*^{(4)} = -\gamma_1 \ldots \gamma_4 = -\sigma_3 \otimes \sigma_0 \tag{4.69}$$
$$c^{(4)} = -i\sigma_0 \otimes \sigma_2$$

We decompose

$$T_{\mu\nu a} \gamma^a = T_{\mu\nu-} \gamma^- + T_{\mu\nu+} \gamma^+ \tag{4.70}$$

in terms of

$$T_{\mu\nu-} = (T_{\mu\nu5} - iT_{\mu\nu6}) \quad \gamma^- = \tfrac{1}{2}(\gamma^5 + i\gamma^6) \quad \gamma^- \gamma_*^{56} = -\gamma^-$$
$$T_{\mu\nu+} = (T_{\mu\nu5} + iT_{\mu\nu6}) \quad \gamma^+ = \tfrac{1}{2}(\gamma^5 - i\gamma^6) \quad \gamma^+ \gamma_*^{56} = +\gamma_+ \tag{4.71}$$

with $\gamma_*^{56} = -i\gamma_{56}$. Since $T_{\mu\nu a}$ is of negative 6d chirality, the $T_{\mu\nu\pm}$ has $\mp$ 4d chirality. We define

$$T_{\mu\nu}^{(4)} \equiv T_{\mu\nu+} - T_{\mu\nu-} = 2iT_{\mu\nu6} \tag{4.72}$$

In terms of the 4d spinors the generalized conformal Killing equation (2.1) takes form

$$D_\mu \varepsilon - \tfrac{1}{16} T_{\rho\sigma}^{(4)} \gamma^{\rho\sigma} \gamma_\mu \varepsilon = \gamma_\mu \eta \tag{4.73}$$

Other 6d-4d notational definitions are

$$\Phi^\pm = \tfrac{1}{2}(\Phi^5 \mp i\Phi^6), \qquad\qquad \Phi_\pm = (\Phi^5 \pm i\Phi^6)$$
$$T_{\mu\nu a} \gamma^{\mu\nu} \Phi^a \varepsilon_+^{(6)} = T_{\mu\nu}^{(4)} \gamma^{\mu\nu} (\Phi^+ \varepsilon_-^{(4)} - \Phi^- \varepsilon_+^{(4)}), \qquad \Phi^a \gamma_a \eta = 2\Phi^- \eta_-^{(4)} - 2\Phi^+ \eta_+^{(4)} \tag{4.74}$$

Appendix 2: Supersymmetry Algebra

The Off-Shell Closure of the Supersymmetry on the Vector Multiplet

Here we explicitly compute δ^2 on vector multiplet for δ defined by:

$$
\begin{aligned}
\delta A_m &= \tfrac{1}{2}\lambda^i \gamma_m \varepsilon_i \\
\delta\lambda^i &= -\tfrac{1}{4}F_{mn}\gamma^{mn}\varepsilon^i + Y^i{}_j \varepsilon^j + \Phi_a \gamma^a \eta^i + \tfrac{1}{8}T_{\mu\nu a}\Phi^a \gamma^{\mu\nu}\varepsilon^i \qquad (4.75) \\
\delta Y^{ij} &= -\tfrac{1}{2}(\varepsilon^{(i}\,\slashed{D}\lambda^{j)})
\end{aligned}
$$

provided that spinors (ε, η) with $\eta = \tfrac{1}{4}\slashed{D}\varepsilon$ satisfy generalized conformal Killing equations (2.3).

We find a contribution of several terms in $\delta^2\lambda^i$. In the flat space we drop the terms proportional to $D_\mu\varepsilon$, η and T and find

$$
\delta^2_{\text{flat}}\lambda^i = \delta(-\tfrac{1}{4}F_{mn}\gamma^{mn}\varepsilon^i) + \delta(Y^i{}_j\varepsilon^j) \qquad (4.76)
$$

with

$$
\begin{aligned}
\delta\left(-\tfrac{1}{4}F_{mn}\gamma^{mn}\varepsilon^i\right) \\
= -\tfrac{1}{4}D_p\left(\lambda^j\gamma_q\varepsilon_j\right)\gamma^{pq}\varepsilon^i &= \tfrac{1}{4}\left(\varepsilon\slashed{D}\lambda\right)\varepsilon^i - \tfrac{1}{4}\left(\varepsilon^j\gamma_q D_p\lambda_j\right)\gamma^p\gamma^q\varepsilon^i \overset{(4.51)}{=} \\
= \tfrac{1}{4}\left(\varepsilon\slashed{D}\lambda\right)\varepsilon^i + \tfrac{1}{4}\left(\varepsilon\gamma^q\varepsilon\right)D_q\lambda^i &- \tfrac{1}{8}\left(\varepsilon\gamma_q\varepsilon\right)\gamma^q\slashed{D}\lambda^i \overset{(4.55)}{=} \tfrac{1}{4}\left(\varepsilon\gamma^q\varepsilon\right)D_q\lambda^i \\
+ \tfrac{1}{4}\left(\varepsilon\slashed{D}\lambda\right)\varepsilon^i &- \tfrac{1}{2}\left(\varepsilon^j\,\slashed{D}\lambda^i\right)\varepsilon_j
\end{aligned}
$$
$$
(4.77)
$$

and together with the $\delta(Y^i{}_j\varepsilon^j)$ we find

$$
\delta^2_{\text{flat}}\lambda^i = \tfrac{1}{4}(\varepsilon\gamma^q\varepsilon)D_q\lambda^i + \tfrac{1}{4}(\varepsilon^j\,\slashed{D}\lambda_j)\varepsilon^i - \tfrac{1}{2}(\varepsilon^{[j}\,\slashed{D}\lambda^{i]})\varepsilon_j \overset{(4.49)}{=} \tfrac{1}{4}(\varepsilon\gamma^q\varepsilon)D_q\lambda^i
$$
$$
(4.78)
$$

Next we account for $D_\mu\varepsilon$ and η terms, still keeping $T = 0$. The transformation would be complete on conformally flat space. The $\delta^2_{\text{cflat}}\lambda$ acquires new contributions

$$
\delta^2_{\text{cflat}}\lambda^i = \delta^2_{\text{flat}}\lambda^i + \textbf{term}_c \qquad (4.79)
$$

where

$$
\begin{aligned}
\textbf{term}_c &= -\tfrac{1}{4}(\lambda\gamma_q\gamma_\mu\eta)\gamma^{\mu q}\varepsilon^i + \tfrac{1}{2}(\lambda\gamma_a\varepsilon)\gamma^a\eta^i \overset{(4.53)\text{ on }\gamma_q\gamma^q}{=} \\
&= -\tfrac{1}{2}(\varepsilon\gamma_q\lambda)\gamma^q\eta + \tfrac{1}{4}(\eta\gamma_{\mu q}\varepsilon)\gamma^{\mu q}\lambda + (\eta\varepsilon)\lambda + (\eta\lambda)\varepsilon
\end{aligned}
$$
$$
(4.80)
$$

Then we expand the middle term in 4d indices

$$\tfrac{1}{4}(\eta\gamma_{\mu q}\varepsilon)\gamma^{\mu q}\lambda = +\tfrac{1}{8}(\eta\gamma_{\mu\nu}\varepsilon)\gamma^{\mu\nu}\lambda + \tfrac{1}{8}(\eta\gamma_{pq}\varepsilon)\gamma^{pq}\lambda - \tfrac{1}{8}(\eta\gamma_{ab}\varepsilon)\gamma^{ab}\lambda \qquad (4.81)$$

and apply Fierz identity (4.38) to the first and the last term in (4.80) to find

$$\begin{aligned}
-\tfrac{1}{2}(\varepsilon^j\gamma_q\lambda_j)\gamma^q\eta^i &= -\tfrac{3}{4}(\varepsilon^j\eta^i)\lambda_j + \tfrac{1}{8}(\varepsilon^j\gamma_{pq}\eta^j)\gamma^{pq}\lambda_j \\
(\eta^j\lambda_j)\varepsilon^i &= \tfrac{1}{4}(\eta^j\varepsilon^i)\lambda_j - \tfrac{1}{8}(\eta^j\gamma_{pq}\varepsilon^i)\gamma^{pq}\lambda_j
\end{aligned} \qquad (4.82)$$

All $\gamma_{pq}\gamma^{pq}$ terms are cancelled using (4.49) and the scalar terms are simplified as

$$(\eta\varepsilon)\lambda - \tfrac{3}{4}(\varepsilon^j\eta^i)\lambda_j + \tfrac{1}{4}(\eta^j\varepsilon^i)\lambda_j = \tfrac{3}{4}(\eta\varepsilon)\lambda + (\eta^{(i}\varepsilon^{j)})\lambda_j \qquad (4.83)$$

and the contribution from the non-flat but conformally flat terms is

$$\mathbf{term_c} = +\tfrac{1}{8}(\eta\gamma_{\mu\nu}\varepsilon)\gamma^{\mu\nu}\lambda - \tfrac{1}{8}(\eta\gamma_{ab}\varepsilon)\gamma^{ab}\lambda + \tfrac{3}{4}(\eta\varepsilon)\lambda + (\eta^{(i}\varepsilon^{j)})\lambda_j \qquad (4.84)$$

Then we compute the T-terms in

$$\delta^2\lambda^i = \delta^2_{\text{cflat}}\lambda^i + \mathbf{term_T} \qquad (4.85)$$

and find

$$\begin{aligned}
\mathbf{term_T} &= \tfrac{1}{64}(\varepsilon\gamma_\mu \mathcal{T}\gamma_q\lambda)\gamma^{\mu q}\varepsilon + \tfrac{1}{16}T_{\mu\nu a}(\varepsilon\gamma^a\lambda)\gamma^{\mu\nu}\varepsilon \overset{(4.53) \text{ on } \gamma_q\gamma^q,(5.35)}{=} \\
&= -\tfrac{1}{64}(\varepsilon\gamma_\mu\mathcal{T}\gamma_q\varepsilon)\gamma^{\mu q}\lambda + \tfrac{1}{64}(\varepsilon\gamma_q\lambda)\gamma^\mu\gamma^q\mathcal{T}\gamma_\mu\varepsilon + \tfrac{1}{32}(\varepsilon\gamma_a\lambda)\mathcal{T}\gamma^a\varepsilon = \\
&= -\tfrac{1}{64}(\varepsilon\gamma_\mu\mathcal{T}\gamma_q\varepsilon)\gamma^{\mu q}\lambda + \tfrac{1}{32}(\varepsilon\gamma_p\lambda)\mathcal{T}\gamma^p\varepsilon \overset{(4.53) \text{ on } \gamma_p\gamma^p}{=} \\
&= -\tfrac{1}{64}(\varepsilon\gamma_\mu\mathcal{T}\gamma_q\varepsilon)\gamma^{\mu q}\lambda - \tfrac{1}{64}(\varepsilon\gamma_q\varepsilon)\mathcal{T}\gamma^q\lambda \overset{(4.59)}{=} \tfrac{1}{32}T_{\mu\nu a}(\varepsilon\gamma^a\varepsilon)\gamma^{\mu\nu}\lambda
\end{aligned} \qquad (4.86)$$

The $\mathbf{term_T}$ can be combined with the $\mathbf{term_c}$:

$$\tfrac{1}{8}(\eta\gamma_{\mu\nu}\varepsilon)\gamma^{\mu\nu}\lambda + \tfrac{1}{32}T_{\mu\nu a}(\varepsilon\gamma^a\varepsilon)\gamma^{\mu\nu}\lambda = \tfrac{1}{16}D_\mu(\varepsilon\gamma_\nu\varepsilon)\gamma^{\mu\nu}\lambda \qquad (4.87)$$

so that finally

$$\delta^2_{\varepsilon,\eta}\lambda^i = \tfrac{1}{4}(\varepsilon\gamma^m\varepsilon)D_m\lambda^i + \tfrac{1}{16}D_\mu(\varepsilon\gamma_\nu\varepsilon)\gamma^{\mu\nu}\lambda^i - \tfrac{1}{8}(\eta\gamma_{ab}\varepsilon)\gamma^{ab}\lambda^i + \tfrac{3}{4}(\eta\varepsilon)\lambda^i + (\eta^{(i}\varepsilon^{j)})\lambda_j \qquad (4.88)$$

The variation $\delta^2_{\varepsilon,\eta}Y^{ij}$ and $\delta^2_{\varepsilon,\eta}A_m$ are computed similarly.

The Invariance of the Lagrangian

The 4d $\mathcal{N} = 2$ supersymmetric Lagrangian for vector multiplet in curved background for vanishing fermionic fields of Weyl multiplet is proportional to (2.11)

$$\frac{1}{2}F_{mn}F^{mn} + \lambda^i\gamma^m D_m\lambda_i + (\frac{1}{6}R + M)\Phi_a\Phi^a - 2Y_{ij}Y^{ij} - F^{\mu\nu}T_{\mu\nu a}\Phi^a$$
$$+ \frac{1}{4}T_{\mu\nu a}T^{\mu\nu b}\Phi^a\Phi_b \tag{4.89}$$

The trace is implicitly implied in all terms. To check the invariance under (2.8) we first consider the flat background with $D_\mu\varepsilon = 0, \eta = 0, T = 0, M = 0$. After that we will add the variational terms in conformally flat background, and finally we will add the remaining T-terms. We find modulo total derivative

$$\delta_{\text{flat}}(\frac{1}{2}F^{mn}F_{mn}) = -(\varepsilon\gamma^n\lambda)D^m F_{mn}$$
$$\delta_{\text{flat}}(\lambda\gamma^m D_m\lambda) = \frac{1}{2}(\lambda\gamma^m D_m F_{pq}\gamma^{pq}\varepsilon) + 2Y^i{}_j(\varepsilon^j\slashed{D}\lambda_i) \overset{(4.60)}{=} \tag{4.90}$$
$$= (\lambda\gamma^n\varepsilon)D^m F_{mn} - 2Y_{ij}(\varepsilon^j\slashed{D}\lambda^i)$$
$$\delta_{\text{flat}}(-2Y_{ij}Y^{ij}) = 2Y_{ij}(\varepsilon^i\slashed{D}\lambda^j)$$

that all terms add to zero. In conformally flat background the new terms appear in the variation of fermionic kinetic term and the coupling of scalars to the curvature

$$\delta_{\text{cflat}}((\frac{1}{6}R + M)\Phi_a\Phi^a) = (\frac{1}{6}R + M)(\lambda\gamma^a\Phi_a\varepsilon)$$
$$\delta_{\text{cflat}}(\lambda\gamma^m D_m\lambda) = \delta_{\text{flat}}(\lambda\gamma^m D_m\lambda) + \mathbf{term_c} \tag{4.91}$$

where

$$\mathbf{term_c} = -2(\lambda[\slashed{D}\gamma^a\Phi_a\eta^i]) + \frac{1}{2}(\lambda\gamma^\mu F_{pq}\gamma^{pq}D_\mu\varepsilon) \overset{(2.1)}{=}$$
$$= -2(\lambda\gamma^{ma}F_{ma}\eta) + 2\Phi_a(\lambda\gamma^a\slashed{D}\eta^i) + 2(\lambda F_{ma}\gamma^{ma}\eta) = 2(\lambda\gamma^a\Phi_a\slashed{D}\eta^i) \overset{(2.1)}{=}$$
$$= -(\frac{1}{6}R + M)(\lambda\gamma^a\Phi_a\varepsilon) \tag{4.92}$$

so all terms in (4.91) cancel when added together.

Next we consider the remaining T-terms for a generic background. We set

$$\mathcal{L} = \mathcal{L}_{\text{cflat}} + \mathcal{L}_T \tag{4.93}$$

where

$$\mathcal{L}_T = -F^{\mu\nu}T_{\mu\nu a}\Phi^a + \frac{1}{4}T_{\mu\nu a}T^{\mu\nu b}\Phi^a\Phi_b \tag{4.94}$$

and we find

$$\delta(\mathcal{L}_T) = (\lambda\gamma^\nu\varepsilon)[D^\mu T_{\mu\nu a}]\Phi^a$$
$$+ \underbrace{(\lambda\gamma^\nu\varepsilon)T_{\mu\nu a}F^{\mu a}}_{①} - \underbrace{\frac{1}{2}(\lambda\gamma^a\varepsilon)F^{\mu\nu}T_{\mu\nu a}}_{②} + \underbrace{\frac{1}{4}(\lambda\gamma^a\varepsilon)T_{\mu\nu a}T_{\mu\nu b}\Phi^b}_{③} \tag{4.95}$$

In the variation of the fermionic action the new terms are

$$\delta(\lambda\gamma^m D_m\lambda) = \delta_{\text{cflat}}(\lambda\gamma^m D_m\lambda) + \textbf{term}_{\text{T1}} + \textbf{term}_{\text{T2}} \tag{4.96}$$

where $\textbf{term}_{\text{T1}}$ comes from T-terms in generalized conformal Killing equation (2.3) and $\textbf{term}_{\text{T2}}$ comes from the T-term in the variation $\delta_{\varepsilon,\eta}\lambda$ (2.8)

$$\textbf{term}_{\text{T1}} = 2\Phi_a(\lambda\gamma^a \slashed{D}\eta)_T + \tfrac{1}{32}(\lambda\gamma^\mu \slashed{F} T \gamma_\mu\varepsilon) \tag{4.97}$$

Then we find

$$\tfrac{1}{32}(\lambda\gamma_\mu \slashed{F} \tfrac{1}{16} T \gamma^\mu\varepsilon) = \underbrace{\tfrac{1}{2}(\lambda\gamma^a\varepsilon)F^{\mu\nu}T_{\mu\nu a}}_{\textcircled{2}} - \underbrace{\tfrac{1}{2}(\lambda\gamma^\nu\varepsilon)T_{\mu\nu a}F^{\mu a}}_{\textcircled{1}}$$
$$+ \underbrace{\tfrac{1}{2}(\lambda\gamma^{\nu ab}\varepsilon)T_{\mu\nu a}F^{\mu b}}_{\textcircled{4}} \tag{4.98}$$

and

$$\textbf{term}_{\text{T2}} = -\tfrac{1}{4}\lambda\slashed{D}(T_{\mu\nu a}\Phi^a\gamma^{\mu\nu}\varepsilon) =$$
$$= -\tfrac{1}{4}(\lambda\gamma^\rho\gamma^{\mu\nu}\varepsilon)[D_\rho T_{\mu\nu a}]\Phi^a \underbrace{-\tfrac{1}{2}(\lambda\gamma^\nu\varepsilon)T_{\mu\nu a}F_{\mu a}}_{\textcircled{1}}$$
$$\underbrace{-\tfrac{1}{4}(\lambda\gamma^{\rho\mu\nu}\varepsilon)T_{\mu\nu a}F_{\rho a}}_{\textcircled{4}} \underbrace{-\tfrac{1}{4}(\lambda\gamma^a\varepsilon)T_{\mu\nu a}T_{\mu\nu b}\Phi^b}_{\textcircled{3}} \tag{4.99}$$

Using (4.61) all TF terms cancel between (4.99) and (4.95) and finally the $[DT]\Phi$ terms in (4.95), (4.97), (4.99) cancel as well as

$$(4.95): \quad (\lambda\gamma^\nu\varepsilon)[D^\mu T_{\mu\nu a}]\Phi^a$$
$$(4.97): \quad \tfrac{1}{2}(D^\mu T_{\mu\nu a})\Phi_b(\lambda\gamma^b\gamma^{\nu a}\varepsilon) = -\tfrac{1}{2}(\lambda\gamma^\nu\varepsilon)D^\mu T_{\mu\nu a}\Phi^a + \tfrac{1}{2}(\lambda\gamma^{\nu ab}\varepsilon)\Phi^b D^\mu T_{\mu\nu a}$$
$$(4.99): \quad -\tfrac{1}{4}\lambda D_\rho(T_{\mu\nu a})\Phi^a\gamma^\rho\gamma^{\mu\nu}\varepsilon = -\tfrac{1}{2}(\lambda\gamma^\mu\varepsilon)D^\mu T_{\mu\nu a}\Phi^a - \tfrac{1}{4}(\lambda\gamma^{\mu\nu\rho}\varepsilon)\Phi^a D_\rho T_{\mu\nu a}$$
$$\tag{4.100}$$

References

[1] Sklyanin, E., Faddeev, L.: Quantum mechanical approach to completely integrable field theory models. Sov. Phys. Dokl. **23**, 902–904 (1978)

[2] Drinfeld, V.G.: Hopf algebras and the quantum Yang-Baxter equation. Dokl. Akad. Nauk SSSR **283**(5), 1060–1064 (1985)

[3] Jimbo, M.: A q-difference analogue of $U(\mathfrak{g})$ and the Yang-Baxter equation. Lett. Math. Phys. **10**(1), 63–69 (1985). doi:10.1007/BF00704588

[4]   Seiberg, N., Witten, E.: Electric-magnetic duality, monopole condensation, and confine-
        ment in N = 2 supersymmetric Yang-Mills theory. Nucl. Phys. **B426**, 19–52 (1994).
        arXiv:hep-th/9407087 [hep-th]
[5]   Atiyah, M.F., Bott, R.: The moment map and equivariant cohomology. Topology **23**(1),
        1–28 (1984). doi:10.1016/0040-9383(84)90021-1
[6]   Witten, E.: Supersymmetry and Morse theory. J. Differ. Geom. **17**, 661–692 (1982)
[7]   Witten, E.: Topological sigma models. Commun. Math. Phys. **118**, 411 (1988)
[8]   Witten, E.: Topological quantum field theory. Commun. Math. Phys. **117**, 353 (1988)
[9]   Nekrasov, N.A.: Seiberg-Witten prepotential from instanton counting. Adv. Theor. Math.
        Phys. **7**, 831–864 (2004). arXiv:hep-th/0206161 [hep-th]. To Arkady Vainshtein on his
        60th anniversary
[10]   Losev, A., Nekrasov, E., Shatashvili, S.L.: Issues in topological gauge theory. Nucl. Phys.
        B534, 549–611 (1998) arXiv:hep-th/9711108 [hep-th]
[11]   Moore, G.W., Nekrasov, N., Shatashvili, S.: Integrating over Higgs branches. Commun.
        Math. Phys. **209**, 97–121 (2000) arXiv:hep-th/9712241 [hep-th]
[12]   Lossev, A., Nekrasov, N., Shatashvili, S.L.: Testing Seiberg-Witten solution.
        arXiv:hep-th/9801061 [hep-th]
[13]   Losev, A., Moore, G.W., Nekrasov, N., Shatashvili, S.: Four-dimensional avatars of two-
        dimensional RCFT. Nucl. Phys. Proc. Suppl. **46**, 130–145 (1996). arXiv:hep-th/9509151
[14]   Faddeev, L., Reshetikhin, N.Y., Takhtajan, L.: Quantization of Lie groups and Lie algebras.
        Leningr. Math. J. **1**, 193–225 (1990)
[15]   Nekrasov, N., Pestun, V.: Seiberg-Witten geometry of four dimensional N = 2 quiver
        gauge theories. arXiv:1211.2240 [hep-th]
[16]   Nekrasov, N., Pestun, V., Shatashvili, S.: Quantum geometry and quiver gauge theories.
        arXiv:1312.6689 [hep-th]
[17]   Nekrasov, N.: On the BPS/CFT correspondence, February 3, 2004. http://www.science.
        uva.nl/research/itf/strings/stringseminar2003-4.html. Lecture at the string theory group
        seminar, University of Amsterdam
[18]   Nakajima, H.: Gauge theory on resolutions of simple singularities and simple Lie algebras.
        Int. Math. Res. Not. **2**, 61–74 (1994). doi:10.1155/S1073792894000085
[19]   Nakajima, H.: Quiver varieties and Kac-Moody algebras. Duke Math. J. **91**(3), 515–560
        (1998). doi:10.1215/S0012-7094-98-09120-7
[20]   Vafa, C., Witten, E.: A strong coupling test of S duality. Nucl. Phys. **B431**, 3–77 (1994).
        arXiv:hep-th/9408074 [hep-th]
[21]   Losev, A.S., Marshakov, A., Nekrasov, N.A.: Small instantons, little strings and free fermi-
        ons. arXiv:hep-th/0302191 [hep-th]
[22]   Nekrasov, N., Okounkov, A.: Seiberg-Witten theory and random partitions.
        arXiv:hep-th/0306238 [hep-th]
[23]   Pestun, V.: Localization of gauge theory on a four-sphere and supersymmetric Wilson
        loops. arXiv:0712.2824 [hep-th]
[24]   Gomis, J., Okuda, T., Pestun, V.: Exact results for 't Hooft loops in gauge theories on S^4.
        arXiv:1105.2568 [hep-th]
[25]   Kapustin, A., Willett, B., Yaakov, I.: Exact results for Wilson loops in superconformal
        Chern-Simons theories with matter. JHEP **1003**, 089 (2010). arXiv:0909.4559 [hep-th]
[26]   Benini, F., Cremonesi, S.: Partition functions of N = (2,2) gauge theories on S^2 and vortices.
        arXiv:1206.2356 [hep-th]
[27]   Doroud, N., Gomis, J., Le Floch, B., Lee, S.: Exact results in D = 2 supersymmetric gauge
        theories. JHEP **1305**, 093 (2013). arXiv:1206.2606 [hep-th]
[28]   Kallen, J., Qiu, J., Zabzine, M.: The perturbative partition function of supersymmet-
        ric 5D Yang-Mills theory with matter on the five-sphere. JHEP **1208**, 157 (2012).
        arXiv:1206.6008 [hep-th]
[29]   Kim, H.-C., Kim, S.: M5-branes from gauge theories on the 5-sphere. JHEP **1305**, 144
        (2013). arXiv:1206.6339 [hep-th]

[30] Hama, N., Hosomichi, K., Lee, S.: SUSY gauge theories on squashed three-spheres. JHEP **1105**, 014 (2011). arXiv:1102.4716 [hep-th]

[31] Imamura, Y., Yokoyama, D.: N = 2 supersymmetric theories on squashed three-sphere. Phys. Rev. **D85**, 025015 (2012). arXiv:1109.4734 [hep-th]

[32] Hama, N., Hosomichi, K.: Seiberg-Witten theories on ellipsoids. JHEP **1209**, 033 (2012). arXiv:1206.6359 [hep-th]

[33] Alday, L.F., Gaiotto, D., Tachikawa, Y.: Liouville correlation functions from four-dimensional gauge theories. Lett. Math. Phys. **91**, 167–197 (2010). arXiv:0906.3219 [hep-th]

[34] Festuccia, G., Seiberg, N.: Rigid supersymmetric theories in curved superspace. JHEP **1106**, 114 (2011). arXiv:1105.0689 [hep-th]

[35] de Wit, B., Van Holten, J., Van Proeyen, A.: Transformation rules of N = 2 supergravity multiplets. Nucl. Phys. **B167**, 186 (1980)

[36] de Wit, B., Van Holten, J., Van Proeyen, A.: Structure of N = 2 supergravity. Nucl. Phys. **B184**, 77 (1981)

[37] de Wit, B., Lauwers, P., Van Proeyen, A.: Lagrangians of $N = 2$ supergravity - matter systems. Nucl. Phys. **B255**, 569 (1985)

[38] de Wit, B., Lauwers, P., Philippe, R., Van Proeyen, A.: Noncompact N = 2 supergravity. Phys. Lett. **B135**, 295 (1984)

[39] Mohaupt, T.: Black hole entropy, special geometry and strings. Fortsch. Phys. **49**, 3–161 (2001). arXiv:hep-th/0007195 [hep-th]

[40] Freedman, D.Z., Van Proeyen, A.: Supergravity. Cambridge University Press, Cambridge (2012). doi:10.1017/CBO9781139026833

[41] Gupta, R.K., Murthy, S.: All solutions of the localization equations for N = 2 quantum black hole entropy. JHEP **1302**, 141 (2013). arXiv:1208.6221 [hep-th]

[42] Dabholkar, A., Gomes, J., Murthy, S.: Quantum black holes, localization and the topological string. JHEP **1106**, 019 (2011). arXiv:1012.0265 [hep-th]

[43] Klare, C., Zaffaroni, A.: Extended supersymmetry on curved spaces. JHEP **1310**, 218 (2013). arXiv:1308.1102 [hep-th]

[44] Proeyen, A.V.: Lectures on N = 2 supergravity. http://itf.fys.kuleuven.be/~toine/LectParis.pdf

[45] Dumitrescu, T.T., Festuccia, G., Seiberg, N.: Exploring curved superspace. JHEP **1208**, 141 (2012). arXiv:1205.1115 [hep-th]

[46] Nakajima, H., Yoshioka, K.: Lectures on instanton counting, ArXiv Mathematics e-prints (2003). arXiv:math/0311058

[47] Nosaka, T., Terashima, S.: Supersymmetric gauge theories on a squashed four-sphere. JHEP **1312**, 001 (2013). arXiv:1310.5939 [hep-th]

[48] Vartanov, G., Teschner, J.: Supersymmetric gauge theories, quantization of moduli spaces of flat connections, and conformal field theory. arXiv:1302.3778 [hep-th]

[49] Nekrasov, N., Witten, E.: The omega deformation, branes, integrability, and Liouville theory. JHEP **1009**, 092 (2010). arXiv:1002.0888 [hep-th]

[50] Kinney, J., Maldacena, J.M., Minwalla, S., Raju, S.: An Index for 4 dimensional super conformal theories. Commun. Math. Phys. **275** 209–254 (2007). arXiv:hep-th/0510251 [hep-th]

[51] Romelsberger, C.: Counting chiral primaries in N = 4 superconformal field theories. Nucl. Phys. **B747**, 329–353 (2006). arXiv:hep-th/0510060 [hep-th]

[52] Dolan, F., Osborn, H.: Applications of the superconformal index for protected operators and q-hypergeometric identities to N = 1 dual theories. Nucl. Phys. **B818**, 137–178 (2009). arXiv:0801.4947 [hep-th]

[53] Dolan, F., Spiridonov, V., Vartanov, G.: From 4d superconformal indices to 3d partition functions. Phys. Lett. **B704**, 234–241 (2011). arXiv:1104.1787 [hep-th]

[54] Gadde, A., Rastelli, L., Razamat, S.S., Yan, W.: Gauge theories and Macdonald polynomials. Commun. Math. Phys. **319**, 147–193 (2013). arXiv:1110.3740 [hep-th]

[55] Witten, E.: Two-dimensional gauge theories revisited. J. Geom. Phys. **9**, 303–368 (1992). arXiv:hep-th/9204083 [hep-th]

[56] Atiyah, M.F., Jeffrey, L.: Topological Lagrangians and cohomology. J. Geom. Phys. **7**(1), 119–136 (1990). doi:10.1016/0393-0440(90)90023-V

[57] Cordes, S., Moore, G.W., Ramgoolam, S.: Lectures on 2-d Yang-Mills theory, equivariant cohomology and topological field theories. Nucl. Phys. Proc. Suppl. **41**, 184–244 (1995). arXiv:hep-th/9411210 [hep-th]

[58] Pestun, V.: Localization of the four-dimensional $N = 4$ SYM to a two-sphere and 1/8 BPS Wilson loops, 49pp. arXiv:0906.0638 [hep-th]

[59] Atiyah, M.F.: Elliptic Operators and Compact Groups. Lecture Notes in Mathematics, vol. 401. Springer, Berlin (1974)

[60] Kennedy, A.: Clifford algebras in two omega dimensions. J. Math. Phys. **22**, 1330–1337 (1981)

Line Operators in Supersymmetric Gauge Theories and the 2d-4d Relation

Takuya Okuda

Abstract Four-dimensional gauge theories with $\mathcal{N} = 2$ supersymmetry admit half-BPS line operators. We review the exact localization methods for analyzing these operators. We also review the roles they play in the relation between four- and two-dimensional field theories, and explain how the two-dimensional CFT can be used to obtain the quantitative results for 4d line operators. This is a contribution to the special LMP volume on the 2d-4d relation, edited by J. Teschner.

1 Introduction

Gauge theory, a fundamental description of nature in our current understanding of particle physics, remains a central subject in theoretical physics. Any quantum field theory with gauge fields possesses a set of universal observables, namely Wilson-'t Hooft line operators, also known as loop operators. The Wilson loop $\mathrm{Tr}\, P \exp(i \oint A)$ exhibits an area law in a confining vacuum. A magnetic analog, the 't Hooft loop, is a disorder operator defined by a singular boundary condition of the gauge field. A Higgs phase can be characterized by a 't Hooft loop obeying an area law. More generally, the behavior of mixed Wilson-'t Hooft operators can be used to classify the vacuum structures of gauge theories [1]. Quantitative understanding of these operators in a non-abelian gauge theory such as QCD is an important open problem.

Four-dimensional theories with extended supersymmetry admit BPS line operators, which represent infinitely massive BPS particles. While they have no known role as order parameters for low-energy physics, the BPS line operators serve as useful probes of various dualities. BPS Wilson loops in $\mathcal{N} = 4$ super Yang-Mills theory were first introduced in [2, 3], and for many years they were studied mostly in the context of the AdS/CFT correspondence. In [4] it was found that the perturbative ladder diagram contributions to the expectation value of a half-BPS circular Wilson

---

A citation of the form [V:x] refers to article number x in this volume.

T. Okuda (✉)
University of Tokyo, Komaba, Meguro-ku, Tokyo 153-8902, Japan
e-mail: takuya@hep1.c.u-tokyo.ac.jp

© Springer International Publishing Switzerland 2016
J. Teschner (ed.), *New Dualities of Supersymmetric Gauge Theories*,
Mathematical Physics Studies, DOI 10.1007/978-3-319-18769-3_7

loop in $\mathcal{N} = 4$ theory with $SU(N)$ gauge group reproduce the large 't Hooft coupling result from AdS. The ladder diagram contributions can be neatly packaged into a Gaussian matrix model, and the authors of [4] conjectured that the matrix model computes the Wilson loop vev in the large N limit. Based on a conformal anomaly, [5] further conjectured that the agreement should hold to all orders in $1/N$ and in the 't Hooft coupling. The agreement was finally proved to be exact in the paper [6], where general $\mathcal{N} = 2$ theories were also treated.

This article reviews the recent developments in the study of BPS line operators in 4d $\mathcal{N} = 2$ gauge theories. There are two main ideas: localization and the 2d-4d correspondence. The former, whose modern version was invented by Pestun [6] building upon earlier works [7, 8], can be applied to line operators in various geometries to obtain exact results. The latter, in particular the AGT correspondence [9], can be used to compute the expectation values of line operators by 2d CFT techniques.

This article is organized as follows. In Sect. 2 we will review the definition of Wilson-'t Hooft line operators and the classification of charges, as well as their counterparts in two dimensions. In Sect. 3 we will review the localization methods applied to line operators. Section 4 is devoted to explaining the 2d CFT techniques used to compute 4d observables involving line operators. The line operators exhibit interesting algebraic structures, which are closely related to a quantization of the Hitchin moduli space. These matters will be reviewed in Sect. 5. The appendix summarizes some relevant facts.

Our emphasis is on the intrinsic UV dynamics of 4d line operators. The exact computation of disorder line operators, which was made possible by localization and was inspired by the 2d-4d relation, is a remarkable progress. The 2d theories themselves also display very rich physics. Other review articles in this volume discuss closely related subjects from different angles.

2   Charges of Line Operators

In this section we review the classification of BPS line operators in an $\mathcal{N} = 2$ gauge theory with gauge group G. We begin by considering all line operators allowed by the Lie algebra of G and the matter content. We will also explain the basic correspondence between line operators and closed curves on a Riemann surface. Then we will review the recent progress on the discrete choice one must make to fully specify a quantum field theory, and how it relates to the spectrum of line operators.

2.1   Definition and Charges of 4d Line Operators

Let us recall some basics of Lie algebras and set notation. (See for example the appendix of [10] for a useful summary.) We denote by t the Cartan subalgebra of G, and by t^* the dual of t. The roots and weights of G take values in t^*, and generate the root lattice Λ_r and the weight lattice Λ_w respectively. We have $\Lambda_r \subset \Lambda_w$. We define the coroot lattice $\Lambda_{cr} \subset \mathfrak{t}$ to be the dual of Λ_w, and the coweight lattice $\Lambda_{cw} \subset \mathfrak{t}$ to be the dual of Λ_r.

Wilson line operators that preserve some supersymmetry were first introduced in the context of AdS/CFT correspondence [2, 3]. For an $\mathcal{N} = 2$ gauge theory on $\mathbb{R}^4$, let us focus on the half-BPS Wilson operators along a straight line or a circle, defined as

$$W_R = \mathrm{Tr}_R P \exp\left[\oint (i A + \mathrm{Re}\,\phi\,ds)\right], \tag{1}$$

where ϕ is a complex scalar in the vector multiplet and ds is the line element determined by the metric. There exist more general curves and scalar couplings that preserve some amount of supersymmetry; their classification may be possible by extending the methods of [11]. Charges of supersymmetric Wilson operators are classified by irreducible representations R, or equivalently their highest weights $w \in \Lambda_w$. Physically, Wilson operators represent the worldline of an electrically charged BPS particle with infinite mass.

The magnetic analogue, 't Hooft operators, were originally introduced to classify the low-energy behavior of non-conformal gauge theories [1], and represent the trajectory of an infinitely heavy magnetic monopole in spacetime. These operators were generalized to preserve a half of supersymmetry in [12]. A 't Hooft operator is a *disorder operator*, meaning that it is defined by a singular boundary condition on the fields in the path integral. In the present case, we define the operator $T(B)$ by demanding that we integrate over the field configurations with Dirac monopole singularities

$$F \sim \frac{B}{4}\epsilon_{ijk}\frac{x^i}{r^3}dx^k \wedge dx^j = -\frac{B}{2}\sin\theta d\theta \wedge d\varphi, \quad \phi \sim i\frac{B}{2r}. \tag{2}$$

In writing this, we assumed that the theta angles of the gauge theory are zero; if they are not we need to excite $\mathrm{Re}\,\phi$ as well as electric components of the field strength on top of (2), for the gauge Noether charge to vanish [13] and supersymmetry to be preserved. We have introduced (x^1, x^2, x^3) and (r, θ, φ), locally-defined Cartesian and spherical coordinates in the directions orthogonal to the trajectory. The 't Hooft operator preserves the same set of supercharges as the Wilson operator (1) when placed along the same curve. The magnetic charge B is constrained by the Dirac quantization condition. Namely, the gauge potential $A \sim -(B/2)(1 - \cos\theta)d\varphi$ has a Dirac string singularity along $\theta = \pi$. For the Dirac string to be unphysical the matter fields must be single-valued. Thus, if we denote by $\langle\,,\rangle$ the natural pairing between coweights and weights, the magnetic charge B must satisfy

$$\langle B, w \rangle \in \mathbb{Z} \tag{3}$$

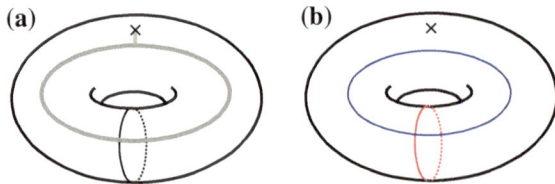

Fig. 1 **a** A one-punctured torus $C_{1,1}$ is decomposed into a pair of pants with one leg degenerate. A trivalent graph Γ is drawn on the decomposed surface. **b** The *red curve*, a pants leg, has $(p, q) = (0, 1)$ and corresponds to the minimal Wilson operator in $\mathcal{N} = 2^*$ theory. The *blue curve* with $(p, q) = (1, 0)$ corresponds to the minimal 't Hooft operator

for the highest weight $w \in \Lambda_w$ of any irreducible representation in which a matter field transforms. The coweights B satisfying (3) form a lattice Λ_m, the dual of the lattice generated by roots and the weights of matter representations.

More generally, dyonic operators, or mixed Wilson-'t Hooft operators, are classified by pairs $(B, w) \in \Lambda_m \times \Lambda_w$ modulo the action of the Weyl group. The weight w may be interpreted as the highest weight of a representation corresponding to the Wilson loop for the subgroup unbroken by B, for an appropriate choice of representative (B, w).

It is illuminating to consider the class S theories of type A_1, reviewed in the Appendix and [V:2]. A weakly coupled description of such a theory is specified by a choice of pants decomposition and a trivalent graph Γ drawn on $C_{g,n}$. (See Fig. 1a for an example.) The universal covering of the gauge group G is $\tilde{G} = SU(2)^{3g-3+n} =: \prod_i SU(2)_i$, where $i = 1, \ldots, 3g - 3 + n$ labels the internal edges of Γ. The external edges labeled by $e = 3g - 2 + n, \ldots, 3g - 3 + 2n$ correspond to the factors in the flavor group $G_F = SU(2)^n =: \prod_e SU(2)_e$. In addition to the electric and magnetic charges for G, it turns out to be convenient to allow line operators to have the magnetic charges for G_F by coupling the theory, via the Cartan of G_F, to non-dynamical gauge multiplets of the form (2). Thus we consider the coweights $\vec{p} = (p_1, \ldots, p_{3g-3+2n}) \in \Lambda_{cw}(\tilde{G} \times G_F) = \mathbb{Z}^{3g-3+2n}$ of the extended group and the weights $\vec{q} = (q_1, \ldots, q_{3g-3+n}) \in \Lambda_w(\tilde{G}) = \mathbb{Z}^{3g-3+n}$ of the gauge group. Each trivalent vertex of the graph Γ corresponds to a half-hypermultiplet with its scalars Φ_{jkl} transforming in the trifundamental representation of $SU(2)_j \times SU(2)_k \times SU(2)_l$, where j, k, l correspond to either a gauge or flavor symmetry and do not need to be distinct. For Φ_{jkl} to be single-valued around a Dirac string, the coweight must satisfy

$$p_j + p_k + p_l \in 2\mathbb{Z}. \tag{4}$$

The conditions (4), imposed for all triplets (j, k, l) corresponding to vertices, define the lattice $\Lambda_m \subset \Lambda_{cw}$. We also identify the charges related by the Weyl group $\mathbb{Z}_2^{3g-3+2n}$. Thus the charges of the line operators in this theory are classified by the set of integers $(\vec{p}, \vec{q})$, subject to the conditions (4) for each trivalent vertex, with the charges identified if they are related by the Weyl group action, i.e., $(p_i, q_i) \mapsto (-p_i, -q_i)$ for some internal edge i or $p_e \mapsto -p_e$ for an external edge *e*. Identification by the Weyl group action is equivalent to requiring that p_i, $p_e \geq 0$, and also that $q_i \geq 0$ if $p_i = 0$. Ordinary Wilson-'t Hooft operators, without non-dynamical fields involved, correspond to $(\vec{p}, \vec{q})$ with $p_e = 0$ for all external edges e.

2.2 Correspondence of Charges and Curves

We are ready to describe the basic 2d/4d relation for the charges of line operators in the A_1-theories [14]. We fix one pants decomposition and use it to describe all homotopy classes of closed curves on $C_{g,n}$ without self-intersection, as well as homotopy classes of arcs connecting punctures without self-intersection. We allow the curve γ to have multiple components, but we assume that no component is homotopic to a point or a curve arbitrarily close to a puncture. Let $\gamma_1, \ldots, \gamma_{3g-3+n}$ be pairwise disjoint connected curves without self-intersection whose complement is a pants decomposition of $C_{g,n}$; these are known as pants legs. Also let $\gamma_{3g-3+n+1}, \ldots, \gamma_{3g-3+2n}$ be simple closed curves near the punctures. In order to describe the correspondence, we define the intersection number $p_j = \#(\gamma \cap \gamma_j)$ for $1 \leq j \leq 3g - 3 + 2n$ to be the minimum of the number, without a sign, of intersection points as γ and γ_j vary among non-self-intersecting curves in their respective isotopy classes. We also need a notion of twisting number q_j for $1 \leq j \leq 3g - 3 + n$. Roughly, q_j counts how many times γ winds around in a direction parallel to γ_j. We refer the reader to [14] for a precise definition. The crucial mathematical fact for the correspondence is the following theorem.

Dehn's Theorem. Let $C_{g,n}$ be an oriented punctured Riemann surface of genus g and negative Euler characteristic with n punctures. Let us define a map

$$\gamma \mapsto (\#(\gamma \cap \gamma_j); q_j) \in (\mathbb{Z}_{\geq 0})^{3g-3+2n} \times \mathbb{Z}^{3g-3+n} \tag{5}$$

which assigns, to each isotopy class of closed curves without self-intersection or arcs connecting punctures without self-intersection, its intersection number $p_j = \#(\gamma \cap \gamma_j)$ with γ_j ($1 \leq j \leq 3g - 3 + 2n$) and its twisting number q_j with respect to γ_j ($1 \leq j \leq 3g - 3 + n$). Note that the intersection and twisting numbers depend only on the homotopy class of γ. With this definition the map is injective, and its image in (5) is

$$\{(p_1, p_2, \ldots, p_{3g-3+2n}; q_1, q_2, \ldots, q_{3g-3+n})$$
$$| \text{ if } p_j = 0 \text{ then } q_j \geq 0, \text{ and } p_j + p_k + p_l \in 2\mathbb{Z}$$
$$\text{when } \gamma_j \cup \gamma_k \cup \gamma_l \text{ is the boundary of a pair of pants}\}.$$

The integers p_j, q_j are known as the Dehn-Thurston parameters of γ.

It is immediate to recognize (6) as the same data that classify the line operator charges in the A_1-theory corresponding to $C_{g,n}$. This is the most basic 2d-4d correspondence involving line operators [14]. An example in Fig. 1b shows curves on the one-punctured torus corresponding to line operators in the $SU(2)$ $\mathcal{N} = 2^*$ theory. The rules for the action of the modular groupoid on the Dehn-Thurston parameters are explicitly known [15]. Since this action is interpreted as S-duality according to the 2d/4d correspondence, the line operators with charges $(\vec{p}, \vec{q})$ are believed to transform according to the same rules.

2.3 Spectrum of Line Operators and Discrete Theta Angles

The specification of a gauge theory, before picking a spacetime geometry, requires several discrete choices. The choice can be phrased in terms of line operators [16, 17].

On $\mathbb{R}^4$ for some purposes one considers all "line operators" that are allowed by the Lie algebra of the gauge group and the matter content. This is useful in the classification of massive phases of a gauge theory [1] by representations of the 't Hooft commutation relation

$$W \cdot T = e^{2\pi i/N} T \cdot W \tag{6}$$

for fundamental Wilson (W) and 't Hooft (T) loops. Here the gauge group has the Lie algebra of $SU(N)$, and we consider two closed curves C_W and C_T that are contained and Hopf-linked in a constant time slice. We place T on C_T while we displace W infinitesimally from C_W forward and backward in time, so that the two sides of (6) arise as operator products that are differently time-ordered. Massive phases such as confining and Higgs vacua arise as representations of (6). Even though C_W and C_T are linked within the three-dimensional slice, C_T and displaced C_W are, for dimensional reasons, not linked in the ambient spacetime. The relation $W \cdot T \neq T \cdot W$ means that the two operators cannot both be genuine line operators. If the gauge group is $SU(N)$, W is a genuine loop operator that is invariant under all gauge transformations, even when it is placed along a homotopically non-trivial curve. On the left hand side of (6) we can link W with the surface swept by the Dirac strings that extend from T in the future time direction. Then $W \cdot T$ picks up the phase $e^{2\pi i/N}$, relative to $T \cdot W$, from the holonomy around the Dirac string. One cannot continuously deform one configuration to the other without W hitting the Dirac sheet of T. Thus T is a boundary of a surface operator defined by the Dirac sheet. For gauge group $SU(N)/\mathbb{Z}_N$ with a zero theta angle, T is a genuine line operator and W is a boundary of a surface operator [18].

For a general gauge group G and two pairs of charges $(B_1, w_1), (B_2, w_2) \in \Lambda_{\mathrm{cw}} \times \Lambda_{\mathrm{w}}$, the product of two would-be loop operators acquires a phase

$$\exp\left(2\pi i \langle B_2, w_1 \rangle - 2\pi i \langle B_1, w_2 \rangle\right) \tag{7}$$

when one operator moves around along a surface that links the other. The correlation function of two genuine loop operators is well-defined only if the phase vanishes, in which case they are *mutually local*. In a consistent theory line operators must be mutually local, and their spectrum must be *maximal* in the sense that one cannot add more line operators without violating mutual locality [16, 17]. The center Z of the universal covering $\tilde{G}$ of G, and its dual Z^*, are isomorphic to $\Lambda_{cw}/\Lambda_{cr}$ and Λ_w/Λ_r, respectively. The phase (7) depends only on the elements $z_1, z_2 \in Z \times Z^*$ that correspond to the two loop operators. A maximal mutually local spectrum then translates to a maximal isotropic subgroup of $Z \times Z^*$.

The authors of [17] showed that the choice of a maximal coisotropic subgroup is equivalent to the choice of what they call *discrete theta angles*. These parameters give, in the Euclidian path integral, non-trivial phases dependent on the topological classes of gauge bundles. Correspondence with line operators arises because the discrete magnetic flux through the $\mathbb{S}^2$ surrounding a line operator induces, combined with discrete theta angles, a discrete electric charge, much as in the usual Witten effect [13]. For A_1 theories of class $\mathcal{S}$ corresponding to a Riemann surface C with no puncture, the discrete choice corresponds to a maximal coisotropic subgroup Δ of $H^1(C, \mathcal{C})$, where $\mathcal{C}$ is the center of the simply connected group with Lie algebra $\mathfrak{g}$ [16, 19].

3 Exact Results for Line Operators by Localization

Line operators with electric and magnetic charges constructed in the previous section fit nicely the framework of supersymmetric localization [7]. In this section we review the exact localization computation of loop/line operators in 4d $\mathcal{N} = 2$ theories on $\mathbb{S}^4$ and other geometries. Since details on the $\mathbb{S}^4$ localization can be found in [6] as well as in the review article [V:6], we restrict ourselves to the bare basics and focus on the features specific to 't Hooft operators. The results will be successfully matched with 2d computations in Sect. 4.2.

3.1 Localization for Wilson loops on $\mathbb{S}^4$

The general procedure in a localization computation consists of the following steps [6].

1. Pick a supercharge Q that annihilates the operator one wants to evaluate. If Q^2 contains terms that vanish on shell, i.e., vanish by the equation of motion, add auxiliary fields so that such terms do not appear. Then Q^2 is a linear combination of bosonic symmetry generators.
2. Choose a Q^2-invariant functional V such that the bosonic terms of $Q \cdot V$ are positive-semidefinite. Add $tQ \cdot V$ to the action where t is a constant. Find the saddle points of the path integral $\int e^{-S_{cl} - tQ \cdot V}$ in the limit $t \to +\infty$, in other words find the configurations such that $Q \cdot V = 0$.
3. Evaluate the classical action S_{cl} and the inserted operator at the saddle point.

4. Compute the fluctuation determinant at the saddle points. This involves gauge-fixing and the inclusion of ghost fields. Either expand fields in the eigenmodes of kinetic operators, or use the equivariant index theorem to compute the determinant.
5. Sum and integrate the above contributions over all the saddle points.

It is possible to define $\mathcal{N} = 2$ non-conformal supersymmetry on $\mathbb{S}^4$, and the steps above were carried out in [6] to compute the partition functions and the Wilson loop vevs for $\mathcal{N} = 2$ gauge theories on $\mathbb{S}^4$.

One of the key steps is the computation of the fluctuation determinant. This is the ratio $\det \Delta_o / (\det \Delta_e)^{1/2}$, where (Δ_e, Δ_o) are the differential operators acting on bosons and fermions in $Q \cdot V$. On $\mathbb{S}^4$ we choose $V = \sum(\text{fermion}) \cdot \overline{Q}(\text{fermion})$, and (Δ_e, Δ_o) can be expressed in terms of simpler differential operators in V. Supersymmetry implies many cancellations among the eigenvalues, and one can show that the determinant is given by

$$\frac{\det \Delta_o}{(\det \Delta_e)^{1/2}} = \left(\frac{\det_{\mathrm{coker}\mathcal{D}} \mathcal{Q}^2}{\det_{\mathrm{ker}\mathcal{D}} \mathcal{Q}^2} \right)^{1/2}, \tag{8}$$

where $\mathcal{D}$ is a differential operator in V, and we recall that $\mathcal{Q}^2$ is a sum of bosonic symmetry generators. Schematically,

$$\mathcal{Q}^2 \sim J + R + a + m, \tag{9}$$

where J, R, a, and m generate an isometry, an R-symmetry rotation, a gauge transformation, and a flavor symmetry transformation. Despite the huge cancellations the fluctuation determinant (8) is still an infinite product and takes the form $\prod_j w_j^{c_j/2}$ with $c_j = \pm 1$. The weights w_j and the signs c_j can be read off from the equivariant index

$$\mathrm{ind}\,\mathcal{D} \equiv \mathrm{Tr}_{\mathrm{ker}\mathcal{D}} e^{\mathcal{Q}^2} - \mathrm{Tr}_{\mathrm{coker}\mathcal{D}} e^{\mathcal{Q}^2} = \sum c_j e^{w_j}, \tag{10}$$

which can be computed by the Atiyah-Singer index theory. In particular, the fixed point formula expresses the index as a sum of contributions from the fixed points of the isometry J. Thus the fluctuation determinant for each saddle point configuration, computed by the index theory, naturally factorizes into the contributions from the fixed points, namely the north and south poles of $\mathbb{S}^4$.

At the north pole, the differential operator $\mathcal{D}$ acts on the vector multiplet fields as the linearization of the anti-self-duality equations, which govern instantons on $\mathbb{C}^2$:

$$\Omega^1(\mathrm{ad}\,E) \xrightarrow{\;D^\dagger \oplus (1+*)D\;} \Omega^0(\mathrm{ad}\,E) \oplus \Omega^{2+}(\mathrm{ad}\,E). \tag{11}$$

On the hypermultiplet $\mathcal{D}$ acts as the Dirac operator. The structure at the south pole is similar, with anti-instantons replacing instantons.

Non-perturbative contributions arise as small instantons and anti-instantons localized at the north and south poles respectively. More precisely, these are the Q^2-fixed points on the instanton moduli space. Let us denote by $Z_{\text{1-loop}}^{\text{pole}}(ia, im_f)$ the north pole contribution to the fluctuation determinant in the topologically trivial backgrounds. The variable a, taking values in the Cartan subalgebra $\mathfrak{t}$, parametrizes the saddle point configurations and is identified with the background value of a vector multiplet scalar Re ϕ. The parameters m_f denote the masses of matter hypermultiplets. A topologically non-trivial configuration at the north pole contributes the universal factor $Z_{\text{1-loop}}^{\text{pole}}(ia, im_f)$, accompanied by an extra rational function of a and m_f. The sum of the rational functions over the invariant instanton configurations is the instanton partition function Z_{inst} [8] with the omega deformation parameters specialized to the values $\epsilon_1 = \epsilon_2 = 1/r$, where r is the radius of $\mathbb{S}^4$. The contributions from the south pole have a similar structure.

Factorization of the determinants implies that the total partition function takes the form

$$Z_{\mathbb{S}^4} = \langle 1 \rangle_{\mathbb{S}^4} = \int_{\mathfrak{t}} da \, \left| Z^{\text{pole}}(a) \right|^2, \tag{12}$$

where the integral is taken over the Cartan subalgebra $\mathfrak{t}$, and

$$Z^{\text{pole}}(a) = Z_{\text{cl}}(ia, \tau) \, Z_{\text{1-loop}}^{\text{pole}}(ia, im_f) \, Z_{\text{inst}}(ia, r^{-1} + im_f; r^{-1}, r^{-1}; \tau). \tag{13}$$

We factorized the classical part by hand: $e^{-S_{\text{cl}}} = |Z_{\text{cl}}|^2$. The precise expressions of various factors in (13) in a similar convention can be found in [20]. The instanton partition function $Z_{\text{inst}}(a, m_f; \epsilon_1, \epsilon_2; \tau)$ defined in [8] arises as a sum of the rational functions over Q-invariant instanton configurations localized at each pole. To compute the vev of the Wilson loop defined by (1) with an integral along the equator, we only need to evaluate it in each saddle point as indicated in step above:

$$\langle W_R \rangle_{\mathbb{S}^4} = \int_{\mathfrak{t}} da \, \left| Z^{\text{pole}}(a) \right|^2 \text{Tr}_R e^{2\pi i r a}. \tag{14}$$

In particular, this reduces to the Gaussian matrix model for $\mathcal{N} = 4$ theory, proving the conjecture [4, 5] mentioned in the introduction.

3.2 Instanton/Monopole Correspondence

We now review a similar localization calculation for a 't Hooft operator [20]. A nice technical tool is a correspondence between singular monopoles on $\mathbb{R}^3$ and $U(1)$-invariant instantons on a Taub-NUT space, discovered by Kronheimer [21]. For our purposes, it is enough to specialize to the single-center Taub-NUT space with metric

$$ds^2 = V(d\rho^2 + \rho^2 d\theta^2 + \rho^2 \sin^2 \theta d\varphi^2) + V^{-1}(d\psi + \omega)^2, \tag{15}$$

where $V = l + 1/(2\rho)$, $\omega = (1/2)(1 - \cos\theta)d\varphi$, $l > 0$ is a constant, and (ρ, θ, φ) are the polar coordinates for $\mathbb{R}^3$. This is a circle fibration over the flat $\mathbb{R}^3$. The variable ψ has periodicity 2π in our convention. From the three-dimensional fields (A, Φ) with singularities

$$A \sim -\frac{B}{2}(1 - \cos\theta)d\varphi, \quad \Phi \sim \frac{B}{2\rho} \tag{16}$$

near the origin, we construct a four-dimensional gauge connection

$$\mathcal{A} \equiv g\left(A + \Phi\frac{d\psi + \omega}{V}\right)g^{-1} - igdg^{-1} \tag{17}$$

and its curvature $\mathcal{F} = d\mathcal{A} + i\mathcal{A} \wedge \mathcal{A}$. Here $g = e^{iB\psi}$ is a singular gauge transformation. The singularities in A and Φ cancel out in (17), and we obtain a smooth four-dimensional gauge field $\mathcal{A}$.

The four-dimensional field $\mathcal{A}$ is invariant under the $U(1)_K$ action $\psi \to \psi + \text{const.}$, which acts on the circle fiber as well as the gauge bundle. The correspondence states that the Bogomolny equations

$$D_i\Phi = \frac{1}{2}\epsilon_{ijk}F_{jk} \quad (i, j, k = 1, 2, 3) \tag{18}$$

on $\mathbb{R}^3$ are equivalent to the anti-self-dual equations

$$\mathcal{F} + *_4\mathcal{F} = 0. \tag{19}$$

Since the Taub-NUT space is isomorphic to $\mathbb{C}^2$ as a complex manifold, we can use instantons on $\mathbb{C}^2$ to perform calculations for 't Hooft operators.

3.3 Localization For 't Hooft Loops on $\mathbb{S}^4$

Let us consider a supersymmetric 't Hooft loop $T(B)$, specified by the coweight B and placed along a large circle of $\mathbb{S}^4$, which we refer to as the equator. See Fig. 2. Since the 't Hooft operator is a disorder operator, we need to evaluate the path integral with the boundary conditions (2) which affect the saddle point configurations. We introduce a convenient set of coordinates, in which the standard round metric on $\mathbb{S}^4$ of radius r is given by

$$ds^2 = r^2\frac{\left(1 - \frac{|\vec{x}|^2}{4r^2}\right)^2}{\left(1 + \frac{|\vec{x}|^2}{4r^2}\right)^2}d\tau^2 + \frac{\sum_{i=1}^3 dx_i^2}{\left(1 + \frac{|\vec{x}|^2}{4r^2}\right)^2}. \tag{20}$$

Fig. 2 Instanton, monopole and anti-instanton field configurations

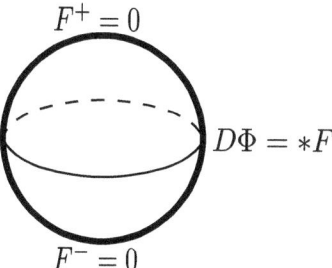

This is Weyl-equivalent to the metric on $\mathbb{S}^1 \times \mathbb{H}^3$, where $\mathbb{H}^3$ is the three-dimensional hyperbolic space. In terms of the Weyl-rescaled fields on $\mathbb{S}^1 \times \mathbb{H}^3$, the only $\mathcal{Q}$-invariant configurations, smooth away from the operator and subject to the boundary condition (2), are given by

$$F_{jk} = -\frac{B}{2}\epsilon_{ijk}\frac{x_i}{|\vec{x}|^3}, \quad \text{Re}\,\phi = \frac{a}{1 + \frac{|\vec{x}|^2}{4r^2}}, \quad \text{Im}\,\phi = \frac{B}{2|\vec{x}|}. \tag{21}$$

We again assumed that $\vartheta = 0$. Note that $a \in \mathfrak{t}$ is the only unfixed parameter which we must integrate over. In this background, the usual classical action diverges due to the infinite mass of the Dirac monopole. Suitable boundary terms [20, 22] cancel the divergence and make the action finite. The values of ϕ at the north and south poles are now shifted. The basic effect of the 't Hooft loop is to shift the argument a of $Z^{\text{pole}}(a)$ by $iB/2r$.

The one-loop determinant receives extra contributions. The differential operator $\mathcal{D}$ in V is now modified, and $\text{ind}\,\mathcal{D}$ receives contributions not only from the north and south poles, but also from the equator. In the neighborhood of the equator, which can be approximated by $\mathbb{S}^1 \times \mathbb{R}^3$, $\mathcal{D}$ acts on the vector multiplet as the differential in the complex defined on $\mathbb{R}^3$:

$$D_{\text{Bogo}} : \Omega^1(\text{ad}\,E) \oplus \Omega^0(\text{ad}\,E) \to \Omega^0(\text{ad}\,E) \oplus \Omega^1(\text{ad}\,E). \tag{22}$$

The arrow involves the dual of the gauge transformation and the linearization of the Bogomolny equation (18). The instanton/monopole correspondence above can be extended to the correspondence between the $U(1)_K$-invariant sections of the self-dual complex (11) and the sections of the Bogomolny complex (22). Thus the index of the latter can be obtained from that of the former by averaging over the $U(1)_K$ action. On the hypermultiplet, $\mathcal{D}$ acts as D_{DH}, the Dirac operator with a coupling to the Higgs field Φ. The one-loop contribution from the equator $Z^{\text{eq}}_{\text{1-loop}}(ia, im_f, B)$ can then be read off from $\text{ind}(D_{\text{Bogo}}) + \text{ind}(D_{\text{DH}})$ by taking into account also the Fourier modes along the $\mathbb{S}^1$.

There are also extra non-perturbative contributions. Recall that zero-size instantons and anti-instantons localized at the north and south poles provide non-perturbative saddle points, even though the configurations are singular. Similarly, infinitesimal dynamical monopoles, which get attached to the Dirac monopole defining the 't

Hooft loop and screen the magnetic charge, also provide non-perturbative saddle points. The saddle point configurations are invariant under $\mathcal{Q}$, and hence also under $\mathcal{Q}^2$. Thus the saddle points are the fixed points, with respect to a certain group action, in the moduli space $\mathcal{M}(B)$ of solutions of the Bogomolny equations on $\mathbb{R}^3$ with a Dirac singularity.

The moduli space $\mathcal{M}(B)$ has components labeled by $v \in \Lambda_{\mathrm{cr}} + B$ with $|v| \leq |B|$. All the fixed points in $\mathcal{M}(B, v)$ take the form of the 't Hooft background (21) except that B is replaced by v. The classical contribution depends only on v and is universal among the fixed points in $\mathcal{M}(B, v)$. We also need to include the fluctuation determinant from each fixed point. By factoring out $Z_{\text{1-loop}}^{\text{eq}}(ia, im_f, v)$, we denote the sum of such determinants by

$$Z^{\text{eq}}(a; B, v) \equiv Z_{\text{1-loop}}^{\text{eq}}(ia, im_f, v) Z_{\text{mono}}^{\text{eq}}(ia, im_f, B, v) \equiv \sum_{\substack{\text{fixed points} \\ \text{in } \mathcal{M}(B,v)}} \prod_j w_j^{c_j/2}. \quad (23)$$

This equation defines $Z_{\text{mono}}^{\text{eq}}$.

Collecting all the contributions (see Fig. 2), the result for the 't Hooft loop expectation value on $\mathbb{S}^4$ is

$$\langle T(B) \rangle_{\mathbb{S}^4} = \int_{\mathfrak{t}} da \sum_v |Z^{\text{pole}}(a + iv/2r)|^2 Z^{\text{eq}}(a; B, v). \quad (24)$$

For example, the vev of the minimal 't Hooft loop $T \equiv L_{1,0}$ in the $SU(2)$ $\mathcal{N} = 2^*$ theory is given by

$$\langle T \rangle_{\mathbb{S}^4} = \sum_{v=\pm 1/2} \int_{-\infty}^{\infty} |Z^{\text{pole}}(a + iv/2r)|^2 \frac{\cosh^{1/2}(\pi r(2a + m)) \cosh^{1/2}(\pi r(2a - m))}{\cosh(2\pi ra)}. \quad (25)$$

This has no bubbling contribution. See [20] for examples with non-trivial bubbling, as well as examples with dyonic charges.

3.4 Other Geometries

3.4.1 $\mathbb{S}^1 \times_b \mathbb{R}^3$

As we saw above, the geometry in the neighborhood of the equatorial $\mathbb{S}^1$ in the four-sphere is essentially $\mathbb{S}^1 \times \mathbb{R}^3$. This suggests that the contributions intrinsic to the loop operators are most naturally formulated on $\mathbb{S}^1 \times \mathbb{R}^3$ itself rather than on $\mathbb{S}^4$. The main advantage of $\mathbb{S}^1 \times \mathbb{R}^3$ is that we can cleanly introduce an analog of omega deformation parameter [8]. In particular the fundamental definition of $Z_{\text{mono}}^{\text{eq}}$, the bubbling contributions, is given in this geometry, just as the instanton partition functions are defined on flat $\mathbb{C}^2$ in the omega background. While there are other geometries that admit an omega deformation, the vevs of 't Hooft operators in such geometries are expressed in terms of the quantities defined on $\mathbb{S}^1 \times \mathbb{R}^3$. Another motivation to consider the omega-deformed product $\mathbb{S}^1 \times \mathbb{R}^3$ comes from the study of the IR dynamics in the same set-up ([16], [V:3]).

We wish to evaluate the expectation values of line operators wrapping the $\mathbb{S}^1$ in $\mathbb{S}^1 \times \mathbb{R}^3$. To preserve SUSY, the metric need not be a direct product; $\mathbb{R}^3$ may be fibered over $\mathbb{S}^1$. If we regard the circle as the time direction, the line operator, an infinitely heavy particle, modifies the theory and defines a Hilbert space $\mathcal{H}_L(\mathbb{R}^3)$. The fibration of $\mathbb{R}^3$ is accounted for by the insertion of the operator $e^{2\pi i b^2 J_3}$, where J_3 generates rotations about the 3-axis. It should be accompanied by $e^{2\pi i b^2 I_3}$ for supersymmetry, where I_3 is a generator of the $SU(2)$ R-symmetry group. We will denote such a space by $\mathbb{S}^1 \times_b \mathbb{R}^3$ to indicate the twist. Recall that an omega deformation is a SUSY-preserving modification of a theory by (equivariant) parameters (ϵ_1, ϵ_2) for the action of the $U(1) \times U(1)$ isometry [8]. In our case the action $U(1) \times U(1)$ acts as the rotation of the $\mathbb{S}^1$ factor and a spatial rotation in $\mathbb{R}^3$. The rotation commutes with supersymmetry if the ratio of the rotation angles is $\epsilon_1/\epsilon_2 = b^2$. We can also introduce flavor symmetry generators F_f and their dual variables $\boldsymbol{m}_f$, which play the role of masses. The expectation value of a line operator (more generally the correlation function of such operators) can be represented as a supersymmetric trace

$$\langle L \rangle_{\mathbb{S}^1 \times_b \mathbb{R}^3} = \mathrm{Tr}_{\mathcal{H}_L(\mathbb{R}^3)}(-1)^F e^{-2\pi R H} e^{2\pi i b^2 (J_3 + I_3)} e^{-2\pi i \boldsymbol{m}_f F_f}. \tag{26}$$

This quantity, in particular the one-loop and bubbling contributions $Z_{\text{1-loop}}^{\mathbb{S}^1 \times_b \mathbb{R}^3}$ and $Z_{\text{mono}}^{\mathbb{S}^1 \times_b \mathbb{R}^3}$, can be computed by localization in the same way as for the equator contributions to $\langle L \rangle_{\mathbb{S}^4}$. The main difference is that in the current case the isometry can act on $\mathbb{S}^1$ and $\mathbb{R}^3$ with a variable ratio $b^2 \in \mathbb{R}$ of rotation angles. The result of localization for the 't Hooft operator T_B is [23]

$$\langle T_B \rangle_{\mathbb{S}^1 \times_b \mathbb{R}^3} = \sum_v e^{2\pi i v \cdot b} Z_{\text{1-loop}}^{\mathbb{S}^1 \times_b \mathbb{R}^3}(a, \boldsymbol{m}_f, b; v) Z_{\text{mono}}(a, \boldsymbol{m}_f, b; B, v), \tag{27}$$

where a and b are respectively the vevs of $\mathrm{Re}\,\phi$ and $\mathrm{Im}\,\phi$ suitably rescaled and complexified by the gauge field. Since the bubbling contribution on this geometry is most fundamental, we simply write $Z_{\text{mono}} \equiv Z_{\text{mono}}^{\mathbb{S}^1 \times_b \mathbb{R}^3}$. Indeed $Z_{\text{mono}}(B, v)$ serve as building blocks for the line operator correlation functions in other geometries. In particular, it is related to the equator contribution on $\mathbb{S}^4$ as $Z_{\text{mono}}^{\text{eq}}(a, \boldsymbol{m}_f; B, v) = Z_{\text{mono}}(ra, r\boldsymbol{m}_f + 1/2, \lambda = 1; B, v)$. The shift in mass appears because the curved metric affects the periodicity of spinors.

For the minimal 't Hooft operator in the $SU(2)$ $\mathcal{N} = 2^*$ theory,

$$\langle T \rangle_{\mathbb{S}^1 \times_b \mathbb{R}^3} = \left(e^{2\pi i b} + e^{-2\pi i b} \right) \left(\frac{\sin(2\pi a + \pi \boldsymbol{m})\sin(2\pi a - \pi \boldsymbol{m})}{\sin\left(2\pi a + \frac{\pi}{2}b^2\right)\sin\left(2\pi a - \frac{\pi}{2}b^2\right)} \right)^{1/2}, \tag{28}$$

with $a, b \in \mathbb{C}$. See [23] for more examples.

3.4.2 $\mathbb{S}^1 \times_b \mathbb{S}^3$

Given an $\mathcal{N} = 2$ superconformal theory, one can perform radial quantization on $\mathbb{R}^4$ by regarding the radial direction as time. One can compactify this direction, and the resulting path integral, the partition function of a supersymmetric theory on $\mathbb{S}^1 \times \mathbb{S}^3$, is known as the superconformal index ([24, 25], [V:9]). One can refine this by including line operators [26]. Line operators, wrapping the $\mathbb{S}^1$ and inserted at arbitrary points along a great circle of $\mathbb{S}^3$, preserve some common supersymmetry [26]. In particular, if one starts with a line operator passing through the origin, by radial quantization one ends up with a line operator L at the north pole of $\mathbb{S}^3$, and its conjugate $\bar{L}$ at the south pole. Let us denote by $\mathcal{H}_{L,\bar{L}}(\mathbb{S}^3)$ the Hilbert space on $\mathbb{S}^3$ with such insertions. The index with the line operator insertions can be represented as a supersymmetric trace

$$\langle L \cdot \bar{L} \rangle_{\mathbb{S}^1 \times_b \mathbb{S}^3} = \mathrm{Tr}_{\mathcal{H}_{L,\bar{L}}(\mathbb{S}^3)} (-1)^F e^{2\pi i b^2 (J_L + J_R + I_3)} \prod_f \eta_f^{F_f}. \tag{29}$$

Here J_L and J_R are the Cartan isometry generators for $\mathbb{S}^3$, b plays the role of an omega deformation parameter, and η_f are the flavor chemical potentials. Recall also from (26) that I_3 and F_f are the R- and flavor symmetry generators respectively.

This was computed in [27] by a hybrid method, where the one-loop contributions are computed by counting BPS states on $\mathbb{S}^3$, and the bubbling contributions, which we expect to be localized to the line operators and given by Z_{mono} above, are put by hand. The classical contribution vanishes as it should for an index. The results agree with a prescription proposed in [26], as well as the predictions of S-duality in $\mathcal{N} = 4$ theory. In addition, the Wilson line operator index for $SU(N)$ $\mathcal{N} = 4$ theory in the large N limit was found to agree with the counting of fluctuation modes on the fundamental string (for the fundamental representation) and on the D5-brane (for the anti-fundamental representation).

3.4.3 $\mathbb{S}^4_b$

Since the CFT side has a variable b parametrizing the central charge, it was immediately recognized after the discovery of the AGT correspondence that the localization computations on $\mathbb{S}^4$ should be generalized. This was done in [28] and is reviewed in [V:6]. The metric of the new geometry, the ellipsoid $\mathbb{S}^4_b$, is given by $ds^2 = \sum_{I=0}^4 dX_I^2$, where

$$X_0^2 + b^{-2}(X_1^2 + X_2^2) + b^2(X_3^2 + X_4^2) = r^2. \tag{30}$$

This geometry has an obvious isometry $U(1) \times U(1)$, whose action commutes with supersymmetry if the ratio of rotation angles is $\epsilon_1/\epsilon_2 = b^2$. Setting b to 1 gives back the round $\mathbb{S}^4$ of radius r. The reference [28] formulated $\mathcal{N} = 2$ gauge theory (with reduced SUSY) on this geometry by introducing a background gauge field for R-symmetry and other auxiliary fields in the $\mathcal{N} = 2$ supergravity multiplet. The partition function now takes the form

$$Z_{\mathbb{S}^4_b} = \int da |Z^{\text{pole}}(a, m_f; b; \tau)|^2 \tag{31}$$

where $Z^{\text{pole}}(a, m_f; b; \tau)$ includes the one-loop and instanton contributions, with omega deformation parameters $(\epsilon_1, \epsilon_2) = (b, b^{-1})$. For any fixed value of $X_0 \in (-r, r)$, there are two circles along which we can place BPS loop operators: the circle $\mathbb{S}^1_{(b)}$ ($X_3 = X_4 = 0$) on the 12-plane, and another circle $\mathbb{S}^1_{(1/b)}$ on the 34-plane. The computation of the Wilson loop vev and S-duality suggests that the 't Hooft loop vev is independent of X_0. Let us focus on $\mathbb{S}^1_{(b)}$ at $X_0 = 0$. The vev of the Wilson loop W_R was computed in [28]; we just insert $\text{Tr}_R e^{2\pi i b r a}$ into (31). We emphasize that 't Hooft loops on $\mathbb{S}^4_b$ have not been treated yet at the time of writing. We expect that the essential part of computation is determined by the symmetry generated by Q^2; in particular it generates the $U(1) \times U(1)$ isometry with equivariant parameters (b, b^{-1}). Generalizing (24), we should get

$$\langle T_B \rangle_{\mathbb{S}^4_b} \overset{\text{expected}}{=} \int_t da \sum_v Z^{\text{eq}}\left(a, m_f; b; B, v\right) \left| Z^{\text{pole}}\left(a + ib\frac{v}{2r}, m_f; b; \tau\right) \right|^2, \tag{32}$$

where $Z^{\text{eq}}\left(a, m_f; b; B, v\right) := Z^{\mathbb{S}^1 \times_b \mathbb{R}^3}_{\text{1-loop+mono}}(ibra, ibrm_f + 1/2; b; B, v)$ [23]. We will compare this with the CFT computations in the next section.

3.5   1/8-BPS Wilson Loops in $\mathcal{N} = 4$ Theory and the 2d Yang-Mills

On $\mathbb{R}^4$ or $\mathbb{S}^4$, $\mathcal{N} = 4$ super Yang-Mills has a variety of loop operators that preserve at least one supercharge [11]. The half-BPS Wilson loop given in (1), where we regard $\mathcal{N} = 4$ theory as an $\mathcal{N} = 2$ theory with a massless adjoint hypermultiplet, is one of them. There are different classes of line operators to which the localization method has been applied [29].

These include 1/8-BPS Wilson loops along arbitrary contours on a two-sphere. One can also place a half-BPS 't Hooft loop that links the $\mathbb{S}^2$. Certain local operators can be further inserted on $\mathbb{S}^2$. Localization can be performed using the common supercharge preserved by the operators. The results are rather different from the case in the previous subsection; the path integral has been shown to reduce, up to the assumption that the one-loop determinant is trivial, to another quantum field theory, namely the bosonic two-dimensional Yang-Mills on the $\mathbb{S}^2$. The correlation functions of the operators turn out to be captured by the analogous observables in the two-dimensional theory, as conjectured in [30]. For certain combinations of the operators, the theory further reduces to multi-matrix models. The results have been tested in a variety of ways using AdS/CFT and S-duality. For more details, see [22, 31, 32] and the references therein.

4 CFT Techniques for Line Operators

In this section we review two-dimensional methods for computing the expectation values of loop operators on $\mathbb{S}_b^4$. We mostly consider the A_1-theories of class $\mathcal{S}$ and the corresponding 2d theory, namely Liouville theory. Generalization to higher rank gauge theories and the $SU(N)$ Toda theories will be explained in Sect. 4.3.

4.1 Verlinde Operators

As we saw in Sect. 2.2, line operator charges in an A_1-theory are in a one-to-one correspondence with closed curves on the Riemann surface $C_{g,n}$. In a conformal field theory, one can associate to any closed curve γ an operation on conformal blocks, defined in terms of the monodromy of a degenerate field along γ. This operation, called the *Verlinde operator*, was introduced by E. Verlinde and applied to the characters (torus conformal blocks) of rational CFT's to argue that the modular S-matrix diagonalizes the coefficients in the fusion rule [33]. Moore and Seiberg proved this conjecture, known as the Verlinde formula, by expressing the Verlinde operators in terms of the fusion and braiding moves, which are the basic ingredients for a general modular transformation of conformal blocks [34, 35]. It turns out that the same construction works even for non-rational CFT's such as Liouville and Toda theories. See [V:12] for a more rigorous discussion of the Verlinde operator. As we will elaborate in Sect. 4.3, there is an alternative definition of a Verlinde operator as a topological defect, whose definition may be conceptually cleaner.

A degenerate field is a primary for which the Kac determinant vanishes. As such it has a descendant that is orthogonal to all states and is decoupled [36]. In the standard parametrization of the Liouville central charge $c = 1 + 6Q^2$ ($Q = b + b^{-1}$), primaries V_α have the conformal weight $\Delta(\alpha) = \alpha(Q - \alpha)$. The most basic degenerate fields are V_α with momenta $\alpha = -b/2$ and $\alpha = -1/2b$. In view of the quantum symmetry $b \leftrightarrow b^{-1}$ of Liouville theory, it suffices to consider $V_{-b/2}$. The condition for decoupling can be stated as

$$\left(\partial_z^2 + b^2 T(z)\right) V_{-b/2}(z) = 0 \quad \text{or equivalently} \quad (L_{-1}^2 + b^2 L_{-2}) \cdot V_{-b/2} = 0, \quad (33)$$

where $T(z)$ is the energy-momentum tensor and L_n are the standard Virasoro generators. Using this, one can show that $V_{-b/2}$ and V_α obey the fusion rule [36]

$$[V_{-b/2}][V_\alpha] = [V_{\alpha-b/2}] + [V_{\alpha+b/2}]. \quad (34)$$

In particular, the OPE of two degenerate fields $V_{-b/2}$ contains the vacuum state $V_0 = 1$.

Since (33) involves only a holomorphic coordinate z, the full correlation function with $V_{-b/2}$, as well as the corresponding conformal blocks, obeys the resulting differential equation. In particular the conformal blocks transform linearly (i.e., there is monodromy) when z is transported along a closed curve.

We are now ready to define the Verlinde operator. Let us consider a conformal block $\mathcal{F}$ specified by a trivalent graph Γ on $C_{g,n}$. Pick also a closed curve γ on $C_{g,n}$, and assume that it has only one connected component. We can define an extended conformal block $\hat{\mathcal{F}}(z, z_0)$ by inserting $V_{-b/2}(z)$ and $V_{-b/2}(z_0)$ at two nearby points z and z_0, taking their OPE, and projecting onto the identity state. Let $\Delta(\alpha) = \alpha(Q-\alpha)$ denote the conformal weight of V_α. We can recover the original block $\mathcal{F}$ from $\hat{\mathcal{F}}$ by taking the limit $z \to z_0$:

$$\hat{\mathcal{F}}(z, z_0) \sim \frac{1}{(z - z_0)^{2\Delta(-b/2)}} \mathcal{F}. \tag{35}$$

A priori $\hat{\mathcal{F}}(z, z_0)$ is defined only for z close enough to z_0. We can analytically continue $\hat{\mathcal{F}}$ and transport z along a closed curve homotopic to γ, and *then* take the limit $z \to z_0$

$$\hat{\mathcal{F}}(z, z_0) \sim \frac{c_0}{(z - z_0)^{2\Delta(-b/2)}} \mathcal{L}_\gamma \cdot \mathcal{F}. \tag{36}$$

We included a universal normalization constant c_0, which we will specify. The map $\mathcal{F} \mapsto \mathcal{L}_\gamma \cdot \mathcal{F}$ is the Verlinde operator.

As an illustration, let us take as $\hat{\mathcal{F}}$ a four-point conformal block with two degenerate fields at 0 and z, and two primaries with momentum α at 1 and ∞. The block can be expressed in terms of the Gauss hypergeometric function $_2F_1$:

$$\hat{\mathcal{F}}(z, 0) = \quad \begin{array}{c} \alpha \\ \\ \hline \\ \alpha \end{array} \Bigg\{ \begin{array}{c} -b/2 \\ \\ -b/2 \end{array} = \frac{(1 - z)^{b\alpha}}{z^{2\Delta(-b/2)}} \,_2F_1\left(1 + b^2, 2b\alpha; 2 + 2b^2; z\right) .$$

$$\tag{37}$$

The solid lines carry a generic momentum, the wiggly ones degenerate states, and the dashed line represents the identity state. The monodromy around $z = 1$ can be computed by a hypergeometric identity and yields a linear combination of (37) and the block with internal momentum $-b$. One can check that the coefficient of the former is

$$\frac{e^{2\pi ba} + e^{-2\pi ba}}{e^{\pi ibQ} + e^{-\pi ibQ}}, \tag{38}$$

where we defined a by $\alpha = Q/2 + ia$. We now choose $c_0 = 1/(e^{\pi i bQ} + e^{-\pi i bQ})$. Then the corresponding Verlinde operator is the multiplication by $e^{2\pi ba} + e^{-2\pi ba}$, which is the trace of a matrix in the fundamental representation of $SU(2)$.

More generally the Verlinde operator $\mathcal{L}_\gamma$ as defined above can be computed very explicitly by fusion and braiding moves. These moves relate the conformal blocks assigned to two trivalent graphs that differ by a local modification. A sequence of such modifications allows us to analytically continue the block as a function of z along γ. Since our aim is to transport a degenerate field $V_{-b/2}(z)$, we only need to implement these moves when the modification involves at least one external edge with $V_{-b/2}$. In this special case, the fusion move is expressed in terms of a 2×2 matrix $F_{s_1 s_2}$ ($s_1, s_2 = \pm$). See (58). Since the fusion move, by definition, locally modifies a trivalent graph by replacing an s-channel with a t-channel, it is enough to describe its action on a four-point block:

$$
\begin{array}{c} \alpha_3 \quad\quad -b/2 \\ \\ \hline \alpha_4 \quad \alpha_1 - s_1\tfrac{b}{2} \quad \alpha_1 \end{array}
= \sum_{s_2=\pm} F_{s_1 s_2} \begin{bmatrix} \alpha_3 & -b/2 \\ \alpha_4 & \alpha_1 \end{bmatrix} \times
\begin{array}{c} \alpha_3 \quad\quad -b/2 \\ \\ \alpha_3 - s_2\tfrac{b}{2} \\ \hline \alpha_4 \quad\quad \alpha_1 \end{array}
\tag{39}
$$

We also use the braiding move

$$
\begin{array}{c} \alpha_2 \quad\quad \alpha_3 \\ \\ \alpha_1 \end{array}
= e^{\pi i(\Delta(\alpha_1) - \Delta(\alpha_2) - \Delta(\alpha_3))} \times
\begin{array}{c} \alpha_3 \quad\quad \alpha_2 \\ \\ \alpha_1 \end{array}
\tag{40}
$$

where $\Delta(\alpha) = \alpha(Q - \alpha)$ is the conformal dimension of the operator V_α and the two exchanged vertex operators are rotated by $180°$. If the rotation is in the other direction the phase is the opposite.

The curve for a spin $1/2$ Wilson operator for a gauge $SU(2)$ factor corresponds to an internal edge of the trivalent graph [37, 38]. Indeed the action of the Verlinde operator $\mathcal{L}_W$ along this curve is calculated by the following sequence of moves.

$$
\begin{array}{c} \\ \alpha \quad\quad \alpha \end{array}
\rightarrow
\begin{array}{c} -\tfrac{b}{2} \quad -\tfrac{b}{2} \\ \\ \alpha \quad \alpha' \quad \alpha \end{array}
\rightarrow
\begin{array}{c} \\ \alpha \quad \alpha' \quad \alpha \end{array}
\rightarrow
\begin{array}{c} \\ \alpha \quad \alpha' \quad \alpha \end{array}
\rightarrow
\begin{array}{c} \\ \alpha \quad\quad \alpha \end{array}
\tag{41}
$$

The result is

$$\mathcal{L}_W \cdot \mathcal{F} = (e^{2\pi ba} + e^{-2\pi ba})\,\mathcal{F}. \tag{42}$$

Of course this agrees with (38) divided by c_0, and is the trace of an $SU(2)$ matrix.

More generally, a closed curve that traverses at least one pair of pants corresponds to a line operator with non-zero magnetic charge, and the associated Verlinde operator involves a non-trivial shift in Liouville momenta. For example, the Verlinde operator corresponding to the minimal 't Hooft loop in $\mathcal{N} = 2^*$ theory is given by the following moves.

The momentum α is replaced by $\alpha' = \alpha \pm b/2$. Explicitly we find

$$[\mathcal{L}_{1,0} \cdot \mathcal{F}](\alpha, \alpha_e) = \sum_{\pm} H_{\pm}(\alpha, \alpha_e)\mathcal{F}(\alpha \pm b/2, \alpha_e), \tag{43}$$

where

$$H_{\pm}(\alpha, \alpha_e) = \frac{\Gamma(\pm 2b(\alpha - Q/2))\Gamma(\pm 2b(\alpha - Q/2) + bQ)}{\Gamma(\pm 2b(\alpha - Q/2) + b\alpha_e)\Gamma(\pm 2b(\alpha - Q/2) - b\alpha_e + bQ)}. \tag{44}$$

See [37, 38] for many more examples of Verlinde operators in Liouville theory.

4.2 Comparison with Gauge Theory

Let us now compare the CFT results with the gauge theory results in Sect. 3. We will write $\vec{\alpha} = (\alpha_j)_{j=1}^{3g-3+n}$.

According the AGT correspondence [9], extended to generic b [28], the $\mathbb{S}_b^4$ partition function (31) of the A_1-theory is the Liouville correlation function on $C_{g,n}$:

$$Z_{\mathbb{S}_b^4} = \int [d\alpha] C(\vec{\alpha}) \overline{\mathcal{F}(\vec{\alpha})} \mathcal{F}(\vec{\alpha}). \tag{45}$$

Here $C(\vec{\alpha})$ is an appropriate product of the three-point functions and $\mathcal{F}(\vec{\alpha})$ is the conformal block. A conjecture put forward in [37, 38], again generalized to arbitrary b, is that the expectation value of a loop operator $L(\vec{p}, \vec{q})$ on $\mathbb{S}_b^4$ is given by the "expectation value" of the Verlinde operator $\mathcal{L}_\gamma$:

$$\langle L(\vec{p}, \vec{q}) \rangle_{\mathbb{S}_b^4} = \int [d\alpha] C(\vec{\alpha}) \overline{\mathcal{F}(\vec{\alpha})} \mathcal{L}_\gamma \cdot \mathcal{F}(\vec{\alpha}). \tag{46}$$

In order to compare gauge theory and CFT, it is natural to adopt a different normalization of conformal blocks [20, 23]:

$$\mathcal{B}(\vec{\alpha}) := C(\vec{\alpha})^{1/2}\mathcal{F}(\vec{\alpha}). \tag{47}$$

This normalization is also natural for the quantization of the Hitchin system and for the interpretation of the fusion move as an analog of the $6j$-symbols [39, 40]. The new block $\mathcal{B}$ is identified with Z^{pole}, which contains not only the instanton contributions, but also the one-loop contributions from the north pole. We define the action of a Verlinde operator on $\mathcal{B}$ as

$$(\mathbb{L}_\gamma \cdot \mathcal{B})(\vec{\alpha}) := C(\vec{\alpha})^{1/2}(c_0 \mathcal{L}_\gamma \cdot \mathcal{F})(\vec{\alpha}) \tag{48}$$

by absorbing c_0 into $\mathbb{L}_\gamma$. In terms of $\mathcal{B}$ and $\mathbb{L}_\gamma$, (46) becomes

$$\langle L(\vec{p}, \vec{q})\rangle_{\mathbb{S}^4} = \int [d\alpha]\,\overline{\mathcal{B}(\vec{\alpha})}\,\mathbb{L}_\gamma \cdot \mathcal{B}(\vec{\alpha}). \tag{49}$$

The Verlinde operator corresponding to a Wilson operator is still multiplicative:

$$\mathbb{L}_W \cdot \mathcal{B} = (e^{2\pi ba} + e^{-2\pi ba})\mathcal{B}. \tag{50}$$

For Verlinde operators that involve shifts in $\vec{\alpha}$, the action is modified. For example, the Verlinde operator $\mathbb{L}_{1,0}$ on the one-punctured torus computed in (43) and (44) becomes

$$\mathbb{L}_{1,0} = \sum_{s=\pm 1} e^{i\frac{s}{4}b\partial_a}\left(\prod_\pm \frac{\cosh(2\pi ba \pm \pi bm)}{\sinh(\pm 2\pi ba + \frac{\pi}{2}ib^2)}\right)^{1/2} e^{i\frac{s}{4}b\partial_a}, \tag{51}$$

where $\alpha = Q/2 + ia$, $\alpha_e = Q/2 + im$.

For the pure Wilson operator, the equality (49) immediately follows from Pestun's computation of the Wilson loop vev [6] and its generalization [28] reviewed in Sect. 3. One can also confirm that the gauge theory result (25) for the minimal 't Hooft loop in $\mathcal{N} = 2^*$ theory agrees with the CFT result (49) combined with (51) for $b = 1$. The expected expression (32) for the 't Hooft loop on $\mathbb{S}_b^4$ is consistent with (49). Many more tests of the correspondence were made in [23, 38].

The exact calculation of the disorder operators such as 't Hooft loops, and its verification by independent methods, is one of the important advances that became possible by localization and the 2d-4d correspondence.

4.3 Higher Rank Gauge Groups and Toda Theories

The higher rank ($N > 2$) A_{N-1}-type theories of class $\mathcal{S}$ are quiver theories that involve $SU(n)$ gauge groups ($2 \le n \le N$) as well as non-Lagrangian theories whose non-Abelian flavor symmetries are gauged ([41], [V:2]). In a higher rank extension of the AGT relation [9, 42], these theories correspond to the $SU(N)$ Toda CFT on a Riemann surface C. Liouville theory is a special case ($N = 2$) of Toda theory. The Toda theory possesses an extended chiral algebra, the W_N-algebra.

The complete dictionary between gauge theory line operators and geometric objects on C has not been developed yet for $N > 2$. Still, for simple theories such as $SU(N)$ $\mathcal{N} = 2^*$ theory and the $SU(N)$ theory with $N_F = 2N$ flavors, one expects from brane constructions that the dictionary for minimal Wilson and 't Hooft operators is essentially the same as in the $N = 2$ case. The W_N-algebra possesses representations with various degeneracy conditions. Using the so-called semi-degenerate fields V_μ, one can construct Verlinde operators [43–45].

For the Wilson loops in the fundamental and anti-fundamental representations, the Verlinde operator was calculated in [44] by the monodromy of a degenerate block given in terms of a generalized hypergeometric function, of which (37) is a special case. The Verlinde operators for the minimal 't Hooft loops in $\mathcal{N} = 2^*$ theory and the conformal SQCD were computed in [45], by determining the relevant fusion move matrices from the monodromy of generalized hypergeometric functions. The agreement between the 4d and 2d results reviewed in Sect. 4.2 extends to the higher rank case.

The authors of [43] expressed the Verlinde operator for the Wilson loop curve in terms of the fusion and braiding moves. General identities that follow from the axioms of CFT imply that such a Verlinde operator inserts $S_{\mu,\alpha}/S_{0,\alpha}$ into the Toda version of (45), where μ is the semi-degenerate representation, α is the generic representation propagating across the curve, and S denotes the modular S-matrix. (See also [46].) This turns out to be the same as the insertion of a so-called topological defect along the curve. The latter is a one-dimensional object in CFT, defined by the condition that the holomorphic and anti-holomorphic generators of the chiral algebra commute with it. Topological defects were originally constructed in [47] for rational CFT and in [48] for Liouville theory. Since the Verlinde operators and the topological defects transform in the same way under the modular groupoid, they must be identical. The definition of a topological defect is local and does not require introducing conformal blocks. It is also possible to compute 't Hooft loops using topological defects in Liouville theory [49]. Another feature of a higher-rank Toda theory is that Verlinde operators/topological defects can be defined on networks with trivalent vertices [43, 50]. The identification of the corresponding line operators is an interesting open problem.

5 Line Operator Algebras and the Hitchin Moduli Space

The previous two sections concerned the expectation values and correlation functions of line operators. It turns out that the line operators also possess interesting algebraic structures.

5.1 Operator Product Expansion from SUSY Quantum Mechanics

Let us consider a general $\mathcal{N} = 2$ theory on $\mathbb{R} \times I \times \mathcal{C}$, where I is an interval and $\mathcal{C}$ (not to be confused with C) is a Riemann surface. With a suitable twist along $\mathcal{C}$, the theory depends on the complex structure of $\mathcal{C}$ but not on its Kähler structure [51, 52]. It is also independent of the gauge coupling and can be analyzed at weak coupling. A Wilson-'t Hooft operator along $\mathbb{R}$, inserted at a point $(s, z) \in I \times \mathcal{C}$, is annihilated by a fermionic charge Q which is a scalar along $\mathcal{C}$. The correlator of two such operators L_i ($i = 1, 2$) depends holomorphically on the complex coordinates z_i on $\mathcal{C}$, and is (locally) independent of the positions s_i on I. We impose suitable boundary conditions at the two ends of I.

At low energies, the theory reduces to an $\mathcal{N} = 2$ quantum mechanics whose target space is the moduli space of solutions of the Bogomolny equations on $I \times \mathcal{C}$, possibly with Dirac monopole singularities of charge B_i at (s_i, z_i). The data for $\mathcal{N} = 2$ quantum mechanics also include a holomorphic vector bundle determined by the electric charges of L_i, as follows from the construction of dyonic operators in Sect. 2.1. In simple cases the moduli space can be described rather explicitly. The BPS Hilbert space is the L^2 Dolbeault cohomology of the vector bundle.

Let us set $z_1 = z_2$ and take $s_1 \neq s_2$. The moduli space in the limit $s_1 \to s_2$ develops a singularity. In simple examples, the singularity corresponds to shrinking exceptional divisors. The L^2 cohomology then splits into several parts, one with elements localized to the smooth part of the moduli space, and the others with support in the vicinity of a divisor. The latter correspond to smaller magnetic charges, and is a manifestation of the phenomenon "monopole bubbling" [52].

This method was applied in [52–56] to compute the operator product expansion (OPE) of line operators in the form

$$L(B_1, w_1) \cdot L(B_2, w_2) = L(B_1 + B_2, w_1 + w_2) + \sum_j (-1)^{s_j} L(B_j, w_j), \quad (52)$$

where $(-1)^{s_j}$ are signs, and $(B_k, w_k) \in \Lambda_{\mathrm{m}} \times \Lambda_{\mathrm{w}}$ in the notation of Sect. 2.1. The signs arise because we weight the BPS Hilbert space by the fermion number. Various checks have been made by S-duality.

The results in [56] should be compared with the two-dimensional methods developed in [50, 57–60] for higher-rank theories. It appears that more work is needed to have a unified view on the algebra of line operators for higher-rank class $\mathcal{S}$ theories.

5.2 Non-commutative Algebra of Line Operators

The set-up $\mathbb{R} \times I \times \mathcal{C}$ above can be identified with $\mathbb{S}^1 \times_b \mathbb{R}^3$ ($=\mathbb{S}^1 \times_b \mathbb{R} \times \mathbb{R}^2$) without omega deformation ($b = 0$) by the identification $(\mathbb{R}, I, \mathcal{C}) \to (\mathbb{S}^1, \mathbb{R}, \mathbb{R}^2)$.

Recall also that the latter geometry with general b is the effective geometry in the neighborhood of the loop operator in $\mathbb{S}_b^4$ and $\mathbb{S}^1 \times_b \mathbb{S}^3$. Thus the algebra of line operators on $\mathbb{S}^1 \times_b \mathbb{R}^3$ captures the counterparts in other geometries.

The OPE of two operators on $\mathbb{S}^1 \times_b (\mathbb{R} \times \mathbb{R}^2)$ with $b \neq 0$ depends on the ordering along the $\mathbb{R}$. Namely, $L_1 \cdot L_2$ (for $s_1 > s_2$) in general does not equal $L_2 \cdot L_1$ (for $s_1 < s_2$). This is because the Poynting vectors in the two cases contribute to the trace (26) with opposite signs of angular momentum J_3 [16]. The OPE takes the form

$$L(B_1, w_1) \cdot L(B_2, w_2) = e^{s_{12}\pi i b^2} L(B_1 + B_2, w_1 + w_2) + \sum_j c_j(b) L(B_j, w_j), \quad (53)$$

where $s_{12} = \langle B_2, w_1 \rangle - \langle B_1, w_2 \rangle$ is a symplectic pairing and the coefficients c_j ($j \neq 1, 2$) depend on b. In fact the localization analysis shows that $\langle L_1 \cdot L_2 \rangle_{\mathbb{S}^1 \times_b \mathbb{R}^3} = \langle L_1 \rangle_{\mathbb{S}^1 \times_b \mathbb{R}^3} * \langle L_2 \rangle_{\mathbb{S}^1 \times_b \mathbb{R}^3}$, where $*$ is the Moyal product:

$$(f * g)(\boldsymbol{a}, \boldsymbol{b}) \equiv \left. e^{i \frac{b^2}{4\pi}(\partial_b \cdot \partial_{a'} - \partial_a \cdot \partial_{b'})} f(\boldsymbol{a}, \boldsymbol{b}) g(\boldsymbol{a}', \boldsymbol{b}') \right|_{\boldsymbol{a}'=\boldsymbol{a}, \boldsymbol{b}'=\boldsymbol{b}}. \quad (54)$$

This product is associative but non-commutative, and is associated with the holomorphic symplectic structure $\Omega = d\boldsymbol{a} \wedge d\boldsymbol{b}$ with $\hbar = b^2/2\pi$ [23]. Also, the relation between $\mathbb{S}_b^4$ and $\mathbb{R}^1 \times_b \mathbb{R}^3$, together with the AGT correspondence, suggests that the corresponding Verlinde operator $\mathbb{L}$ acting on the normalized conformal blocks $\mathcal{B}$ in (47) is the Weyl ordering of $\langle L \rangle_{\mathbb{S}^1 \times_b \mathbb{R}^3}$ viewed as a function of $\boldsymbol{a}$ and $\boldsymbol{b}$ [23].

5.3 Quantization of the Hitchin Moduli Space

The various 4d geometries considered in Sect. 3 admit a natural action of $U(1) \times U(1)$ isometries. In the tubular neighborhood of a supersymmetric loop operator, or two such operators very close to each other, the local geometry can be effectively approximated by $\mathbb{S}^1 \times_b \mathbb{R}^3$, where one $U(1)$ rotates the $\mathbb{S}^1$ and the other acts as rotations about the 3-axis. Since the algebra of supersymmetric line operators is a UV property of the theory, it suffices to analyze the theory on $\mathbb{S}^1 \times_b \mathbb{R}^3$.

For class $\mathcal{S}$ theories, line operators are intimately related to functions on the Hitchin moduli space. One can see this most clearly as follows [16]. Recall that a class $\mathcal{S}$ theory is specified by a simply laced Lie algebra $\mathfrak{g}$ and a Riemann surface C ([41, 61], [V:2]). For simplicity we assume that there is no puncture. For $b = 0$ such a theory on $\mathbb{S}^1 \times_b \mathbb{R}^3$ is simply the 6d $\mathcal{N} = (0, 2)$ theory on $\mathbb{S}^1 \times \mathbb{R}^3 \times C$, topologically twisted along infinitesimally small C. If instead the size of C is much bigger than the radius of the $\mathbb{S}^1$, a better description is the 5d maximally supersymmetric Yang-Mills theory on $\mathbb{R}^3 \times C$. The condition for preserving supersymmetry is precisely the Hitchin equations on C [16, 62]:

$$F_{z\bar{z}} = [\varphi_z, \bar{\varphi}_{\bar{z}}], \quad D_{\bar{z}}\varphi_z = 0, \quad D_z\bar{\varphi}_{\bar{z}} = 0, \quad (55)$$

where φ_z and its complex conjugate $\bar{\varphi}_{\bar{z}}$ arise from two real scalars via twisting. The space of solutions modulo gauge transformations taking values in the simply connected group $\tilde{G}$ with Lie algebra $\mathfrak{g}$ is the Hitchin moduli space $\mathcal{M}_H(\tilde{G}, C)$. The Coulomb moduli space of the 5d theory on $\mathbb{R}^3 \times C$ is believed to be the quotient $\mathcal{M}_H(\tilde{G}, C)/\Delta$ by a discrete group Δ. Here Δ is a subgroup of the group of flat line bundles whose structure group is the center $\mathcal{C}$ of $\tilde{G}$, and can be identified with the maximal coisotropic subgroup of $H^1(C, \mathcal{C})$ denoted by the same symbol in Sect. 2.3 [16, 19]. The line operators in the 4d theory arise from surface operators in the 6d theory wrapping a curve (or a trivalent network) on C. The surface operators descend to Wilson loops for a complex gauge field obtained from (A_z, φ_z) in the 5d theory. We note that the twist along C eliminates dependence of BPS observables on the scale of the metric on C, and that the 5d theory is IR free. It is then natural to expect [16] that the correlation functions of BPS line operators on $\mathbb{S}^1 \times_{b=0} \mathbb{R}^3$ are given by the classical holonomies on C. This expectation was shown to be consistent with wall-crossing [16] in the 4d IR theories, and was also directly demonstrated [23] for a few examples by noting that the parameters (a, b) in Sect. 3.4.1 are the complexification of the Fenchel-Nielsen coordinates on the Hitchin moduli space.

In order to see that the omega deformation $\mathbb{S}^1 \times_b \mathbb{R}^3$ induces non-commutativity, one approach is to reduce the theory by the action of $U(1) \times U(1)$ to two dimensions [63]. This can be done for a topologically twisted theory, and in the limit that the orbits of the action become small, the reduced 2d theory is the $\mathcal{N} = (4, 4)$ sigma model with target space $\mathcal{M}_H(\tilde{G}, C)/\Delta$. The 4d geometry reduces to a half plane, and line operators get inserted along the boundary. The presence of a B-field accounts for non-commutativity [64].

If we reduce $\mathbb{S}_b^4$ by the action of $U(1) \times U(1)$ above, the neighborhood of the equator $\mathbb{S}_b^3$ of $\mathbb{S}_b^4$ reduces to a two-dimensional strip, as considered in [63]. By topologically twisting the 4d theory, one obtains a two-dimensional sigma model. The line operators along the circle $\mathbb{S}_{(b)}^1$ define the boundary chiral ring $\mathcal{A}_b$ on the left boundary, while those along $\mathbb{S}_{(1/b)}^1$ define another ring $\mathcal{A}_{1/b}$ on the right boundary. The A_1-theory on $\mathbb{S}_b^4$ realizes the quantization of the Hitchin moduli space with a Hilbert space; the two rings act on the Hilbert space of conformal blocks. If we included all the operators labeled by $\Lambda_m \times \Lambda_w/(\text{Weyl group})$, $\mathcal{A}_b$ and $\mathcal{A}_{1/b}$ would not commute because the two circles $\mathbb{S}_{(b)}^1$ and $\mathbb{S}_{(1/b)}^1$ are linked inside the $\mathbb{S}_b^3$ in the constant time slice $\{X_0 = 0\}$ [1], as explained in Sect. 2.3:

$$L_{\gamma_1}^{(b)} \cdot L_{\gamma_2}^{(1/b)} = (-1)^{\langle \gamma_1, \gamma_2 \rangle} L_{\gamma_2}^{(1/b)} \cdot L_{\gamma_1}^{(b)}, \tag{56}$$

where we denote by γ_j the corresponding charges $(\vec{p}_j, \vec{q}_j)$, and $\langle \gamma_1, \gamma_2 \rangle$ is a symplectic product. Indeed as explained in Sect. 2.3, we must restrict to a maximal mutually local subset of $\Lambda_m \times \Lambda_w/(\text{Weyl group})$ such that $\langle \gamma_1, \gamma_2 \rangle$ is even for any pair of line operators. Then $\mathcal{A}_b$ and $\mathcal{A}_{1/b}$ commute with each other, as they should because they are chiral rings on two separate boundary components. As explained above such a restriction modifies the target space from $\mathcal{M}_H(C, SU(2))$ to its quotient by a finite group Δ [16].

These are manifestations of the relation between the A_{N-1}-gauge theories and quantization of $\mathcal{M}_H(C, SU(N))$ associated with the curve C. See [V:3] for discussions and references. The connection between the gauge theory and the Hitchin system can also be used to study line operators from the IR point of view, where a different class of Darboux coordinates naturally appears ([16], [V:3]). For some theories the non-commutative algebra of line operators can be computed using IR quiver quantum mechanics [65]. An important open problem is the comparison of the algebraic relations obtained in different approaches.

Acknowledgments I am grateful to my collaborators for enlightening discussions on this subject. I also thank N. Drukker and the referee in this project for useful comments on drafts. This research is supported in part by Grant-in-Aid for Young Scientists (B) No. 23740168 and by Grant-in-Aid for Scientific Research (B) No. 25287049.

Appendix: Summary of Relevant Facts

Class $\mathcal{S}$ Theories of Type A_1

The low-energy theory in the world-volume of two M5-branes is a six-dimensional $\mathcal{N} = (0, 2)$ supersymmetric theory with no known Lagrangian description. An A_1-theory of class $\mathcal{S}$ is believed to arise by compactifying the six-dimensional theory on the Riemann surface $C_{g,n}$ of genus g with n punctures, with each puncture carrying a codimension-two defect of the $(0, 2)$ theory ([41, 61], [V:2]). The A_1-theories of class $\mathcal{S}$ provide basic examples of 2d-4d correspondence.

A weakly coupled description of such a theory may be encoded in a choice of decomposition of $C_{g,n}$ into $3g - 3 + n$ pairs of pants, and a trivalent graph Γ drawn on $C_{g,n}$. Each pair of pants contains one vertex, and three edges come out through distinct boundary components (pants legs). An example is shown in Fig. 1a. We allow a pants leg to degenerate to a puncture. The graph Γ has $3g - 3 + n$ internal edges and n external edges ending on the punctures. The field content in this description of the $\mathcal{N} = 2$ theory can be read off from Γ by associating to each internal edge an $SU(2)$ gauge group and to each vertex eight half-hypermultiplets in the trifundamental representation of the $SU(2)^3$ group associated to the three attached edges. When the edge is external the $SU(2)$ symmetry corresponds to a flavor symmetry. A change of pants decomposition and Γ corresponds to a S-duality transformation.

Liouville Theory

Liouville field theory is formally defined by the path integral over a single real field ϕ weighted by e^{-S}, where

$$S = \frac{1}{4\pi} \int_C \left(\partial^\mu \phi \partial_\mu \phi + 4\pi \mu e^{2b\phi} + QR\phi \right). \tag{57}$$

Here R is the scalar curvature, and $Q = b + 1/b$ parametrizes the central charge $c = 1 + 6Q^2$. The "cosmological constant" μ can be absorbed into a shift of ϕ, and affects the theory in a very mild way. Liouville theory is a non-rational CFT, meaning that it contains infinitely many representations of the Virasoro algebra. The spectrum of representations is continuous, and the conformal dimension Δ is parametrized by the Liouville momentum $\alpha \in Q/2 + i\mathbb{R}_{\geq 0}$ as $\Delta = \alpha(Q - \alpha)$. We denote the corresponding primary field by V_α.

Fusion move coefficients $F_{s_1 s_2} = F_{s_1 s_2}\left[\begin{smallmatrix} \alpha_3 & -b/2 \\ \alpha_4 & \alpha_1 \end{smallmatrix} \right]$ in (39) are explicitly known:

$$F_{++} = \frac{\Gamma(b(2\alpha_1 - b))\Gamma(b(b - 2\alpha_3) + 1)}{\Gamma(b(\alpha_1 - \alpha_3 - \alpha_4 + b/2) + 1)\Gamma(b(\alpha_1 - \alpha_3 + \alpha_4 - b/2))}, \tag{58}$$

etc.

References

[1]   't Hooft, G.: On the phase transition towards permanent quark confinement. Nucl. Phys. **B138**, 1 (1978)
[2]   Maldacena, J.M.: Wilson loops in large N field theories. Phys. Rev. Lett. **80**, 4859–4862 (1998). arXiv:hep-th/9803002
[3]   Rey, S.-J., Yee, J.-T.: Macroscopic strings as heavy quarks in large N gauge theory and anti-de sitter supergravity. Eur. Phys. J. **C22**, 379–394 (2001). arXiv:hep-th/9803001
[4]   Erickson, J., Semenoff, G., Zarembo, K.: Wilson loops in N = 4 supersymmetric Yang-Mills theory. Nucl. Phys. **B582**, 155–175 (2000). arXiv:hep-th/0003055
[5]   Drukker, N., Gross, D.J.: An Exact prediction of N = 4 SUSYM theory for string theory. J. Math. Phys. **42**, 2896–2914 (2001). arXiv:hep-th/0010274
[6]   Pestun, V.: Localization of gauge theory on a four-sphere and supersymmetric Wilson loops. Commun. Math. Phys. **313**, 71–129 (2012). arXiv:0712.2824
[7]   Witten, E.: Topological quantum field theory. Commun. Math. Phys. **117**, 353 (1988)
[8]   Nekrasov, N.A.: Seiberg-Witten prepotential from instanton counting. Adv. Theor. Math. Phys. **7**, 831–864 (2004). arXiv:hep-th/0206161
[9]   Alday, L.F., Gaiotto, D., Tachikawa, Y.: Liouville correlation functions from four-dimensional gauge theories. Lett. Math. Phys. **91**, 167–197 (2010). arXiv:0906.3219
[10] Gukov, S., Witten, E.: Gauge theory, ramification, and the geometric Langlands program. arXiv:hep-th/0612073
[11] Dymarsky, A., Pestun, V.: Supersymmetric Wilson loops in N = 4 SYM and pure spinors. JHEP **1004**, 115 (2010). arXiv:0911.1841
[12] Kapustin, A.: Wilson-'t Hooft operators in four-dimensional gauge theories and S-duality. Phys. Rev. **D74**, 025005 (2006). arXiv:hep-th/0501015
[13] Witten, E.: Dyons of charge e theta/2 pi. Phys. Lett. **B86**, 283–287 (1979)

[14]   Drukker, N., Morrison, D.R., Okuda, T.: Loop operators and S-duality from curves on
        Riemann surfaces. JHEP **0909**, 031 (2009). arXiv:0907.2593
[15]   Penner, R.C.: The action of the mapping class group on curves in surfaces. Enseign. Math.
        (2) **30**(1–2), 39–55 (1984)
[16]   Gaiotto, D., Moore, G.W., Neitzke, A.: Framed BPS states. arXiv:1006.0146
[17]   Aharony, O., Seiberg, N., Tachikawa, Y.: Reading between the lines of four-dimensional
        gauge theories. JHEP **1308**, 115 (2013). arXiv:1305.0318
[18]   Kapustin, A., Seiberg, N.: Coupling a QFT to a TQFT and duality. JHEP **1404**, 001 (2014).
        arXiv:1401.0740
[19]   Tachikawa, Y.: On the 6d origin of discrete additional data of 4d gauge theories. JHEP
        1405, 020 (2014). arXiv:1309.0697
[20]   Gomis, J., Okuda, T., Pestun, V.: Exact results for 't Hooft loops in gauge theories on S^4.
        JHEP **1205**, 141 (2012). arXiv:1105.2568
[21]   Kronheimer, P.: Monopoles and Taub-NUT metrics. M.Sc. thesis Oxford University, 1986,
        available on the author's home page
[22]   Giombi, S., Pestun, V.: The 1/2 BPS 't Hooft loops in N = 4 SYM as instantons in 2d
        Yang-Mills. J. Phys. **A46**, 095402 (2013). arXiv:0909.4272
[23]   Ito, Y., Okuda, T., Taki, M.: Line operators on $\mathbb{S}^1 \times \mathbb{R}^3$ and quantization of the Hitchin
        moduli space. JHEP **1204**, 010 (2012). arXiv:1111.4221
[24]   Romelsberger, C.: Counting chiral primaries in N = 1, d = 4 superconformal field theories.
        Nucl. Phys. **B747**, 329–353 (2006). arXiv:hep-th/0510060
[25]   Kinney, J., Maldacena, J.M., Minwalla, S., Raju, S.: An index for 4 dimensional super
        conformal theories. Commun. Math. Phys. **275**, 209–254 (2007). arXiv:hep-th/0510251
[26]   Dimofte, T., Gaiotto, D., Gukov, S.: 3-manifolds and 3d indices. arXiv:1112.5179
[27]   Gang, D., Koh, E., Lee, K.: Line operator index on $S^1 \times S^3$. JHEP **1205**, 007 (2012).
        arXiv:1201.5539
[28]   Hama, N., Hosomichi, K.: Seiberg-Witten theories on ellipsoids. JHEP **1209**, 033 (2012).
        arXiv:1206.6359
[29]   Pestun, V.: Localization of the four-dimensional N = 4 SYM to a two-sphere and 1/8 BPS
        Wilson loops. JHEP **1212**, 067 (2012). arXiv:0906.0638
[30]   Drukker, N., Giombi, S., Ricci, R., Trancanelli, D.: Wilson loops: from four-dimensional
        SYM to two-dimensional YM. Phys. Rev. **D77**, 047901 (2008). arXiv:0707.2699
[31]   Giombi, S., Pestun, V.: Correlators of local operators and 1/8 BPS Wilson loops on S**2
        from 2d YM and matrix models. JHEP **1010**, 033 (2010). arXiv:0906.1572
[32]   Giombi, S., Pestun, V.: Correlators of Wilson loops and local operators from multi-matrix
        models and strings in AdS. JHEP **1301**, 101 (2013). arXiv:1207.7083
[33]   Verlinde, E.P.: Fusion rules and modular transformations in 2D conformal field theory.
        Nucl. Phys. **B300**, 360 (1988)
[34]   Moore, G.W., Seiberg, N.: Polynomial equations for rational conformal field theories. Phys.
        Lett. **B212**, 451 (1988)
[35]   Moore, G.W., Seiberg, N.: Naturality in conformal field theory. Nucl. Phys. **B313**, 16 (1989)
[36]   Belavin, A., Polyakov, A.M., Zamolodchikov, A.: Infinite conformal symmetry in two-
        dimensional quantum field theory. Nucl. Phys. **B241**, 333–380 (1984)
[37]   Alday, L.F., Gaiotto, D., Gukov, S., Tachikawa, Y., Verlinde, H.: Loop and surface oper-
        ators in N = 2 gauge theory and Liouville modular geometry. JHEP **1001**, 113 (2010).
        arXiv:0909.0945
[38]   Drukker, N., Gomis, J., Okuda, T., Teschner, J.: Gauge theory loop operators and Liouville
        theory. JHEP **1002**, 057 (2010). arXiv:0909.1105
[39]   Teschner, J., Vartanov, G.: 6j symbols for the modular double, quantum hyperbolic
        geometry, and supersymmetric gauge theories. Lett. Math. Phys. **104**, 527–551 (2014).
        arXiv:1202.4698
[40]   Vartanov, G., Teschner, J.: Supersymmetric gauge theories, quantization of moduli spaces
        of flat connections, and conformal field theory. arXiv:1302.3778
[41]   Gaiotto, D.: N = 2 dualities. JHEP **1208**, 034 (2012). arXiv:0904.2715

[42]   Wyllard, N.: A(N-1) conformal Toda field theory correlation functions from conformal N
        = 2 SU(N) quiver gauge theories. JHEP **0911**, 002 (2009). arXiv:0907.2189
[43]   Drukker, N., Gaiotto, D., Gomis, J.: The virtue of defects in 4D gauge theories and 2D
        CFTs. JHEP **1106**, 025 (2011). arXiv:1003.1112
[44]   Passerini, F.: Gauge theory Wilson loops and conformal Toda field theory. JHEP **1003**, 125
        (2010). arXiv:1003.1151
[45]   Gomis, J., Le Floch, B.: 't Hooft operators in gauge theory from Toda CFT. JHEP **1111**,
        114 (2011). arXiv:1008.4139
[46]   Wu, J.-F., Zhou, Y.: From Liouville to Chern-Simons, alternative realization of Wilson loop
        operators in AGT duality. arXiv:0911.1922
[47]   Petkova, V., Zuber, J.: Generalized twisted partition functions. Phys. Lett. **B504**, 157–164
        (2001). arXiv:hep-th/0011021
[48]   Sarkissian, G.: Defects and permutation branes in the Liouville field theory. Nucl. Phys.
        B821, 607–625 (2009). arXiv:0903.4422
[49]   Petkova, V.: On the crossing relation in the presence of defects. JHEP **1004**, 061 (2010).
        arXiv:0912.5535
[50]   Bullimore, M.: Defect networks and supersymmetric loop operators. arXiv:1312.5001
[51]   Johansen, A.: Twisting of $N = 1$ SUSY gauge theories and heterotic topological theories.
        Int. J. Mod. Phys. **A10**, 4325–4358 (1995). arXiv:hep-th/9403017
[52]   Kapustin, A., Witten, E.: Electric-magnetic duality and the geometric Langlands program.
        Commun. Number Theory Phys. **1**, 1–236 (2007). arXiv:hep-th/0604151
[53]   Kapustin, A.: Holomorphic reduction of N = 2 gauge theories, Wilson-'t Hooft operators,
        and S-duality. arXiv:hep-th/0612119
[54]   Kapustin, A., Saulina, N.: The algebra of Wilson-'t Hooft operators. Nucl. Phys. **B814**,
        327–365 (2009). arXiv:0710.2097
[55]   Saulina, N.: A note on Wilson-'t Hooft operators. Nucl. Phys. **B857**, 153–171 (2012).
        arXiv:1110.3354
[56]   Moraru, R., Saulina, N.: OPE of Wilson-'t Hooft operators in N=4 and N=2 SYM with
        gauge group G = PSU(3). arXiv:1206.6896
[57]   Xie, D.: Higher laminations, webs and N = 2 line operators. arXiv:1304.2390
[58]   Xie, D.: Aspects of line operators of class S theories. arXiv:1312.3371
[59]   Cirafici, M.: Line defects and (framed) BPS quivers. JHEP **1311**, 141 (2013).
        arXiv:1307.7134
[60]   Saulina, N.: Spectral networks and higher web-like structures. arXiv:1409.2561
[61]   Gaiotto, D., Moore, G.W., Neitzke, A.: Wall-crossing, Hitchin systems, and the WKB
        approximation. arXiv:0907.3987
[62]   Bershadsky, M., Johansen, A., Sadov, V., Vafa, C.: Topological reduction of 4-d SYM to
        2-d sigma models. Nucl. Phys. **B448**, 166–186 (1995). arXiv:hep-th/9501096
[63]   Nekrasov, N., Witten, E.: The omega deformation, branes, integrability, and Liouville the-
        ory. JHEP **1009**, 092 (2010). arXiv:1002.0888
[64]   Kapustin, A.: A-branes and noncommutative geometry. arXiv:hep-th/0502212
[65]   Cordova, C., Neitzke, A.: Line defects, tropicalization, and multi-centered quiver quantum
        mechanics. arXiv:1308.6829

Surface Operators

Sergei Gukov

Abstract We give an introduction and a broad survey of surface operators in 4d gauge theories, with a particular emphasis on aspects relevant to AGT correspondence. One of the main goals is to highlight the boundary between what we know and what we don't know about surface operators. To this end, the survey contains many open questions and suggests various directions for future research. Although this article is mostly a review, we did include a number of new results, previously unpublished.

1 What Is a Surface Operator?

Surface operators (a.k.a. surface defects) in a four-dimensional gauge theory are operators supported on two-dimensional submanifolds in the space-time manifold M. They are particular examples of non-local operators in quantum field theory (QFT) that play the role of "thermometers" in a sense that, when introduced in the Feynman path integral, their correlation functions provide us with valuable information about the physics of a QFT in question (phases, non-perturbative phenomena, etc.).

In general, non-local operators can be classified by dimension (or, equivalently, codimension) of their support, which in four dimensions clearly can range from zero to four, so that we have the following types of operators:

- codimension 4: the operators of codimension 4 are the usual local operators $\mathcal{O}(p)$ supported at a point $p \in M$. These are the most familiar operators in this list, which have been extensively studied e.g. in the context of the AdS/CFT correspondence. Typical examples of local operators can be obtained by considering gauge-invariant combinations of the fields in the theory, e.g. $\mathcal{O}(p) = \mathrm{Tr}(\phi^n \ldots)$.

S. Gukov (✉)
California Institute of Technology, Pasadena, CA 91125, USA
e-mail: gukov@theory.caltech.edu

S. Gukov
Walter Burke Institute for Theoretical Physics, California Institute of Technology,
Pasadena, CA 91125, USA

© Springer International Publishing Switzerland 2016
J. Teschner (ed.), *New Dualities of Supersymmetric Gauge Theories*,
Mathematical Physics Studies, DOI 10.1007/978-3-319-18769-3_8

223

- codimension 3: line operators. Important examples of such operators are Wilson and 't Hooft operators, which are labeled, respectively, by a representation, R, of the gauge group, G, and by a representation $^L R$ of the dual gauge group $^L G$.
- codimension 2: surface operators. These are perhaps least studied among the operators and defects listed here, and will be precisely our main subject.
- codimension 1: domain walls and boundaries.

After giving the reader a basic idea about different types of non-local operators classified by (co)dimension of their support, perhaps it is worth mentioning that some of them—usually called "electric"—can be constructed directly from elementary fields present in the path integral formulation of the theory. In the above classification, we already mentioned examples of such operators that are actually local, i.e. any gauge-invariant combination of elementary fields gives an example. Among non-local operators, a typical example of "electric" operators is a Wilson line operator labeled by a representation R of the gauge group G:

$$W_R(K) = \mathrm{Tr}_R \, \mathrm{Hol}_K(A) = \mathrm{Tr}_R \left(P \exp \oint A \right) \qquad (1.1)$$

Another type of operators, called "magnetic" (a.k.a. disorder operators) can not be defined via (algebraic) combinations of elementary fields and calls for alternative definitions, which will be considered below and which will be crucial for defining surface operators.

A surface operator in four-dimensional gauge theory is an operator supported on a 2-dimensional submanifold $D \subset M$ in the space-time manifold M. In other words, according to the above classification, it is an operator whose dimension and codimension are both equal to 2:

$$4 = 2 + 2 \qquad (1.2)$$

This simple equation illustrates how the dimension of the space-time manifold M splits into the tangent and normal spaces to the support, D, of the surface operator. Note, that 2 also happens to be the degree of the differential form F, the curvature of the gauge field A. This basic fact and Eq. (1.2) make surface operators somewhat special in the context of 4d gauge theory.

Indeed, since the degree of the 2-form F matches the dimension of the tangent as well as normal space to $D \subset M$, we can either write an integral

$$\exp \left(i\eta \int_D F \right) \qquad (1.3)$$

which defines an electric surface operator analogous to (1.1) in abelian $U(1)$ gauge theory, or write a relation

$$F = 2\pi\alpha\delta_D + \cdots \qquad (1.4)$$

where δ_D is a 2-form delta-function Poincaré dual to D. In (1.3) and (1.4) we used the basic fact that, respectively, dimension and codimension of the surface operator (or, to be more precise, its support) equals the degree of the differential form F. These relations define magnetic (resp. electric) surface operators in abelian 4d gauge theory—with any amount of supersymmetry, including $\mathcal{N} = 2$ that will be of our prime interest in this note—and admit a simple generalization to non-abelian theories that will be discussed shortly.

Already at this stage, however, it is a good idea to pause and ask the following questions that shall guide us in the exploration of surface defects:

- How can one define surface operators?
- What are they classified by?
- Are there supersymmetric surface operators?
- What are the correlation functions of surface operators?
- What is the OPE algebra of line operators in the presence of a surface operator?
- How do surface operators transform under dualities?

The answer to many of these questions is not known at present, except in some special cases. One such special case is that of abelian gauge theory with gauge group $G \cong U(1)^r = \mathbb{L}$, where all of the above questions can be answered:

- By combining the above constructions (1.3) and (1.4) for each $U(1)$ factor in G one can produce a surface operator that, in general, preserves some part of the gauge group, $\mathbb{L} \subseteq G$.
- The resulting surface operators are labeled by a *discrete* choice of $\mathbb{L} \subseteq G$ and two sets of *continuous* parameters

$$(\alpha, \eta) \in \mathbb{T} \times \mathbb{T}^\vee \tag{1.5}$$

where $\mathbb{T} = G/\mathbb{L}$ and $\mathbb{T}^\vee$ is its dual.
- They are compatible with any amount of supersymmetry and define half-BPS surface operators in SUSY gauge theories.
- Physically, the world-volume D of such a surface operator can be interpreted as a "visible" Dirac string for a dyon with electric and magnetic charges (η, α) that do not obey Dirac quantization condition.
- A remarkable property of abelian 4d gauge theory is that it enjoys electric-magnetic duality, even in the absence of supersymmetry [1, 2]. This duality exchanges the role of α and η:

$$(\alpha, \eta) \rightarrow (\eta, -\alpha) \tag{1.6}$$

- A novel feature of surface operators is that they are labeled not only by discrete but also by continuous parameters. A kink-like configuration within the surface operator that represents an adiabatic change of continuous parameters along a closed loop in the parameter space (1.5) represents a Wilson-'t Hooft line operator. In other words, line operators correspond to closed loops in the space of continuous parameters and are labeled by elements of

$$\pi_1(\{\text{parameters}\}) \tag{1.7}$$

Many of these aspects have analogues in non-abelian gauge theory, where essential features may look similar though dressed with lots of quantum and non-perturbative effects, which potentially can not only affect the details, but also lead to new physics. As one might anticipate, such effects are under better control in supersymmetric theories and in situations where surface operators preserve some fraction of supersymmetry.

1.1 Construction of Surface Operators

Now, once we have presented the basic idea of what a surface operator is, we can elaborate on various points, starting with the definition. In the standard formulation of quantum field theory, based on a Feynman path integral, there are several (often, equivalent) ways to define surface operators [3, 4]:

- as singularities or boundary conditions for the gauge filed A_μ (and, possibly, other fields) along a surface D in four-dimensional space-time;
- as a coupled 2d-4d systems, namely a 2d theory supported on D with a flavor symmetry group G that is gauged upon coupling to 4d theory on M.

The latter option, in turn, is often subdivided into two large classes of models where the 2d theory on D is either (a) gauge theory itself, or (b) non-linear sigma-model. Clearly, these two classes do not exhaust all possibilities and, yet, there are models which belong to both. A prominent example of such a model that has the advantage of being looked at from several viewpoints is a 2d sigma-model with target space $\mathbb{CP}^1 = \mathbb{C}^2 /\!/ U(1)$ that can be equivalently described as a GLSM with $U(1)$ gauge group. It defines a surface operator in 4d gauge theory with gauge group $G = SU(2)$ that is a symmetry of $\mathbb{CP}^1$ (and for which $\mathbb{C}^2$ is the defining two-dimensional representation).

As for the first way of defining surface operators, we already saw examples in (1.3) and (1.4) where one did not need to introduce any additional 2d degrees of freedom. In particular, the disorder operator (1.4) has an obvious analogue in a non-abelian gauge theory with a general gauge group G. Namely, one can define operators supported on a surface D by requiring the gauge field A (and, possibly, other fields) to have a prescribed singularity along D:

$$\text{Hol}_\ell(A) \in \mathfrak{C} \tag{1.8}$$

where ℓ is a small loop that links surface $D \subset M$ in the space-time 4-manifold M, and $\mathfrak{C}$ is a fixed conjugacy class in the gauge group G (or, possibly, its complexification $G_{\mathbb{C}}$). The latter option, $\mathfrak{C} \subset G_{\mathbb{C}}$, is realized in $\mathcal{N} \geq 2$ supersymmetric gauge theory, where the gauge field A combines with a Higgs field ϕ in a complex combination $\mathcal{A} = A + i\phi$ (see e.g. Fig. 1 for a list of complex conjugacy classes in $SO(7)$ and $Sp(6)$ gauge theory).

Fig. 1 Surface operators shown in *red* and labeled by ∗ appear to spoil electric-magnetic duality between $SO(7)$ and $Sp(6)$ gauge theories. In order to restore a nice match, one has to introduce a larger class of surface operators

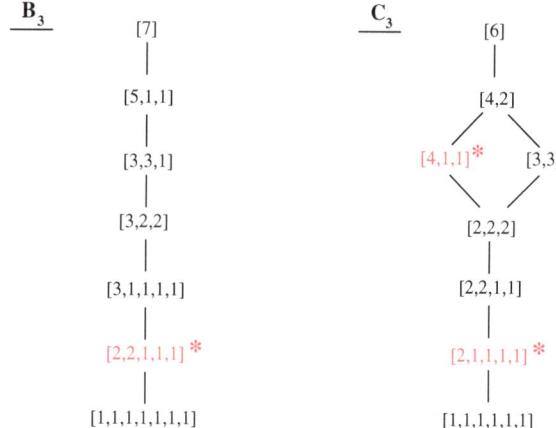

1.2 Classification of Surface Operators

A careful definition of surface operators essentially gives an answer to the question about classification of surface operators.

In general, parameters of surface operators can be divided into discrete data and continuous parameters. In a way, the former is analogous to the choice[1] of a representation that labels line operators, cf. (1.1), while the latter are a novel feature of surface operators. Moreover, it turns out that understanding these continuous parameters is the key to addressing other important questions about the properties of surface operators. For example, the non-commutative structure of line operators supported on a surface operator—that will be discussed in Sect. 3—is described by the fundamental group (1.7) of the suitable (sub)space of continuous parameters.

In our previous discussion we already saw examples of both discrete and continuous parameters. In (1.5), the parameters α and η are examples of continuous parameters, whereas the choice of the subgroup $\mathbb{L} \subseteq G$, called the Levi subgroup, preserved by the surface operator along D is a typical example of the discrete parameter. Although we introduced these parameters in the simplest (abelian) examples, they have immediate analogues in a very broad class of surface operators known at present. The number of continues parameter can vary, usually from 0 to the rank of the gauge group G (multiplied by $\mathcal{N}$). The surface operators which do not have continuous parameters at all are usually called *rigid*.

The classification problem consists of making a list of discrete and continues parameters that label surface operators in a given gauge theory. At present, this is an open problem, which is very far from satisfactory solution. One might hope to

---

[1]For example, in $\mathcal{N} = 2^*$ theory with gauge group $G = SU(N)$ this choice includes the choice of a partition of N. When G is a classical group of Cartan type B, C, or D, the choice of partition must satisfy certain conditions, as illustrated in Fig. 1. In particular, the transformation of surface operators under electric-magnetic duality becomes a rather non-trivial matter in non-abelian gauge theories.

make more progress by imposing additional conditions, e.g. focusing on SUSY gauge theories and requiring surface operators to preserve some fraction of supersymmetry. Thus, one might hope that the general construction of [4] is not too far from a complete classification of half-BPS surface operators in the maximally supersymmetric gauge theory in four dimensions. The next natural step is the classification of half-BPS surface operators in $\mathcal{N} = 2$ gauge theories (that will be of our main interest here), followed by $\frac{1}{4}$-BPS surface operators and surface operators in $\mathcal{N} = 1$ and $\mathcal{N} = 0$ gauge theories.[2]

Let us illustrate how this construction and classification of half-BPS surface operators works in the simplest case, namely in $\mathcal{N} = 4$ super-Yang-Mills, which can be viewed as a special case of $\mathcal{N} = 2$ gauge theory with a massless hypermultiplet in the adjoint representation of the gauge group G. (Its deformation by turning on the mass parameter $m \neq 0$ for the adjoint hypermultiplet is usually called $\mathcal{N} = 2^*$ theory.) Much like in our preliminary discussion around (1.4), we can produce a large class of half-BPS surface operators which break the gauge group down to a Levi subgroup $\mathbb{L} \subset G$ and which also break the global R-symmetry group,

$$SO(6)_R \to SO(4) \times SO(2) \tag{1.9}$$

by introducing a singularity for the gauge field that corresponds to the monodromy (1.8) and for two components of the Higgs field, say $\varphi = \phi_1 + i\phi_2$,

$$A = \alpha d\theta + \cdots, \tag{1.10}$$

$$\varphi = \frac{1}{2}(\beta + i\gamma)\frac{dz}{z} + \cdots \tag{1.11}$$

Here, $z = x^2 + ix^3 = re^{i\theta}$ is a local complex coordinate, normal to the surface $D \subset M$, and the dots stand for less singular terms. In order to obey the supersymmetry equations [3], the parameters α, β, and γ must take values in the $\mathbb{L}$-invariant part of $\mathfrak{t}$, the Lie algebra of the maximal torus $\mathbb{T}$ of G. Moreover, gauge transformations shift values of α by elements of the cocharacter lattice, Λ_{cochar}. Hence, α takes values in $\mathbb{T} = \mathfrak{t}/\Lambda_{\text{cochar}}$.

In addition to the classical (or "geometric") parameters (α, β, γ), the surface operators of this type are also labeled by quantum parameters (1.3), the "theta angles" of the two-dimensional theory on $D \subset M$. It is easy to see that parameters η take values in the $\mathbb{L}^\vee$-invariant part of the maximal torus of the Langlands/GNO dual group $G^\vee$. We can summarize all this by saying that maximally supersymmetric ($\mathcal{N} = 4$) super-Yang-Mills theory admits a large class of surface operators labeled by a choice[3] of the Levi subgroup $\mathbb{L} \subset G$ and continuous parameters

$$(\alpha, \beta, \gamma, \eta) \in \left(\mathbb{T} \times \mathfrak{t} \times \mathfrak{t} \times \mathbb{T}^\vee\right)/\mathcal{W} \tag{1.12}$$

---

[2]See [5, 6] for discussion of $\frac{1}{4}$-BPS surface operators in 4d $\mathcal{N} = 2$ gauge theories.

[3]In a theory with gauge group $G = SU(N)$ this choice is equivalent to a choice of a partition of N.

invariant under the Weyl group $\mathcal{W}_{\mathbb{L}}$ of $\mathbb{L}$. Similar surface operators exist in $\mathcal{N} = 2$ supersymmetric gauge theories; the only difference is that they don't have parameters β and γ.

These surface operators naturally correspond to the so-called Richardson conjugacy classes in the complexified gauge group $G_{\mathbb{C}}$, cf. (1.8) and, in a theory with gauge group $G = SU(N)$, cover all half-BPS surface operators which correspond to singularities with simple poles.

1.3 Surface Operators in 4d $\mathcal{N} = 2$ Gauge Theory

The construction described in the end of the previous section can be easily generalized to define half-BPS surface operators in $\mathcal{N} = 2$ gauge theories, see e.g. [7–11] and subsequent work. As a result, one finds a fairly large class of surface operators labeled by the Levi subgroup $\mathbb{L} \subseteq G$ and continuous parameters $(\alpha, \eta) \in (\mathbb{T} \times \mathbb{T}^{\vee})/\mathcal{W}$, which in $\mathcal{N} = 2$ theories conveniently unify into holomorphic combinations

$$t = \eta + \tau\alpha \tag{1.13}$$

where τ is the coupling (matrix) of the $\mathcal{N} = 2$ gauge theory.

A novel feature of such half-BPS surface operators in $\mathcal{N} = 2$ theories—compared to maximally supersymmetric Yang-Mills or abelian ($\mathcal{N} = 0$) theories without supersymmetry discussed above—is that one must be wary of quantum effects, which can not only renormalize the values of various parameters but also change the nature of a surface operator altogether. In other words, defined as a singularity for the gauge field (and, possibly, other fields) as described in Sect. 1.1, a surface operator is defined at a given energy scale in the 4d theory. It can be a UV theory, or an IR theory, or some effective theory at intermediate energy scale. An interesting question, then, is to study what becomes of such surface operator at other energy scales and/or regimes of parameters.

In order to answer such questions, it is often helpful to use another definition of surface operators described in Sect. 1.1. Namely, one can define a surface operator supported on $D \subset M$ by introducing additional 2d degrees of freedom along D, with their own Lagrangian and a global symmetry group G that becomes gauged upon coupling to 4d degrees of freedom. Of course, if 4d gauge theory in question has matter fields Q, they too can be coupled to 2d degrees of freedom supported on D in a gauge invariant manner. As explained in [3, 4], integrating out 2d degrees of freedom leaves behind a singularity (obviously, supported on D) in the field equations of the four-dimensional theory:

$$F_{23} - QQ^{\dagger} = 2\pi\delta^2(\vec{x})\mu_1 \tag{1.14}$$
$$D_{\bar{z}}Q = \pi\delta^2(\vec{x})(\mu_2 + i\mu_3)$$

Table 1 Half-BPS surface operators in SUSY gauge theories can be described as coupled 2d-4d systems with suitable amount of supersymmetry in 2d theory

4d theory on M	2d theory on D	Superconformal symmetry
$\mathcal{N} = 4$	$\mathcal{N} = (4, 4)$	$PSU(2, 2\mid 4) \rightarrow PSU(1, 1\mid 2) \times PSU(1, 1\mid 2) \times U(1)$
$\mathcal{N} = 2$	$\mathcal{N} = (2, 2)$	$SU(2, 2\mid 2) \rightarrow SU(1, 1\mid 1) \times SU(1, 1\mid 1) \times U(1)$
$\mathcal{N} = 1$	$\mathcal{N} = (0, 2)$	$SU(2, 2\mid 1) \rightarrow SU(1, 1\mid 1) \times SL(2, \mathbb{R}) \times U(1)$

The last column is only relevant to superconformal theories and describes the symmetry breaking pattern due to surface operator

where $\mathcal{N} = 4$ SYM corresponds to a special case when Q transforms in the adjoint representation of the gauge group G.

Which two-dimensional theories can one use in this construction? In general, any 2d theory will do as long as it has a symmetry group G that can be gauged and as long as it is free of anomalies. In fact, coupling 2d degrees of freedom to 4d gauge theory even allows one to experiment with anomalous 2d theories where anomalies can be canceled by the inflow from the four-dimensional bulk [12].

When one aims to build a surface operator that preserves certain symmetries of the four-dimensional gauge theory, the 2d theory on the defect must be chosen accordingly, so that it also enjoys the desired symmetries. For instance, if the goal is to build a half-BPS surface operator in a supersymmetric 4d gauge theory, the 2d theory on D must have at least half of the supercharges present in 4d, as illustrated in Table 1.

A simple way to achieve this is to take 2d theory to be a sigma-model with the desired supersymmetry and a target space X that has a symmetry group G. Of course, depending the on the desired amount of supersymmetry, the space X may also need to be Kähler or hyper-Kähler for applications to $\mathcal{N} = 2$ and $\mathcal{N} = 4$ gauge theory, respectively. Large class of such targets that have all the desired properties are coadjoint orbits (or, via the exponentiation map, conjugacy classes $X = \mathfrak{C}$) and their complexifications. Indeed, they admit Kähler and hyper-Kähler metrics, respectively, in addition to a G-action that one needs for coupling to 4d degrees of freedom. In the $\mathcal{N} = 2$ case, let μ_1 be the moment map for the action of G on the Kähler target space X and, similarly, in the $\mathcal{N} = 4$ theory let $\vec{\mu} = (\mu_1, \mu_2, \mu_3)$ be the hyper-Kähler moment map for the action of G on X.

Then, integrating out 2d degrees of freedom in these cases leads to surface operators defined as singularities (1.14), where the holonomy of the (complexified) gauge field is required to be in a fixed conjugacy class, cf. (1.8). This provides a link between two ways of defining surface operators described in Sect. 1.1, namely, as singularities and as coupled 2d-4d systems.

Supersymmetry also often tightly constrains the geometry of the surface $D \subset M$. A popular example is $D = \mathbb{R}^2$ linearly embedded in $M = \mathbb{R}^4$ which breaks the Lorentz symmetry as, cf. (1.2):

$$SO(1, 3) \rightarrow SO(1, 1)_{01} \times SO(2)_{23} \qquad (1.15)$$

where, for concreteness, we chose the surface operator to be oriented along the (x^0, x^1) plane. Since surface operators break Lorentz symmetry, they must break at least part of the supersymmetry and some of the R-symmetries. Thus, in $\mathcal{N} = 4$ gauge theory the R-symmetry breaking pattern is (1.9). Similarly, a generic $\mathcal{N} = 2$ gauge theory has R-symmetry group $SU(2)_R \times U(1)_r$, of which $U(1)_r$ may be broken by quantum effects. A half-BPS surface operator further breaks $SU(2)_R$ down to $U(1)_R$.

Of particular interest to us, especially in applications to the AGT correspondence [13] will be half-BPS surface operators in superconformal gauge theories. The conformal group in four dimensions is $SO(4, 2) \sim SU(2, 2)$, and a surface operator oriented along the (x^0, x^1) plane breaks it down to a subgroup, cf. (1.15):

$$SO(2, 2) \times U(1)_{23} \subset SO(4, 2) \qquad (1.16)$$

Here, $SO(2, 2) \cong SL(2, \mathbb{R})_L \times SL(2, \mathbb{R})_R$ is the conformal group in two dimensions and $U(1)_{23}$ is the rotation symmetry in the (x^2, x^3) plane transverse to the surface operator.

The analogous symmetry breaking patterns in supersymmetric theories are summarized in Table 1. In particular, the superconformal symmetry group of 4d $\mathcal{N} = 2$ gauge theory is $SU(2, 2|2)$. Its bosonic subgroup is $S[U(2, 2) \times U(2)] \sim SU(2, 2) \times SU(2)_R \times U(1)_r$, where $SU(2, 2)$ is the familiar conformal group and $SU(2)_R \times U(1)_r$ is the R-symmetry group mentioned earlier. Apart from the conformal symmetry (1.16), a half-BPS surface operator also preserves $U(1)_L \times U(1)_R \subset SU(2)_R \times U(1)_r$ part of the R-symmetry group and four (out of eight) supercharges $\mathcal{Q}^2_-, \tilde{\mathcal{Q}}^1_-, \mathcal{Q}^1_+, \tilde{\mathcal{Q}}^2_+$ of the four dimensional theory. The bosonic subgroup $SL(2, \mathbb{R})_L \times U(1)_L$ combines with the supersymmetries $\mathcal{Q}^2_-, \tilde{\mathcal{Q}}^1_-$ to form $SU(1, 1|1)_L$. Similarly, the remaining charges generate $SU(1, 1|1)_R$, so that in total a half-BPS surface operator in $\mathcal{N} = 2$ superconformal theory preserves $SU(1, 1|1)_L \times SU(1, 1|1)_R \times U(1)_e$ subgroup of $SU(2, 2|2)$, where $U(1)_e$ is the commutant of the embedding.

1.4 Their Role in AGT Correspondence

Now we are ready to review the role of surface operators in the 2d-4d correspondence [13] that relates Liouville conformal block on a Riemann surface C and the equivariant instanton partition function [14] of the class $\mathcal{S}$ gauge theory [15, 16] labeled by the Riemann surface C:

Fig. 2 Upon the hemispherical stereographic projection on two copies of $\mathbb{R}^4$, surface operators on S^4 factorize into a two surface operators, a north and a south half, glued together at the equator

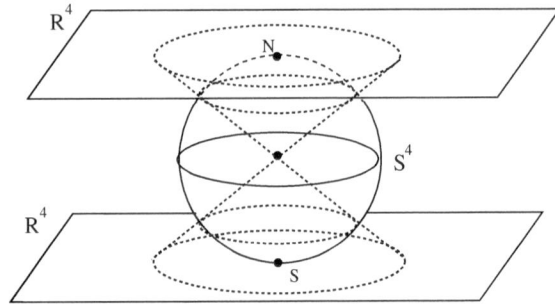

$$Z^{\text{inst}}(a, \tau, \epsilon) = Z^{\text{Liouv}}(\alpha, q, b)$$
$$\epsilon_1 : \epsilon_2 = b : 1/b \tag{1.17}$$
$$\exp(2\pi i \tau_{\text{UV}}) = q$$
$$a = \alpha - Q/2$$

where $Q = b + 1/b$ is the standard notation in the literature on Liouville theory. Note, the left-hand side of this dictionary involves supersymmetric gauge theory in four dimensions, whereas the right-hand side is about a *non-supersymmetric* 2d theory. There is a similar version of this correspondence [17] that relates superconformal index of the 4d $\mathcal{N} = 2$ gauge theory $T[C]$ labeled by C and a certain deformation of 2d non-supersymmetric Yang-Mills theory on C which we shall briefly discuss in Sect. 4.

The conformal block in (1.17) represents a "chiral half" of the full Liouville correlation function, which has the form of an integral of the absolute value squared of a conformal block and also admits a nice interpretation in 4d gauge theory as a partition function [18] on a 4-sphere S^4. Indeed, dividing S^4 into the northern and southern hemispheres illustrated in Fig. 2 corresponds to the chiral decomposition of the Liouville CFT correlation functions into "left-moving" and "right-moving" chiral halves.

The Ω-deformation of the Euclidean $\mathcal{N} = 2$ gauge theory on $M = \mathbb{R}^4$, used in the definition of the instanton partition function, involves the subgroup of the rotation symmetry

$$SO(2) \times SO(2) \subset SO(4) \tag{1.18}$$

which is precisely the part of the symmetry preserved by a surface operators, cf. (1.15). Therefore, following [14], one can introduce Ω-deformation and the partition function in the presence of a surface operators:

$$Z_{k,\mathbf{m}}^{\text{inst}}(\epsilon_1, \epsilon_2) = \oint_{\mathcal{M}_{k,\mathbf{m}}} 1 \tag{1.19}$$

Here, $\mathcal{M}_{k,\mathbf{m}}$ is the moduli space of "ramified instantons" on $M \backslash D$ labeled by the ordinary instanton number $k := c_2(E)$ and the monopole number

$$\mathfrak{m} = \frac{1}{2\pi} \int_D F \qquad \text{(“monopole number”)} \qquad (1.20)$$

that measures the magnetic charge of the gauge bundle E restricted to D. Then, the path integral of the Ω-deformed $\mathcal{N} = 2$ gauge theory in the presence of a surface operator of Levi type $\mathbb{L}$ gives the generating function

$$Z^{\text{inst}}(a, \Lambda, \epsilon; \mathbb{L}, z) = \sum_{k=0}^{\infty} \sum_{\mathfrak{m} \in \Lambda_{\mathbb{L}}} \Lambda^{2Nk} e^{iz \cdot \mathfrak{m}} Z_{k,\mathfrak{m}}^{\text{inst}}(a, \epsilon) \qquad (1.21)$$

where the coefficients $Z_{k,\mathfrak{m}}^{\text{inst}}$ are precisely the integrals (1.19).

The basic surface operator (with next-to-maximal $\mathbb{L} = S[U(1) \times U(N-1)]$) is labeled by a single complex parameter $z = \eta + i\alpha$ that takes values in $\mathbb{C}$. Incorporating this surface operator in the instanton partition function on $\mathbb{R}^4_{\epsilon_1,\epsilon_2}$ or $S^4_{\epsilon_1,\epsilon_2}$ on the Liouville side corresponds to inserting a degenerate primary operator at a point $z \in C$,

$$\Phi_{2,1}(z) = e^{-(b/2)\phi(z)} \qquad (1.22)$$

There are several ways to argue for this identification:

- using higher-dimensional constructions (that will be discussed below),
- using the "semi-classical limit" $\epsilon_{1,2} \to 0$,
- studying line operators within the surface operator,
- using tests based on direct computations of both sides.

Since higher-dimensional constructions and line operators will be discussed in Sects. 2 and 3, respectively, let us make a few comments on the semi-classical limit $\epsilon_{1,2} \to 0$. One of the main results of [14] is that, in this limit, the (logarithm of the) instanton partition function has a second order pole, whose coefficient is the Seiberg-Witten prepotential $\mathcal{F}(a_i)$. This matches the structure of the Liouville conformal block in the limit $\hbar^2 = \epsilon_1 \epsilon_2 \to 0$. The insertion of a degenerate field does not affect the leading singularity, but leads to a new first-order pole

$$Z^{\text{Liouv}} \sim \exp\left(-\frac{\mathcal{F}(a_i)}{\hbar^2} + \frac{b\mathcal{W}(a_i, z)}{\hbar} + \cdots\right) \qquad (1.23)$$

which has an elegant translation to the language of 4d $\mathcal{N} = 2$ gauge theory.

Indeed, using basic properties of the AGT correspondence, one can identify the function $\mathcal{W}(a_i, z)$ with the an integral [10]:

$$\mathcal{W} = \int_{p_*}^{p} \lambda_{SW} \qquad (1.24)$$

along some path on the Seiberg-Witten curve, starting at some reference point p_* (see Fig. 3). This is precisely how the insertion of a surface operator modifies the instanton partition function:

Fig. 3 The effective twisted
superpotential $\mathcal{W}$ can be
expressed as an integral over
an open path on the
Seiberg-Witten curve Σ

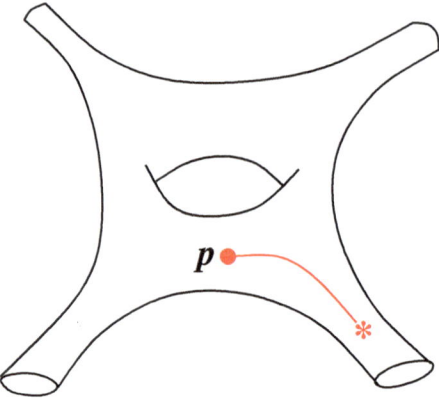

$$Z^{\text{inst}} \sim \exp\left(-\frac{\mathcal{F}(a_i)}{\epsilon_1 \epsilon_2} + \frac{\mathcal{W}(a_i, t)}{\epsilon_1} + \cdots\right) \quad (1.25)$$

The leading singularities here come from the "regularized" equivariant volume contributions of the 4d bulk degrees of freedom supported on $M = \mathbb{R}^4_{\epsilon_1, \epsilon_2}$ and the 2d contribution of a surface operator supported on $D = \mathbb{R}^2_{\epsilon_1}$:

$$\text{Vol}(\mathbb{R}^4_{\epsilon_1, \epsilon_2}) = \int_{\mathbb{R}^4_{\epsilon_1, \epsilon_2}} 1 = \frac{1}{\epsilon_1 \epsilon_2}, \quad \text{Vol}(\mathbb{R}^2_{\epsilon_1}) = \int_{\mathbb{R}^2_{\epsilon_1}} 1 = \frac{1}{\epsilon_1} \quad (1.26)$$

Naturally, here we are more interested in the contribution of a surface operator. The function $\mathcal{W}(a_i, z)$ that depends on both the Coulomb branch parameters of the 4d theory as well as continuous parameters of the surface operator has a simple physical interpretation: it is the effective twisted superpotential of the 2d $\mathcal{N} = (2, 2)$ theory on D. The relation

$$d\mathcal{W} = \eta \, da + \alpha \, da_D \quad (1.27)$$

tells us that the IR parameters (α, η) coincide with the points on the Jacobian of the Seiberg-Witten curve. Indeed, differentiating with respect to a,

$$\partial_a \mathcal{W} = \eta + \tau \alpha \quad (1.28)$$

and using (1.24) we conclude that the map

$$t_i = \frac{\partial \mathcal{W}}{\partial a^i} = \int_{p_*}^{p} \frac{\partial \lambda}{\partial a^i} = \int_{p_*}^{p} \omega_i \quad (1.29)$$

is precisely the Abel-Jacobi map from a Riemann surface to its Jacobian.

Note, the shifts of $\mathcal{W}$ by $n_e a + n_m a_D$ correspond to the monodromies of α and η. In Sect. 3 we relate them to line operators localized within a surface operator.

2 Surface Operators from Higher Dimensions

Four-dimensional $\mathcal{N} = 2$ gauge theories—that are of our prime interest here in view of the AGT correspondence—can be realized in a variety of higher-dimensional models, that include 6d $(0, 2)$ theory, type II string theory, and M-theory. Even though such constructions involve more sophisticated higher-dimensional systems, they often shed light on strongly coupled gauge dynamics and help understand various aspects of $\mathcal{N} = 2$ gauge theories, such as the Seiberg-Witten exact solution [19, 20] and Nekrasov's (K-theoretic) instanton partition function [14].

For example, a nice heuristic derivation of the AGT correspondence (1.17) follows from the $(0, 2)$ superconformal theory in six dimensions, which combines the 2-manifold C (where the Liouville theory lives) and the 4-manifold $M = \mathbb{R}^4_{\epsilon_1, \epsilon_2}$ (where the 4d $\mathcal{N} = 2$ gauge theory lives):

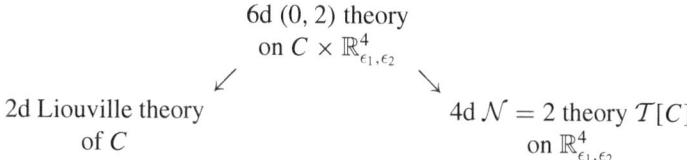

Here, the two sides of the AGT correspondence (1.17) are simply the two ways of reducing the 6d theory, either on a 2-manifold C or on a 4-manifold $M = \mathbb{R}^4_{\epsilon_1, \epsilon_2}$ (or $M = S^4_{\epsilon_1, \epsilon_2}$). In order to preserve supersymmetry, the former must be accompanied by a partial topological twist [21], whereas the latter involves deformed supersymmetry algebra that can be conveniently understood via coupling to the corresponding off-shell supergravity theory [22].

The 6d $(0, 2)$ theory itself admits surface operators (a.k.a. codimension-4 defects) which, upon reduction on C, give rise to surface operators in 4d $\mathcal{N} = 2$ theory $T[C]$. The existence of such surface operators can be deduced by realizing 6d $(0, 2)$ theory itself on the world-volume of N five-branes supported on $C \times M \times \{\text{pt}\}$ in 11d M-theory on $T^*C \times M \times \mathbb{R}^3$. And, in order to reduce to a surface operator on M, the codimension-4 defect of the six-dimensional theory must be supported at a point on C. From the viewpoint of the 4d gauge theory, its position $z \in C$ becomes a continuous parameter that labels half-BPS surface operator.

Note, six-dimensional $(0, 2)$ superconformal theory also has codimension-2 defects that can also produce half-BPS surface operators in four dimensions upon wrapping all of the Riemann surface C [23]. Since codimension-2 defects carry G-bundles, such surface operators are naturally labeled by points $x \in \text{Bun}_G(C)$. These surface operators are dual to the surface operators that arise from codimension-4 defects [24].

2.1 Brane Constructions

In addition, there exist various string constructions of surface operators. The one relevant to our discussion here is based on the brane realization of $\mathcal{N} = 2$ gauge theory in type IIA string theory [25], where basic surface operators (with next-to-maximal $\mathbb{L}$) can be described by introducing semi-infinite D2-branes [10]:

$$
\begin{array}{lll}
\text{NS5}: & 012345 \\
\text{D4}: & 0123\ \ 6 \\
\text{D2}: & 01 \quad\ \ 7
\end{array}
\tag{2.1}
$$

Lifting this configuration to M-theory, we obtain a M5-brane with world-volume $\mathbb{R}^4 \times \Sigma$ and a M2-brane (ending on the M5-brane) with world-volume $\mathbb{R}^2 \times \mathbb{R}_+$. Here, $D = \mathbb{R}^2$ is the support of the surface operator in the four-dimensional space-time $M = \mathbb{R}^4$, and Σ is the Seiberg-Witten curve of the $\mathcal{N} = 2$ gauge theory (Fig. 4).

In this construction, the M2-brane is localized along Σ (the choice of the point $t \in \Sigma$ corresponds to the IR parameters of the surface operator) and has a semi-infinite extent along the direction x^7, as described in (2.1).

Similar construction can be used to define UV surface operators in 4d $\mathcal{N} = 2$ superconformal theories obtained from compactifications of 6d $(0, 2)$ fivebrane theory on a UV Riemann surface C.

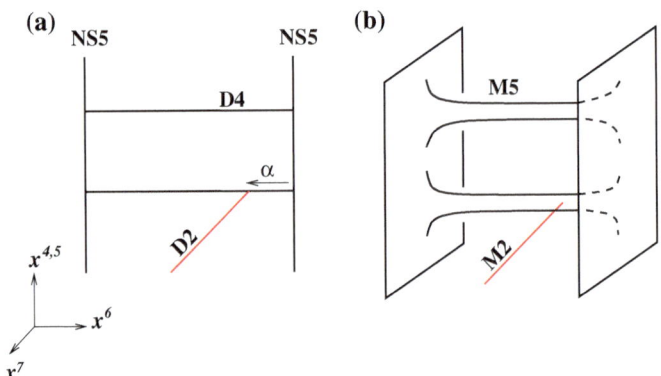

Fig. 4 The brane construction of $\mathcal{N} = 2$ super Yang-Mills theory with a half-BPS surface operator in type IIA string theory (**a**) and its M-theory lift (**b**)

Fig. 5 $U(1)$ toric geometry with a single Lagrangian brane

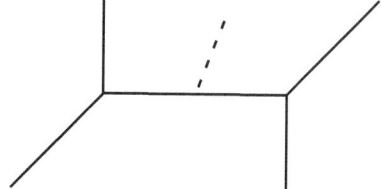

2.2 Geometric Engineering

Let us consider a four-dimensional $\mathcal{N} = 2$ gauge theory that can be geometrically engineered via type IIA string "compactification" on a Calabi-Yau space CY_3. In other words, we take the ten-dimensional space-time to be $M \times CY_3$, where M is a 4-manifold (where $\mathcal{N} = 2$ gauge theory lives) and CY_3 is a suitable Calabi-Yau space [26]. We recall that such CY_3 is non-compact and toric, and that its toric polygon coincides with the Newton polygon of the Seiberg-Witten curve Σ. As in most of our applications, one can simply take $M = \mathbb{R}^4$.

Aiming to reproduce half-BPS surface operators supported on $D = \mathbb{R}^2$, we need an extra object that breaks part of the Lorentz symmetry (along $M = \mathbb{R}^4$) and half of the supersymmetry. It is easy to see that D4-branes supported on supersymmetric 3-cycles in CY_3 provide just the right candidates [27]. Indeed, if the world-volume of a D4-brane is $\mathbb{R}^2 \times L$, where

$$\begin{array}{ccc} \text{space-time:} & \mathbb{R}^4 \times & CY_3 \\ & \cup & \cup \\ \text{D4-brane:} & \mathbb{R}^2 \times & L \end{array} \tag{2.2}$$

and L is a special Lagrangian submanifold of X, then such a D4-brane preserves exactly the right set of symmetries and supersymmetries as the half-BPS surface operators discussed in Sect. 1.3.

A nice feature of this construction is that it is entirely geometric: all the parameters of a surface operators (discrete and continuous) are encoded in the geometry of $L \subset CY_3$. In particular, among the different choices of L we should be able to find those which correspond to half-BPS surface operators of Levi type $\mathbb{L}$ with the continuous parameters α and η. When $\mathbb{L} = S[U(1) \times U(N-1)]$ is the next-to-maximal Levi subgroup, the corresponding surface operator is geometrically engineered [28] by a simple Lagrangian submanifold $L \cong S^1 \times \mathbb{R}^2$ invariant under the toric symmetry of CY_3 (see Fig. 5).

The space of IR parameters of such a surface operator is the algebraic curve Σ which is mirror to the original Calabi-Yau 3-fold CY_3 via local mirror symmetry.[4] Equivalently, from the viewpoint of CY_3 these continuous parameters are open string moduli of L corrected by world-sheet disk instantons. Just like open string moduli

---

[4]In the case of non-compact toric Calabi-Yau 3-folds mirror symmetry (often called "local mirror symmetry") relates enumerative invariants of CY_3 with complex geometry of a Riemann surface Σ.

become parameters of the surface operator in 4d $\mathcal{N} = 2$ gauge theory, the gauge theory itself is determined by closed string moduli, which are Kähler parameters of CY_3. (Note, a non-compact toric Calabi-Yau 3-fold CY_3 is rigid, i.e. has no complex structure deformations.)

4d gauge theory	geometry of CY_3
parameters of 4d $\mathcal{N} = 2$ theory (a_i, m_i, Λ)	closed string moduli (Kähler moduli Q)
Ω-background ϵ_1 and ϵ_2	string coupling/graviphoton $q_1 = e^{\epsilon_1}$ and $q_2 = e^{\epsilon_2}$
surface operator	Lagrangian submanifold
parameters of surface operator	open string moduli
$\Lambda_{\mathbb{L}}$	$H_1(L; \mathbb{Z})/\text{torsion}$

(2.3)

The geometric realization of a half-BPS surface operator (2.2) allows to express many interesting partition functions in terms enumerative invariants of the pair (CY_3, L). For example, the instanton partition function of the 4d $\mathcal{N} = 2$ gauge theory relevant to the AGT correspondence (1.17) and its variant with a half-BPS surface operator (1.21) both find a natural home on the right-hand side of the dictionary (2.3) as so-called "closed" and "open" BPS partition functions, respectively. To be more precise, the K-theoretic instanton counting on M is captured by counting *refined BPS invariants* on CY_3:

$$Z_K^{\text{inst}}(\Lambda, a_i; q_1, q_2) = Z_{\text{BPS}}^{\text{closed}}(Q; q_1, q_2). \tag{2.4}$$

Similarly, in the presence of a surface operator the K-theoretic analogue of (1.21) is equal to the generating function of open (as well as closed) refined BPS invariants of the pair (CY_3, L):

$$Z_{K-\text{theory}}^{\text{inst}}(\Lambda, a_i, z; q_1, q_2) = Z^{\text{BPS}}(Q, z; q_1, q_2). \tag{2.5}$$

An important special case of this relation is the limit $\Lambda \to 0$ (i.e. $Q_\Lambda \to 0$) and $(q_1, q_2) \to (q, 1)$. On the gauge theory side, this decouples the four-dimensional theory from the surface operator, and counts vortices on the surface operator with respect to two-dimensional rotations (but not R-charge). The resulting partition function counts only 2d vortices and not 4d instantons:

$$Z_{K-\text{theory}}^{\text{vortex}}(z, a_i; q) = Z_{\text{BPS}}^{\text{open}}(Q_\Lambda = 0, Q_{a_i}, z; q, 1). \tag{2.6}$$

For example, in the case of CY_3 shown in Fig. 5 that engineers $U(1)$ gauge theory, this limit corresponds to a degeneration upon which (CY_3, L) is replaced by $(\mathbb{C}^3, L)$.

2.3 Surface Operators and BPS States

Upon lift to M-theory, the BPS states in the system (2.2) are represented by membranes, with and without boundary, as illustrated in (2.11) and (2.14) below. They are completely localized along M and besides their support in CY_3 have non-trivial extent only along the "eleventh" dimension of M-theory, which can be treated as "time":

$$
\begin{array}{cccccc}
\text{space-time:} & \mathbb{R} & \times & M & \times & CY_3 \\
& \| & & \cup & & \cup \\
\text{M5-brane:} & \mathbb{R} & \times & D & \times & L \\
& \| & & \cup & & \\
\text{M2-brane:} & \mathbb{R} & \times & \{\text{pt}\} & \times & \Sigma_g
\end{array}
\qquad (2.7)
$$

In the five-dimensional gauge theory on $\mathbb{R} \times M$ such BPS states (open or closed) all look like particles. Therefore, one can equivalently talk about BPS particles in 5d theory with a surface operator supported on $\mathbb{R} \times D$. This system is very similar to our original 2d-4d system and can be related to that via reduction along one of the dimensions of M. For either system, one can introduce the space of BPS states that can move in 4d (resp. 5d) bulk as well as the space of BPS states localized on 2d (resp. 3d) surface operator, $\mathcal{H}_{\mathrm{BPS}}^{\mathrm{bulk}}$ and $\mathcal{H}_{\mathrm{BPS}}^{\mathrm{surface}}$. The space $\mathcal{H}_{\mathrm{BPS}}^{\mathrm{bulk}}$ depends only on the 4d/5d gauge theory on M (resp. $\mathbb{R} \times M$), whereas the space $\mathcal{H}_{\mathrm{BPS}}^{\mathrm{surface}}$ depends on both 4d/5d gauge theory as well as the surface operator.

What is the relation between $\mathcal{H}_{\mathrm{BPS}}^{\mathrm{bulk}}$ and $\mathcal{H}_{\mathrm{BPS}}^{\mathrm{surface}}$? The geometric engineering (2.7) can teach us an important lesson and help to answer this question. It has been known for a long time that $\mathcal{H}_{\mathrm{BPS}}^{\mathrm{bulk}}$ form an algebra [29], and recently it was further conjectured [30] that the space of BPS states localized on a surface operator forms a representation of this algebra

$$
\boxed{
\begin{array}{c}
\text{BPS states on a surface operator} : \ \mathcal{H}_{\mathrm{BPS}}^{\mathrm{surface}} \\
\circlearrowleft \\
\text{BPS states in 4d/5d gauge theory} : \ \mathcal{H}_{\mathrm{BPS}}^{\mathrm{bulk}}
\end{array}
}
\qquad (2.8)
$$

Indeed, the space of BPS states in bulk gauge theory is graded by a charge lattice Γ, which in the context of geometric engineering can be identified with even cohomology of local toric Calabi-Yau manifold:

$$
\Gamma = H^{\mathrm{even}}(CY_3; \mathbb{Z}). \qquad (2.9)
$$

Then, as explained in [29], two BPS states of the bulk theory, $\mathcal{B}_1$ and $\mathcal{B}_2$, of charge $\gamma_1, \gamma_2 \in \Gamma$ can form a bound state, $\mathcal{B}_{12}$ of charge $\gamma_1 + \gamma_2$, as a sort of "extension" of $\mathcal{B}_1$ and $\mathcal{B}_2$,

$$
0 \rightarrow \mathcal{B}_2 \rightarrow \mathcal{B}_{12} \rightarrow \mathcal{B}_1 \rightarrow 0, \qquad (2.10)
$$

thereby defining a product on $\mathcal{H}_{\mathrm{BPS}}^{\mathrm{bulk}}$:

$$\mathcal{H}_{\mathrm{BPS}}^{\mathrm{bulk}} \otimes \mathcal{H}_{\mathrm{BPS}}^{\mathrm{bulk}} \longrightarrow \mathcal{H}_{\mathrm{BPS}}^{\mathrm{bulk}}$$

$$(\mathcal{B}_1 , \mathcal{B}_2) \quad \mapsto \quad \mathcal{B}_{12}$$

(2.11)

Mathematical candidates for the algebra $\mathcal{H}_{\mathrm{BPS}}^{\mathrm{bulk}}$ include variants of the Hall algebra [31], which by definition encodes the structure of the space of extensions (2.10):

$$[\mathcal{B}_1] \cdot [\mathcal{B}_2] = \sum_{\mathcal{B}_{12}} |0 \to \mathcal{B}_2 \to \mathcal{B}_{12} \to \mathcal{B}_1 \to 0| \, [\mathcal{B}_{12}] \qquad (2.12)$$

In the present case, the relevant algebras include the motivic Hall algebra [32], the cohomological Hall algebra [33], and its various ramifications, e.g. cluster algebras. In Sect. 3.1 we will also discuss algebras of line operators localized on a surface operator that preserve the same amount of symmetry and supersymmetry as BPS states discussed here. In fact, line operators can be viewed as infinite mass limits of BPS states discussed here; this viewpoint explains many similarities between algebras of BPS states discussed here and algebras of line operators discussed in Sect. 3.

In the last line of (2.11) we illustrate the process of bound formation in the context of geometric engineering (2.7), where from the Calabi-Yau viewpoint each BPS state in the bulk gauge theory is represented by a closed membrane on $\Sigma_g \subset CY_3$. Similarly, BPS states localized on a surface operator in the system (2.7) correspond to *open* membranes with boundary on L:

$$\mathcal{H}_{\mathrm{BPS}}^{\mathrm{bulk}} = \mathcal{H}_{\mathrm{BPS}}^{\mathrm{closed}} \quad (=\text{refined closed BPS states}) \qquad (2.13)$$
$$\mathcal{H}_{\mathrm{BPS}}^{\mathrm{surface}} = \mathcal{H}_{\mathrm{BPS}}^{\mathrm{open}} \quad (=\text{refined open BPS states})$$

Specifically, the BPS states discussed here are, in fact, the so-called *refined* BPS states: besides grading by the charge lattice Γ, their space has an additional $\mathbb{Z}$-grading by the difference between $U(1)_{23}$ and $U(1)_R$ symmetries. From the viewpoint of a surface operator, this symmetry behaves in many ways as non-R flavor symmetry and plays an important role in [28, 34]. We will return to the role of this symmetry in Sect. 4.

By analogy with (2.11), when a bulk BPS state $\mathcal{B}_1^{\mathrm{bulk}} \in \mathcal{H}_{\mathrm{BPS}}^{\mathrm{bulk}}$ forms a bound state with a BPS state localized on a surface operator $\mathcal{B}_2^{\mathrm{surface}} \in \mathcal{H}_{\mathrm{BPS}}^{\mathrm{surface}}$ we obtain another BPS state localized on a surface operator $\mathcal{B}_{12}^{\mathrm{surface}} \in \mathcal{H}_{\mathrm{BPS}}^{\mathrm{surface}}$:

$$(\mathcal{B}_1^{\text{bulk}} , \mathcal{B}_2^{\text{surface}}) \quad \mapsto \quad \mathcal{B}_{12}^{\text{surface}}$$

$$(2.14)$$

This defines an action of the algebra of bulk BPS states on the space of BPS states localized on a surface operator. For example, when CY_3 is the total space of the $\mathcal{O}(-1) \oplus \mathcal{O}(-1)$ bundle over $\mathbb{CP}^1$ and L is defined by a knot in S^3 (cf. Sect. 2.4), the space of BPS states localized on a surface operator can be identified with a homological knot invariant,

$$\mathcal{H}_{\text{BPS}}^{\text{surface}} \cong \mathcal{H}_{\text{knot}} \qquad (2.15)$$

and the action (2.8) defines a plethora of anti-commuting operators (i.e. differentials) acting on this space.

2.4 Relation to 3d-3d Correspondence and Integrable Systems

The fivebrane configuration (2.7) encountered in the previous section has several interesting interpretations. We already discussed the five-dimensional point of view: in gauge theory on $\mathbb{R} \times M$ the fivebrane defines a codimension-2 defect supported on $\mathbb{R} \times D$. Likewise, from the vantage point of CY_3 it defines a defect supported on a special Lagrangian submanifold L and relates BPS state count to enumerative invariants of the pair (CY_3, L).

Here, we briefly comment on another interpretation of the system (2.7), from the viewpoint of the fivebrane observer on $R \times D$. It leads to yet another, equivalent description of physics—including the spectrum of BPS objects localized on a surface operator—in terms of 3d $\mathcal{N} = 2$ theory that in general depends on both L and CY_3. Particular choices of the Calabi-Yau 3-fold that have been extensively studied in the literature and play an important role in many applications include $CY_3 \cong \mathbb{C}^3$, T^*L, and the conifold geometry. In particular, since neighborhood of any special Lagrangian submanifold L looks like the total space of the cotangent bundle, the choice $CY_3 \cong T^*L$ is especially canonical and depends only on L. In this case, the effective 3d $\mathcal{N} = 2$ theory on $\mathbb{R} \times D$ also depends only on the 3-manifold L (and the total number of M5-branes), so that we get a correspondence

$$L \quad \leadsto \quad T[L], \qquad (2.16)$$

often called 3d-3d correspondence. In our presentation, we tried to emphasize its similarity to the study of surface operators. Indeed, compactification of the system (2.7) on a circle, obtained by replacing $\mathbb{R}$ with S^1, yields the familiar construction (2.2) of a half-BPS surface operator in 4d gauge theory. Moreover, there are many parallels between the space of SUSY vacua in the theory $T[L]$ on a circle and the space of vacua in the surface operator theory. Both are described by algebraic equations

$$\mathcal{M}_{\text{SUSY}} = \{A_i = 0\} \tag{2.17}$$

which play the role of Ward identities for line operators in 3d $\mathcal{N} = 2$ theory [35]. Specifically, for 3d $\mathcal{N} = 2$ theories (2.16) labeled by 3-manifolds, the algebraic relations $A_i = 0$ define the moduli space of complex flat connections on L:

$$\mathcal{M}_{\text{SUSY}}(T[L]) = \mathcal{M}_{\text{flat}}(L) \tag{2.18}$$

Besides this basic property, there are many other elements of the dictionary between 3-manifolds and 3d $\mathcal{N} = 2$ gauge theories that are described in [35] and summarized in a companion contribution to this volume [V:11].

As we reviewed in Sect. 1.4, for applications to the AGT correspondence one is interested in turning on the Ω-background, so that $M = \mathbb{R}^4_{\epsilon_1,\epsilon_2}$ and $D = \mathbb{R}^2_{\epsilon_1}$. This has the following effect on the surface operator theory or 3d theory $T[L]$, where the role of ϵ_1 and ϵ_2 is clearly very different. The Ω-deformation *along* the surface operator controlled by the parameter ϵ_1 has the effect of "quantizing" the system, i.e. replacing the polynomials A_i by their non-commutative deformation

$$A_i \xrightarrow{\quad \epsilon_1 \neq 0 \quad} \widehat{A}_i \tag{2.19}$$

so that classical equations $A_i = 0$ are replaced by the Schrodinger-like equations $\widehat{A}_i Z = 0$. In a particular class of models where $A(x, y) = 0$ realize spectral curves of integrable systems, deformation by ϵ_1 leads to Baxter equations of the corresponding integrable systems [6]. For general values of the S^1 radius, the integrable systems in question are *trigonometric* (also called *hyperbolic* in some of the literature), whose prominent examples include the XXZ spin chain and the trigonometric Ruijsenaars model.

The role of ϵ_2 is very different. Turning on $\epsilon_2 \neq 0$ (while keeping $\epsilon_1 = 0$) does not make A_i non-commutative and leads to the Nekrasov-Shatashvili duality [36] between $\mathcal{N} = 2$ theory on $S^1 \times D$ and, in general, a different integrable system. The relation between the two integrable systems is some sort of spectral duality [6] which in the present physical setup clearly corresponds to exchanging the role of ϵ_1 and ϵ_2 (or, equivalently, the support of surface operator inside $M = \mathbb{R}^4_{\epsilon_1,\epsilon_2}$). Note, the two relations with integrable systems invoke rather different aspects, e.g. one goes via Baxter equation, as was mentioned earlier, while the Nekrasov-Shatashvili correspondence goes via Bethe equations. Conversely, Bethe equations are not manifest in a duality with $\epsilon_2 = 0$, while Baxter equations are not manifest in a duality with $\epsilon_1 = 0$.

3 Surface Operators and Line Operators

Line operators remain, even at present time, the most familiar and better understood representatives in the list of non-local operators in Sect. 1. They have a wide range of applications, from supersymmetric gauge theories—where they play an important role in computations of partition functions *a la* [18, 37, 38] as discussed e.g. in a companion contribution to this volume [V:7] —to phase structure of non-supersymmetric gauge theories, where they serve as excellent order parameters (cf. Sect. 5). This justifies the study of line operators in their own right.

Here, we will focus on rather specific aspects of line operators that have to do with how they interact with surface operators. In fact, the main aspect we wish to discuss is that, in the presence of a surface operator, the OPE algebra of line operators becomes non-commutative. And, then, we shall give some examples of such non-commutative structure and explain its simple geometric interpretation. It is useful to keep in mind the parallel discussion of BPS states confined to a surface operator in Sect. 2.3: in both cases we deal with one-dimensional world-lines within a surface operator that lead to a non-commutative algebra. Moreover, the rotation symmetry in two space-time dimensions transverse to the surface operator gives rise to a deformation of the algebra by the parameter $q = e^{\hbar}$.

First, let us consider a four-dimensional gauge theory on a space-time manifold M without surface operators. (As usual, for concreteness, one can keep in mind a simple example of $M = \mathbb{R}^4$.) It is well known that line operators form an algebra—very similar to the algebra of BPS states discussed earlier—with the product

$$L_1 \times L_2 \sim \sum_i V_i L_i \tag{3.1}$$

given by the operator product expansion (OPE). In many familiar examples, that include topological and supersymmetric theories, this product is commutative simply because one can continuously exchange positions of line operators by moving them around each other in four dimensional space.

One important aspect of the product (3.1) is that its coefficients V_i are, in fact, vector spaces. This aspect is not yet about surface operators *per se*, but does become more pronounced in the presence of surface operators, as we explain shortly. In many applications, V_i's can be replaced by numbers, especially in situations where only dimensions $v_i = \dim V_i$ are relevant to a particular application in question. This happens, for instance, when line operators are compactified (either effectively or explicitly) on a circle. Then, the OPE product of the resulting "loop operators" has the form of a typical OPE of local operators

$$\mathcal{O}_1 \times \mathcal{O}_2 \sim \sum_i v_i \mathcal{O}_i \tag{3.2}$$

with numerical coefficients v_i.

Fig. 6 Line operators
confined to a surface
operator do not commute

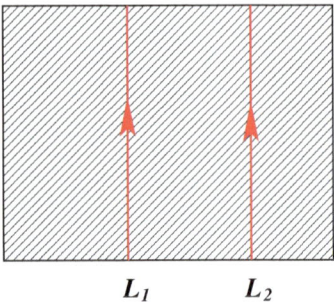

Fig. 6 Line operators confined to a surface operator do not commute

L_1 L_2

Now, let us imagine that line operators L_1 and L_2 are stuck to a two-dimensional subspace $D \subset M$, as illustrate in Fig. 6. For example, as in our previous discussion, turning on the Ω-background in directions transverse to D inside M will not allow line operators to move off from the surface D without breaking the $SO(2)_{23}$ rotation symmetry of the transverse space ($\cong \mathbb{R}_{\hbar}^2$). What this means is the following: The rotation symmetry $SO(2)_{23}$ makes V_i's into graded vector spaces, graded by the angular momentum h_{23}. And, therefore, the operator product expansion (3.1) has a "refinement" with graded vector spaces as coefficients, as long as line operators are confined within the surface D. In other words, the product (3.1) is commutative, but its graded version in general is not.

A more dramatic way to make the product (3.1) non-commutative is to introduce surface operators supported on $D \subset M$. This has several important ramifications. First, it breaks the 4d Poincaré invariance and, therefore, does not allow to naively move around line operators L_1 and L_2 in three transverse directions. Moreover, in the presence of a surface operator there can exist additional line operators which are supported on the surface operator and can not move into the rest of the 4-manifold M. Since such line operators are confined to the surface $D \subset M$, they can not be passed through each other without encountering a singularity. As a result, the OPE algebra of such line operators in general is non-commutative.

For example, in applications to AGT correspondence, one can consider line operators localized within a surface operator in 4d $\mathcal{N} = 2$ theory $\mathcal{T}[C, \mathfrak{g}]$, where C is a Riemann surface (possibly with punctures). From the six-dimensional perspective reviewed in Sect. 2, these are line operators localized on the 2-dimensional worldsheet $p \times D \times \{0\}$ of a surface operator (=codimension-4 defect) in the six-dimensional $(2, 0)$ theory on $C \times D \times \mathbb{R}_{\hbar}^2$, where

$$M = D \times \mathbb{R}_{\hbar}^2 \tag{3.3}$$

is the 4d space-time of the $\mathcal{N} = 2$ gauge theory. It was argued in [3] that such line operators generate an affine Hecke algebra H_{aff} of type $\mathfrak{g}$ with parameter $q = e^{\hbar}$. Note that this affine Hecke algebra is "local on C." In other words, it does not depend on the details of the Riemann surface C away from the point p. This observation will be useful to us in what follows since for the purpose of deriving the non-commutative algebra associated with the presence of surface operator (=ramification at $p \in C$) one can take C to be something simple, e.g. a torus or a disk. For instance, in the basic case $G = SU(2)$ the affine Hecke algebra is generated by T, X, and X^{-1}, which obey the relations (see e.g. [39]):

$$(T + 1)(T - q) = 0$$
$$TX^{-1} - XT = (1 - q)X \qquad (3.4)$$
$$XX^{-1} = X^{-1}X = 1$$

The affine Hecke algebra has two close cousins: the affine Weyl group $\mathcal{W}_{\text{aff}}$, which corresponds to the limit $q \to 1$, and its "categorification", the affine braid group B_{aff}. They too have a simple physical interpretation which, moreover, offers an intuitive explanation of the non-commutative product of line operators within a surface operator. It follows from a compactification of our 2d-4d system on a circle, which can be achieved e.g. by taking $D = S^1 \times \mathbb{R}$. From the six-dimensional perspective, reduction on S^1 gives the maximally supersymmetric Yang-Mills, and a further reduction on C yields a 3d $\mathcal{N} = 4$ sigma-model with target space $\mathcal{M}_H(G, C)$, the moduli space of Higgs bundles on C (also known as the 'Hitchin moduli space') [40, 41]. The presence of a surface operator introduces ramification at $p \in C$, so that in the present case $\mathcal{M}_H$ is the moduli space of ramified Higgs bundles [3].

In the sigma-model on $\mathcal{M}_H$, line operators correspond to functors acting on branes (or, boundary conditions). According to (1.7), these functors form a group which often can be identified with the fundamental group of the (sub)space of parameters of a surface operator. Indeed, kinks on a surface operators are nothing but line operators. In order to see this, consider a kink corresponding to an (adiabatic) variation of the continuous parameters of a surface operator. On the one hand, it traverses a closed loop in the space of parameters. On the other hand, it is localized in one dimension (="space") and extended along the other dimension (="time") on D, just like line operators illustrated in Fig. 6.

Line operators preserving certain symmetry and supersymmetry correspond to varying particular parameters (that don't break these symmetries). For example, in applications to the geometric Langlands correspondence [42], the Galois side corresponds to the B-model of $\mathcal{M}_H$. Branes and boundary conditions that preserve this particular supersymmetry are described by the derived category of coherent sheaves on $\mathcal{M}_H$ and their charges are described by the K-theory. Therefore, depending on whether one is interested in D-branes (as objects of the derived category of coherent sheaves on $\mathcal{M}_H$) or in D-brane charges (classified by K-theory) one finds the following groups acting the on the K-theory/derived category of the moduli space of ramified Higgs bundles[5]:

---

[5] For simplicity, here we consider only one ramification point $p \in C$. For the case of ramification at several points, one finds several group actions, one for each ramification point.

Claim [3]: *affine Weyl group* $\mathcal{W}_{\text{aff}}$ *acts on* $K(\mathcal{M}_H)$
    affine Hecke algebra H_{aff} *acts on* $K^{\mathbb{C}^*}(\mathcal{M}_H)$
    affine braid group B_{aff} *acts on* $D^b(\mathcal{M}_H)$

This result can be regarded as a categorification of the affine Hecke algebra, which in the local version of $\mathcal{M}_H$ was also obtained by Bezrukavnikov [43] using a "noncommutative counterpart" of the Springer resolution $\widetilde{\mathcal{N}} \to \mathcal{N}$. The action of $\mathcal{W}_{\text{aff}}$ and B_{aff} in the first and the last part of this claim can be understood as the monodromy action in the space of parameters of the surface operator (1.7).

For example, let us illustrate how this group action arises at the level of D-brane charges, which are classified by $K(\mathcal{M}_H)$. The space of D-brane charges $K(\mathcal{M}_H)$ varies as the fiber of a flat bundle over the space of parameters away from the points where $\mathcal{M}_H$ develops singularities. Since for the purposes of this question we are interested only in the geometry of $\mathcal{M}_H$, we can ignore the "quantum" parameter η. Hence, the relevant parameters are (α, β, γ), which take values in the space, cf. (1.12):

$$(\alpha, \beta, \gamma) \in \left(\mathfrak{t} \times \mathfrak{t} \times \mathfrak{t}\right)/\mathcal{W}_{\text{aff}} \tag{3.5}$$

Moreover, $\mathcal{M}_H$ becomes singular precisely for those values of (α, β, γ) which are fixed by some element of $\mathcal{W}_{\text{aff}}$. The set of such points is at least of codimension three in $\mathfrak{t}^3$ (since it takes three separate conditions to be satisfied for (α, β, γ) to be fixed by some element of $\mathcal{W}_{\text{aff}}$). Therefore, the space of regular values of $(\alpha, \beta, \gamma) \in \mathfrak{t}^3$ where $\mathcal{M}_H$ is non-singular is connected and simply-connected, and since $\mathcal{W}_{\text{aff}}$ acts freely on this space, the fundamental group of the quotient is

$$\pi_1\left(\{(\alpha, \beta, \gamma)\}^{\text{reg}}\right) = \mathcal{W}_{\text{aff}} \tag{3.6}$$

This is the group that acts on D-brane charges, that is on $K(\mathcal{M}_H)$. In a similar way, one can deduce the action of the affine braid group B_{aff} on $D^b(\mathcal{M}_H)$ as the fundamental group of the Kähler moduli space. Indeed, for the B-model in complex structure J the complexified Kähler parameters are $\eta + i\beta$, and from (1.12) one finds:

$$\pi_1\left(\{(\beta, \eta)\}^{\text{reg}}\right) = B_{\text{aff}} \tag{3.7}$$

The same results can be derived more directly in the description of surface operators as 2d-4d coupled systems.

3.1 Line Operators and Hecke Algebras

In Sect. 1 we explained that a surface operator in 4d gauge theory can be equivalently defined as a 2d sigma-model supported on D and with a target space X that has G-action. Here we use this description of surface operators to explain more directly how algebra of line operators localized on a surface operator leads to the affine Hecke algebra H_{aff} or its close cousins W_{aff} and B_{aff}. This particular approach offers an alternative derivation of the results in [3] and to the best of our knowledge has not appeared in the literature. For concreteness and in order to keep things simple we shall restrict our attention to gauge theory with $G = SU(2)$.

When the surface operator is described by a sigma-model with target space X, line operators (which act on branes on X) in turn can be viewed as branes on $X \times X$. In the language of derived category, this means that an object $\mathcal{Z} \in D^b(X \times X)$ called the "kernel" defines an exact functor $\Phi_{\mathcal{Z}} : D^b(X) \to D^b(X)$, such that

$$\Phi_{\mathcal{Z}}(\mathcal{E}) := p_{2*}(\mathcal{Z} \otimes p_1^*(\mathcal{E})) \tag{3.8}$$

where p_1 (resp. p_2) is the projection to the first (resp. the second) factor.[6] We can define a product of two line operators $\mathcal{A}$ and $\mathcal{B}$ by bringing them together (as in Fig. 6) which leads to a composition of the transforms $\Phi_{\mathcal{A}}$ and $\Phi_{\mathcal{B}}$. This gives a new transform

$$\Phi_{\mathcal{B}\star\mathcal{A}} \cong \Phi_{\mathcal{A}} \circ \Phi_{\mathcal{A}} \tag{3.9}$$

with the kernel

$$\mathcal{B} \star \mathcal{A} = p_{13*}(p_{12}^*\mathcal{A} \otimes p_{23}^*\mathcal{B}) \tag{3.10}$$

where p_{ij} are the obvious projection maps from $X \times X \times X$ to $X \times X$. In particular, the diagonal $\Delta_X : X \hookrightarrow X \times X$ gives the identity. The product (3.10) is associative

$$\mathcal{C} \star (\mathcal{B} \star \mathcal{A}) \cong (\mathcal{C} \star \mathcal{B}) \star \mathcal{A} \tag{3.11}$$

We are interested in the case where $X = \mathcal{N}$ is the nilpotent cone for $SL(2, \mathbb{C})$ or its Springer resolution $\tilde{\mathcal{N}}$. This is a special case of a larger class of examples where X is (the minimal resolution) of the Kleinian quotient singularity $\mathbb{C}^2/\Gamma$ for a finite subgroup $\Gamma \subset SL(2, \mathbb{C})$. In this case, there is an equivalence (the derived McKay correspondence):

$$D^b(X) \cong D_\Gamma^b(\mathbb{C}^2) \tag{3.12}$$

The category $D_\Gamma^b(\mathbb{C}^2)$ has simple objects

$$S_i = \rho_i \otimes \mathcal{O}_p \tag{3.13}$$

where ρ_i are irreducible representations of Γ and $\mathcal{O}_p$ is the skyscraper sheaf supported at the origin of $\mathbb{C}^2$. These are precisely the fractional branes on $\mathbb{C}^2/\Gamma$. In the derived category of the minimal resolution X, the simple objects (3.13) are represented by [44],

---

[6]To be more precise, the pull-back p_1^* is left-derived and the push-forward p_{2*} is right-derived.

$$S_0 = \mathcal{O}_{\sum C_i} \tag{3.14}$$
$$S_i = \mathcal{O}_{C_i}(-1)[1]$$

where C_i are the exceptional divisors.

An important feature of fractional branes is that, in the derived category of X, they are described by spherical objects and, therefore, according to the results of Seidel and Thomas [45], define twist functors T_i which generate the braid group $Br(\Gamma)$. As the name suggests, an object $\mathcal{E} \in D^b(X)$ is called d-spherical if $\mathrm{Ext}^*(\mathcal{E}, \mathcal{E})$ is isomorphic to $H^*(S^d, \mathbb{C})$ for some $d > 0$,

$$\mathrm{Ext}^i(\mathcal{E}, \mathcal{E}) = \begin{cases} \mathbb{C} & \text{if } i = 0 \text{ or } d \\ 0 & \text{otherwise} \end{cases} \tag{3.15}$$

A spherical B-brane defines a twist functor $T_{\mathcal{E}} \in \mathrm{Auteq}(D^b(X))$ which, for any $\mathcal{F} \in D^b(X)$, fits into exact triangle

$$\mathrm{Hom}^*(\mathcal{E}, \mathcal{F}) \otimes \mathcal{E} \longrightarrow \mathcal{F} \longrightarrow T_{\mathcal{E}}(\mathcal{F}) \tag{3.16}$$

where the first map is evaluation. The functor $T_{\mathcal{E}}$ can be written as a Fourier-Mukai transform (3.8) associated with the brane $\mathcal{Z}$ on $X \times X$,

$$\mathcal{Z} = \mathrm{Cone}\left(\mathcal{E}^\vee \boxtimes \mathcal{E} \to \mathcal{O}_{\Delta_X}\right) \tag{3.17}$$

where $\mathcal{E}^\vee$ denotes the dual complex, Δ_X is the diagonal in $X \times X$, and $\mathcal{E} \boxtimes \mathcal{F} = p_2^* \mathcal{E} \otimes p_1^* \mathcal{F}$ is the exterior tensor product. At the level of cohomology, the twist functor $T_{\mathcal{E}}$ acts as,

$$x \mapsto x + (v(\mathcal{E}) \cdot x)\, v(\mathcal{E})$$

where $v(\mathcal{E}) = ch(\mathcal{E})\sqrt{Td(X)} \in H^*(X)$ is the D-brane charge (the Mukai vector) of $\mathcal{E}$. Summarizing, "spherical branes" (spherical objects in $D^b(X)$) lead to autoequivalences of $D^b(X)$. What is the group they generate?

Given an A_n chain of spherical objects, that is a collection of spherical objects $\mathcal{E}_1, \ldots, \mathcal{E}_n$ which satisfy the condition

$$\sum_k \dim \mathrm{Ext}^k(\mathcal{E}_i, \mathcal{E}_j) \begin{cases} 1 & |i - j| = 1 \\ 0 & |i - j| > 1 \end{cases} \tag{3.18}$$

with some minor technical assumptions Seidel and Thomas [45] showed that the corresponding twist functors $T_{\mathcal{E}_i}$ generate an action of the braid group Br_{n+1} on $D^b(X)$. More generally, a chain of spherical objects associated with Γ gives rise to the action of the braid group $Br(\Gamma)$ on B-branes. The generators of $Br(\Gamma)$ correspond to vertices of the affine Dynkin diagram of Γ and obey the relations

$$T_i T_j T_i = T_j T_i T_j \qquad (3.19)$$

if the vertices i and j are connected by an edge, and $T_i T_j = T_j T_i$ otherwise. In particular, in the situation we are interested in, namely when X is the minimal resolution quotient singularity $\mathbb{C}^2/\Gamma$, the braid group $Br(\Gamma)$ is the essential part of the group of autoequivalences of X. Specifically, the group of autoequivalences of X is [46, 47]:

$$\text{Auteq}(D^b(X)) = \mathbb{Z} \times (\text{Aut}(\Gamma) \ltimes Br(\Gamma)) \qquad (3.20)$$

where the first factor is generated by the shift functor $[1]$, the group $\text{Aut}(\Gamma)$ is the group of symmetries of the affine Dynkin diagram associated to Γ, and $Br(\Gamma)$ is the braid considered above.

In particular, for $\Gamma = \mathbb{Z}_2$ which is relevant to the $SU(2)$ gauge theory, the group $\text{Auteq}(D^b(X))$ is generated by the functors $T_{\pm}$ and R (and the shift functor, of course). Indeed, in the present case there are two spherical objects (two fractional branes (3.13)) which lead to the twist functors T_+ and T_-. These two are exchanged by R, the generator of $\text{Aut}(\Gamma) \cong \mathbb{Z}_2$, so that in total we obtain the group generated by $T_{\pm}$ and R which obey the relations

$$T_+ R = R T_- \qquad (3.21)$$
$$R^2 = 1$$

Notice, for $\Gamma = \mathbb{Z}_2$ there are no braid relations of the form (3.19). Nevertheless, we still shall refer to the resulting group as the affine braid group of type $\hat{A}_1$.

To make contact with our earlier discussion, we can identify the autoequivalences that generate the braid group with the monodromies in the category of B-branes around special points in the Kähler moduli space of X. In the case at hand, there are three such points: (i) the large volume limit, (ii) the "conifold limit" (where $X = \mathbb{C}^2/\mathbb{Z}_2$ with zero B-field), and (iii) the orbifold limit (where $X = \mathbb{C}^2/\mathbb{Z}_2$ with $B = \frac{1}{2}$). A monodromy around each of these points defines a Fourier-Mukai transform associated with a certain brane on $X \times X$. Following [48], we denote these branes, respectively, as $\mathcal{L}$, $\mathcal{K}$, and $\mathcal{G}$. The corresponding transforms will be denoted by $\Phi_{\mathcal{L}}$, $\Phi_{\mathcal{K}}$, and $\Phi_{\mathcal{G}}$.

As we shall see below, the monodromy $\Phi_{\mathcal{L}}$ has infinite order and, therefore, is related to the generator X in (3.4) or the generator $T_{\pm}$ in (3.21). On the other hand, while the monodromies around the conifold point and the orbifold point are both of order 2 at the level of K-theory charges, in the derived category we have

$$\Phi_{\mathcal{G}}^2 = 1, \quad \Phi_{\mathcal{K}}^2 \neq 1 \qquad (3.22)$$

Therefore, $\Phi_{\mathcal{G}}$ which comes from the quantum $\mathbb{Z}_2$ symmetry of $\mathbb{C}^2/\mathbb{Z}_2$ should be identified with the generator R in (3.21) which has a similar origin. The monodromy around the orbifold point is a composition of the monodromies around the conifold point and the large radius limit. Therefore, from (3.9) we get

$$\mathcal{G} = \mathcal{K} \star \mathcal{L} \tag{3.23}$$

Now, let us describe explicitly $\mathcal{L}$, $\mathcal{K}$, and $\mathcal{G}$, and verify (3.22). The monodromy around the large radius limit is always associated with

$$\mathcal{L} = \mathcal{O}_{\Delta_X}(1) \tag{3.24}$$

Since D-brane wrapped on the exceptional divisor C becomes massless at the conifold point, the monodromy around the conifold point is the twist functor $T_{\mathcal{E}}$ associated with the spherical object $\mathcal{E} = \mathcal{O}_C$. Therefore,

$$\mathcal{K} = \left(\mathcal{O}_C^{\vee} \boxtimes \mathcal{O}_C \to \mathcal{O}_{\Delta_X} \right) \tag{3.25}$$

and using (3.23) we get

$$\mathcal{G} = \left(\mathcal{O}_C(-1)^{\vee} \boxtimes \mathcal{O}_C \to \mathcal{O}_{\Delta_X}(1) \right) \tag{3.26}$$

Now, to verify (3.22) we can either compute how the functors $\Phi_{\mathcal{L}}$, $\Phi_{\mathcal{K}}$, and $\Phi_{\mathcal{G}}$ act on simple branes, such as the 0-brane $\mathcal{O}_p$, or to study their composition using (3.9). In particular, computing $\Phi_{\mathcal{K}}^n(\mathcal{O}_p)$ we can verify that $\Phi_{\mathcal{K}}$ is indeed of infinite order. Similarly, we find

$$\mathcal{G} \star \mathcal{G} = \mathcal{O}_{\Delta_X}$$

which is the first relation in (3.22).

The transforms $\Phi_{\mathcal{L}}$, $\Phi_{\mathcal{K}}$, and $\Phi_{\mathcal{G}}$ are autoequivalences of $D^b(X)$. In fact, they generate the entire group (3.21) which we found earlier by looking at the fractional branes on $\mathbb{C}^2/\mathbb{Z}_2$. This can be shown by explicitly matching the generators. First, one of the generators $T_{\pm}$ is the twist functor associated with the spherical object $\mathcal{E} = \mathcal{O}_C$. Without loss of generality, we assume that this generator is T_+. According to (3.25), it should be identified with the monodromy around the conifold point, $\Phi_{\mathcal{K}}$. Similarly, the order-2 generator R should be identified with the monodromy around the orbifold point $\Phi_{\mathcal{G}}$ which is also of order 2 and has a similar origin (both come from the quantum $\mathbb{Z}_2$ symmetry of $\mathbb{C}^2/\mathbb{Z}_2$). Summarizing,

$$\Phi_{\mathcal{K}} \longleftrightarrow T_+^{-1} \tag{3.27}$$
$$\Phi_{\mathcal{G}} \longleftrightarrow R$$

The remaining generator can be expressed as a product of these two, cf. (3.21) and (3.23). In particular, we conclude that the monodromy around the large volume limit $\Phi_{\mathcal{L}}$ should be identified with $T_+ R = R T_-$.

As we mentioned earlier, at the level of K-theory charges the OPE algebra of line operators that we are considering should reduce to the affine Weyl group $\mathcal{W}_{aff}$ or affine Hecke algebra H_{aff}. From (3.24)–(3.26) it is easy to see that the monodromies $\Phi_{\mathcal{L}}$, $\Phi_{\mathcal{K}}$, and $\Phi_{\mathcal{G}}$ act on the charges of D0, D2, and D4 branes as

$$M_{\mathcal{L}} = \begin{pmatrix} 1 & 0 & 0 \\ 1 & 1 & 0 \\ \frac{1}{2} & 1 & 1 \end{pmatrix}, \quad M_{\mathcal{K}} = \begin{pmatrix} 1 & 0 & 0 \\ 0 & -1 & 0 \\ 0 & 0 & 1 \end{pmatrix}, \quad M_{\mathcal{G}} = \begin{pmatrix} 1 & 0 & 0 \\ -1 & -1 & 0 \\ \frac{1}{2} & 1 & 1 \end{pmatrix} \tag{3.28}$$

It is easy to verify that

$$M_{\mathcal{G}} = M_{\mathcal{K}} M_{\mathcal{L}} \tag{3.29}$$

in agreement with (3.23), and that both $M_{\mathcal{K}}$ and $M_{\mathcal{G}}$ are of order two:

$$M_{\mathcal{K}}^2 = 1, \quad M_{\mathcal{G}}^2 = 1 \tag{3.30}$$

Via the identification (3.27) this implies that, in addition to the relation $R^2 = 1$ which is already included in (3.21), we need to impose an extra condition $T_+^2 = 1$ which, of course, implies $T_-^2 = 1$ as well:

$$T_i^2 = 1 \tag{3.31}$$

Therefore, at the level of K-theory charges, the group generated by the monodromies is a semidirect product of $\mathbb{Z}_2$ (generated by T_+) and $\mathbb{Z}$ (generated by T_+R). This is precisely the affine Weyl group $\mathcal{W}_{\text{aff}}$ for $G = SU(2)$.

More generally, instead of the quadratic relations (3.31), we can consider imposing extra relations

$$T_i^2 = (q^{1/2} - q^{-1/2})T_i + 1 \tag{3.32}$$

on all the generators $T_i \in B_{\text{aff}}$ which correspond to the simple reflections in $\mathcal{W}_{\text{aff}}$. As we discussed earlier, this should lead to the affine Hecke algebra H_{aff}. In the example we are considering, we require T_+ (and T_-) to obey (3.32). Furthermore, motivated by the specialization to the affine Weyl group considered above, we introduce the notations

$$T = q^{1/2}T_+ \tag{3.33}$$
$$X = T_+R = RT_-$$

Then, the quadratic constraint (3.32) on T_+ implies a similar constraint on T,

$$T^2 = (q-1)T + q \tag{3.34}$$

which is precisely one of the relations in the affine Hecke algebra (3.4). Moreover, from (3.32) we obtain

$$T_\pm^{-1} = T_\pm + (q^{-1/2} - q^{1/2})$$

which can be used to find

$$X^{-1} = T_-R + (q^{-1/2} - q^{1/2})R \tag{3.35}$$

Now, using (3.21), (3.33), and (3.35), it is easy to check that X and T satisfy

$$TX^{-1} - XT = (1 - q)X \qquad (3.36)$$

which is precisely the second relation in the affine Hecke algebra (3.4). Therefore, we verified that imposing extra relations (3.32) on the generators of the affine braid group leads to the affine Hecke algebra.

Conversely, starting with (3.32) and using the identifications (3.33), it easy to verify that $T_{\pm}$ and R obey (3.21) with the additional relations (3.32). For example,

$$R = q^{-1/2}TX + (q^{-1/2} - q^{1/2})X \qquad (3.37)$$

so that after a little algebra we get $R^2 = 1$.

Finally, we conclude our discussion of line operators confined to a surface operator with one more deformation of the Hecke algebra that depends on two deformation parameters, q and t. This deformation is called the *double affine Hecke algebra*, or DAHA for short. It already made an appearance in the physical literature [49, 50] on refined BPS states and knot invariants that we mentioned in Sect. 2.3 and cries out for an interpretation either as algebra of BPS states or algebra of line operators.

In fact, a convenient starting point for defining DAHA (which for simplicity we explain in the basic case of $G = SU(2)$) is the orbifold fundamental group of the elliptic curve quotient, cf. (3.7):

$$\pi_1^{\mathrm{orb}}(\{E\backslash 0\}/\mathbb{Z}_2) \cong \pi_1^{\mathrm{orb}}(\{E \times E\backslash\mathrm{diag}\}/\mathbb{Z}_2) \qquad (3.38)$$

generated by X, Y, and T with the relations

$$\begin{aligned}
TXT &= X^{-1} \\
TY^{-1}T &= Y \\
Y^1 X^{-1} YXT^2 &= 1
\end{aligned} \qquad (3.39)$$

Deforming the last relation to $Y^1 X^{-1} YXT^2 = t^{-1/2}$ gives the so-called *elliptic braid group* B_{ell}. Furthermore, imposing by now familiar quadratic Hecke relation as in (3.4), (3.32), or (3.34) with another deformation parameter q leads to the complete definition of DAHA:

$$\mathrm{DAHA} = \mathbb{C}[B_{\mathrm{ell}}]/((T - q^{1/2})(T + q^{-1/2})) \qquad (3.40)$$

as a quotient of the group algebra of B_{ell}. The operator Y in this algebra is called the difference Dunkl operator.

Comparing the fundamental group (3.38)–(1.7), (1.5) and (1.12), we see that it classifies monodromies of the parameters $(\alpha, \eta) \in [(\mathbb{T} \times \mathbb{T}^\vee)/W]^{\mathrm{reg}}$ of a surface operator in $SU(2)$ gauge theory. In other words, it classifies line operators that correspond to monodromies of the parameters α and η, which are precisely the parameters of the basic surface operator in $\mathcal{N} = 2$ gauge theory, cf. Sect. 1.3. Of course, such surface operators can be also embedded in $\mathcal{N} = 4$ gauge theory (with the same gauge group) where (3.38) means classifying monodromies of α and η, while keeping β and γ fixed. This is similar to what we encountered in (3.7), except that now the relevant problem involves the B-model of $\mathcal{M}_H$ in complex structure I, where $\beta + i\gamma$ is the complex structure parameter, while $\eta + i\alpha$ represents the Kähler modulus (stability condition). See [3] for more details.

Therefore, we expect that the algebra of line operators confined to a surface operator in $\mathcal{N} = 2$ gauge theory (or in $\mathcal{N} = 4$ gauge theory with supersymmetry of type B_I) is intimately related, if not equal, to the double affine Hecke algebra. It would be interesting to tackle a similar physical realization of quantum affine algebras and Kac-Moody algebras acting on the equivariant K-theory of certain quiver varieties constructed by Nakajima [51, 52]. It is natural to expect that this action can be lifted to an action of the fundamental group of the Kähler moduli space on the derived category of the quiver variety involved in this construction.

4 Superconformal Index

The AGT correspondence has a sister that, on the one hand, is simpler, but in another respect is more mysterious. It relates another observable (partition function) of the 4d $\mathcal{N} = 2$ theory $T[C]$ to the partition function of a non-supersymmetric 2d theory on the Riemann surface C,

$$\mathcal{I}_{4d}(T[C]) = Z_{2d}(C) \tag{4.1}$$

where $\mathcal{I}_{4d}(T[C])$ is a superconformal index of the theory $T[C]$, defined as

$$\mathcal{I}_{4d}(\mathfrak{p}, \mathfrak{q}, \mathfrak{t}) = \mathrm{Tr}(-1)^F \mathfrak{p}^{h_{23}-r} \mathfrak{q}^{h_{01}-r} \mathfrak{t}^{R+r}. \tag{4.2}$$

For a theory with weakly coupled Lagrangian description the index is computed by a matrix integral:

$$\mathcal{I}_{4d}(\mathfrak{p}, \mathfrak{q}, \mathfrak{t}) = \int [dU] \exp\left(\sum_{n=1}^{\infty} \sum_j \frac{1}{n} f^{(j)_{4d}}(\mathfrak{p}^n, \mathfrak{q}^n, \mathfrak{t}^n) \chi_{R_j}(U^n, V^n) \right). \tag{4.3}$$

Here, U and V denote elements of gauge and flavor groups, respectively. The invariant Haar measure integral $\int [dU]$ imposes the Gauss law over the Fock space. The sum is over different $\mathcal{N} = 2$ supermultiplets appearing in the Lagrangian, with R_j being the representation of the jth multiplet under gauge and flavor group, and χ_{R_j} the character of R_j. The function $f^{(j)}$ is called single letter index. It is equal to either f_{4d}^{vector} or $f_{4d}^{\text{half-hyper}}$ depending on whether the jth multiplet is $\mathcal{N} = 2$ vector multiplet or half-hypermultiplet [17]:

Table 2 Embedding of 2d $\mathcal{N} = (2, 2)$ algebra into 4d $\mathcal{N} = 2$

4d	2d
$\{Q_-^1, (Q_-^1)^\dagger\} = E + h_{01} + h_{23} - (2R + r)$	Broken
$\{Q_+^1, (Q_+^1)^\dagger\} = E + h_{01} - h_{23} - (2R + r)$	$\{\mathcal{G}_L^+, (\mathcal{G}_L^+)^\dagger\} = 2H_L - J_L$
$\{Q_-^2, (Q_-^2)^\dagger\} = E - h_{01} + h_{23} + (2R - r)$	$\{\mathcal{G}_R^+, (\mathcal{G}_R^+)^\dagger\} = 2H_R + J_R$
$\{Q_+^2, (Q_+^2)^\dagger\} = E + h_{01} - h_{23} + (2R - r)$	Broken
$\{\widetilde{Q}_-^1, (\widetilde{Q}_-^1)^\dagger\} = E - h_{01} - h_{23} - (2R - r)$	$\{\mathcal{G}_R^-, (\mathcal{G}_R^-)^\dagger\} = 2H_R - J_R$
$\{\widetilde{Q}_+^1, (\widetilde{Q}_+^1)^\dagger\} = E + h_{01} + h_{23} - (2R - r)$	Broken
$\{\widetilde{Q}_-^2, (\widetilde{Q}_-^2)^\dagger\} = E - h_{01} - h_{23} + (2R + r)$	Broken
$\{\widetilde{Q}_+^2, (\widetilde{Q}_+^2)^\dagger\} = E + h_{01} + h_{23} + (2R + r)$	$\{\mathcal{G}_L^-, (\mathcal{G}_L^-)^\dagger\} = 2H_L + J_L$

$$f_{4d}^{\text{vector}} = \frac{-\mathfrak{p} - \mathfrak{q} - \mathfrak{t} + 2\mathfrak{p}\mathfrak{q} + \mathfrak{p}\mathfrak{q}/\mathfrak{t}}{(1 - \mathfrak{p})(1 - \mathfrak{q})} \qquad f_{4d}^{\text{half-hyper}} = \frac{\sqrt{\mathfrak{t}} - \mathfrak{p}\mathfrak{q}/\sqrt{\mathfrak{t}}}{(1 - \mathfrak{p})(1 - \mathfrak{q})}. \quad (4.4)$$

Now let us incorporate half-BPS surface operators supported on (x^0, x^1) plane in a four-dimensional space-time. As discussed in Sect. 1.3 (cf. Table 1), such surface operators preserve $SU(1, 1|1)_L \times SU(1, 1|1)_R \times U(1)_e$ subgroup of the superconformal symmetry group, which is basically the superconformal symmetry of a 2d $\mathcal{N} = (2, 2)$ theory on the (x^0, x^1) plane. The standard bosonic generators of this two-dimensional superconformal algebra can be easily identified via embedding into 4d $\mathcal{N} = 2$ algebra summarized in Table 2:

$$H_{L,R} = \frac{1}{2} (E \pm h_{01}), \qquad (4.5)$$

$$J_{L,R} = h_{23} + (2R \pm r)$$

Luckily, the unbroken part of the symmetry and supersymmetry suffices for defining the superconformal index (4.2) even in the presence of surface operators, with all of the fugacities. Moreover, since half-BPS surface operators can be defined via coupling to 2d $\mathcal{N} = (2, 2)$ theory supported on D, it gives a very convenient way of computing the index: one simply needs to add the contribution of 2d degrees of freedom, namely the so-called "flavored elliptic genus" of the 2d $\mathcal{N} = (2, 2)$ system [6]:

$$\mathcal{I}_{2d}(a_j; q, t) = \text{Tr}(-1)^F q^{H_L + \frac{1}{2}J_L} t^{-J_L} \prod_j a_j^{f_j}. \quad (4.6)$$

As the ordinary elliptic genus [53], it depend on the "Jacobi variables" q and t that can be identified with the 4d fugacities $(\mathfrak{p}, \mathfrak{q}, \mathfrak{t})$ by means of the embedding in Table 2:

$$q = \mathfrak{q}, \qquad t = \mathfrak{p}\mathfrak{q}/\mathfrak{t}, \qquad e = \mathfrak{p}^2/\mathfrak{t}. \quad (4.7)$$

Much like the basic building blocks of 4d $\mathcal{N} = 2$ theories are vector- and hyper-multiplets whose contributions to the index are summarized in (4.4), the basic building blocks of 2d $\mathcal{N} = (2, 2)$ theories are chiral and vector multiplets. Their contributions to the flavored elliptic genus, respectively, are

$$\mathcal{I}_{2d \text{ chiral}} = \frac{\theta(at; q)}{\theta(a; q)} \tag{4.8}$$

and

$$\mathcal{I}_{2d\text{vector}}^{U(n)} = \left(\frac{(q; q)^2}{\theta(t; q)}\right)^n \prod_{i \neq j} \left((1 - \frac{a_i}{a_j}) \frac{\theta(ta_i/a_j; q)}{\theta(a_i/a_j; q)}\right)^{-1} \tag{4.9}$$

where $\theta(x; q) := (x; q)(q/x; q)$ and $(x; q) = \prod_{i=0}^{\infty}(1 - xq^i)$. With these basic tools one can easily compute the index of any 2d-4d coupled system that, as in Sect. 1, describes a fairly generic surface operator. One interesting question, that still remains open since the pioneering work [17], is the identification of 2d TQFT whose partition function on a Riemann surface C matches the index in (4.1).

For further discussion of the superconformal index with surface operators see e.g. [54–58].

5 Surface Operators as Order Parameters

Finally, going back to the origins, we wish to explain that surface operators can serve as order parameters, in particular, they can distinguish the IR phases of 4d gauge theories. Typically, the information captured by surface operators is roughly equivalent to the spectrum of line operators in the low-energy theory [59].

Extending standard arguments that show how 't Hooft and Wilson line operators exhibit "area law" in Higgs and confining phases, respectively, one can quickly conclude that surface operators can exhibit a "volume law" in phases that admit domain walls which can end on surface operators. Not much is known about such peculiar domain walls, and exploring this direction would be an excellent research topic. In particular, by studying the spectra of domain walls in 4d supersymmetric gauge theories one might hope to learn whether surface operators can detect interesting phases not distinguished by Wilson and 't Hooft operators, cf. [60].

Here, we present a slightly different mechanism for the "volume law" behavior due to thermal effects. This material is new and has not appeared in the previous literature. It will also give us an excellent opportunity to illustrate how surface operators are described in the holographic dual of gauge theory, which is a convenient way to study thermal physics. As usual, in order to study 4d gauge theory at finite temperature T, we compactify the time direction on a circle of circumference $\beta = 2\pi/T$ and study the theory on a space-time manifold $M = S^1_\beta \times S^3$ with thermal (anti-periodic) boundary conditions on fermions. Following [61], we can study this system using a holographic dual description, which is available for many 4d gauge theories.

For concreteness, let us focus on $\mathcal{N} = 2$ gauge theory with a massless adjoint hypermultiplet or, equivalently, $\mathcal{N} = 4$ super-Yang-Mills, for which the gravity dual is especially simple and well studied [62–64]. After all, at finite temperature the precise details of the spectrum and interactions are expected to be less important and we expect that results should apply to more general $\mathcal{N} = 2$ gauge theories. Surface operators in this theory are well understood and have a simple description in the holographic dual [3, 4, 65].

Specifically, the holographic dual of $\mathcal{N} = 4$ super-Yang-Mills is type IIB string theory on $X_5 \times S^5$, where the 5-manifold X_5 is either a "thermal AdS"

$$X_5 \cong B^4 \times S^1 \qquad \text{(low temperature)}$$

in the low temperature phase, or the Schwarzschild black hole on AdS space

$$X_5 \cong S^3 \times B^2 \qquad \text{(high temperature)}$$

in the high temperature phase [61]. Note, that both of these manifolds are bounded by $S^1 \times S^3$, which is precisely M where the boundary gauge theory lives.

Now, we can introduce surface operators supported on $D \subset M$. For generic values of the continuous parameters α and η, in the holographic dual such surface operators are represented by D3-branes with four-dimensional world-volume

$$Q \times S^1 \subset X_5 \times S^5$$

where $Q \subset X_5$ is a volume-minimizing 3-dimensional submanifold bounded by $D = \partial Q$, and S^1 is a great circle in the S^5. Indeed, notice that such a D3-brane probe breaks the isometry/superconformal symmetry precisely as described in Table 1.

There are two qualitatively different choices of D, which correspond to spatial surface operators with $D \subset S^3$ or temporal surface operators with $D = \gamma \times S^1_\beta$, for some closed path $\gamma \subset S^3$. In the low temperature confining phase we have

$$\langle \mathcal{O}_{\text{temporal}} \rangle = 0$$

since S^1_β is not contractible in X_5, and so there is no minimal submanifold Q bounded by D. On the other hand, spatial surface operators exhibit the area law in this phase:

$$\langle \mathcal{O}_{\text{spatial}} \rangle \sim e^{-\text{Area}(D)}$$

As we decrease the value of β, the theory undergoes a phase transition to a deconfining phase [61] with $X_5 \cong S^3 \times B^2$, which does admit minimal submanifolds $Q \cong \gamma \times B^2$ bounded by temporal surface operators. Hence, in this high temperature phase we have

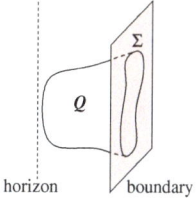

Fig. 7 A surface operator of the boundary theory is represented by a D3-brane in the holographic dual

$$\langle \mathcal{O}_{\text{temporal}} \rangle \neq 0 \tag{5.1}$$

and the spatial surface operators exhibit a "volume law":

$$\langle \mathcal{O}_{\text{spatial}} \rangle \sim e^{-\text{Volume}(D)} \tag{5.2}$$

since in the AdS black hole geometry the warp factor is bounded from below, as illustrated in Fig. 7. Since finite temperature breaks supersymmetry explicitly and makes scalars and fermions massive, this behavior is expected to be generic, in particular, present in pure gauge theory or more general $\mathcal{N} = 2$ gauge theories.

Also note, that in the limit $\beta \to 0$ the theory reduces to a pure (non-supersymmetric) three-dimensional Yang-Mills theory, which is expected to exhibit confinement and a mass gap. Since under this reduction a temporal surface operator turns into a line operator (supported on γ) in the 3d gauge theory, the behavior (5.1) is certainly expected.

References

[1]   Witten, E.: On S duality in Abelian gauge theory. Selecta Math. **1**, 383 (1995). arXiv:hep-th/9505186

[2]   Verlinde, E.P.: Global aspects of electric—magnetic duality. Nucl. Phys. **B455**, 211–228 (1995). arXiv:hep-th/9506011

[3]   Gukov, S., Witten, E.: Gauge theory, ramification, and the geometric Langlands program. arXiv:hep-th/0612073

[4]   Gukov, S., Witten, E.: Rigid surface operators. Adv. Theor. Math. Phys. **14**, 87–178 (2010). arXiv:0804.1561

[5]   Koh, E., Yamaguchi, S.: Holography of BPS surface operators. JHEP **0902**, 012 (2009). arXiv:0812.1420

[6]   Gadde, A., Gukov, S., Putrov, P.: Walls, lines, and spectral dualities in 3d gauge theories. arXiv:1302.0015

[7]   Gukov, S.: Gauge theory and knot homologies. Fortsch. Phys. **55**, 473–490 (2007). arXiv:0706.2369

[8]   Koh, E., Yamaguchi, S.: Surface operators in the Klebanov-Witten theory. JHEP **0906**, 070 (2009). arXiv:0904.1460

[9]   Tan, M.-C.: Surface operators in N = 2 Abelian gauge theory. JHEP **0909**, 047 (2009). arXiv:0906.2413

[10] Alday, L.F., Gaiotto, D., Gukov, S., Tachikawa, Y., Verlinde, H.: Loop and surface operators in N = 2 gauge theory and Liouville modular geometry. JHEP **1001**, 113 (2010). arXiv:0909.0945

[11] Gaiotto, D.: Surface operators in N = 2 4d gauge theories. JHEP **1211**, 090 (2012). arXiv:0911.1316

[12] Callan, C.G., Harvey, J.A.: Anomalies and fermion zero modes on strings and domain walls. Nucl. Phys. **B250**, 427 (1985)

[13] Alday, L.F., Gaiotto, D., Tachikawa, Y.: Liouville correlation functions from four-dimensional gauge theories. Lett. Math. Phys. **91**, 167–197 (2010). arXiv:0906.3219

[14] Nekrasov, N.A.: Seiberg-Witten prepotential from instanton counting. Adv. Theor. Math. Phys. **7**, 831–864 (2004). arXiv:hep-th/0206161

[15] Gaiotto, D.: N = 2 dualities. JHEP **1208**, 034 (2012). arXiv:0904.2715

[16] Gaiotto, D., Moore, G.W., Neitzke, A.: Wall-crossing, Hitchin systems, and the WKB approximation. arXiv:0907.3987

[17] Gadde, A., Rastelli, L., Razamat, S.S., Yan, W.: Gauge theories and Macdonald polynomials. Commun. Math. Phys. **319**, 147–193 (2013). arXiv:1110.3740

[18] Pestun, V.: Localization of gauge theory on a four-sphere and supersymmetric Wilson loops. Commun. Math. Phys. **313**, 71–129 (2012). arXiv:0712.2824

[19] Seiberg, N., Witten, E.: Electric-magnetic duality, monopole condensation, and confinement in N = 2 supersymmetric Yang-Mills theory. Nucl. Phys. **B426**, 19–52 (1994). arXiv:hep-th/9407087

[20] Seiberg, N., Witten, E.: Monopoles, duality and chiral symmetry breaking in N = 2 supersymmetric QCD. Nucl. Phys. **B431**, 484–550 (1994). arXiv:hep-th/9408099

[21] Bershadsky, M., Vafa, C., Sadov, V.: D-branes and topological field theories. Nucl. Phys. **B463**, 420–434 (1996). arXiv:hep-th/9511222

[22] Festuccia, G., Seiberg, N.: Rigid supersymmetric theories in curved superspace. JHEP **1106**, 114 (2011). arXiv:1105.0689

[23] Alday, L.F., Tachikawa, Y.: Affine SL(2) conformal blocks from 4d gauge theories. Lett. Math. Phys. **94**, 87–114 (2010). arXiv:1005.4469

[24] Frenkel, E., Gukov, S., Teschner J.: In preparation

[25] Witten, E.: Solutions of four-dimensional field theories via M theory. Nucl. Phys. **B500**, 3–42 (1997). arXiv:hep-th/9703166

[26] Katz, S.H., Klemm, A., Vafa, C.: Geometric engineering of quantum field theories. Nucl. Phys. **B497**, 173–195 (1997). arXiv:hep-th/9609239

[27] Ooguri, H., Vafa, C.: Knot invariants and topological strings. Nucl. Phys. **B577**, 419–438 (2000). arXiv:hep-th/9912123

[28] Dimofte, T., Gukov, S., Hollands, L.: Vortex counting and Lagrangian 3-manifolds. Lett. Math. Phys. **98**, 225–287 (2011). arXiv:1006.0977

[29] Harvey, J.A., Moore, G.W.: On the algebras of BPS states. Commun. Math. Phys. **197**, 489–519 (1998). arXiv:hep-th/9609017

[30] Gukov, S., Stosic, M.: Homological algebra of Knots and BPS states. arXiv:1112.0030

[31] Schiffmann, O.: Lectures on Hall algebras. arXiv:math/0611617

[32] Kontsevich, M., Soibelman, Y.: Stability structures, motivic Donaldson-Thomas invariants and cluster transformations. arXiv:0811.2435

[33] Kontsevich, M., Soibelman, Y.: Cohomological Hall algebra, exponential Hodge structures and motivic Donaldson-Thomas invariants. arXiv:1006.2706

[34] Gukov, S., Schwarz, A.S., Vafa, C.: Khovanov-Rozansky homology and topological strings. Lett. Math. Phys. **74**, 53–74 (2005). arXiv:hep-th/0412243

[35] Dimofte, T., Gaiotto, D., Gukov, S.: Gauge theories labelled by three-manifolds. arXiv:1108.4389

[36] Nekrasov, N.A., Shatashvili, S.L.: Supersymmetric vacua and Bethe Ansatz. Nucl. Phys. Proc. Suppl. **192–193**, 91–112 (2009). arXiv:0901.4744

[37] Gomis, J., Okuda, T., Pestun, V.: Exact results for 't Hooft loops in gauge theories on S^4. JHEP **1205**, 141 (2012). arXiv:1105.2568

[38] Gang, D., Koh, E., Lee, K.: Line operator index on $S^1 \times S^3$. JHEP **1205**, 007 (2012). arXiv:1201.5539

[39] Chriss, N., Ginzburg, V.: Representation Theory and Complex Geometry. Birkhauser Boston Inc., Boston (1997)

[40] Harvey, J.A., Moore, G.W., Strominger, A.: Reducing S duality to T duality. Phys. Rev. **D52**, 7161–7167 (1995). arXiv:hep-th/9501022

[41] Bershadsky, M., Johansen, A., Sadov, V., Vafa, C.: Topological reduction of 4-d SYM to 2-d sigma models. Nucl. Phys. **B448**, 166–186 (1995). arXiv:hep-th/9501096

[42] Kapustin, A., Witten, E.: Electric-magnetic duality and the geometric Langlands program. Commun. Number Theory Phys. **1**, 1–236 (2007). arXiv:hep-th/0604151

[43] Bezrukavnikov, R.: Noncommutative counterparts of the Springer resolution. arXiv:math/0604445

[44] Kapranov, M., Vasserot, E.: Kleinian singularities, derived categories and Hall algebras. arXiv:math/9812016

[45] Seidel, P., Thomas, R.: Braid group actions on derived categories of coherent sheaves. Duke Math. J. **108**, 37 (2001)

[46] Ishii, A., Uehara, H.: Autoequivalences of derived categories on the minimal resolutions of A_n-singularities on surfaces. arXiv:math/0409151

[47] Bridgeland, T.: Stability conditions and Kleinian singularities. arXiv:math/0508257

[48] Aspinwall, P.S.: D-branes on Calabi-Yau manifolds. arXiv:hep-th/0403166

[49] Aganagic, M., Shakirov, S.: Knot homology from refined Chern-Simons theory. arXiv:1105.5117

[50] Cherednik, I.: Jones polynomials of torus knots via DAHA. arXiv:1111.6195

[51] Nakajima, H.: Quiver varieties and Kac-Moody algebras. Duke Math. J. **91**(3), 515–560 (1998)

[52] Nakajima, H.: Quiver varieties and finite-dimensional representations of quantum affine algebras. J. Am. Math. Soc. **14**(1), 145–238 (2001)

[53] Witten, E.: Elliptic genera and quantum field theory. Commun. Math. Phys. **109**, 525 (1987)

[54] Nakayama, Y.: 4D and 2D superconformal index with surface operator. JHEP **1108**, 084 (2011). arXiv:1105.4883

[55] Gaiotto, D., Rastelli, L., Razamat, S.S.: Bootstrapping the superconformal index with surface defects. arXiv:1207.3577

[56] Iqbal, A., Vafa, C.: BPS degeneracies and superconformal index in diverse dimensions. arXiv:1210.3605

[57] Gadde, A., Gukov, S.: 2d index and surface operators. arXiv:1305.0266

[58] Bullimore, M., Fluder, M., Hollands, L., Richmond, P.: The superconformal index and an elliptic algebra of surface defects. arXiv:1401.3379

[59] Gukov, S., Kapustin, A.: Topological quantum field theory, nonlocal operators, and gapped phases of gauge theories. arXiv:1307.4793

[60] Cachazo, F., Seiberg, N., Witten, E.: Phases of N = 1 supersymmetric gauge theories and matrices. JHEP **0302**, 042 (2003). arXiv:hep-th/0301006

[61] Witten, E.: Anti-de Sitter space, thermal phase transition, and confinement in gauge theories. Adv. Theor. Math. Phys. **2**, 505–532 (1998). arXiv:hep-th/9803131

[62] Maldacena, J.M.: The Large N limit of superconformal field theories and supergravity. Adv. Theor. Math. Phys. **2**, 231–252 (1998). arXiv:hep-th/9711200

[63] Witten, E.: Anti-de Sitter space and holography. Adv. Theor. Math. Phys. **2**, 253–291 (1998). arXiv:hep-th/9802150

[64] Gubser, S., Klebanov, I.R., Polyakov, A.M.: Gauge theory correlators from noncritical string theory. Phys. Lett. **B428**, 105–114 (1998). arXiv:hep-th/9802109

[65] Gomis, J., Matsuura, S.: Bubbling surface operators and S-duality. JHEP **0706**, 025 (2007). arXiv:0704.1657

The Superconformal Index of Theories of Class $\mathcal{S}$

Leonardo Rastelli and Shlomo S. Razamat

Abstract We review different aspects of the superconformal index of $\mathcal{N} = 2$ superconformal theories of class $\mathcal{S}$. In particular we discuss the relation of the index of class $\mathcal{S}$ theories to topological QFTs and integrable models, and review how this relation can be harnessed to completely determine the index. This is part of a combined review on $2d$-$4d$ relations, edited by J. Teschner.

Keywords Conformal field theory · Supersymmetry · Class $\mathcal{S}$ · Topological field theory

1 Introduction

This volume surveys the $4d/2d$ relations that arise in the study of class $\mathcal{S}$, the set of four-dimensional $\mathcal{N} = 2$ supersymmetric field theories obtained by compactification of a six-dimensional $(2, 0)$ theory on a punctured Riemann surface $\mathcal{C}$.[1] There is an extensive $4d/2d$ dictionary relating several protected observables of the four-dimensional theory $T[\mathcal{C}]$ to observables of certain natural theories defined on the associated surface $\mathcal{C}$. In this chapter we focus on the *superconformal index* of $T[\mathcal{C}]$

---

[1] See [V:2] in this volume for a general introduction to class $\mathcal{S}$.

A citation of the form [V:x] refers to article number x in this volume.

L. Rastelli (✉)
C. N. Yang Institute for Theoretical Physics, Stony Brook University, Stony Brook, NY 11794-3840, USA
e-mail: leonardo.rastelli@gmail.com

S.S. Razamat
Institute for Advanced Study, Einstein Dr., Princeton, NJ 08540, USA
e-mail: srtachyon@gmail.com

S.S. Razamat
NHETC, Rutgers University, Piscataway, NJ 08854, USA

S.S. Razamat
Department of Physics, Technion, 32000 Haifa, Israel

© Springer International Publishing Switzerland 2016
J. Teschner (ed.), *New Dualities of Supersymmetric Gauge Theories*,
Mathematical Physics Studies, DOI 10.1007/978-3-319-18769-3_9

and on its re-interpretation as a topological quantum field theory (TQFT) living on $\mathcal{C}$. We shall restrict our discussion to the subset of theories that enjoy conformal invariance, for which the general index is well-defined.

The superconformal index, or index for short, encodes some detailed information about the protected spectrum of a superconformal field theory. By construction, it is invariant under exactly marginal deformations of the SCFT. A basic item of the $4d/2d$ dictionary equates the conformal manifold of $T[\mathcal{C}]$ (i.e., the space of its exactly marginal gauge couplings) with the complex structure moduli of $\mathcal{C}$. We should then expect on general grounds that the index is computed by a TQFT living on $\mathcal{C}$. A concrete description of this TQFT as an explicit $2d$ theory is only available for certain specializations of the general index, in particular the so-called Schur index corresponds to q-deformed two-dimensional Yang-Mills theory in the zero-area limit. The TQFT viewpoint is however very fruitful also in the general case. As different TQFT correlators compute the indices of different $4d$ theories, we are led to study consistency conditions in *theory space*. This turns out to be a very effective strategy, which allows for the complete determination of the general index for theories of class $\mathcal{S}$.

2 The Superconformal Index

Let us introduce the main character of this review. To a superconformal field theory in d space-time dimensions one can associate its superconformal index [1, 2], which is nothing but the Witten index of the theory in radial quantization, refined to keep track of a maximal set of commuting conserved quantum numbers $\{C_i\}$,

$$\mathcal{I}(\mu_i) = \mathrm{Tr}\,(-1)^F \prod_i \mu_i^{C_i} \, e^{-\beta\delta}\,, \quad \delta := \{\mathcal{Q}, \mathcal{Q}^\dagger\}. \qquad (2.1)$$

The trace is taken over the Hilbert space of the radially quantized theory on $\mathbb{S}^{d-1}$, F is the fermion number and $\mathcal{Q}$ a chosen Poincaré supercharge. In a given theory, the index is thus a function of the "fugacities" $\{\mu_i\}$ that couple to the conserved charges $\{C_i\}$. The conserved charges are chosen as to commute with each other, with the chosen supercharge $\mathcal{Q}$ and with its conjugate (conformal) supercharge $\mathcal{Q}^\dagger$. If the theory is unitary, which we shall always assume, then $\delta := \{\mathcal{Q}, \mathcal{Q}^\dagger\} \geqslant 0$. By a familiar argument, the index counts (with signs) cohomology classes of $\mathcal{Q}$. Indeed the Hilbert space decomposes into the subspace of states with $\delta = 0$, which are automatically killed by both $\mathcal{Q}$ and $\mathcal{Q}^\dagger$ (these are the "harmonic representatives" of the cohomology classes), and the subspace with $\delta \neq 0$, where one can choose a basis such that all states belong to a pair $(\psi, \mathcal{Q}\psi)$, with $\mathcal{Q}^\dagger\psi = 0$. The paired states have the same charges $\{C_i\}$ but opposite statistics, so their combined contribution to the trace vanishes. Since the trace in (2.1) receives contributions only from the harmonic representatives, the index is in fact independent of β. The states with $\delta = 0$

are annihilated by some of the supercharges and as such they belong to shortened representation of the superconformal algebra.

As the energy (conformal dimension) of a generic long multiplet of the superconformal algebra is lowered to the unitarity bound, the long multiplet breaks up into a direct sum of short multiplets, containing states with $\delta = 0$, but by continuity their total contribution to the index is zero. So even within the $\delta = 0$ subspace there may be fermion/boson cancellations between states with the same charges $\{C_i\}$, associated to recombinations of short multiplets into long ones. In fact one can equivalently characterize the index as the most general invariant that counts short multiplets, up to the equivalence relation setting to zero combinations of short multiplets that have the right quantum numbers to recombine into long ones [2]. It follows, at least formally, that the index is invariant under changes of continuous parameters of the theory preserving superconformal invariance, i.e. it is constant over the *conformal manifold* of the theory. As the exactly marginal couplings are varied, long multiplets may split into short ones or short multiplets recombine into long ones, but this is immaterial for the index. In other contexts, the formal independence of supersymmetric indices on continuous parameters is known to fail, leading to rich wall-crossing phenomena. In our case, however, we are dealing with theories that have a discrete spectrum of states, and such that the subspaces with fixed values of the quantum numbers $\{C_i\}$ are *finite-dimensional*, so the formal argument is completely rigorous. The index is thus truly invariant under exactly marginal deformations preserving the full superconformal algebra of the model.

The superconformal index can be defined for theories in various spacetime dimensions, and with different amounts of superconformal symmetry. We have given the "Hamiltonian" definition in terms of a trace formula, but the index has an equivalent "Lagrangian" interpretation as a supersymmetric partition function on $\mathbb{S}^{d-1} \times \mathbb{S}^1$, with twisted boundary conditions around the "temporal" $\mathbb{S}^1$ to incorporate the dependence on the various fugacities. See [V:6] in this volume for more details on this approach. Viewed as a partition function, the index makes sense for non-conformal theories, though in those cases it should be more properly referred to as a *supersymmetric* index. One can show that such a partition function is independent on the RG scale, so that the superconformal index of a theory realized as the IR fixed point of some RG flow can be often computed using the non-conformal UV starting point of the flow [1, 3, 4]. Examples where this is a very useful strategy include $\mathcal{N} = 1$ gauge theories in four dimensions, and susy gauge theories in three and two dimensions. The partition function interpretation is also useful to obtain the index in the presence of various BPS defects, by the techniques of supersymmetric localization. In this review we will mostly stick to the trace interpretation of the index, and localization will not play a role. We will determine the index of the $\mathcal{N} = 2$ SCFTs of class $\mathcal{S}$ (even in the presence of certain BPS defects) by a more abstract algebraic viewpoint. A direct localization approach would not be an option since these theories do not generally admit a known Lagrangian description.

We now specialize to the case of interest, namely $\mathcal{N} = 2$ superconformal theories in four dimensions. The $\mathcal{N} = 2$ superconformal index depends on three superconformal fugacities (p, q, t) and on any number of fugacities $\{a_i\}$ associated to flavor symmetries (which, by definition, commute with the superconformal algebra),[2]

$$\mathcal{I}(p,q,t;a_i) := \mathrm{Tr}(-1)^F \left(\frac{t}{pq}\right)^r p^{j_{12}} q^{j_{34}} t^R \prod_i a_i^{f_i} e^{-\beta\delta_{2\dot{-}}}, \qquad (2.2)$$

where

$$2\delta_{2\dot{-}} := \{\tilde{\mathcal{Q}}_{2\dot{-}}, \tilde{\mathcal{Q}}_{2\dot{-}}^\dagger\} = E - 2j_2 - 2R + r. \qquad (2.3)$$

We will always assume that

$$|p| < 1, \quad |q| < 1, \quad |t| < 1, \quad |a_i| = 1, \quad \left|\frac{pq}{t}\right| < 1. \qquad (2.4)$$

Our notations are as follows. We denote by E the conformal hamiltonian (dilatation generator), by j_1 and j_2 the Cartan generators of of the $SU(2)_1 \times SU(2)_2$ isometry group of $\mathbb{S}^3$, by R and r the Cartan generators of the $SU(2)_R \times U(1)_r$ the superconformal R-symmetry. We have also defined $j_{12} := j_2 - j_1$ and $j_{34} := j_2 + j_1$, which generate rotations in two orthogonal planes (thinking of $\mathbb{S}^3$ as embedded in $\mathbb{R}^4$). Finally $\{f_i\}$ are the flavor symmetry generators. In our conventions, we label the supercharges as

$$\mathcal{Q}_\alpha^\mathcal{I} \quad \tilde{\mathcal{Q}}_{\mathcal{I}\dot{\alpha}} \quad \mathcal{S}_\mathcal{I}^\alpha \quad \tilde{\mathcal{S}}^{\mathcal{I}\dot{\alpha}}, \qquad (2.5)$$

where $\alpha = \pm$ is an $SU(2)_1$ index, $\dot{\alpha} = \pm$ an $SU(2)_2$ index and $\mathcal{I} = 1, 2$ an $SU(2)_R$ index. We have $\mathcal{S} = \mathcal{Q}^\dagger$ and $\tilde{\mathcal{S}} = \tilde{\mathcal{Q}}^\dagger$. Writing an explicit trace formula for the index involves a choice of supercharge. With no loss of generality, we chose in (2.2) to count cohomology classes of $\tilde{\mathcal{Q}}_{2\dot{-}}$, which has quantum numbers $E = R = -r = \frac{1}{2}$, $(j_1, j_2) = (0, -\frac{1}{2})$. The states that to this index are the "harmonic representatives" satisfy $\delta \equiv \delta_{2\dot{-}} = 0$. All other choices of a Poincaré supercharge would give an equivalent index [2].

In Appendix 2 we review the shortening conditions of the $\mathcal{N} = 2$ superconformal algebra and the recombination rules of short multiplets into long ones. Explicit formulae for the index of individual short multiplets are given in Appendix B of [6] and will not be repeated here. It is important to keep in mind that knowledge of the index alone is in general not sufficient to completely reconstruct the spectrum of short representations of a given theory. Schematically, the issue is the following [8]. Suppose that two short multiplets, S_1 and S_2, can recombine to form a long multiplet L_1,

$$S_1 \oplus S_2 = L_1, \qquad (2.6)$$

---

[2]In this review we follow the conventions of [5]. In comparing with [6, 7], the only significant change is $j_1 \to -j_1$ in the definitions of j_{12} and j_{34}. The conventions for labeling supercharges are also slightly different in these two sets of references, but notations aside all of them choose "same" supercharge to define the general index (i.e. the supercharge with quantum numbers $E = R = -r = \frac{1}{2}$, $(j_1, j_2) = (0, -\frac{1}{2})$).

and similarly that S_2 can recombine with a third short multiplet S_3 to give another long multiplet L_2,

$$S_2 \oplus S_3 = L_2. \tag{2.7}$$

By construction, the index evaluates to zero on long multiplets, so

$$\mathcal{I}(S_1) = -\mathcal{I}(S_2) = \mathcal{I}(S_3). \tag{2.8}$$

The index cannot distinguish between the two multiplets S_1 and S_3. (Note that S_2 is distinguished from $S_1 \sim S_3$ by the overall sign.) A detailed discussion of equivalence classes of multiplets that have the same $\mathcal{N} = 2$ superconformal index can be found in Sect. 5.2 of [8].

2.1 Free Field Combinatorics

The simplest examples of conformal quantum field theories are free theories. In a free theory, the general local operator is obtained from normal ordering of the elementary fields, and its quantum numbers including the conformal dimension take their classical "engineering" values. By the state/operator map, local operators inserted at the origin are in one-to-one correspondence with states. Enumerating states reduces then to the simple combinatorial problem of enumerating all possible composite "words" (or "multi-particles") built out of the elementary "letters" (or "single-particles"), which are the elementary fields and their space-time derivatives.

For our purposes, we are interested in enumerating states with $\delta = 0$, and since in a free theory the value of δ of a composite operator is simply the sum of the values of δ of its elementary letters, we may from the start restrict to the letters with $\delta = 0$. The letters contributing the index of the free $\mathcal{N} = 2$ hypermultiplet and of the free vector multiplet $\mathcal{N} = 2$ are shown in Table 1. One immediately finds the following single-particle indices (i.e., the indices computed over the set of single-particle states):

$$\mathcal{I}_H^{s.p.} = t^{\frac{1}{2}} \frac{1 - \frac{pq}{t}}{(1 - p)(1 - q)} (a + a^{-1}) \chi_\Lambda(\mathbf{x}), \tag{2.9}$$

$$\mathcal{I}_V^{s.p.} = -\frac{q}{1 - q} - \frac{p}{1 - p} + \frac{\frac{pq}{t} - t}{(1 - p)(1 - q)}. \tag{2.10}$$

Here a is a $U(1)$ fugacity under which the two half-hypers have opposite charges and $\chi_\Lambda(\mathbf{x})$ is the character of the representation of some global symmetry. The multi particle-indices are given by the plethystic exponentials of the single-particle ones. In particular the index of a free hypermultiplet in a bi-fundamental representation of $SU(n) \times SU(n)$, which will play an important role in our discussion, is given by

Table 1 Contributions to the index from single-particle (letter) operators of the two basic $\mathcal{N} = 2$ multiplets: the vector multiplet and the hypermultiplet

Letters	E	j_1	j_2	R	r	$\mathcal{I}(p,q,t)$
ϕ	1	0	0	0	-1	pq/t
$\lambda^1_{\pm}$	$\frac{3}{2}$	$\pm\frac{1}{2}$	0	$\frac{1}{2}$	$-\frac{1}{2}$	$-p, -q$
$\bar{\lambda}_{2\dot{+}}$	$\frac{3}{2}$	0	$\frac{1}{2}$	$\frac{1}{2}$	$\frac{1}{2}$	$-t$
$\bar{F}_{\dot{+}\dot{+}}$	2	0	1	0	0	pq
$\partial_{-\dot{+}}\lambda^1_{+} + \partial_{+\dot{+}}\lambda^1_{-} = 0$	$\frac{5}{2}$	0	$\frac{1}{2}$	$\frac{1}{2}$	$-\frac{1}{2}$	pq
Q	1	0	0	$\frac{1}{2}$	0	$\sqrt{t}$
$\bar{\psi}_{\dot{+}}$	$\frac{3}{2}$	0	$\frac{1}{2}$	0	$-\frac{1}{2}$	$-pq/\sqrt{t}$
$\partial_{\pm\dot{+}}$	1	$\pm\frac{1}{2}$	$\frac{1}{2}$	0	0	p, q

We denote by $(\phi, \bar{\phi}, \lambda^{\mathcal{I}}_{\alpha}, \bar{\lambda}_{\mathcal{I}\dot{\alpha}}, F_{\alpha\beta}, \bar{F}_{\dot{\alpha}\dot{\beta}})$ the components of the adjoint $\mathcal{N} = 2$ vector multiplet, by $(Q, \bar{Q}, \psi_{\alpha}, \bar{\psi}_{\dot{\alpha}})$ the components of the $\mathcal{N} = 1$ chiral multiplet, and by $\partial_{\alpha\dot{\alpha}}$ the spacetime derivatives

$$
\mathcal{I}_H(a, \mathbf{x}, \mathbf{y}; p, q, t) = \text{PE}\left[t^{\frac{1}{2}} \frac{1 - \frac{pq}{t}}{(1-p)(1-q)} (a + a^{-1}) \left(\sum_{i=1}^{n} x_i \right) \left(\sum_{j=1}^{n} y_j \right) \right]
$$

$$
= \prod_{i,j=1}^{n} \Gamma(t^{\frac{1}{2}} a \, x_i \, y_j \, ; \, p, q) \, \Gamma(t^{\frac{1}{2}} (a \, x_i \, y_j)^{-1}; \, p, q). \quad (2.11)
$$

We collect the definitions of the plethystic exponenential, elliptic Gamma function, and related objects in Appendix 1.

Conversely, one of the hallmarks of a free theory is the fact that the plethystic log of the index is simple. For example, formally analogous to the counting problem in free field theory is the counting problem for large N theories. It often happens that the conformal gauge theories come in families labeled by the rank of the gauge group and in the limit of large rank they have a dual description in terms of supergravity in *AdS* backgrounds [2]. In such cases the operators counted by the index are dual to free supergravity modes. Thus, taking the limit of large N the index reduces again to a simple plethystic exponential of the towers of single trace operators dual to the finite number of free supergravity fields.

2.2 Gauging

After we dealt with free theories, let us turn to interacting models. In general, we should not expect any simple combinatorial description of the set of local operators in an interacting theory. An important exception are the superconformal field theories that admit a Lagrangian description, which by definition are continuously connected to free field theories by turning off the gauge couplings. Since the index is independent of exactly marginal deformations, we may as well compute it in the free limit (setting to zero all gauge couplings). The only effect of the gauging is the Gauss law constraint, i.e. the projection onto gauge invariant states.

More generally, starting from a SCFT $\mathcal{T}$, we can obtain a new superconformal field theory $\mathcal{T}_G$ by gauging a subgroup G of the flavor symmetry of $\mathcal{T}$, provided of course that the gauge coupling beta functions vanish. If the index of $\mathcal{T}$ is known, we find the index of $\mathcal{T}_G$ by multiplying by the index of a vector multiplet in the adjoint representation of G, and then integrating over G with the invariant Haar measure to enforce the projecting over gauge singlets,

$$\mathcal{I}[\mathcal{T}_G] = \int [d\mathbf{z}]_G \, \mathcal{I}_V(\mathbf{z}) \, \mathcal{I}[\mathcal{T}](\mathbf{z}). \qquad (2.12)$$

In fact we can treat the index $\mathcal{I}_{\mathcal{T}}(\mathbf{z})$ as a "black-box": it might be the index of a collection of free hypermultiplets, the index of a gauge theory, or the index of an interacting theory for which we do not know a useful description in terms of a Lagrangian. Whenever a flavor symmetry is gauged in four dimensions, the effect on the index is simply to introduce the vector multiplet and project onto gauge-invariant states.[3]

In all known examples, conformal manifolds of $\mathcal{N} = 2$ SCFTs are parametrized by gauge couplings. It is tempting to speculate that the most general $\mathcal{N} = 2$ SCFT is obtained by gauging a set of elementary building blocks, each of which is an isolated theory with no exactly marginal couplings. The simplest of such an elementary building block is the free hypermultiplet theory. We will encounter below several other examples of building blocks with no known Lagrangian description. Determining the index of such isolated theories would appear to be very challenging. Fortunately, for theories of class $\mathcal{S}$ we can leverage the additional structure of generalized S-duality. Let us turn to a concrete illustration.

3 Interlude: Duality and the Index of E_6 SCFT

In this section, we will sketch how to determine the index of a canonical example of isolated non-Lagrangian theory, the SCFT with E_6 flavor symmetry of Minahan and Nemeschansky [9]. The general idea is to couple the isolated theory to some extra stuff, and use dualities to relate the larger theory to a more tractable model.

For the case at hand, we exploit Argyres-Seiberg duality [10]. On one side of the duality we have an $SU(3)$ SYM with $N_f = 6$ flavors. On the other side of the duality we have a hypermultiplet in the fundamental representation of gauged $SU(2)$ under which also a strongly-coupled theory with E_6 flavor symmetry [9] is charged.

---

[3]In other dimensions the situation can be slightly more involved. For example, in three dimensions a gauge theory contains local monopole operators which have to be introduced into the index computations along with the vector multiplets.

The $SU(2)$ gauged group is a sub-group of the E_6 flavor symmetry. By the rules of computing the index reviewed in the previous section, this duality can be written as equality of two integrals [11],

$$\int [d\mathbf{z}]_{G=SU(3)}\, \mathcal{I}_V^{SU(3)}(\mathbf{z})\, \mathcal{I}_H^{(1)}(\mathbf{z}, \mathbf{y}, \mathbf{x}, a, b) = \tag{3.1}$$

$$\int [d z]_{G=SU(2)}\, \mathcal{I}_V^{SU(2)}(z)\, \mathcal{I}_H^{(2)}(z, (a/b)^{3/2})\, \mathcal{I}_{E_6}(\mathbf{x}, \mathbf{y}, \{z(ab)^{-\frac{1}{2}}, z^{-1}(ab)^{-\frac{1}{2}}\}).$$

Here $\mathcal{I}_{E_6}(\mathbf{x}_1, \mathbf{x}_2, \mathbf{x}_3)$ is the unknown index of the theory with E_6 flavor symmetry with $\{\mathbf{x}_i\}$ being the fugacities for $SU(3)^3$ maximal subgroup of E_6; $\mathbf{y}$ and $\mathbf{x}$ are $SU(3)$ fugacities and a, b are two $U(1)$ fugacities. The quantity $\mathcal{I}_H^{(1)}(\mathbf{z}, \mathbf{y}, \mathbf{x}, a, b)$ represents the index of a collection of hypermultiplets in the bi-fundamental representation of flavor of $SU(3)^2$ and the gauged $SU(3)$, whereas $\mathcal{I}_H^{(2)}(z, (a/b)^{3/2})$ is a fundamental hypermultiplet of $SU(2)$. The powers of $U(1)$ fugacities a and b on the right-hand side of the equality are a consequence of the details of the map of global symmetries between the two duality frames. In general from equalities of integrals of this sort one cannot extract the precise values of the integrands. However, in this particular case the integral on the right-hand side is invertible and just by assuming the Argyres-Seiberg duality as manifested for the index in (3.1) one can explicitly deduce the index $\mathcal{I}_{E_6}$. Schematically, this inversion procedure takes the following form

$$\mathcal{I}_{E_6}(\mathbf{x}, \mathbf{y}, \{c(ab)^{-\frac{1}{2}}, c^{-1}(ab)^{-\frac{1}{2}}\}) = \oint_C \frac{d h}{2\pi i h}\, \Delta(h, c) \int [d\mathbf{z}]_{G=SU(3)}\, \mathcal{I}_V^{SU(3)}(\mathbf{z})\, \mathcal{I}_H^{(1)}(\mathbf{z}, \mathbf{y}, \mathbf{x}, a, b).$$

Here C is a well-defined integration contour and Δ is a specific inversion kernel [12]. Physically, the fact that the integral is invertible means that the extra hyper-multiplet introduced while gauging a sub-group of the E_6 symmetry adds enough structure so that the information about the protected spectrum of the E_6 theory itself, a-priori lost after gauging, can be still recovered.

We thus are able to completely fix the superconformal index of a theory not connected to a free theory by a continuous parameter. The trick is to *enlarge* the theory with the bigger theory admitting an alternative description which *can* be connected to a free theory by continuous deformation. This basic idea will be behind the general procedure we will outline in the next sections.

Before turning to the general discussion of class $\mathcal{S}$ theories, let us illustrate in this concrete example what kind of physical information can be extracted from the index. Explicitly computing (3.2) one obtains to the lowest orders in the series expansion in fugacities

$$\mathcal{I}_{E_6} = PE[\mathcal{I}_u]\, PE[\mathcal{I}_H(\chi_{78})]\, PE[\mathcal{I}_T] \times$$

$$\left(1 - (t^2 - p q t + t^2(p + q))\,(\chi_{650} + 1) - \frac{p^3 q^3}{t^2} + p q t\, \chi_{78} + \cdots\right). \tag{3.2}$$

In the first line we have the protected multiplets appearing in this theory: $\mathcal{I}_u$ is the Coulomb branch multiplet (the dual of $u = Tr\phi^3$ of the $SU(3)$ gauge theory), $\mathcal{I}_H$ is the Higgs branch generator, X, in **78** of the E_6 global symmetry, and $\mathcal{I}_T$ is the stress-energy multiplet. The quantum numbers of these multiplets are different from free fields. Moreover, on the second line we have *constraints* appearing removing some of the contributions generated on the first line: these constraints are the footprint of the non-trivial dynamics of the theory. For example one constraint encoded here is

$$[X \otimes X]_{650\oplus 1} = 0, \tag{3.3}$$

which is the Joseph's relation discussed in [13].

4 Derivation of the Index for Theories of Class $\mathcal{S}$

In this section we will determine the index for all theories of class $\mathcal{S}$. Broadly speaking, we will be using the same kind of physical input as in the previous section, namely knowledge of the index for Lagrangian theories and the assumption of generalized S-duality. We will however exploit these ingredients in a different way, arriving at a particularly elegant and uniform description of the general index. For simplicity we focus on the basic index (the $\mathbb{S}^3 \times \mathbb{S}^1$ partition function), and to the simplest class $\mathcal{S}$ theories of type A. Several generalizations will be mentioned in Sect. 6.

4.1 Class $\mathcal{S}$

A lightening review of class $\mathcal{S}$ is in order. A $4d$ superconformal field theory of class $\mathcal{S}$ is specified the following data[4]:

- A choice of the type $\mathfrak{g}$ of the $(2, 0)$ theory, where $\mathfrak{g} = \{A_n, D_n, E_6, E_7, E_8\}$ is a simply-laced Lie algebra.
- A choice of UV curve $\mathcal{C}_{g,s}$, where g indicates the genus and s the number of punctures of the curve. Only the complex structure moduli of $\mathcal{C}_{g,s}$ matter. They are interpreted as the exactly marginal gauge couplings of the $4d$ SCFT.
- Each puncture corresponds to a codimension two defect of the $(2, 0)$ theory. We restrict to the so-called regular defects, which are labelled by a choice of embedding $\Lambda : \mathfrak{su}(2) \to \mathfrak{g}$. The centralizer $\mathfrak{h} \subset \mathfrak{g}$ of the image of Λ in $\mathfrak{g}$ is the flavor symmetry associated to the defect. All in all, the theory enjoys at least[5] the flavor symmetry algebra $\oplus_{i=1}^s \mathfrak{h}_i$.

---

[4]These are the "basic" theories. A larger list is obtained by allowing for "irregular" punctures. Further possibilities arise by decorating the UV curve with outer automorphisms twist lines , see [14].

[5]In some special cases, the symmetry is enhanced by additional generators which are not naturally assigned to any puncture.

We will label the corresponding $4d$ SCFT as $T[\mathfrak{g}; \mathcal{C}_{g,s}; \{\Lambda_i\}]$. We summarize the basics of class $\mathcal{S}$ dictionary in (Table 2).

From now on we will restrict our discussion to class $\mathcal{S}$ theories of type A, $\mathfrak{g} = A_{n-1}$. The embeddings $\Lambda : \mathfrak{su}(2) \to \mathfrak{su}(n)$ are in one-to-one correspondence with partitions of n, $[n_1^{\ell_1}, n_2^{\ell_2}, \ldots n_k^{\ell_k}]$ with $\sum_i \ell_i n_i = n$ and $n_i > n_{i+1}$, which indicate how the fundamental representation of $\mathfrak{su}(n)$ decomposes under representations of $\Lambda(\mathfrak{su}(2))$. For the trivial embedding $\Lambda = 0$, associated to the partition $[1^n]$, we have maximal flavor symmetry $\mathfrak{h} = \mathfrak{su}(n)$ and the corresponding puncture is called *maximal*. The other extreme case case is the principal embedding, associated to the partition $[n]$, leading to $\mathfrak{h} = 0$ (no flavor symmetry), so the puncture is effectively deleted. Another important case is the subregular embedding, associated to the partition $[n - 1, 1]$, which leads to $\mathfrak{h} = \mathfrak{u}(1)$, the smallest non-trivial flavor symmetry, so the corresponding puncture is called a *minimal* puncture.[6]

The surface $\mathcal{C}_{g,s}$ can be assembled by gluing together three-punctured spheres, or "pairs of pants" (viewed as three-vertices) and cylinders (viewed as propagators). Each cylinder is associated to a simple gauge group factor of the $4d$ SCFT, with the plumbing parameter interpreted as the corresponding marginal gauge coupling. The degeneration limit of the surface where one cylinder becomes very long corresponds to the weak coupling limit of that gauge group. Cutting a cylinder is interpreted as "ungauging" an $SU(n)$ gauge group, leaving behind two maximal punctures, each carrying $SU(n)$ flavor symmetry. Conversely, gluing two maximal punctures corresponds to gauging the diagonal subgroup of their $SU(n) \times SU(n)$ flavor symmetry. The basic building blocks of class $\mathcal{S}$ are thus the theories associated to three-punctured spheres, $T_n^{\Lambda_1, \Lambda_2, \Lambda_3} := T[\mathfrak{su}(n); \mathcal{C}_{0,3}; \Lambda_1 \Lambda_2 \Lambda_3]$. These are isolated SCFTs with no tunable couplings, in harmony with the fact that three-punctured spheres carry no complex structure moduli. Most of them have no known Lagrangian description.

Table 2 The basic class $\mathcal{S}$ dictionary

$4d$ theory $T[\mathcal{C}]$	Riemann surface $\mathcal{C}$
Conformal manifolds	Complex structure moduli of $\mathcal{C}$
$SU(n)$ gauge group with coupling τ	Cylinder with sewing parameter $q = \exp(2\pi i \tau)$
Flavor-symmetry factor $H \subset SU(n)$	Puncture labelled by $SU(2) \to SU(n)$ with commutant H
Weakly-coupled frame	Pair-of-pant decomposition of $\mathcal{C}$
Generalized S-duality	Moore-Seiberg groupoid of $\mathcal{C}$
Partition function on S^4	Correlator in Liouville/Toda on $\mathcal{C}$
Superconformal index	Correlator in a TQFT on $\mathcal{C}$

---

[6]Throughout this review we will often associate punctures with flavor symmetry factors. For theories of type A this association is well motivated (although there can be two different punctures with same flavor symmetry), but one has to remember that for type D and E theories one can have non-trivial punctures with no flavor symmetry associated with them.

An important exception is the theory associated to two maximal and one minimal puncture, $T_n^{[1^n][1^n][n-1,1]}$, which is identified with the free hypermultiplet in the bifundamental representation of $SU(n) \times SU(n)$.

Different pairs-of-pants decompositions of the UV curve correspond to different weakly coupled descriptions of the same SCFT, related by generalized S-dualities. The Moore-Seiberg groupoid of the UV curve is thus identified with the S-duality groupoid of the SCFT.

4.2 TQFT Interpretation of the Index

The index of $T[\mathfrak{g}; \mathcal{C}_{g,s}; \{\Lambda_i\}]$ is a function of the superconformal fugacities (p, q, t) and of the flavor fugacities $\mathbf{a}_i$, $i = 1, \ldots s$, associated to the Cartan generators of the global symmetry group $H_1 \otimes \cdots \otimes H_s$, but it is independent of the complex structure moduli of the UV curve $\mathcal{C}_{g,s}$. We can thus regard the index as a correlator of a TQFT defined on the UV curve [15],

$$\mathcal{I}^g[p, q, t; \mathbf{a}_i] = \langle \mathcal{O}(\mathbf{a}_1) \ldots \mathcal{O}(\mathbf{a}_s) \rangle_{\mathcal{C}_{g,s}}, \tag{4.1}$$

where we have formally introduced "local operators" $\mathcal{O}(\mathbf{a}_i)$ associated to the punctures. This is natural, because the index enjoys the kind of factorization property expected for a TQFT correlator. Given a pair-of-paints decomposition of $\mathcal{C}_{g,s}$ we may cut an internal cylinder and disconnect the surface into the two surfaces[7] $\mathcal{C}_{g_1,s_1+1}$ and $\mathcal{C}_{g_2,s_2+1}$, with $g_1 + g_2 = g$ and $s_1 + s_2 = s$. By applying the general gauging prescription (2.12), we have the "factorization" formula[8]

$$\mathcal{I}^g[\mathbf{a}_1, \ldots \mathbf{a}_s] = \int [d\mathbf{b}]_G \ \mathcal{I}^{g_1}[\mathbf{a}_j, \mathbf{b}] \, \mathcal{I}_V(\mathbf{b}) \, \mathcal{I}^{g_2}[\mathbf{b}, \mathbf{a}_k], \quad j \in S_1, \ k \in S_2, \tag{4.2}$$

where S_1 and S_2 are the set of indices labeling the punctures on the two components, with $S_1 \cup S_2 = \{1, \ldots s\}$. As the index is invariant under generalized S-dualities, one must obtain the same answer by applying the factorization formula in different channels. This is the essential property that must be satisfied by a $2d$ TQFT correlator.

To make the connection with the standard treatment of $2d$ TQFT more explicit, let us make a change of basis, from a continuous to a discrete set of operators. For simplicity we restrict to the case where all punctures are maximal, carrying the full flavor symmetry $\mathfrak{g}$. The operator $\mathcal{O}(\mathbf{a})$ is labelled by the flavor fugacity $\mathbf{a}$ dual to the Cartan subalgebra of $\mathfrak{g}$. Consider now a complete set of Weyl invariant functions $\{\psi_\alpha(\mathbf{a})\}$, where the label α runs over the finite-dimensional irreps of G, and define the discrete set of operators $\mathcal{O}_\alpha$ by the integral transform

---

[7]This is the generic situation. The remaining possibility is that cutting the cylinder yields the connected surface $\mathcal{C}_{g-1,s+2}$. This case can be treated analogously.

[8]We'll often omit the dependence on the superconformal fugacities to avoid cluttering.

$$\mathcal{O}_\alpha := \int [d\mathbf{a}]_G \, \mathcal{I}_V(\mathbf{a}) \, \psi_\alpha(\mathbf{a}) \mathcal{O}(\mathbf{a}). \tag{4.3}$$

It is convenient to choose the $\{\psi_\alpha(\mathbf{a})\}$ to be orthonormal under the propagator measure,

$$\int [d\mathbf{a}]_G \, \mathcal{I}_V(\mathbf{a}) \psi_\alpha(\mathbf{a}) \psi_\beta(\mathbf{a}) = \delta_{\alpha\beta}. \tag{4.4}$$

In this discrete basis, the factorization property reads simply

$$\mathcal{I}^g_{\alpha_1,\dots\alpha_s} = \mathcal{I}^{g_1}_{\{\alpha_j\}\beta} \, \mathcal{I}^{g_2}_{\beta\,\{\alpha_k\}}, \tag{4.5}$$

where the repeated index β is summed over. It is then clear that the general correlator on an surface of arbitrary topology can be obtained by successive contractions of the three-point correlator, i.e. the index of the three-punctured sphere, $\mathcal{I}^{g=0}_{\alpha_1\alpha_2\alpha_3} =: C_{\alpha_1\alpha_2\alpha_3}$. These "TQFT structure constants" $C_{\alpha_1\alpha_2\alpha_3}$ are symmetric functions of the three labels α_i and must satisfy the associativity constraint that follows from demanding that factorization of $\mathcal{I}^{g=0}_{\alpha_1\alpha_2\alpha_3\alpha_4}$ in two different ways must yield the same result,

$$C_{\alpha_1\alpha_2\beta} \, C_{\beta\alpha_3\alpha_4} = C_{\alpha_1\alpha_3\gamma} \, C_{\gamma\alpha_2\alpha_4}. \tag{4.6}$$

This condition is in fact sufficient to ensure independence of the general correlator on any specific choice of pair-of-pants decomposition. The structure that we have just described is very close to the standard axiomatic description of $2d$ TQFTs, but with the caveat that in the mathematical literature the state-space of the TQFT is usually taken to be finite-dimensional, whereas we have the infinite-dimensional space of finite-dimensional irreps of $\mathfrak{g}$.

It is a simple linear algebra fact that one may always[9] perform a further change of basis to a preferred discrete basis, in which associativity relations (4.6) become trivial (see Appendix A of [6] for an explicit example). This is the so-called Frobenius basis, which is still orthonormal under the propagator measure and is such that the structure constants have the diagonal structure

$$C_{\lambda_1\lambda_2\lambda_3} = C_\lambda \, \delta_{\lambda\lambda_1} \, \delta_{\lambda\lambda_2} \, \delta_{\lambda\lambda_3}. \tag{4.7}$$

In the Frobenius basis the non-vanishing components of the index associated to $\mathcal{C}_{g,s}$ take the very simple form

$$\mathcal{I}^g_{\lambda\dots\lambda} = C^{2g-2+s}_\lambda, \tag{4.8}$$

which just follows from the observation that $\mathcal{C}_{g,s}$ can be built by gluing $(2g-2+s)$ three-punctured sphere, and that the contractions of indices implementing the gluings are all trivial in this basis. Going back to the continuous fugacity basis,

---

[9]Here we should mention that since the state-space of the QFT obtained from the index is infinite dimensional there might be in principle issues of converges when changing basis. Such complication though do not actually arise in practice in the index computations.

$$\mathcal{I}^g[\mathbf{a}_1, \ldots \mathbf{a}_s] = \sum_\lambda C_\lambda^{2g-2+s} \, \psi_\lambda(\mathbf{a}_1) \ldots \psi_\lambda(\mathbf{a}_s). \tag{4.9}$$

In summary, the task of evaluating the general index is reduced to the task of finding the Frobenius basis $\{\psi_\lambda(\mathbf{a})\}$ and the structure constants C_λ.

4.3 Bootstrapping the Index

The structure just outlined is so constraining that it essentially fixes the index of class $\mathcal{S}$ theories, when supplemented with the extra physical input about the special cases that have a Lagrangian description [7].

We focus on A_{n-1} theories. Let us first aim to find the index for theories containing only maximal punctures. For $n > 2$, none of these theories have a Lagrangian description. Nevertheless, their index must obey compatibility conditions that follow by gluing in an extra three-punctured sphere of type $T_n^{[1^n][1^n][n-1,1]}$, which is identified with the free hypermultiplet theory in the bifundamental of $SU(n) \times SU(n)$. The physical input mentioned above is then

$$\mathcal{I}[T_n^{[1^n][1^n][n-1,1]}] = \mathcal{I}_H(a, \mathbf{x}, \mathbf{y}), \tag{4.10}$$

where the explicit expression of $\mathcal{I}_H$ is given in (2.11). Recall that a is the $U(1)$ fugacity associated with minimal puncture while $\mathbf{x}, \mathbf{y}$ the $SU(n)$ fugacities associated with the two maximal punctures.

Let the index of $T[\mathcal{C}]$ with all maximal punctures be some unknown function[10] $\mathcal{I}_C(\mathbf{x}_i)$, symmetric under permutations of the arguments $\mathbf{x}_i$, $i = 1, \ldots s$. We construct a larger theory with s maximal and one minimal puncture by gluing in a free hypermultiplet. The resulting index is given by

$$\mathcal{I}(a, \mathbf{x}_1, \mathbf{x}_2 \cdots \mathbf{x}_s) = \int [d\mathbf{z}] \, \mathcal{I}_V(\mathbf{z}) \, \mathcal{I}_H(a, \mathbf{x}_1, \mathbf{z}) \, \mathcal{I}_C(\mathbf{z}^{-1}, \mathbf{x}_2, \ldots, \mathbf{x}_s). \tag{4.11}$$

While in the above expression $\mathbf{x}_1$ appears to be treated asymmetrically from $\mathbf{x}_2, \ldots \mathbf{x}_s$, generalized S-duality (the TQFT structure of the index) demands that the integral be invariant under permutations of all the $\mathbf{x}_i$. Remarkably, this will be sufficient to determine the function $\mathcal{I}_C$. To reach this conclusion, we take an apparent detour and study the analytical properties of the integral as a function of the $U(1)$ fugacity a.

One can show that the integral has simple poles for

$$p^r q^s t^{\frac{n}{2}} a^{-n} = 1, \tag{4.12}$$

---

[10]The dependence on the superconformal fugacities (p, q, t) is again left implicit.

where r and s non-negative integers. To see this one notices that the poles in $\mathbf{z}$ in the integrand move around when one varies a. At the special values (4.12) pairs of poles pinch the integration contours and cause the whole integral to diverge. A toy example of this mathematical phenomenon is as follows. Consider ($|t\,a|,\ |t\,b|,\ |t\,c| < 1$)

$$\oint \frac{dz}{2\pi i z} \oint \frac{dy}{2\pi i y} \frac{1}{t\,a - y} \frac{1}{t\,b - z} \frac{1}{t^{-1}c^{-1} - z\,y} = \oint \frac{dz}{2\pi i z} \frac{1}{t\,b - z} \frac{1}{t^{-1}c^{-1} - z\,t\,a}$$
$$= \frac{t\,c}{1 - t^3\,a\,b\,c}.$$

We have a pole at $t^3 a\,b\,c = 1$. This can be viewed as the pole in y at $t\,a$ colliding with pole in y at $t^{-1}c^{-1}z^{-1}$ simultaneously with pole in z at $t\,b$ colliding with the pole at $t^{-1}c^{-1}y^{-1}$.

The residues of the poles (4.12) are easy to compute. This residue gets contributions in the $\mathbf{z}$ contour integrals only from the finite number of poles that pinch the integration contours. The simplest case is the residue at $t^{\frac{n}{2}}a^{-n} = 1$,

$$\mathcal{I}_V\,\mathrm{Res}_{t^{\frac{n}{2}}a^{-n}\to 1}\mathcal{I}(a, \mathbf{x}_1, \mathbf{x}_2, \ldots) = \mathcal{I}_C(\mathbf{x}_1, \mathbf{x}_2, \ldots), \qquad (4.13)$$

where $\mathcal{I}_V$ is the index of $U(1)$ $\mathcal{N} = 2$ vector multiplet. So picking up the residue at $a^2 t^{-1} = 1$ has the effect of "deleting" the extra $U(1)$ puncture. A slightly more involved calculation gives the residue at $q t^{\frac{n}{2}}a^{-n} = 1$,

$$\mathcal{I}_V\,\mathrm{Res}_{q t^{\frac{n}{2}}a^{-n}\to 1}\mathcal{I}(a, \mathbf{x}, \mathbf{y}, \ldots) = \qquad\qquad\qquad\qquad (4.14)$$
$$= \frac{\theta(t; p)}{\theta(q^{-1}; p)} \sum_{i=1}^{n} \prod_{j\neq i} \frac{\theta(\frac{t}{q}x_i/x_j; p)}{\theta(x_j/x_i; p)} \mathcal{I}_C(\{x_i \to q^{-\frac{1}{2}}x_i,\ x_{j\neq i} \to q^{\frac{1}{2}}x_j\}, \mathbf{y}, \ldots)$$
$$=: \frac{\theta(t; p)}{\theta(q^{-1}; p)} \mathfrak{S}_{(r=0, s=1)}(\mathbf{x})\,\mathcal{I}_C(\mathbf{x}, \mathbf{y}, \ldots).$$

We see that the residue is computed by the action on $\mathcal{I}_C$ of an interesting difference operator, which we have named $\mathfrak{S}_{(r=0, s=1)}(\mathbf{x})$, shifting the values of the fugacity $\mathbf{x}$. The residues can be easily computed for general values of r and s in (4.12), and are again given by acting on $\mathcal{I}_C$ with certain difference operators $\mathfrak{S}_{(r,s)}(\mathbf{x})$ which we will not write explicitly. The operators $\mathfrak{S}_{(r,s)}(\mathbf{x})$ all commute with each other and are self-adjoint under the propagator measure.

As we have already observed, there is nothing special about the puncture labelled by $\mathbf{x}$. What singled out $\mathbf{x}$ in the above calculation is the choice of a pair-of-pants decomposition where the punctured labelled by $\mathbf{x}$ belongs to the three-punctured sphere associated to the free hypermultiplet theory. A different pair-of-pants decomposition would single out a different puncture. By generalized S-duality, acting with $\mathfrak{S}_{(r,s)}$ on different punctures must give the same answer:

$$\mathfrak{S}_{(r,s)}(\mathbf{x}_k)\,\mathcal{I}_C(\mathbf{x}_1, \ldots, \mathbf{x}_s) = \mathfrak{S}_{(r,s)}(\mathbf{x}_\ell)\,\mathcal{I}_C(\mathbf{x}_1, \ldots, \mathbf{x}_s) \qquad (4.15)$$

Fig. 1 Two different pair of pants decompositions corresponding to two different S-duality frames of the field theory. In the two duality frames the minimal puncture labeled by a sits in a pair-of-pants with a different maximal puncture. The index computed in the two frames should give the same result

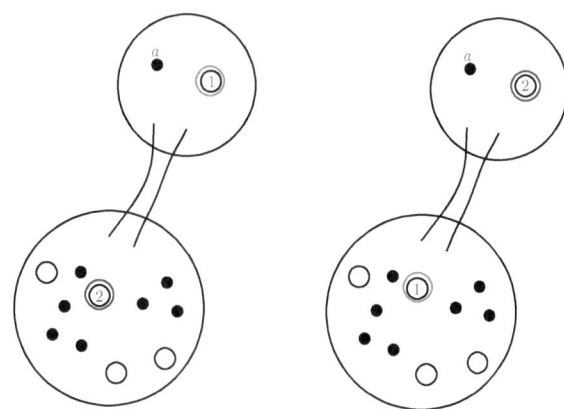

for any choice of $k, \ell = 1, \ldots s$. This is the basic relation that allows to fix the index (Fig. 1).

Consider a complete basis of simultaneous eigenfunctions of the difference operators,

$$\mathfrak{S}_{(r,s)}(\mathbf{x}) \, \psi_\lambda(\mathbf{x}) = \mathcal{E}_\lambda^{(r,s)} \, \psi_\lambda(\mathbf{x}). \tag{4.16}$$

If the eigenvalues are non-degenerate (as can indeed be checked to be case), these functions are automatically orthogonal under the propagator measure, and can be normalized to be orthonormal. The punchline is now simply stated: this is precisely the Frobenius basis introduced in the previous section for the TQFT of the index. Indeed, expanding the index associated to the three-punctured sphere as

$$\mathcal{I}(\mathbf{x}_1, \mathbf{x}_2, \mathbf{x}_3) = \sum_{\lambda_1, \lambda_2, \lambda_3} C_{\lambda_1 \lambda_2 \lambda_3} \, \psi_{\lambda_1}(\mathbf{x}_1) \, \psi_{\lambda_2}(\mathbf{x}_2) \, \psi_{\lambda_3}(\mathbf{x}_3), \tag{4.17}$$

we see from (4.15) and the assumption of non-degenerate eigenvalues that the structure constants can be non-vanishing only for $\lambda_1 = \lambda_2 = \lambda_3$.

The eigenfunctions ψ_λ are not known in closed analytic from for general values of the superconformal fugacities (q, p, t), but there are well-defined algorithms to find them as series expansions (see e.g. [16]). Moreover, as we will see in detail in the following section, closed analytic forms are available for special limits of the superconformal fugacities.

To complete the computation, it remains to determine the structure constants C_λ. First, expanding the index of the free hypermultiplet theory as

$$\mathcal{I}_H(a, \mathbf{x}, \mathbf{y}) = \sum_\lambda \phi_\lambda(a) \, \psi_\lambda(\mathbf{x}) \, \psi_\lambda(\mathbf{y}), \tag{4.18}$$

we define the functions $\phi_\lambda(a)$ associated to the minimal puncture. The functions ψ_λ are chosen to be orthonormal under the vector multiplet measure but functions

ϕ_λ do not have natural normalization properties at this level of the discussion and their normalization is defined by (4.18).[11] Second, we consider the theory associated to the sphere with two maximal and $n - 1$ minimal punctures.[12] This theory has two equivalent descriptions, depicted respectively in the top and bottom pictures in Fig. 2: (i) It can be obtained by gluing to the basic non-Lagrangian building block $T_n^{[1^n][1^n][1^n]}$ a *superconformal tail* [17], which is Lagrangian quiver SCFT with flavor symmetry $SU(n - 1) \times U(1)^{n-1}$. (ii) It can be obtained in a completely Lagrangian setup as a linear quiver. For the index this implies the following equality:

$$\sum_\lambda \psi_\lambda(\mathbf{x}) \psi_\lambda(\mathbf{y}) \prod_{i=1}^{n-1} \phi_\lambda(b_i) = \sum_\lambda C_\lambda \, \psi_\lambda(\mathbf{x}) \psi_\lambda(\mathbf{y}) \int [d\mathbf{z}] \, \Delta(\mathbf{z}; \{b_i\}) \, \psi_\lambda(\{\mathbf{z}, b\}),$$

(4.19)

where $\mathbf{z}$ is an $SU(n - 1)$ fugacity and an appropriate function of the b_i fixed by matching the $U(1)$ symmetries on the two sides. The function $\Delta(\mathbf{z}; \{b_i\})$ can be easily calculated from the superconformal tail. Since all quantities are known except the structure constants C_λ, this relation allows to fix them explicitly. This completes the derivation of the index of class S theories of type A, with maximal and minimal punctures.

To include punctures of general type Λ, we need more general superconformal tails. For each Λ, there exists a minimal integer $n(\Lambda)$ such that the theory associated to one maximal puncture, one puncture of type Λ and $n(\Lambda)$ minimal punctures can be described by a Lagrangian quiver gauge theory [17]. This can in fact be viewed as a *definition* of the puncture of type Λ. By equating the abstract definition of the index of such a theory, namely

$$\sum_\lambda \psi_\lambda(\mathbf{x}) \, \phi_\lambda^\Lambda(\mathbf{y}_\Lambda) \prod_{i=1}^{n(\Lambda)} \phi_\lambda(b_i),$$

(4.20)

with the explicit integral expression of the same index given by Lagrangian quiver description we can determine the factor $\phi_\lambda^\Lambda(\mathbf{y}_\Lambda)$ associated to the puncture of type Λ.

In summary, we have described an algorithm that determines the superconformal index for all theories of class S with regular punctures. The index takes an elegant general form in terms of structure constants $C_\lambda(p, q, t)$ and of "wavefunctions" $\{\phi_\lambda^{\Lambda_i}(\mathbf{y}_{\Lambda_i}; p, q, t)\}$ associated to the punctures,[13]

---

[11]The same will hold for functions ϕ_λ^Λ associated to general punctures we will define later in this section.

[12]We take $n > 2$ as the $n = 2$ case is trivial. For $n = 2$ there is no distinction between minimal and maximal punctures. The basic building block T_2 is identified with a free hypermultiplet in the trifundamental representation of $SU(2)^3$. The structure constants can then be obtained directly by expanding the free hypermultiplet index.

[13]Comparing with (4.9), we have reabsorbed some factors of C_λ into wavefunctions, by setting a new normalization for the wave function of the maximal puncture, $\phi_\lambda^{[1^n]} := C_\lambda \psi_\lambda$.

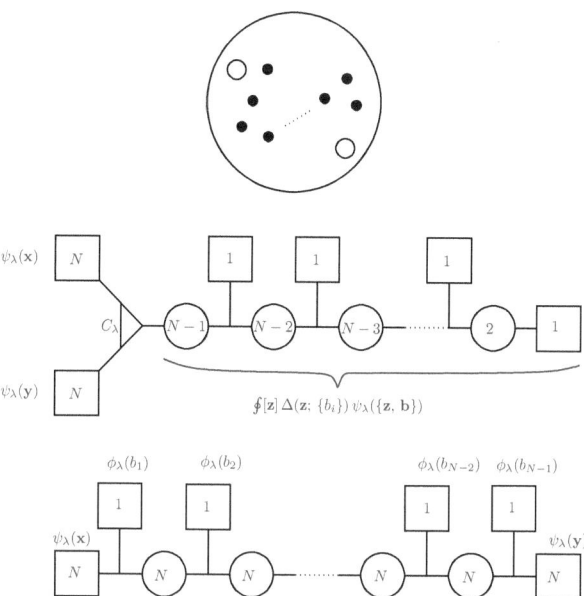

Fig. 2 One can determine the structure constants C_λ of the A_{n-1} theories by studying the theory associated to a sphere with two maximal and $n - 1$ minimal punctures (*top picture*). In one duality frame (*middle picture*) this is given by a T_n theory, involving C_λ, coupled to a "superconformal tail" quiver. In another duality frame (*bottom picture*) this is given by a linear quiver with an $SU(n)^{n-2}$ gauge group, where each $SU(n)$ is coupled to $2n$ hypermultiplets. For $n = 3$, the equivalence of the two frames is the celebrated Argyres-Seiberg duality, whose consequences for the index of T_3 ($\equiv$ the E_6 SCFT) have already been explored in Sect. 3

$$\mathcal{I} = \sum_\lambda C_\lambda^{2g-2} \prod_{j=1}^{s} \phi_\lambda^{\Lambda_i}(\mathbf{y}_{\Lambda_i}), \qquad (4.21)$$

where the sum is over the set of finite-dimensional irreps of $\mathfrak{g} = \mathfrak{su}(n)$.

A caveat is in order. Not every possible choice of Riemann surface decorated by a choice of $\{\Lambda_i\}$ at the punctures corresponds to a physical SCFT. An indication that a choice of decorated surface may be unphysical is if the sum in (4.21) diverges, which happens when the flavor symmetry is "too small". There are subtle borderline cases where the sum diverges, but the theory is perfectly physical—this can happen when the theory has additional "accidental" flavor symmetries not associated to punctures. An example of such a theory is the rank two E_6 SCFT. These cases have to be treated with more care [18].

We will discuss how to calculate explicit expressions for the wavefunctions and structure constants in the next section. In the rest of this section we offer two viewpoints that illuminate the structure of the result, the first related to Higgsing and the second to dimensional reduction.

4.4 Higgsing: Reduced Punctures and Surface Defects

The index is a meromorphic function of flavor and superconformal fugacities, with a rich structure of poles. A large class of these poles has a nice physical interpretation [7].

Consider a schematic version of the index,

$$\mathcal{I}(a, b) = \text{Tr}(-1)^F a^f t^R, \tag{4.22}$$

where f and R are two conserved charges. Let us assume that $\mathcal{I}$ has a pole in fugacity a,

$$\mathcal{I} = \frac{\widetilde{\mathcal{I}}(a, t)}{1 - a^{f_{\mathcal{O}}} t^{R_{\mathcal{O}}}}. \tag{4.23}$$

It is natural to associate the pole to a bosonic operator $\mathcal{O}$, with charges $f = f_{\mathcal{O}}$ and $R = R_{\mathcal{O}}$, such that an infinite tower of composites of the form $\mathcal{O}^n$ contribute to the index. In the simplest case, $\mathcal{O}$ is the generator of a ring spanned by $\{\mathcal{O}^n\}$, and the pole appears by resumming the geometric sum,

$$1 + a^{f_{\mathcal{O}}} t^{R_{\mathcal{O}}} + (a^{f_{\mathcal{O}}} t^{R_{\mathcal{O}}})^2 + \cdots \tag{4.24}$$

In more complicated cases, there can be several generators obeying non-trivial relations, which are encoded in the numerator of (4.23). The residue at $a^{f_{\mathcal{O}}} t^{R_{\mathcal{O}}} = 1$ is given by $\widetilde{\mathcal{I}}(t^{-R_{\mathcal{O}}/f_{\mathcal{O}}}, t)$, which can be interpreted as

$$\text{Tr}'(-1)^F t^{\bar{R}}, \quad \bar{R} := R - \frac{R_{\mathcal{O}}}{f_{\mathcal{O}}} f, \tag{4.25}$$

where the prime on the trace indicates that we are omitting the infinite set of states with $\bar{R} = 0$, which are of course the states responsible for the pole in the first place. The shifted charge $\bar{R}$ is the linear combination of charges preserved in a background where $\mathcal{O}$ has acquired a non-zero vacuum expectation value (vev). In a path integral representation of the index as the $\mathbb{S}^3 \times \mathbb{S}^1$ partition function, the divergence at $a^{f_{\mathcal{O}}} t^{R_{\mathcal{O}}} = 1$ arises from the integration over a bosonic zero mode, which heuristically we identify with $\langle \mathcal{O} \rangle$. Following this intuition, we expect the *residue* to be controlled by the behavior of theory "at infinity" in the moduli space parametrized by $\langle \mathcal{O} \rangle$, that is, by the properties of the IR theory reached at the endpoint of the the RG flow triggered by giving $\mathcal{O}$ a vev. We interpret $\widetilde{\mathcal{I}}$ as the index of this IR fixed point.

Reducing punctures
As a first application of these ideas, let us obtain more directly the index in the presence of punctures of general type, taking as starting point the index with maximal punctures. The idea is that the theory with a partially-closed puncture can be obtained from the theory with a full puncture by partially higgsing the full $\mathfrak{su}(n)$

flavor symmetry, and flowing to the IR.[14] The role of the operator $\mathcal{O}$ that featured the above general discussion is played by the *moment map* operator μ. The moment map is the superconformal primary of the supermultiplet that contains the flavor symmetry current, and thus transforms in the adjoint representation of $\mathfrak{su}(n)$.[15] Given an embedding $\Lambda : \mathfrak{su}(2) \to \mathfrak{su}(n)$, we choose the vev of μ to be

$$\langle \mu \rangle = \Lambda(t^-) \in \mathrm{adj}_{\mathfrak{su}(n)}, \tag{4.26}$$

where t^- is the lowest weight of $\mathfrak{su}(2)$. The flavor symmetry is broken down to the centralizer of Λ in $\mathfrak{su}(n)$, which we call $\mathfrak{g}_\Lambda$. We expect to find poles in the wavefunction $\phi_\lambda(\mathbf{a})$ in correspondence to each component of μ that receives a vev. Extracting the residues with respect to such poles should give the wave function $\phi_\lambda^\Lambda(\mathbf{x}_\Lambda)$ associated to the reduced puncture. More precisely, the symmetry breaking also generates Goldstone modes that give a decoupled free sector, and we should remove their contribution if we are interested in the interacting IR SCFT. Finally we should remember to redefine charges, following the general principle outlined in (4.25). In our case, the vev for μ breaks the $SU(2)_R$ symmetry, however a linear combination $\bar{R}$ of the original R Cartan generator and of flavor Cartan generators is preserved; we expect this symmetry to enhance in the IR to the full non-abelian $SU(2)_{\bar{R}}$ of the interacting fixed point.[16] All in all, we have the prescription

$$G^\Lambda(\mathbf{a}_\Lambda)\, \phi_\lambda^\Lambda(\mathbf{a}_\Lambda) = \mathrm{Res}_{\mathbf{a} \to \mathrm{fug}_\Lambda(\mathbf{a}_\Lambda, t)}\, \phi_\lambda(\mathbf{a}), \tag{4.27}$$

where the prefactor G^Λ, which is easily computable, accounts for the contribution to the index of the Goldstone bosons induced by the symmetry breaking. The fugacity replacement $\mathbf{a} \to \mathrm{fug}_\Lambda(\mathbf{a}_\Lambda, t)$ can be obtained with a little representation theory. Any representation $\mathfrak{R}$ of $\mathfrak{g} = \mathfrak{su}(n)$ decomposes as

$$\mathfrak{R} = \bigoplus_j \mathcal{R}_j^{(\mathfrak{R})} \otimes V_j, \tag{4.28}$$

where $\mathcal{R}_j^{(\mathfrak{R})}$ is some (generally reducible) representation of $\mathfrak{g}_\Lambda$ and V_j the spin j representation of $\mathfrak{su}(2)$. Then $\mathrm{fug}_\Lambda(\mathbf{a}_\Lambda, t)$ is the solution for $\mathbf{a}$ in the character decomposition equation,[17]

---

[14]The equivalence between the realization of general punctures by superconformal tails (as sketched in the previous subsection) and the higgsing procedure that we are about to implement is explained in Sect. 12.5 of [19].

[15]The moment map is also an $SU(2)_R$ triplet and $U(1)_r$ singlet. We consider the highest $SU(2)_R$ weight (which has $R = 1$), since it is the component that contributes to the index.

[16]It might be that the vev actually preserves the diagonal subgroup of the UV $su(2)$ R-symmetry and some $su(2)$ subgroup of the flavor symmetry. In such a case there is no need for the IR enhancement of the R-symmetry. We thank C. Beem, D. Gaiotto, and A. Neitzke for pointing this out to us.

[17]The solution is unique up to the action of the Weyl group.

$$\chi_f^g(\mathbf{a}) = \sum_j \chi_{\mathcal{R}_j^{(f)}}^{g_\Lambda}(\mathbf{a}_\Lambda) \chi_{V_j}^{su(2)}(t^{\frac{1}{2}}, t^{-\frac{1}{2}}). \tag{4.29}$$

One can check that (4.27) reproduces the wavefunctions obtained using superconformal tails by the method outlined in the previous subsection. Let us give a couple
of simple examples.

Taking $g = su(2)$ and $\Lambda : su(2) \to su(2)$ the principal embedding, which in this
case is just the identity map, the centralizer is of course trivial and (4.29) reads

$$a + a^{-1} = t^{\frac{1}{2}} + t^{-\frac{1}{2}}, \tag{4.30}$$

which has the two solutions $a = t^{\frac{1}{2}}, t^{-\frac{1}{2}}$, related by the action of the Weyl group
$a \leftrightarrow a^{-1}$. Since we are interested in the vev of the $su(2)$ lowest weight μ^- of the
moment map, whose contribution to the index is $a^{-2}t$, we should pick $a = t^{\frac{1}{2}}$; the
other solution $t^{-\frac{1}{2}}$ would be associated to the $su(2)$ highest weight μ^+. The lesson
(which generalizes) is that if we are interested in giving a vev to specific operator,
we should fix a representative of the Weyl orbit. Extracting the residue at $a^2 t^{-1} = 1$
will give the index of the IR theory at the end of the RG flow triggered by $\langle \mu^- \rangle$,
times the contribution from the free Goldstone bosons. In this case, the Goldstone
bosons consist of a free hypermultiplet in the fundamental of the flavor $su(2)$. Both
the flavor and R symmetry are broken by the vev, but the combination $\bar{R} = R + f/2$
is preserved. Under the new $SU(2)_{\bar{R}}$, the scalars of the free hypermultiplet transform
as $3 + 1$, with the singlet corresponding to the states responsible for the divergence.
Extracting the pole is precisely equivalent to omitting this singlet states. Setting
$a = t^{\frac{1}{2}}$ in (2.9) we see that under this new charge assignment the non-singlet states
of the free hypermultiplet give a contribution to the index exactly equal to the inverse
of the index of a free $U(1)$ vector multiplet, so the Goldstone boson factor in (4.27)
is $G = \mathcal{I}_V^{-1}$. All in all, we have derived from general principles the following
prescription to close an $su(2)$ puncture,

$$\mathcal{I}_V \operatorname{Res}_{a^{-2}t \to 1} \phi_\lambda^{[1^2]}(a) = \phi_\lambda^{[2]} \equiv 1. \tag{4.31}$$

In the last equality we have just reminded ourselves that the wavefunction of a fully
closed puncture is identically equal to one. One can check (4.31) using the expression
for $\phi_\lambda^{[1^2]}$ derived by the methods of the previous subsection.

A sightly more involved example is $g = su(3)$ and Λ the subregular embedding,
corresponding to the partition $[2, 1]$. The centralizer is $g_\Lambda = u(1)$. If a_1, a_2, a_3 with
$a_1 a_2 a_3 = 1$ are the $su(3)$ fugacities, and b the $u(1)$ fugacity, (4.29) takes the form

$$a_1 + a_2 + a_3 = b(t^{\frac{1}{2}} + t^{-\frac{1}{2}}) + b^{-2}. \tag{4.32}$$

The only solution (up to the action of the Weyl group, which permutes the a_i) is $a_1 = t^{\frac{1}{2}}b, a_2 = t^{-\frac{1}{2}}b, a_3 = b^{-2}$. Extracting the residue and removing the contribution of the Goldstone bosons accomplishes the reduction of the full puncture to the minimal puncture.

Surface defects

Next, we would like to interpret in a similar light the poles (4.12) that played such a crucial role in the previous subsection. Recall the basic setup: we "glued" the bifundamental hypermultiplet theory $T_n^{[1^n][1^n][n-1,1]}$ to a general theory $T[\mathcal{C}]$, connecting a maximal puncture of one theory with a maximal puncture of the other theory by gauging the diagonal $SU(n)$ symmetry. We then extracted residues with respect to the fugacity a for the $U(1)$ global symmetry of the hypermultiplet. This is the $U(1)$ baryon symmetry, under which the complex scalars q and $\tilde{q}$ have charge -1 and $+1$ respectively. It is then clear that the operator associated to the simplest pole, at $a^{-n}t^{\frac{n}{2}} = 1$, is the baryon operator $B = \det q$. Giving a vev to B higgses the $SU(n)$ gauge group, triggering an RG flow whose IR endpoint is the original theory $T[\mathcal{C}]$ and a collection of decoupled free fields [7]. This explains (4.13).[18]

By the same logic, the poles at $p^r q^s t^{\frac{n}{2}} a^{-n} = 1$ are naturally associated to holomorphic derivatives of the baryon operator in the 12 and 34 planes, $\partial_{12}^r \partial_{34}^s \det q$. We expect the residue at these poles to describe the IR physics of the flow triggered by a spacetime-dependent vev of the form $\langle B \rangle \sim z^r w^s$. Consider first the $r = 0, s \neq 0$ case. Away from the $w = 0$ plane, the endpoint of the flow is still $T[\mathcal{C}]$. However some extra degrees of freedom survive at $w = 0$, which we interpret as a *surface defect* for $T[\mathcal{C}]$ extended in the 12 plane. Similarly, the endpoint of the flow with $r \neq 0, s = 0$ is $T[\mathcal{C}]$ decorated with an extra surface defect extended in the 34 plane. In the general case with $rs \neq 0$ both type of defects will be present. In the $\mathbb{S}^3 \times \mathbb{S}^1$ geometry, these surface defects fill the "temporal" $\mathbb{S}^1$ and the two maximal circles inside the $\mathbb{S}^3$ fixed by the j_{12} and j_{34} rotations, respectively. This proposal has been checked [20] in a set of examples where $T[\mathcal{C}]$ admits a Lagrangian description, and surface defects can be added by coupling the $4d$ SCFT to a $(2, 2)$ sigma model; the index can then be independently evaluated by localization techniques, confirming the prescription that we have just outlined.

In summary, we have found a physical interpretation for the difference operators $\mathfrak{S}_{(r,s)}$: their action on the index of $T[\mathcal{C}]$ yields the index of the same theory decorated by some extra surface defect [7]. Since the difference operators act "locally" on the generalized quiver, we should associate them to special punctures of the UV curve. This agrees with the M-theory picture, where the surface defects correspond to M2 branes localized on the UV curve. Acting with a difference operator on a given flavor fugacity corresponds pictorially to colliding the special puncture with a flavor puncture. The location of the special punctures on the UV curve is immaterial, so collision of the same special puncture with different flavor punctures is bound to give the same result—which is a restatement of (4.15).

This description of surface defects bears a striking kinship with the analogous picture that arises in the AGT correspondence [21, 22]. The introduction of surface defects in the $\mathbb{S}^4$ partition function is accomplished by the insertion of special,

---

[18]For $n = 2$, the $U(1)$ baryon symmetry enhances to $SU(2)$, $B \equiv \mu^-$ (the lowest weight component of the moment map), and (4.13) is precisely equivalent to (4.31).

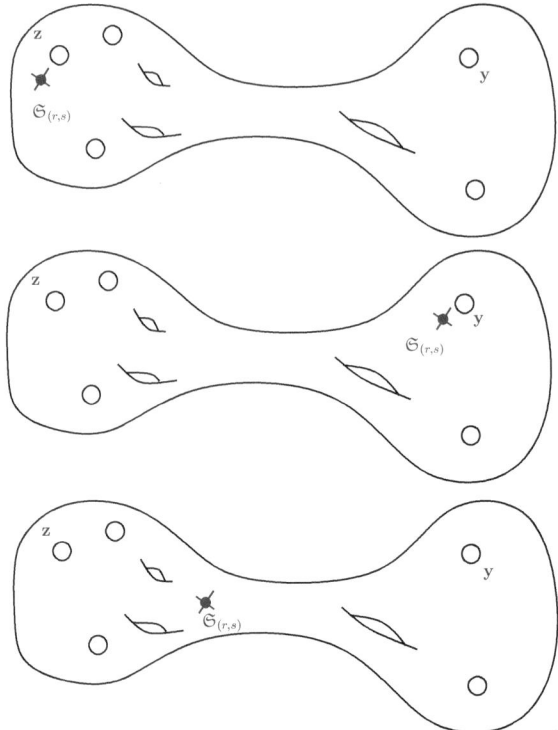

Fig. 3 The difference operators $\mathfrak{S}_{(r,s)}$, which compute residues and introduce surface defects, can be visualized as special punctures on the UV curve. The action of $\mathfrak{S}_{(r,s)}$ on a flavor fugacity is interpreted as the collision of the special puncture with a flavor puncture. We can act on different punctures and obtain the same result for the index (*top* and *middle pictures*). We can also define the action of $\mathfrak{S}_{(r,s)}$ on a long tube (*bottom picture*), by cutting open a cylinder, acting on one of the open punctures and gluing the surface back. S-duality guarantees that this is a well-defined procedure. In this way we can introduce the special punctures $\mathfrak{S}_{(r,s)}$ on a UV curve with no flavor punctures at all

semi-degenerate operators in the Toda CFT correlator defined on the UV curve. These operators are the key to the solution of Liouville theory by the conformal bootstrap [23]: considering their fusion with normalizable vertex operators one can derive functional equations that admit a unique solution. Similarly, we have special punctures in our 2d TQFT that insert surface defects in the $\mathbb{S}^3 \times \mathbb{S}^1$ partition function. Their fusion with ordinary flavor punctures leads to the topological bootstrap equations (4.31), which uniquely fix the superconformal index (Fig. 3).

4.5 Reduction to 3d

The index of theories of class $\mathcal{S}$ has a very definite structure (4.21). This structure is natural since it is a manifestation of the 2d TQFT nature of the index of the

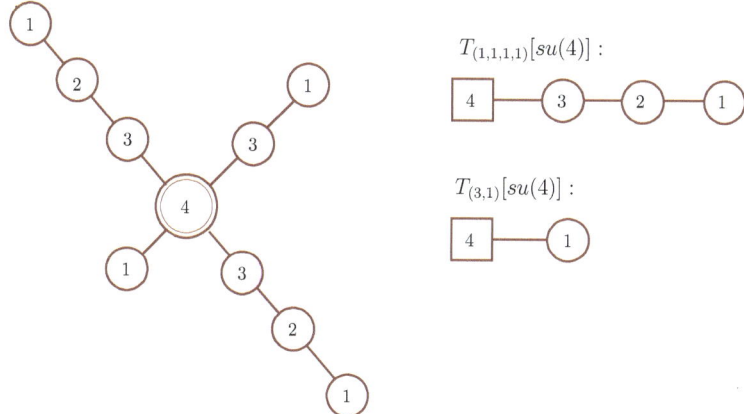

$T_{(1,1,1,1)}[su(4)]$:

$T_{(3,1)}[su(4)]$:

Fig. 4 On the *left* we have an example of a star-shaped quiver mirror of the A_3 theory corresponding to a sphere with four punctures, two of which are maximal and one is minimal. On the *right* the quiver theories for $\mathcal{T}_{(1,1,1,1)}[su(4)]$ corresponding to the maximal puncture and $\mathcal{T}_{(3,1)}[su(4)]$ corresponding to minimal puncture are depicted

theories at hand as was anticipated in Sect. 4.2. It is however an important question to understand better the physical meaning of the different ingredients entering (4.21). For example, we would like to gain more insight into the physical significance of the eigenfunctions $\phi_\lambda^{\Lambda_i}(\mathbf{y}_{\Lambda_i})$ and the eigenvalues $\mathcal{E}_\lambda^{(r,s)}$. Let us consider here a very informative 3d interpretation of (4.21).[19]

We can consider theories of class $\mathcal{S}$ on $\mathcal{M}_3 \times \mathbb{S}^1$ with $\mathcal{M}_3$ some three dimensional manifold. Upon reduction on the $\mathbb{S}^1$ we obtain a 3d theory on $\mathcal{M}_3$. The $\mathcal{N} = 2$ class $\mathcal{S}$ theories admitting a known description in terms of a Lagrangian upon dimensional reduction on $\mathbb{S}^1$ are described in terms of the same field content and same gauge and superpotential interaction as the 4d parent theory. The 3d Lagrangians however are not conformal and the theories flow in general to an interacting $\mathcal{N} = 4$ 3d SCFT in the IR. The 4d conformal S-dualities imply IR (Seiberg-like) dualities of the 3d models. Thus the complex moduli of the Riemann surface defining the model in 4d do not translate to physical parameters in 3d: the topology of the surface and the information at the punctures alone are sufficient to completely specify the 3d model. An extremely interesting fact about the class $\mathcal{S}$ theories in 3d is that they possess yet another dual description. All theories of class $\mathcal{S}$ reduced to 3d, with and without known Lagrangian description in 4d, have a mirror description in 3d in terms of a *star-shaped* quiver theory [25].

This mirror symmetry states that a theory corresponding to a Riemann surface with genus g and s punctures of types Λ_i is dual to a quiver theory coupling s linear quivers $\mathcal{T}_{\Lambda_i}[g]$ [26] associated to Lie algebra $g = su(N)$ by gauging the common g with an addition of g $\mathcal{N} = 4$ adjoint hypermultiplet, see Fig. 4 for an example.

---

[19]A 6d physical interpretation of this equation can be also entertained [24] but we will not discuss it in this review.

The dimensional reduction on $\mathbb{S}^1$ can be performed at the level of the index. Here $\mathcal{M}_3$ is $\mathbb{S}^3$ and upon reduction of the index on $\mathbb{S}^1$ we obtain the partition function of the dimensionally reduced theory on a squashed sphere $\mathbb{S}_b^3$ [27–29] (see also [30]). The reduction is done by first parametrizing the fugacities as

$$t = e^{2\pi i r_1(\gamma + \frac{i}{2r_3}(b+b^{-1}))}, \quad z = e^{2\pi i r_1 \sigma}, \quad p = e^{2\pi b r_1/r_3}, \quad q = e^{2\pi b^{-1} r_1/r_3}, \quad (4.33)$$

where r_1 is the radius of $\mathbb{S}^1$ and r_3 is the radius of $\mathbb{S}^3$. Then the radius of $\mathbb{S}^1$, r_1, is sent to zero. The parameter b is the squashing parameter of the sphere.

We have defined the functions $\phi_\lambda^{\Lambda_i}(\mathbf{y}_{\Lambda_i})$ as eigenfunctions of difference operators $\mathfrak{S}_{(r,s)}$ and argued that this operators have a physical interpretation of introducing linked surface defects to the index computation. The surface defects corresponding to $\mathfrak{S}_{(0,s)}$ and $\mathfrak{S}_{(r,0)}$ span the $\mathbb{S}^1$ and one of the two equators of $\mathbb{S}^3$. Upon reduction on the $\mathbb{S}^1$ these become line defects sitting on one of the two equators of $\mathbb{S}_b^3$. When sending $r_1 \to 0$ the difference operators have very simple limit. For example in the A_1 case we have[20]

$$\mathfrak{S}_{(0,1)} \cdot f(z) \quad \to \quad \mathcal{T}(\sigma) \cdot \hat{f}(\sigma) = \quad\quad\quad\quad (4.34)$$

$$= \frac{\sinh \pi b \left(\frac{i(b-b^{-1})}{2} - \gamma + 2\sigma \right)}{\sinh 2\pi b \sigma} \hat{f}(\sigma + \frac{ib}{2}) + \frac{\sinh \pi b \left(\frac{i(b-b^{-1})}{2} - \gamma - 2\sigma \right)}{\sinh -2\pi b \sigma} \hat{f}(\sigma - \frac{ib}{2}),$$

where $\hat{f}(\sigma) = \lim_{r_1 \to 0} f(e^{2\pi i r_1 \sigma})$. Interestingly a set of eigenfunctions of this operator is given by the $\mathbb{S}_b^3$ partition functions of the $\mathfrak{T}[\mathfrak{su}(2)]$ theory. The $\mathfrak{T}[\mathfrak{su}(N)]$ theory has global $\mathfrak{su}(N)_H \times \mathfrak{su}(N)_C$ symmetry with the $\mathfrak{su}(N)_H$ acting on the Higgs branch and $\mathfrak{su}(N)_C$ acting on the Coulomb branch. Turning on real mass parameters, σ_H and σ_C, for the two symmetries the $\mathbb{S}_b^3$ partition function of $\mathfrak{T}[\mathfrak{su}(N)]$ can be denoted by $\phi^{(\gamma,b)}(\sigma_H|\sigma_C)$ and we have the property

$$\mathcal{T}(\sigma_C) \cdot \phi^{(\gamma,b)}(\sigma_H|\sigma_C) = \mathcal{W}(\sigma_H) \, \phi^{(\gamma,b)}(\sigma_H|\sigma_C). \quad\quad (4.35)$$

The eigenvalue $\mathcal{W}(\sigma_H)$ is the expectation value of the Wilson loop for the $\mathfrak{su}(N)_H$ global symmetry. This eigenvalue property of the partition function thus suggests the physical interpretation that the line defect for the gauge symmetry of $\mathfrak{T}[\mathfrak{su}(N)]$ is equivalent to a Wilson line for the global $\mathfrak{su}(N)_H$ symmetry. This fact is not surprising since the $\mathfrak{T}[\mathfrak{su}(N)]$ theories make their appearance as models living on S-duality domain wall separating two S-dual $\mathcal{N} = 4$ $SU(N)$ SYM theories. Since under 4d S-duality defect ('t Hooft) line operators map to Wilson operators our 3d eigenvalue statement is natural.

Further, the $\mathbb{S}_b^3$ partition function of a star shaped quiver mirror dual say to the A_1 theory with genus g and s punctures has the following form,

---

[20]This operator is called the Macdonald operator in math literature and we will shortly encounter a different incarnation of it in 4d index context.

$$Z_{g,s}(\{\sigma_C^{(i)}\}_{i=1}^s) = \int d\sigma_H Z_V(\sigma_H) \ Z_{\mathcal{H}}(\sigma_H)^g \prod_{i=1}^s \phi^{(\gamma,b)}(\sigma_H|\sigma_C^{(i)}). \qquad (4.36)$$

Here $Z_{\mathcal{H}}(\sigma_H)$ is the contribution of an $\mathcal{N} = 4 \ 3d$ adjoint hypermultiplet and $Z_V(\sigma_H)$ is the contribution of the vector. Note the striking structural similarity between (4.21) and (4.36). This is not a coincidence [31]. One can argue that indeed in the $r_1 \to 0$ limit the eigenfunctions $\phi_\lambda^\Lambda(\mathbf{y}_\Lambda)$ reduce to the $\mathbb{S}_b^3$ partition functions of $\mathfrak{T}_\Lambda[\mathfrak{su}(N)]$. The discrete labels of the eigenfunctions, λ_i, become (linear combinations of) the real masses of the symmetry rotating the Coulomb branch of $\mathfrak{T}_\Lambda[\mathfrak{su}(N)]$, $\sigma_C^{(i)}$: roughly, taking the $r_1 \to 0$ limit we should also concentrate on large representations and keep $r_1 \lambda_i$ fixed.

Let us summarize the $3d$ interpretation of the eigenfunctions,

- The difference operators introduce line defects.
- The eigenfunctions are $\mathbb{S}_b^3$ partition functions of $\mathfrak{T}_\Lambda[\mathfrak{su}(N)]$.
- The eigenvalues are expectation values of Wilson loops.
- The existence of the eigenvalue equation follows from $4d$ S-duality through the statement that Wilson and 't Hooft lines are S-dual to each other.[21]

In particular the fact that the index of theories of class $\mathcal{S}$ in $4d$ can be written in the form (4.21) is a $4d$ manifestation of the fact that the dimensionally reduced theories admit a mirror description. *That is the index written as* (4.21) *is a 4d precursor of the 3d mirror symmetry.* The interested reader might consult [33] for more thorough discussion of these issues.

Finally let us also mention that the $3d$ eigenfunctions, $\mathbb{S}_b^3$ partition functions of $\mathfrak{T}_\Lambda[\mathfrak{su}(N)]$, provide a connection between the $4d$ index and the $4d$ $\mathbb{S}^4$ partition functions of theories of class $\mathcal{S}$. As we mentioned $\mathfrak{T}_\Lambda[\mathfrak{su}(N)]$ models are obtained by considering $\mathcal{N} = 4 \ 4d$ theories with a duality domain wall. The kernel which implements the insertion of such duality wall in the $\mathbb{S}^4$ partition function computation is precisely the $\mathbb{S}_b^3$ partition function of $\mathfrak{T}_\Lambda[\mathfrak{su}(N)]$ [34]. In particular the difference operator we obtained by reduction to $3d$ are the same difference operators introducing line defects into Liouville-Toda/$\mathbb{S}^4$ (AGT correspondence [21]) computations [7] (see also [35]).

5 Integrable Models and Limits of the Index

The discussion of the previous section reduces the physical problem of determining the superconformal index of class $\mathcal{S}$ theories to the mathematical problem of finding a complete set of orthonormal eigenfunctions of the difference operators $\mathfrak{S}_{(r,s)}(\mathbf{x})$. Remarkably, these operators are closely related to the Hamiltonians that define a well-known class of integrable models, the elliptic relativistic Ruijsenaars-Schneider (RS) models, *aka* relativistic elliptic Calogero-Moser-Sutherland models.

---

[21] When writing this equation as a difference operator annihilating the partition function, it gives rise actually to the difference operator annihilating holomorphic blocks of the $3d$ partition function [32].

The operator (4.14), $\mathfrak{S}_{(0,1)}(\mathbf{x})$, is related to the basic RS Hamiltonian $\mathcal{H}_1(t, q; p)$ by a similarity transformation,

$$\mathcal{H}_1(t, q; p) = \frac{\theta(q^{-1}; p)}{\theta(t; p)} \frac{1}{\prod_{i \neq j} \Gamma(t z_i/z_j; p, q)} \mathfrak{S}_{(0,1)}(\mathbf{z}) \prod_{i \neq j} \Gamma(t z_i/z_j; p, q). \tag{5.1}$$

Under the same similarity transformation, the propagator measure in the A_{n-1} case becomes

$$\frac{1}{n!} \oint \prod_{i=1}^{n-1} \frac{dz_i}{2\pi i z_i} \prod_{i \neq j} \frac{\Gamma(t z_i/z_j; p, q)}{\Gamma(z_i/z_j; p, q)} \cdots . \tag{5.2}$$

Higher operators, $\mathcal{H}_\ell$, can be constructed as polynomials in $\mathfrak{S}_{(0,s)}$. One can think of the $n - 1$ independent $\mathcal{H}_\ell$ operators as associated to antisymmetric representations of $SU(n)$, whereas $\mathfrak{S}_{(0,s)}$ are associated to symmetric representations. Then by exploiting group theory and the fact that the fundamental representation can be trivially thought as either symmetric or antisymmetric one can translate between $\mathcal{H}_\ell$ and $\mathfrak{S}_{(0,s)}$ (see for example [35]).

The parameters p, q, and t appear in the Hamiltonian $\mathcal{H}_1(t, q; p)$ on different footing: (i) the parameter t plays a role of coupling constant, (ii) q is the shift parameter of the difference operator and can be understood as an exponent of the "speed of light" parameter of the relativistic integrable system, (iii) the integrable model is associated to an elliptic curve parametrized by p. Given an eigenfunction of $\mathcal{H}_1$ dressing it with an arbitrary elliptic function in q a huge class of new eigenfunctions can be obtained. This arbitrariness is lifted by the demand that the eigenfunction we are after diagonalize both operators $\mathfrak{S}_{(0,s)}$ and $\mathfrak{S}_{(s,0)}$ and in particular are symmetric with respect to exchanging p and q.

The RS models have a long history of rich connections with gauge theories in various dimensions (see e.g. [36, 37]). Nevertheless, for general values of (p, q, t) determining the exact eigenfunctions and eigenvalues of the difference operators is still an open problem. For some natural limits of the parameters the eigenfunctions are well known. Curiously, many of the same limits have independent physical interest, because they lead to a supersymmetry enhancement of the $\mathbb{S}^3 \times \mathbb{S}^1$ partition function. One can systematically classify the limits of the index that enjoy enhanced supersymmetry, and relate them to integrable models. We will shortly review some of the salient results in this direction. Physical properties of theories of class $\mathcal{S}$ impose additional constraints on $\phi_\lambda(\mathbf{z})$. For example, since some of the theories have known Lagrangian description the indices can be explicitly computed as integrals of elliptic Gamma functions and the results have to match the expressions evaluated using the eigenfunctions. Exploiting the known expressions for the eigenfunctions for specialized values of the parameters and the additional physical constraints one can set up a perturbative scheme around the known results to compute the eigenfunctions for general values of the parameters [7, 16].

We now turn to discuss several useful limits of the index for which explicit expressions for eigenfunctions are known.

Schur index

The trace formula (2.2) that defines the general index can be written in the following equivalent form (we suppress flavor fugacities to avoid cluttering):

$$\mathcal{I}(q, p, t) = \mathrm{Tr}(-1)^F \, p^{\frac{1}{2}\delta^1_-} \, q^{\frac{1}{2}\delta^1_+} \, t^{R+r} \, e^{-\beta'\delta_{2\dot{-}}}, \tag{5.3}$$

where

$$2\delta^1_+ := \{\mathcal{Q}^1_+, (\mathcal{Q}^1_+)^\dagger\} = E + 2j_1 - 2R - r \geqslant 0 \tag{5.4}$$

$$2\delta^1_- := \{\mathcal{Q}^1_-, (\mathcal{Q}^1_-)^\dagger\} = E - 2j_1 - 2R - r \geqslant 0$$

$$2\delta_{2\dot{-}} := \{\tilde{\mathcal{Q}}_{2\dot{-}}, (\tilde{\mathcal{Q}}_{2\dot{-}})^\dagger\} = E - 2j_2 - 2R + r \geqslant 0.$$

The inequalities follow from unitarity of the representation and will be useful momentarily. The equivalence of (2.2) and (5.3) follows immediately by recalling that only states with $\delta_{2\dot{-}} = 0$ contribute to the trace. The Schur index is the "unrefined" index obtained by setting $q = t$. . One readily observes that on this slice the combination of conserved charges appearing in the trace formula commute with a second supercharge, $\mathcal{Q}^1_-$, in addition to the supercharge $\tilde{\mathcal{Q}}_{2\dot{-}}$ that leaves invariant the general index. As the p dependence is $\mathcal{Q}^1_-$-exact, it drops out, and we are left with a simple expression that depends on q alone,[22]

$$\mathcal{I}_{\mathrm{Schur}} := \mathrm{Tr}(-1)^F \, q^{E-R}. \tag{5.5}$$

The index counts operators with $\delta^1_- = \delta_{2\dot{-}} = 0$, or equivalently

$$\widehat{L}_0 := \frac{E - (j_1 + j_2)}{2} - R = 0, \qquad \mathcal{Z} := j_1 - j_2 + r = 0. \tag{5.6}$$

In fact, the unitarity inequalities in (5.4) give $\widehat{L}_0 \geqslant \frac{|\mathcal{Z}|}{2}$, so the first condition implies the second. We refer to operators obeying $\widehat{L}_0 = 0$ as *Schur operators*. A Schur operator is annihilated by two Poincaré supercharges of *opposite* chiralities ($\mathcal{Q}^1_-$ and $\tilde{\mathcal{Q}}_{2\dot{-}}$ in our conventions). This is a consistent condition because the supercharges have the same $SU(2)_R$ weight, and thus anticommute with each other. No analogous BPS condition exists in an $\mathcal{N} = 1$ supersymmetric theory, because the anticommutator of opposite-chirality supercharges necessarily yields a momentum operator, which annihilates only the identity.

---

[22]In principle the Schur index might make sense also for *non-conformal* $\mathcal{N} = 2$ theories quantized on $\mathbb{S}^3 \times \mathbb{R}$, although we are not aware of a detailed analysis of the requisite deformations needed to define an $\mathcal{N} = 2$ theory on such a curved background (the analysis of [38] might be of help here). The $\mathcal{N} = 1$ analysis of [4] is not sufficient, because the Schur index cannot be understood as a special case of the $\mathcal{N} = 1$ index. Of course, in the non-conformal case one cannot relate $\mathbb{S}^3 \times \mathbb{R}$ to $\mathbb{R}^4$ by a Weyl rescaling and there is no state/operator map.

Table 3 This table summarizes the manner in which Schur operators fit into short multiplets of the $\mathcal{N}=2$ superconformal algebra

Multiplet	$\mathcal{O}_{\text{Schur}}$	$h := \frac{E+(j_1+j_2)}{2}$	r	Lagrangian "letters"
$\hat{\mathcal{B}}_R$	$\Psi^{11...1}$	R	0	$Q, \tilde{Q}$
$\mathcal{D}_{R(0,j_2)}$	$\tilde{Q}^1_+ \Psi^{11...1}_{+...+}$	$R+j_2+1$	$j_2+\frac{1}{2}$	$Q, \tilde{Q}, \tilde{\lambda}^1_+$
$\bar{\mathcal{D}}_{R(j_1,0)}$	$Q^1_+ \Psi^{11...1}_{+...+}$	$R+j_1+1$	$-j_1-\frac{1}{2}$	$Q, \tilde{Q}, \lambda^1_+$
$\hat{\mathcal{C}}_{R(j_1,j_2)}$	$Q^1_+ \tilde{Q}^1_+ \Psi^{11...1}_{+...+...+}$	$R+j_1+j_2+2$	j_2-j_1	$D^n_{++} Q, D^n_{++} \tilde{Q},$ $D^n_{++}\lambda^1_+, D^n_{++}\tilde{\lambda}^1_+$

We use the naming conventions for supermultiplets of Dolan and Osborn [39]. For each supermultiplet, we denote by Ψ the superconformal primary. There is then a single conformal primary Schur operator $\mathcal{O}_{\text{Schur}}$, which in general is obtained by the action of some Poincaré supercharges on Ψ. We list the holomorphic dimension h and $U(1)_r$ charge r of $\mathcal{O}_{\text{Schur}}$ in terms of the quantum numbers (R, j_1, j_2) that label the shortened multiplet (left-most column). We also indicate the schematic form that $\mathcal{O}_{\text{Schur}}$ can take in a Lagrangian theory by enumerating the elementary "letters" from which the operator may be built. We denote by Q and $\tilde{Q}$ the complex scalar fields of a hypermultiplet, by $\lambda^\mathcal{I}_\alpha$ and $\tilde{\lambda}^\mathcal{I}_{\dot{\alpha}}$ the left- and right-moving fermions of a vector multiplet, and by $D_{\alpha\dot{\alpha}}$ the gauge-covariant derivatives. Note that while in a Lagrangian theory Schur operators are built from these letters, the converse is false—not *all* gauge-invariant words of this kind are Schur operators, only the special combinations with vanishing anomalous dimensions

A summary of the different classes of Schur operators, organized according to how they fit in shortened multiplets of the superconformal algebra, is given in Table 3 [5]. The first line lists the half-BPS operators belonging to the Higgs branch $\mathcal{N}=1$ chiral ring, which have $E=2R$ and $j_1=j_2=0$. In a Lagrangian theory, these are operators of the schematic form $QQ\ldots\tilde{Q}\tilde{Q}$. The $SU(2)_R$ highest weight component of the moment map operator μ^{11}, which has $E=2R=2$ (and transforms in the adjoint representation of the flavor group) is in this class. The second and third lines of the table list more general $\mathcal{N}=1$ antichiral (respectively chiral) operators. In a Lagrangian theory they may be obtained by considering gauge-invariant words that contain $\tilde{\lambda}^1_+$ (respectively λ^1_+) in addition to Q and $\tilde{Q}$. Finally the forth line lists the most general class of Schur operators, belonging to supermultiplet obeying less familiar semishortening conditions. An important operator in this class is the Noether current for the $SU(2)_R$ R-symmetry, which belongs to the same superconformal multiplet as the stress-energy tensor and is universally present in any $\mathcal{N}=2$ SCF. Its J^{11}_{++} component, with $E=3$, $R=1$, $j_1=j_2=\frac{1}{2}$, is a Schur operator. Finally, note that the half-BPS operators of the *Coulomb* branch chiral ring (of the form $\text{Tr}\,\phi^k$ in a Lagrangian theory) are *not* Schur operators.

The Schur index earns its name from the fact that the wavefunctions are proportional to Schur polynomials, and simple closed form expressions are available for all the ingredients that enter the TQFT formula for the index (4.21). We will quote the full expressions below in the more general Macdonald limit. The structure constants $C_\lambda(q)$ turn out to be inversely proportional to the *quantum dimension* of the representation λ. One recognizes [40] the TQFT of the index as the zero-area limit[23] of q-deformed $2d$ Yang-Mills theory [43], which can also be understood as an analytic continuation of Chern-Simons theory on $\mathcal{C} \times \mathbb{S}^1$. This observation has been reproduced by a top-down approach [44, 45], starting from the $(2, 0)$ theory on the geometry $\mathbb{S}^3 \times \mathbb{S}^1 \times \mathcal{C}$, first reducing on $\mathbb{S}^1$ to obtain $5d$ YM, and then reducing further on $\mathbb{S}^3$ and using supersymmetric localization to obtain a bosonic gauge theory on $\mathcal{C}$, which is argued to coincide with q-YM.

In q-YM theory, introducing flavor punctures correspond to fixing the holonomies of the gauge fields around the punctures. One can also define additional local operators by fixing the *dual* variables at the punctures [46]—in the language of Chern-Simons theory on $\mathcal{C} \times \mathbb{S}^1$, this corresponds to adding a Wilson loop along the temporal $\mathbb{S}^1$. These operators are the natural candidates to correspond to the surface defects discussed in the previous section [7, 47].

Perhaps the most interesting fact about the Schur index is that it can be viewed as the character of a $2d$ chiral algebra canonically associated to the $4d$ SCFT [5], as we shall review in Sect. 7. A related point is that the Schur index enjoys intriguing modular properties encoding conformal anomalies [48]. For example the indices of a hypermultiplet and vector multiplets in the Schur limit become combinations of theta functions,

$$\mathcal{I}_H = \frac{1}{\theta(q^{\frac{1}{2}} a; q)}, \qquad \Delta_{Haar}(\mathbf{z}) \, \mathcal{I}_V(\mathbf{z}) = \frac{1}{n!} (q; q)^{2n-2} \prod_{i \neq j} \theta(q z_i / z_j; q), \qquad (5.7)$$

which have simple modular properties under

$$q = e^{2\pi i \tau} \quad \rightarrow \quad q' = e^{-\frac{2\pi i}{\tau}}, \qquad z = e^{2\pi i \zeta} \quad \rightarrow \quad z' = e^{\frac{2\pi i \zeta}{\tau}}. \qquad (5.8)$$

Here Δ_{Haar} is the Haar measure and we specialized for concreteness to $SU(n)$ vector field. An index of the gauge theory is given by contour integrals with the integrand built from products of theta functions. The combination of theta functions in in the integrand, $\mathcal{I}_{integ.}(\mathbf{z}; q)$, always forms an elliptic function in the fugacities, $\mathbf{z}$ corresponding to the gauged symmetries,

$$\mathcal{I}_{integ.}(q \, \mathbf{z}; q) = \mathcal{I}_{integ.}(\mathbf{z}; q). \qquad (5.9)$$

The gauge fugacities $\mathbf{z}$ can be thus thought as taking values on a torus with modular parameter τ. The contour integral defining the Schur index of a gauge theory then can be thought of as an integral over a cycle of the torus while the index after modular transformation is given as an integral over the dual cycle. These properties beg the question of the relation of the Schur index to *mock modular forms*, a relation which is yet to be explored (Fig. 5).

---

[23]On a surface of finite (non-zero) area, q-YM is not topological, but it still admits a natural class $\mathcal{S}$ interpretation [41] as the supersymmetric partition function of the $(2, 0)$ theory on $\mathbb{S}^3 \times \mathbb{S}^1 \times \mathcal{C}$ where the UV curve $\mathcal{C}$ is kept of finite area [42].

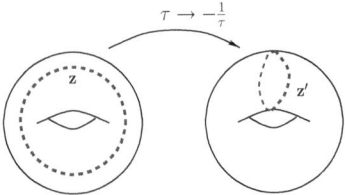

Fig. 5 The Schur index of a gauge theory is given by an integral over fugacities **z** taking value in a torus with modular parameter τ. After modular transformation, $\tau \to -\frac{1}{\tau}$, the index is written as an integral over the dual cycle

Macdonald limit

Taking $p \to 0$ in (5.3) is a well-defined limit, thanks to positive-definiteness of the associated charge δ^1_-. The trace formula reads

$$\mathcal{I}_{\mathrm{Mac}}(q,t) := \mathrm{Tr}_{\mathrm{M}}(-1)^F q^{E-2R-r} t^{R+r} = \mathrm{Tr}_{\mathrm{M}}(-1)^F q^{2j_1} t^{R+r}, \tag{5.10}$$

where the subscript in the trace indicates that we are restricting by hand to the states with $\delta^1_- = 0$. Clearly, we are concentrating on the operators that are also annihilated by the supercharge $\mathcal{Q}^1_-$, in addition to $\widetilde{\mathcal{Q}}_{2\dot{-}}$. These are of course the same as the Schur operators, but we are now refining their counting by keeping track of the quantum number $R + r$. For $q = t$, we recover the Schur index.

This limit is mathematically very interesting. Our difference operators and our integration measure become identical (up to conjugation) to the well-studied Macdonald difference operators and Macdonald measure [49]. The diagonalization problem is completely solved in terms of Macdonald polynomials, a beautiful two-parameter generalization of the Schur polynomials. In the Macdonald limit we set the elliptic curve of the Ruijsenaars-Schneider model, p, to zero the integrable model becomes thus trigonometric (but still relativistic). For example, in the A_1 case after conjugation (5.1) the basic hamiltonian becomes,[24]

$$\mathcal{H}_1 \cdot f(z) \sim \frac{1-tz^2}{1-z^2} f(q^{\frac{1}{2}}z) + \frac{1-tz^{-2}}{1-z^{-2}} f(q^{-\frac{1}{2}}z). \tag{5.11}$$

We are then able to find closed form expressions for the general wavefunctions and for the structure constants [6]. The wavefunction for a general choice of puncture (embedding) and representation now takes the following form,

$$\psi^\Lambda_{\mathfrak{R}}(\mathbf{z}_\Lambda; q,t) = K_\Lambda(\mathbf{z}_\Lambda; q,t) \, P^{\mathfrak{g}}_{\mathfrak{R}}(\mathrm{fug}_\Lambda(\mathbf{z}_\Lambda;t); q,t). \tag{5.12}$$

Here $P^{\mathfrak{g}}_{\mathfrak{R}}(\mathbf{z}; q,t)$ are the Macdonald polynomials labeled by finite dimensional representations $\mathfrak{R}$ of Lie algebra $\mathfrak{g}$ and orthonormal under the Macdonald measure, which,

---

[24] Note that this is the same operator that we obtained in a quite different context of the reduction of the elliptic difference operator $\mathfrak{S}_{(0,1)}$ to three dimensions 4.34.

e.g., for $\mathfrak{g} = \mathfrak{su}(n)$ is given by,

$$\Delta_{q,t}(\mathbf{z}) = \frac{1}{n!} \prod_{i \neq j} \frac{(z_i/z_j; q)}{(t z_i/z_j; q)}. \tag{5.13}$$

The K-factors admit a compact expression as a plethystic exponential [50],

$$K_\Lambda(\mathbf{z}_\Lambda; q, t) = \text{P.E.} \left[\sum_j \frac{t^{j+1}}{1-q} \chi^{\mathfrak{h}_\Lambda}_{\mathcal{R}_j^{(\text{adj})}}(\mathbf{z}_\Lambda) \right], \tag{5.14}$$

where the summation is over the terms appearing in the decomposition of Eq. (4.28) applied to the adjoint representation,

$$\text{adj}_\mathfrak{g} = \bigoplus_j \mathcal{R}_j^{(\text{adj})} \otimes V_j. \tag{5.15}$$

$\chi^{\mathfrak{h}_\Lambda}_{\mathcal{R}_j^{(\text{adj})}}(\mathbf{z})$ is the Schur polynomial of Lie algebra $\mathfrak{h}_\Lambda$ corresponding to representation $\mathcal{R}_j^{(\text{adj})}$. For the maximal puncture, corresponding to the trivial embedding $\Lambda_{\text{max}} \equiv 0$, the wavefunction reads,

$$\psi_{\mathfrak{R}}^{\Lambda_{\text{max}}}(\mathbf{x}; q, t) = K_{\text{max}}(\mathbf{x}; q, t) \, P_{\mathfrak{R}}^{\mathfrak{g}}(\mathbf{x}; q, t), \quad K_{\text{max}}(\mathbf{x}; q, t) := \text{P.E.} \left[\frac{t}{1-q} \chi^{\mathfrak{g}}_{\text{adj}}(\mathbf{x}) \right]. \tag{5.16}$$

At the other extreme, for the principal embedding $\Lambda = \rho$, the decomposition of Eq. (5.15) reads

$$\text{adj}_\mathfrak{g} = \bigoplus_{i=1}^{\text{rank } \mathfrak{g}} V_{d_i - 1}, \tag{5.17}$$

where $\{d_i\}$ are the degrees of invariants of $\mathfrak{g}$, so in particular $d_i = i + 1$ for $\mathfrak{su}(n)$. We then find

$$\psi_{\mathfrak{R}}^{\rho}(q, t) = \text{P.E.} \left[\sum_i^{\text{rank } \mathfrak{g}} \frac{t^{d_i}}{1-q} \right] P_{\mathfrak{R}}^{\mathfrak{g}}(\text{fug}_\rho(t)). \tag{5.18}$$

For $\mathfrak{g} = \mathfrak{su}(n)$, the fugacity assignment associated to the principal embedding takes a particularly simple form,

$$\text{fug}_\rho(t) = (t^{\frac{n-1}{2}}, t^{\frac{n-3}{2}}, \ldots t^{-\frac{n-1}{2}}). \tag{5.19}$$

Provided that,

$$C_{\mathfrak{R}}(q)^{-1} = \psi_{\mathfrak{R}}^{\rho}(q, t), \tag{5.20}$$

we thus obtain an expression for the Macdonald index of any class $\mathcal{S}$ theory with regular punctures,

$$\mathcal{I}_{\text{Mac}}(q,t;\mathbf{x}) = \sum_{\mathfrak{R}} C_{\mathfrak{R}}(q,t)^{2g-2+s} \prod_{i=1}^{s} \psi_{\mathfrak{R}}^{\Lambda_i}(\mathbf{x}_{\Lambda_i};q,t), \qquad (5.21)$$

with all the ingredients explicitly given above.

In the Macdonald limit, the TQFT of the index is recognized as a certain deformation of q-YM, closely related to the *refined* Chern-Simons theory on $\mathcal{C} \times \mathbb{S}^1$ discussed in [51]; the refinement amounts to changing the measure in the path integral of q-YM from Haar to Macdonald.

Hall-Littlewood limit

Proceeding one step further, we can take the $q \to 0$ limit in the Macdonald index. The trace formula reads

$$\mathcal{I}_{\text{HL}}(t) := \text{Tr}_{\text{HL}}(-1)^F t^{R+r}, \qquad (5.22)$$

where we are restricting the trace to states with $\delta_+^1 = \delta_-^1 = 0$. In the $q \to 0$ limit, Macdonald polynomials reduce to the much more manageable Hall-Littlewood (HL) polynomials. The HL index of theories of class $\mathcal{S}$ takes a relatively simple form: it is always a rational function of t.

The HL index receives contributions from operators annihilated by the three supercharges $\mathcal{Q}_{+}^1$, $\mathcal{Q}_{\dot{+}}^1$ and $\mathcal{Q}_{2\dot{-}}$. This is precisely the subset of Schur operators with $j_1 = 0$, corresponding to the $\hat{B}$ and D multiplets, listed in the first two rows of Table 3. Since such *Hall-Littlewood operators* are killed by both spinorial components of $\mathcal{Q}_{\alpha}^1$, they are chiral[25] with respect to an $\mathcal{N} = 1$ subalgebra, and thus form a ring, which is consistent truncation of the full $\mathcal{N} = 1$ chiral ring. In a Lagrangian theory, they are composite operators made with the complex hypermultiplet scalars Q and $\tilde{Q}$ and the λ_+^1 component of the gaugino, but no derivatives. There is a further consistent truncation of the ring to operators with $j_2 = 0$: this is the Higgs branch chiral ring, spanned by the bottom component of the $\hat{B}_R$ multiplets.

For an $\mathcal{N} = 2$ SCFTs associated to a linear quiver, one can show that only the $\hat{B}_R$ multiplets contribute to the HL index. This is the case because the gauginos are in one-to-one correspondence with the F-term constraints on the Higgs branch chiral operators, so their contribution to the index (which comes with a minus sign) is precisely such to enforce those constraints. It follows that for linear quivers the HL index coincides [6] with the Hilbert series of the Higgs branch (see e.g. [52, 53]). The equivalence between the HL index and the Higgs branch Hilbert series appears also to hold for the T_n building blocks (see [6]), and so by the same reasoning it extends to all class $\mathcal{S}$ theories associated to curves of of *genus zero*. One can then use the HL index to compute the Hilbert series of multi-instanton moduli spaces for E_n groups [18, 54], which are quite intricate to compute using other methods (see

---

[25] To be pedantic, *anti*chiral.

e.g. [55]). The HL index and the Higgs Hilbert series are not the same for theories with genus one or higher, where $\mathcal{D}$ multiplets play a role.[26]

Coulomb limit

There is another limit of the index that leads to supersymmetry enhancement: one takes t, $p \to 0$ while keeping q and $\frac{pq}{t}$ fixed. It is called the Coulomb limit because in a Lagrangian theory the hypermultiplet single-particle index (2.9) goes to zero; the only supermultiplets that contribute in this limit are the short multiplets of type $\bar{\mathcal{E}}_{-\ell(0,0)}$ (in the notations of [39]), whose lowest components are the operators of the Coulomb branch chiral ring, of the form $\mathrm{Tr}\,\phi^k$. That there should exist a limit of the general index for which only $\{\bar{\mathcal{E}}_{-\ell(0,0)}\}$ contribute is *a priori* clear from the fact that these multiplets do not appear in any of the recombination rules, so their multiplicities define an index.

In a Lagrangian theory with simple gauge group G, the Coulomb index is given by [6]

$$\mathcal{I}_C = \oint [d\mathbf{z}]_G \, \Delta_{q,\frac{pq}{t}}(\mathbf{z}) = \mathrm{PE}\left[\sum_{\ell \in \exp(G)} \tilde{\mathcal{I}}_{\ell+1}\right], \tag{5.23}$$

where $\exp(G)$ stands for the set of exponents of G, $\Delta_{q,\frac{pq}{t}}(\mathbf{z})$ is the Macdonald measure (5.13) (which arises by taking the Coulomb limit of the usual propagator measure), and $\tilde{\mathcal{I}}_{\ell+1}$ is the index of an individual $\bar{\mathcal{E}}_{-\ell(0,0)}$ multiplet. This is a well-known mathematical equality, going by the name of the the Macdonald central term identity. It can be understood physically as the statement that the Coulomb chiral ring is freely generated by a set of operators in one-to-one correspondence with the Casimir invariants of G, for example $\{\mathrm{Tr}\,\phi^k\}$, $k = 2, \ldots n$ for $G = SU(n)$.[27]

6 Some Generalizations

The discussion in previous sections can be extended and generalized in several ways. We will discuss some of the open problems in Sect. 8, while here let us briefly mention some of the work that has already appeared in the literature.

- In this review we have concentrated on class $\mathcal{S}$ theories of type A. A similar analysis can be performed for theories of type D and E. Following our TQFT intuition the indices should be expressible in terms of a complete set of functions.

---

[26] Assuming that the Higgs branch of the $4d$ theory of class $\mathcal{S}$ is isomorphic to the Higgs branch of the dimensionally reduced theory, we can consider the Coulomb index [33, 56, 57] of the mirror dual theory (see Sect. 4.5). The $3d$ Coulomb index of the mirror coincides with the Hilbert series of the Higgs branch of theories of class $\mathcal{S}$ for any genus. We refer the reader to [33] for further discussion of this issue.

[27] The fact that the Coulomb branch is freely generated is known to be true by inspection for theories of class $\mathcal{S}$ of type A we discuss here, but is not obvious for theories of type D and E: it would be interesting to clarify this issue. We thank Y. Tachikawa for this comment.

The integrable models we discussed here for which the relevant set of functions for the A case is a set of eigenfunctions have natural generalizations to the D and E cases. In particular the eigenfunctions for D and E cases are known in the Macdonald limit. These eigenfunctions have been used to compute indices for the three-punctured spheres D type class S theories [58, 59] and for the E type class S theories [60]. One can also consider indices with outer-automorphism twists around the temporal $\mathbb{S}^1$ as was done in [50].

- Performing a different twist of the $6d$ putting Riemann surface can result in a $4d$ theory with $\mathcal{N} = 1$ supersymmetry rather than $\mathcal{N} = 2$ [61]. The resulting $\mathcal{N} = 1$ theories are closely related to the $\mathcal{N} = 2$ class theories and in particular their indices can be exactly computed resulting in expressions which are very similar to the ones discussed here [62, 63]. The $\mathcal{N} = 1$ theories can be also built using outer-automorphism twists and the corresponding indices can be computed as was done in [64].

- In the process of determining the index we have found it useful to consider indices of theories with surface defects. The theories of interest admit a variety of other supersymmetric defects in presence of which the index can be computed. For example, one can compute the Schur index in presence of supersymmetric line operator wrapping the $\mathbb{S}^1$ [65, 66]. Here the answers are easily obtained in case of Wilson lines but in case of 't Hooft lines the computation is much more involved [66] if one chooses to perform the computation without making use of S-duality. Other examples of extended objects involve domain walls [67] and more general surface defects than discussed here [35, 47].[28]

- Finally let us mention that the dualities satisfied by the theories of class S imply highly non-trivial identities satisfied by the superconformal indices. These identities take usually the form of equalities between different integrals of elliptic Gamma functions and or (infinite) sums of orthogonal functions. To give an example let us write down the index of the $SU(N)$ $\mathcal{N} = 2$ SYM with $2N$ flavors. This theory corresponds to a sphere with two maximal and two minimal punctures and its index is proportional to [11],

$$\oint \prod_{\ell=1}^{N-1} \frac{dz_\ell}{2\pi i z_\ell} \prod_{i \neq j} \frac{\Gamma(\frac{pq}{t} z_i/z_j; p, q)}{\Gamma(z_i/z_j; p, q)} \prod_{i=1}^{N} \prod_{\alpha,\beta=1}^{2N} \Gamma(t^{\frac{1}{2}} (z_i y_\alpha a)^{\pm 1}; p, q) \Gamma(t^{\frac{1}{2}} (z_i^{-1} x_\beta b)^{\pm 1}; p, q).$$

$$(6.1)$$

Here a and b are fugacities for the $U(1)$ symmetries associated with the minimal punctures and $\mathbf{x}$ with $\mathbf{y}$ are fugacities associated with the maximal punctures. The S-duality exchanging the two minimal punctures implies that the above integral is invariant under exchange of a and b. Mathematically this property is not at all obvious and was proven for the $SU(2)$ case in [69]. As far as we know no mathematical proof for higher rank cases exists as of this moment. Another simple

---

[28]The index of theories of class S in presence of codimension two defects of the 6d theory wrapping the Riemann surface [68] has not been analyzed yet.

example of an unproven identity following from S-duality propertied of the index is the equality of the indices of $SO(2n+1)$ and $SP(n)$ $\mathcal{N} = 4$ theories [15].

7 Chiral Algebras and the Schur Index

In this section, we give a brief outline of the structure discovered in [5]. The basic claim is that any $\mathcal{N} = 2$ SCFT admits a closed sector of operators and observables, isomorphic to a two-dimensional chiral algebra. The Schur index is recognized as the character of this chiral algebra,

$$\text{Tr}_{2d} (-1)^F q^{L_0} \equiv \mathcal{I}_{\text{Schur}}(q). \tag{7.1}$$

To understand this surprising claim, we start with the following seemingly innocent observation. The states that contribute to the Schur index can be equivalently characterized as belonging to the cohomology of a single nilpotent supercharge, a linear combination of Poincaré and conformal supercharges,

$$\mathbb{Q} := \mathcal{Q}^1_- + \tilde{\mathcal{S}}^{\dot{-}2}. \tag{7.2}$$

Indeed,

$$\{\mathbb{Q}, \mathbb{Q}^\dagger\} = 2\widehat{L}_0 = E - (j_1 + j_2) - 2R, \tag{7.3}$$

so the harmonic cohomology representatives obey the Schur condition (5.6). By the state/operator map, states are as always in correspondence with local operators inserted at the origin. So Schur operators $\mathcal{O}_{\text{Schur}}(0)$ inserted the origin belong to the cohomology of $\mathbb{Q}$.

What is the cohomology of $\mathbb{Q}$ more generally? One easily shows that $\mathcal{Z}$ defined in (5.6) is $\mathbb{Q}$-exact, so a local operator can be $\mathbb{Q}$-closed only if it lies on the plane fixed by $j_1 - j_2$, which we call the *chiral algebra plane*. We use the complex coordinate z (and its conjugate $\bar{z}$) to parametrize the chiral algebra plane. The global conformal algebra on the chiral algebra plane is the standard $\mathfrak{sl}(2) \times \bar{\mathfrak{sl}}(2)$, with generators L_n and $\bar{L}_n$, for $n = -1, 0, 1$, and is of course a subalgebra of the four-dimensional conformal algebra. For example,

$$L_0 = \frac{E + j_1 + j_2}{2}, \quad \bar{L}_0 = \frac{E - (j_1 + j_2)}{2}. \tag{7.4}$$

It turns out that

$$[\mathbb{Q}, L_n] = 0 \quad \text{but} \quad [\mathbb{Q}, \bar{L}_n] \neq 0, \tag{7.5}$$

so a Schur operator $\mathcal{O}_{\text{Schur}}(z, \bar{z})$ inserted away from the origin is *not* $\mathbb{Q}$-closed. There is however a simple fix. We introduce a twisted algebra $\widehat{\mathfrak{sl}(2)}$ as the diagonal subalgebra of $\bar{\mathfrak{sl}}(2) \times \mathfrak{su}(2)_R$,

$$\widehat{L}_{-1} := \bar{L}_{-1} + \mathcal{R}^-, \qquad \widehat{L}_0 := \bar{L}_0 - \mathcal{R}, \qquad \widehat{L}_{+1} := \bar{L}_{+1} - \mathcal{R}^+. \qquad (7.6)$$

(In retrospect, this explains why the combination of charges in the first equation of (5.6) was denoted by $\widehat{L}_0$). Remarkably, the twisted generators $\widehat{L}_n$ are $\mathbb{Q}$-exact. It follows that starting from a Schur operator inserted at the origin, we can act with twisted translations to obtain a $\mathbb{Q}$-closed operator defined at a generic point $(z, \bar{z})$ on the chiral algebra plane,

$$\mathcal{O}(z, \bar{z}) = e^{zL_{-1} + \bar{z}\widehat{L}_{-1}} \mathcal{O}_{\text{Schur}}(0, 0) e^{-zL_{-1} - \bar{z}\widehat{L}_{-1}}. \qquad (7.7)$$

A Schur operator is necessarily an $\mathfrak{su}(2)_R$ highest weight state, carrying the maximum eigenvalue R of the Cartan. Indeed, if this were not the case, states with greater values of R would have negative $\widehat{L}_0$ eigenvalue, violating unitarity. We denote the whole spin k representation of $\mathfrak{su}(2)_R$ as $\mathcal{O}^{(\mathcal{I}_1 \cdots \mathcal{I}_{2k})}$, with $\mathcal{I}_i = 1, 2$. Then the Schur operator is $\mathcal{O}_{\text{Schur}} = \mathcal{O}^{11 \cdots 1}(0)$, and the twisted-translated operator at any other point is given by

$$\mathcal{O}(z, \bar{z}) := u_{\mathcal{I}_1}(\bar{z}) \cdots u_{\mathcal{I}_{2k}}(\bar{z}) \, \mathcal{O}^{(\mathcal{I}_1 \cdots \mathcal{I}_{2k})}(z, \bar{z}), \qquad u_{\mathcal{I}}(\bar{z}) := (1, \bar{z}). \qquad (7.8)$$

By construction, such an operator is annihilated by $\mathbb{Q}$, and $\mathbb{Q}$-exactness of $\widehat{L}_{-1}$ implies that its $\bar{z}$ dependence is $\mathbb{Q}$-exact. It follows that the cohomology class of the twisted-translated operator defines a purely meromorphic operator,

$$[\mathcal{O}(z, \bar{z})]_{\mathbb{Q}} \quad \rightsquigarrow \quad \mathcal{O}(z). \qquad (7.9)$$

Operators constructed in this manner have correlation functions that are meromorphic functions of the insertion points, and enjoy well-defined meromorphic OPEs at the level of the cohomology. These are precisely the ingredients that define a two-dimensional chiral algebra! The relation (7.1) of the chiral algebra character with the Schur index follows at once by observing that $\widehat{L}_0 = 0$ implies $L_0 = E - R$, so the trace formula (5.5) that defines the Schur index is reproduced.

There is a rich dictionary related properties of the $4d$ SCFT with properties of its associated chiral algebra. Let us briefly mention some universal features of this correspondence:

• The global $\mathfrak{sl}(2)$ symmetry is enhanced to the full Virasoro symmetry, with the $2d$ holomorphic stress tensor $T(z)$ arising from the Schur operator in the $SU(2)_R$ conserved current, $T(z) := [\mathcal{J}_R(z, \bar{z})]_{\mathbb{Q}}$. The $2d$ central charge is given by

$$c_{2d} = -12 c_{4d}, \qquad (7.10)$$

where c_{4d} is one of conformal anomaly coefficients of the $4d$ theory (the one associated to the Weyl tensor squared).

• The global flavor symmetry of the SCFT is enhanced to an affine symmetry in the associated chiral algebra, with the affine current $J(z)$ arising from the moment map operator, $J(z) := [M(z, \bar{z})]_{\mathbb{Q}}$. The $2d$ level is related to the $4d$ level by another universal relation,

$$k_{2d} = -\frac{k_{4d}}{2}. \tag{7.11}$$

- The generators of the HL chiral ring give rise to generators of the chiral algebra. Remarkably, the geometry of the $4d$ Higgs branch is encoded algebraically in vacuum module of the chiral algebra: Higgs branch relations correspond to null states.

Free SCFTs are associated to free chiral algebras. The free hypermultiplet corresponds to the chiral algebra of symplectic bosons $(q, \tilde{q})$, of weights $(\frac{1}{2}, \frac{1}{2})$, while the free vector multiplet corresponds to a (b, c) ghost system of weights $(1, 0)$.

There is also a chiral algebra counterpart of the index gauging prescription (2.12). We start with a SCFT $\mathcal{T}$, whose chiral algebra $\chi[\mathcal{T}]$ is known, and define a new SCFT $\mathcal{T}_G$ by gauging a subgroup of the flavor symmetry, such that the gauge coupling is exactly marginal. A naive guess for finding the chiral algebra associated $\mathcal{T}_G$ is to take the tensor product of $\chi[\mathcal{T}]$ with a $(b^A c_A)$ ghost system in the adjoint representation of G, and restrict to gauge singlets. This would be the direct analog of (2.12), and is indeed the correct answer at zero gauge coupling. But at finite coupling, some of the Schur states are lifted and the chiral algebra must be smaller. There is an elegant prescription to find the quantum chiral algebra: one is instructed to pass to the cohomology of

$$Q_{\text{BRST}} := \oint \frac{dz}{2\pi i} j_{\text{BRST}}(z), \quad j_{\text{BRST}} := c_A \left[J^A - \frac{1}{2} f^{AB}{}_C c_B b^C \right], \tag{7.12}$$

where J^A is the G affine current of $\chi[\mathcal{T}]$. This BRST operator is nilpotent precisely when the $\beta_G = 0$, which amounts to $k_{2d} = -2h^\vee$, where $h^\vee$ is the dual Coxeter number of G. By this prescription, we can in principle find $\chi[\mathcal{T}]$ for any Lagrangian SCFT $\mathcal{T}$.

The chiral algebra contains much more information that the Schur index. The state space of the chiral algebra can be regarded as a "categorification" of the Schur index: it consists of the cohomology classes of $\mathbb{Q}$, whereas the index only counts such cohomology classes with signs, and so it knows about sets of short multiplets that are kinematically allowed to recombine but do not. In addition, there may be multiplets that cannot recombine but nonetheless make accidentally cancelling contributions to the index, and these are also seen in the categorification. And of course, the chiral algebra structure goes well beyond categorification—it is a rich algebraic system that also encodes the OPE coefficients of the Schur operators, and is subject to non-trivial associativity constraints.

For theories of class $\mathcal{S}$, there is a generalized topological quantum field theory that associates to a decorated Riemann surface the corresponding chiral algebra. Associativity of the gluing of Riemann surfaces imposes highly non-trivial requirements on the chiral algebras of the elementary building blocks T_n. Finally, let us mention that the task of reducing the rank of a puncture can be accomplished directly in the chiral algebra setting, by a generalization of quantum Drinfeld-Sokolov reduction. We refer to [5, 70] for a detailed discussion of this very rich structure.

8 Some Open Questions

We conclude by discussing some open problems and possible generalizations of the topics discussed in this review. The main focus of this review was the partition function on $\mathbb{S}^3 \times \mathbb{S}^1$ for $\mathcal{N} = 2$ superconformal theories in four dimensions: generalizations and extensions of our logic can be entertained by relaxing each of these qualifiers.

- **More partition functions**
  A rather natural generalization is to consider indices with the theory quantized on more generic manifolds, i.e. $\mathcal{M}_3 \times \mathbb{S}^1$. For example, one can take $\mathcal{M}_3 = \mathbb{S}^3/\mathbb{Z}_r$, the lens space [71]. The superconformal index discussed in this paper is a special case of the partition functions defined using this sequence of manifolds, $r = 1$. The lens space with $r > 1$ has a non-contractable cycle and thus is sensitive to non-local objects in the theory. In particular, unlike the superconformal index it can distinguish theories differing by choices of allowed line operators and/or by choices of the global structure of the gauge groups. One would expect that as long as the manifold on which the partition function is computed has an $\mathbb{S}^1$ the arguments of this paper can be reiterated. In particular the partition functions in these cases should be computable by a $2d$ TQFT. This has been discussed in the case of the lens space [72, 73], and it would be interesting to extend the analysis to other partition functions, e.g. $\mathbb{T}^2 \times \mathbb{S}^2$ [74, 75].

- **More theories**
  The superconformal index is not yet fixed for *all* $\mathcal{N} = 2$ theories in $4d$. For example, we do not know at the moment how to compute the index depending on the most general set of fugacities for Argyres-Douglas theories and theories corresponding to Riemann surfaces with irregular singularities [76].[29] It would be very interesting to fill this gap in our current understanding. To do so it might be useful to exploit the chiral algebra associated to these theories and its relation to the (Schur) index. Another, related, question is what kind of partition functions can be exactly computed for $\mathcal{N} = 2$ theories which are not superconformal.

- **Properties of the index**
  The indices which we can compute have many interesting properties not all of which were sufficiently well studied. For example, the $4d$ indices have factorization properties [78, 79] similar to the ones studied for the partition functions of $3d$ theories [32, 80, 81]. Another example is that of modular properties the indices have under non linear transformations of some of the chemical potentials [48, 82] (see also [30, 83–85]).

---

[29]See however [77] for some recent discussion.

- **Less supersymmetry and/or other space-time dimensions**
  A very important open question is whether the methodology which allowed us to fix the index of a large class of $\mathcal{N} = 2$ theories can be applied to theories with less supersymmetry and/or theories in different space-time dimensions: this remains to be seen.

- **Relations to mathematics**
  The superconformal index is directly related to different branches of exciting mathematics. To list just couple examples: it is a gold mine for extracting identities satisfied by elliptic hypergeometric integrals; and it is closely related to quantum mechanical integrable systems with their very rich mathematical structure. There is a real chance here for a mutually beneficial dialogue between the mathematics community working on these topics and the physics community.

Acknowledgments It is a great pleasure to thank Chris Beem, Abhijit Gadde, Davide Gaiotto, Madalena Lemos, Pedro Liendo, Wolfger Peelaers, Elli Pomoni, Brian Willett, and Wenbin Yan, for very enjoyable collaboration and countless discussions on the material reviewed here. We thank Davide Gaiotto and Yuji Tachikawa for useful comments on the draft. LR thanks the Simons Foundation and the Solomon Guggenheim Foundation for their generous support. He is grateful to the IAS, Princeton, and to the KITP, Santa Barbara, for their wonderful hospitality during his sabbatical leave. LR is also supported in part by the National Science Foundation under Grant No. NSF PHY1316617. SSR gratefully acknowledges support from the Martin A. Chooljian and Helen Chooljian membership at the Institute for Advanced Study. The research of SSR was also partially supported by National Science Foundation under Grant No. PHY-0969448, and by "Research in Theoretical High Energy Physics" grant DOE-SC00010008. SSR would like to thank KITP, Santa Barbara, and the Simons Center, Stony Brook, for hospitality and support during different stages of this work.

Appendix 1: Plethystics

In this appendix we collect the definitions of some special functions and combinatorial objects used in the bulk of the review. The Pochammer symbol is defined as

$$(z; \, p) := \prod_{\ell=0}^{\infty}(1 - z \, p^{\ell}). \tag{8.1}$$

The theta-function is given by

$$\theta(z; \, p) := (z; \, p) \, (p \, z^{-1}; \, p). \tag{8.2}$$

The plethystic exponential is given by

$$\text{PE}\,[f(x, y, \ldots)] := \exp\left[\sum_{\ell=1}^{\infty} \frac{1}{\ell} f(x^{\ell}, y^{\ell}, \ldots)\right]. \tag{8.3}$$

In particular

$$
\mathrm{PE}[x] = \frac{1}{1-x} , \qquad \mathrm{PE}[-x] = 1 - x . \tag{8.4}
$$

The inverse of the plethystic exponential is the logarithm, given by

$$
\mathrm{PL}\,[f(x, y, \ldots)] := \sum_{\ell=1}^{\infty} \frac{\mu(\ell)}{\ell} \, \ln f(x^{\ell}, y^{\ell}, \ldots), \tag{8.5}
$$

where $\mu(\ell)$ is the Mobius mu-function. Finally the elliptic Gamma function is defined as

$$
\Gamma(z; \, p, \, q) := \mathrm{PE}\left[\frac{z - \frac{pq}{z}}{(1-p)(1-q)} \right] = \prod_{i,j=0}^{\infty} \frac{1 - p^{i+1} q^{j+1} z^{-1}}{1 - p^i q^j z} . \tag{8.6}
$$

Appendix 2: $\mathcal{N} = 2$ Superconformal Representation Theory

In this appendix (adapted from [5]) we review the classification of short representations of the four-dimensional $\mathcal{N} = 2$ superconformal algebra [2, 39, 86].

Short representations occur when the norm of a superconformal descendant state in what would otherwise be a long representation is rendered null by a conspiracy of quantum numbers. The unitarity bounds for a superconformal primary operator are given by

$$
\begin{aligned}
E &\geqslant E_i, & j_i &\neq 0 , \\
E &= E_i - 2 \quad \text{or} \quad E \geqslant E_i , & j_i &= 0,
\end{aligned} \tag{8.7}
$$

where we have defined

$$
E_1 = 2 + 2j_1 + 2R + r , \qquad E_2 = 2 + 2j_2 + 2R - r , \tag{8.8}
$$

and short representations occur when one or more of these bounds are saturated. The different ways in which this can happen correspond to different combinations of Poincaré supercharges that will annihilate the superconformal primary state in the representation. There are two types of shortening conditions, each of which has four incarnations corresponding to an $SU(2)_R$ doublet's worth of conditions for each supercharge chirality:

$$
\begin{aligned}
\mathcal{B}^{\mathcal{I}} : \quad & Q_{\alpha}^{\mathcal{I}} |\psi\rangle = 0 , \quad \alpha = 1, 2 \tag{8.9} \\
\bar{\mathcal{B}}_{\mathcal{I}} : \quad & \tilde{Q}_{\mathcal{I}\dot{\alpha}} |\psi\rangle = 0 , \quad \dot{\alpha} = 1, 2 \tag{8.10}
\end{aligned}
$$

Table 4 Unitary irreducible representations of the $\mathcal{N} = 2$ superconformal algebra

Shortening	Quantum number relations	DO	KMMR
$\varnothing$	$E \geqslant \max(E_1, E_2)$	$\mathcal{A}^\Delta_{R,r(j_1,j_2)}$	$\mathbf{aa}_{\Delta,j_1,j_2,r,R}$
$\mathcal{B}^1$	$E = 2R + r \quad j_1 = 0$	$\mathcal{B}_{R,r(0,j_2)}$	$\mathbf{ba}_{0,j_2,r,R}$
$\bar{\mathcal{B}}_2$	$E = 2R - r \quad j_2 = 0$	$\bar{\mathcal{B}}_{R,r(j_1,0)}$	$\mathbf{ab}_{j_1,0,r,R}$
$\mathcal{B}^1 \cap \mathcal{B}^2$	$E = r \quad R = 0$	$\mathcal{E}_{r(0,j_2)}$	$\mathbf{ba}_{0,j_2,r,0}$
$\bar{\mathcal{B}}_1 \cap \bar{\mathcal{B}}_2$	$E = -r \quad R = 0$	$\bar{\mathcal{E}}_{r(j_1,0)}$	$\mathbf{ab}_{j_1,0,r,0}$
$\mathcal{B}^1 \cap \bar{\mathcal{B}}_2$	$E = 2R \quad j_1 = j_2 = r = 0$	$\hat{\mathcal{B}}_R$	$\mathbf{bb}_{0,0,0,R}$
$\mathcal{C}^1$	$E = 2 + 2j_1 + 2R + r$	$\mathcal{C}_{R,r(j_1,j_2)}$	$\mathbf{ca}_{j_1,j_2,r,R}$
$\bar{\mathcal{C}}_2$	$E = 2 + 2j_2 + 2R - r$	$\bar{\mathcal{C}}_{R,r(j_1,j_2)}$	$\mathbf{ac}_{j_1,j_2,r,R}$
$\mathcal{C}^1 \cap \mathcal{C}^2$	$E = 2 + 2j_1 + r \quad R = 0$	$\mathcal{C}_{0,r(j_1,j_2)}$	$\mathbf{ca}_{j_1,j_2,r,0}$
$\bar{\mathcal{C}}_1 \cap \bar{\mathcal{C}}_2$	$E = 2 + 2j_2 - r \quad R = 0$	$\bar{\mathcal{C}}_{0,r(j_1,j_2)}$	$\mathbf{ac}_{j_1,j_2,r,0}$
$\mathcal{C}^1 \cap \bar{\mathcal{C}}_2$	$E = 2 + 2R + j_1 + j_2 \quad r = j_2 - j_1$	$\hat{\mathcal{C}}_{R(j_1,j_2)}$	$\mathbf{cc}_{j_1,j_2,j_2-j_1,R}$
$\mathcal{B}^1 \cap \bar{\mathcal{C}}_2$	$E = 1 + 2R + j_2 \quad r = j_2 + 1$	$\mathcal{D}_{R(0,j_2)}$	$\mathbf{bc}_{0,j_2,j_2+1,R}$
$\bar{\mathcal{B}}_2 \cap \mathcal{C}^1$	$E = 1 + 2R + j_1 - r = j_1 + 1$	$\bar{\mathcal{D}}_{R(j_1,0)}$	$\mathbf{cb}_{j_1,0,-j_1-1,R}$
$\mathcal{B}^1 \cap \mathcal{B}^2 \cap \bar{\mathcal{C}}_2$	$E = r = 1 + j_2 \quad r = j_2 + 1 \quad R = 0$	$\mathcal{D}_{0(0,j_2)}$	$\mathbf{bc}_{0,j_2,j_2+1,0}$
$\mathcal{C}^1 \cap \bar{\mathcal{B}}_1 \cap \bar{\mathcal{B}}_2$	$E = -r = 1 + j_1 \quad -r = j_1 + 1 \quad R = 0$	$\bar{\mathcal{D}}_{0(j_1,0)}$	$\mathbf{cb}_{j_1,0,-j_1-1,0}$

$$
\mathcal{C}^{\mathcal{I}} : \quad \begin{cases} \epsilon^{\alpha\beta} \mathcal{Q}^{\mathcal{I}}_\alpha \; |\psi\rangle_\beta = 0 \,, & j_1 \neq 0 \\ \epsilon^{\alpha\beta} \mathcal{Q}^{\mathcal{I}}_\alpha \mathcal{Q}^{\mathcal{I}}_\beta \; |\psi\rangle = 0 \,, & j_1 = 0 \end{cases} , \tag{8.11}
$$

$$
\bar{\mathcal{C}}_{\mathcal{I}} : \quad \begin{cases} \epsilon^{\alpha\beta} \tilde{\mathcal{Q}}_{\mathcal{I}\alpha} \; |\psi\rangle_\beta = 0 \,, & j_2 \neq 0 \\ \epsilon^{\alpha\beta} \tilde{\mathcal{Q}}_{\mathcal{I}\alpha} \tilde{\mathcal{Q}}_{\mathcal{I}\beta} \; |\psi\rangle = 0 \,, & j_2 = 0 \end{cases} , \tag{8.12}
$$

The different admissible combinations of shortening conditions that can be simultaneously realized by a single unitary representation are summarized in Table 4, where the reader can also find the precise relations that must be satisfied by the quantum numbers (E, j_1, j_2, r, R) of the superconformal primary operator, as well as the notations used to designate the different representations in [39] (DO) and [2] (KMMR).[30]

At the level of group theory, it is possible for a collection of short representations to recombine into a generic long representation whose dimension is equal to one of the unitarity bounds of (8.7). In the DO notation, the generic recombinations are as follows:

---

[30] We follow the R-charge conventions of DO.

$$\mathcal{A}^{2R+r+2+2j_1}_{R,r(j_1,j_2)} \simeq \mathcal{C}_{R,r(j_1,j_2)} \oplus \mathcal{C}_{R+\frac{1}{2},r+\frac{1}{2}(j_1-\frac{1}{2},j_2)}, \tag{8.13}$$

$$\mathcal{A}^{2R-r+2+2j_2}_{R,r(j_1,j_2)} \simeq \bar{\mathcal{C}}_{R,r(j_1,j_2)} \oplus \bar{\mathcal{C}}_{R+\frac{1}{2},r-\frac{1}{2}(j_1,j_2-\frac{1}{2})}, \tag{8.14}$$

$$\mathcal{A}^{2R+j_1+j_2+2}_{R,j_1-j_2(j_1,j_2)} \simeq \hat{\mathcal{C}}_{R(j_1,j_2)} \oplus \hat{\mathcal{C}}_{R+\frac{1}{2}(j_1-\frac{1}{2},j_2)} \oplus \hat{\mathcal{C}}_{R+\frac{1}{2}(j_1,j_2-\frac{1}{2})} \oplus \hat{\mathcal{C}}_{R+1(j_1-\frac{1}{2},j_2-\frac{1}{2})}. \tag{8.15}$$

There are special cases when the quantum numbers of the long multiplet at threshold are such that some Lorentz quantum numbers in (8.13) would be negative and unphysical:

$$\mathcal{A}^{2R+r+2}_{R,r(0,j_2)} \simeq \mathcal{C}_{R,r(0,j_2)} \oplus \mathcal{B}_{R+1,r+\frac{1}{2}(0,j_2)}, \tag{8.16}$$

$$\mathcal{A}^{2R-r+2}_{R,r(j_1,0)} \simeq \bar{\mathcal{C}}_{R,r(j_1,0)} \oplus \bar{\mathcal{B}}_{R+1,r-\frac{1}{2}(j_1,0)}, \tag{8.17}$$

$$\mathcal{A}^{2R+j_2+2}_{R,-j_2(0,j_2)} \simeq \hat{\mathcal{C}}_{R(0,j_2)} \oplus \mathcal{D}_{R+1(0,j_2)} \oplus \hat{\mathcal{C}}_{R+\frac{1}{2}(0,j_2-\frac{1}{2})} \oplus \mathcal{D}_{R+\frac{3}{2}(0,j_2-\frac{1}{2})}, \tag{8.18}$$

$$\mathcal{A}^{2R+j_1+2}_{R,j_1(j_1,0)} \simeq \hat{\mathcal{C}}_{R(j_1,0)} \oplus \hat{\mathcal{C}}_{R+\frac{1}{2}(j_1-\frac{1}{2},0)} \oplus \bar{\mathcal{D}}_{R+1(j_1,0)} \oplus \bar{\mathcal{D}}_{R+\frac{3}{2}(j_1-\frac{1}{2},0)}, \tag{8.19}$$

$$\mathcal{A}^{2R+2}_{R,0(0,0)} \simeq \hat{\mathcal{C}}_{R(0,0)} \oplus \mathcal{D}_{R+1(0,0)} \oplus \bar{\mathcal{D}}_{R+1(0,0)} \oplus \hat{\mathcal{B}}_{R+2}. \tag{8.20}$$

The last three recombinations involve multiplets that make an appearance in the associated chiral algebra described in this work. Note that the $\mathcal{E}, \bar{\mathcal{E}}, \hat{\mathcal{B}}_{\frac{1}{2}}, \hat{\mathcal{B}}_1, \hat{\mathcal{B}}_{\frac{3}{2}}, \mathcal{D}_0$, $\bar{\mathcal{D}}_0, \mathcal{D}_{\frac{1}{2}}$ and $\bar{\mathcal{D}}_{\frac{1}{2}}$ multiplets can never recombine, along with $\mathcal{B}_{\frac{1}{2},r(0,j_2)}$ and $\bar{\mathcal{B}}_{\frac{1}{2},r(j_1,0)}$.

References

[1] Romelsberger, C.: Counting chiral primaries in N = 1, d = 4 superconformal field theories. Nucl. Phys. **B747**, 329–353 (2006). arXiv:hep-th/0510060

[2] Kinney, J., Maldacena, J.M., Minwalla, S., Raju, S.: An index for 4 dimensional super conformal theories. Commun. Math. Phys. **275**, 209–254 (2007). arXiv:hep-th/0510251

[3] Romelsberger, C.: Calculating the superconformal index and Seiberg duality. arXiv:0707.3702

[4] Festuccia, G., Seiberg, N.: Rigid supersymmetric theories in curved superspace. JHEP **1106**, 114 (2011). arXiv:1105.0689

[5] Beem, C., Lemos, M., Liendo, P., Peelaers, W., Rastelli, L., et al.: Infinite chiral symmetry in four dimensions. arXiv:1312.5344

[6] Gadde, A., Rastelli, L., Razamat, S.S., Yan, W.: Gauge theories and Macdonald polynomials. Commun. Math. Phys. **319**, 147–193 (2013). arXiv:1110.3740

[7] Gaiotto, D., Rastelli, L., Razamat, S.S.: Bootstrapping the superconformal index with surface defects. JHEP **01**, 022 (2013). arXiv:1207.3577

[8] Gadde, A., Pomoni, E., Rastelli, L.: The Veneziano limit of N = 2 superconformal QCD: towards the string dual of $N = 2SU(N_c)$ SYM with $N_f = 2N_c$. arXiv:0912.4918

[9] Minahan, J.A., Nemeschansky, D.: An N = 2 superconformal fixed point with E(6) global symmetry. Nucl. Phys. **B482**, 142–152 (1996). arXiv:hep-th/9608047

[10] Argyres, P.C., Seiberg, N.: S-duality in n = 2 supersymmetric gauge theories. JHEP **0712**, 088 (2007)

[11] Gadde, A., Rastelli, L., Razamat, S.S., Yan, W.: The superconformal index of the E_6 SCFT. JHEP **08**, 107 (2010). arXiv:1003.4244

[12] Spiridonov, V.P., Warnaar, S.O.: Inversions of integral operators and elliptic beta integrals on root systems. Adv. Math. **207**, 91–132 (2006). arXiv:math/0411044

[13] Gaiotto, D., Neitzke, A., Tachikawa, Y.: Argyres-Seiberg duality and the Higgs branch. Commun. Math. Phys. **294**, 389–410 (2010). arXiv:0810.4541

[14] Tachikawa, Y.: N = 2 S-duality via outer-automorphism twists. J. Phys. **A44**, 182001 (2011). arXiv:1009.0339

[15] Gadde, A., Pomoni, E., Rastelli, L., Razamat, S.S.: S-duality and 2d topological QFT. JHEP **03**, 032 (2010). arXiv:0910.2225

[16] Razamat, S.S.: On the N = 2 superconformal index and eigenfunctions of the elliptic RS model. arXiv:1309.0278

[17] Gaiotto, D.: N = 2 dualities. arXiv:0904.2715

[18] Gaiotto, D., Razamat, S.S.: Exceptional indices. JHEP **1205**, 145 (2012). arXiv:1203.5517

[19] Tachikawa, Y.: N = 2 supersymmetric dynamics for dummies. arXiv:1312.2684

[20] Gadde, A., Gukov, S.: 2d index and surface operators. JHEP **1403**, 080 (2014). arXiv:1305.0266

[21] Alday, L.F., Gaiotto, D., Tachikawa, Y.: Liouville correlation functions from four-dimensional gauge theories. Lett. Math. Phys. **91**, 167–197 (2010). arXiv:0906.3219

[22] Alday, L.F., Gaiotto, D., Gukov, S., Tachikawa, Y., Verlinde, H.: Loop and surface operators in N = 2 gauge theory and Liouville modular geometry. JHEP **01**, 113 (2010). arXiv:0909.0945

[23] Teschner, J.: On the Liouville three point function. Phys. Lett. **B363**, 65–70 (1995). arXiv:hep-th/9507109

[24] Gaiotto, D., Rastelli, L., Razamat, S.S.: Un-published

[25] Benini, F., Tachikawa, Y., Xie, D.: Mirrors of 3d Sicilian theories. JHEP **09**, 063 (2010). arXiv:1007.0992

[26] Gaiotto, D., Witten, E.: S-duality of boundary conditions in N = 4 super Yang-Mills theory. arXiv:0807.3720

[27] Dolan, F., Spiridonov, V., Vartanov, G.: From 4d superconformal indices to 3d partition functions. arXiv:1104.1787

[28] Gadde, A., Yan, W.: Reducing the 4d index to the S^3 partition function. JHEP **1212**, 003 (2012). arXiv:1104.2592

[29] Imamura, Y.: Relation between the 4d superconformal index and the S^3 partition function. arXiv:1104.4482

[30] Aharony, O., Razamat, S.S., Seiberg, N., Willett, B.: 3d dualities from 4d dualities. JHEP **1307**, 149 (2013) arXiv:1305.3924

[31] Nishioka, T., Tachikawa, Y., Yamazaki, M.: 3d partition function as overlap of wavefunctions. JHEP **1108**, 003 (2011). arXiv:1105.4390

[32] Beem, C., Dimofte, T., Pasquetti, S.: Holomorphic blocks in three dimensions. arXiv:1211.1986

[33] Razamat, S.S., Willett, B.: Down the rabbit hole with theories of class S. arXiv:1403.6107

[34] Hosomichi, K., Lee, S., Park, J.: AGT on the S-duality wall. JHEP **1012**, 079 (2010). arXiv:1009.0340

[35] Bullimore, M., Fluder, M., Hollands, L., Richmond, P.: The superconformal index and an elliptic algebra of surface defects. arXiv:1401.3379

[36] Gorsky, A.: Integrable many-body systems in the field theories. Theor. Math. Phys. **103**, 681–700 (1995). doi:10.1007/BF02065867

[37] Gorsky, A., Nekrasov, N.: Hamiltonian systems of Calogero type and two-dimensional Yang-Mills theory. Nucl. Phys. **B414**, 213–238 (1994). arXiv:hep-th/9304047

[38] Klare, C., Zaffaroni, A.: Extended supersymmetry on curved spaces. JHEP **1310**, 218 (2013) arXiv:1308.1102

[39] Dolan, F.A., Osborn, H.: On short and semi-short representations for four dimensional superconformal symmetry. Ann. Phys. **307**, 41–89 (2003). arXiv:hep-th/0209056

[40] Gadde, A., Rastelli, L., Razamat, S.S., Yan, W.: The 4d superconformal index from q-deformed 2d Yang-Mills. Phys. Rev. Lett. **106**, 241602 (2011). arXiv:1104.3850

[41] Tachikawa, Y.: 4d partition function on $S^1 \times S^3$ and 2d Yang-Mills with nonzero area. PTEP **2013**, 013B01 (2013). arXiv:1207.3497

[42] Gaiotto, D., Moore, G.W., Tachikawa, Y.: On $6d$ $N = (2, 0)$ theory compactified on a Riemann surface with finite area. arXiv:1110.2657

[43] Aganagic, M., Ooguri, H., Saulina, N., Vafa, C.: Black holes, q-deformed 2d Yang-Mills, and non-perturbative topological strings. Nucl. Phys. **B715**, 304–348 (2005). arXiv:hep-th/0411280

[44] Kawano, T., Matsumiya, N.: 5D SYM on 3D sphere and 2D YM. Phys. Lett. **B716**, 450–453 (2012). arXiv:1206.5966

[45] Fukuda, Y., Kawano, T., Matsumiya, N.: 5D SYM and 2D q-deformed YM. Nucl. Phys. **B869**, 493–522 (2013). arXiv:1210.2855

[46] Witten, E.: On quantum gauge theories in two-dimensions. Commun. Math. Phys. **141**, 153–209 (1991)

[47] Alday, L.F., Bullimore, M., Fluder, M., Hollands, L.: Surface defects, the superconformal index and q-deformed Yang-Mills. arXiv:1303.4460

[48] Razamat, S.S.: On a modular property of N = 2 superconformal theories in four dimensions. JHEP **1210**, 191 (2012). arXiv:1208.5056

[49] Macdonald, I.G.: Symmetric Functions and Hall Polynomials. Oxford University Press, Oxford (1995)

[50] Mekareeya, N., Song, J., Tachikawa, Y.: 2d TQFT structure of the superconformal indices with outer-automorphism twists. JHEP **1303**, 171 (2013). arXiv:1212.0545

[51] Aganagic, M., Shakirov, S.: Knot homology from refined Chern-Simons theory. arXiv:1105.5117

[52] Gray, J., Hanany, A., He, Y.-H., Jejjala, V., Mekareeya, N.: SQCD: a geometric apercu. JHEP **0805**, 099 (2008). arXiv:0803.4257

[53] Hanany, A., Mekareeya, N.: Counting gauge invariant operators in SQCD with classical gauge groups. JHEP **0810**, 012 (2008). arXiv:0805.3728

[54] Hanany, A., Mekareeya, N., Razamat, S.S.: Hilbert series for moduli spaces of two instantons. JHEP **1301**, 070 (2013). arXiv:1205.4741

[55] Keller, C.A., Song, J.: Counting exceptional instantons. JHEP **1207**, 085 (2012). arXiv:1205.4722

[56] Cremonesi, S., Hanany, A., Mekareeya, N., Zaffaroni, A.: Coulomb branch Hilbert series and Hall-Littlewood polynomials. arXiv:1403.0585

[57] Cremonesi, S., Hanany, A., Mekareeya, N., Zaffaroni, A.: Coulomb branch Hilbert series and three dimensional Sicilian theories. arXiv:1403.2384

[58] Lemos, M., Peelaers, W., Rastelli, L.: The superconformal index of class S theories of type D. JHEP **1405**, 120 (2014). arXiv:1212.1271

[59] Chacaltana, O., Distler, J., Trimm, A.: Tinkertoys for the twisted D-series. arXiv:1309.2299

[60] Chacaltana, O., Distler, J., Trimm, A.: Tinkertoys for the E_6 theory. arXiv:1403.4604

[61] Bah, I., Beem, C., Bobev, N., Wecht, B.: Four-dimensional SCFTs from M5-branes. JHEP **1206**, 005 (2012). arXiv:1203.0303

[62] Beem, C., Gadde, A.: The $N = 1$ superconformal index for class S fixed points. JHEP **1404**, 036 (2014). arXiv:1212.1467

[63] Gadde, A., Maruyoshi, K., Tachikawa, Y., Yan, W.: New N = 1 dualities. JHEP **1306**, 056 (2013). arXiv:1303.0836

[64] Agarwal, P., Song, J.: New N = 1 dualities from M5-branes and outer-automorphism twists. JHEP **1403**, 133 (2014). arXiv:1311.2945

[65] Dimofte, T., Gaiotto, D., Gukov, S.: 3-manifolds and 3d indices. arXiv:1112.5179

[66] Gang, D., Koh, E., Lee, K.: Line operator index on $S^1 \times S^3$. arXiv:1201.5539

[67] Gang, D., Koh, E., Lee, K.: Superconformal index with duality domain wall. JHEP **1210**, 187 (2012). arXiv:1205.0069

[68] Alday, L.F., Tachikawa, Y.: Affine SL(2) conformal blocks from 4d gauge theories. Lett. Math. Phys. **94**, 87–114 (2010). arXiv:1005.4469

[69] van de Bult, F.: An elliptic hypergeometric integral with $w(f_4)$ symmetry. Ramanujan J. **25**, 1–20 (2011). arXiv:0909.4793

[70] Beem, C., Peelaers, W., Rastelli, L., van Rees, B.C.: Chiral algebras of class S. arXiv:1408.6522

[71] Benini, F., Nishioka, T., Yamazaki, M.: 4d index to 3d index and 2d TQFT. arXiv:1109.0283

[72] Alday, L.F., Bullimore, M., Fluder, M.: On S-duality of the superconformal index on lens spaces and 2d TQFT. arXiv:1301.7486

[73] Razamat, S.S., Yamazaki, M.: S-duality and the N = 2 lens space index. JHEP **1310**, 048 (2013). arXiv:1306.1543

[74] Closset, C., Dumitrescu, T.T., Festuccia, G., Komargodski, Z.: The geometry of supersymmetric partition functions. JHEP **1401**, 124 (2014). arXiv:1309.5876

[75] Closset, C., Shamir, I.: The $\mathcal{N} = 1$ chiral multiplet on $T^2 \times S^2$ and supersymmetric localization. JHEP **1403**, 040 (2014). arXiv:1311.2430

[76] Xie, D.: General Argyres-Douglas theory. JHEP **1301**, 100 (2013). arXiv:1204.2270

[77] Del Zotto, M., Hanany, A.: Complete graphs, Hilbert series, and the Higgs branch of the 4d N = SCFT's. arXiv:1403.6523

[78] Yoshida, Y.: Factorization of 4d N = 1 superconformal index. arXiv:1403.0891

[79] Peelaers, W.: Higgs branch localization of $\mathcal{N} = 1$ theories on $S^3 \times S^1$. arXiv:1403.2711

[80] Pasquetti, S.: Factorisation of N = 2 theories on the squashed 3-sphere. JHEP **1204**, 120 (2012). arXiv:1111.6905

[81] Cecotti, S., Gaiotto, D., Vafa, C.: tt^* geometry in 3 and 4 dimensions. JHEP **1405**, 055 (2014). arXiv:1312.1008

[82] Spiridonov, V., Vartanov, G.: Elliptic hypergeometric integrals and 't Hooft anomaly matching conditions. JHEP **1206**, 016 (2012). arXiv:1203.5677

[83] Di Pietro, L., Komargodski, Z.: Cardy Formulae for SUSY theories in d = 4 and d = 6. arXiv:1407.6061

[84] Ardehali, A.A., Liu, J.T., Szepietowski, P.: Central charges from the $\mathcal{N} = 1$ superconformal index. arXiv:1411.5028

[85] Buican, M., Nishinaka, T., Papageorgakis, C.: Constraints on chiral operators in $\mathcal{N} = 2$. JHEP 12, 095 (2014). doi:10.1007/JHEP12(2014)095

[86] Dobrev, V., Petkova, V.: All positive energy unitary irreducible representations of extended conformal supersymmetry. Phys. Lett. **B162**, 127–132 (1985)

A Review on SUSY Gauge Theories on S^3

Kazuo Hosomichi

Abstract We review the exact computations in 3D $\mathcal{N} = 2$ supersymmetric gauge theories on the round or squashed S^3 and the relation between 3D partition functions and 4D superconformal indices. This is part of a combined review on the recent developments of the 2d–4d relation, edited by J. Teschner.

1 Introduction

Localization principle has been a powerful tool in the study of supersymmetric field theories which allows one to evaluate certain SUSY-preserving quantities by explicit path integration. It was first applied to 3D SUSY gauge theories on S^3 in [1], where a closed formula for partition function and Wilson loop was obtained for a class of $\mathcal{N} \geq 2$ superconformal Chern-Simons matter theories. With generalization by [2, 3], exact formula is now available for arbitrary 3D $\mathcal{N} = 2$ SUSY gauge theories. The essential idea of localization is that, since nonzero contribution to supersymmetric path integrals arise only from SUSY invariant configurations of bosonic fields called saddle points, infinite dimensional path integrals can be reduced to finite-dimensional integrals over saddle points. It turned out that the analysis of 3D gauge theories on S^3 is much simpler than the case of 4D $\mathcal{N} = 2$ SUSY gauge theories on S^4 [4] (see [V:6] for a review in this volume), due to the absence of saddle points with non-trivial topological quantum numbers.

The exact partition function, which depends on the radius of S^3 as well as some of the coupling constants, is one of the most basic quantities characterizing $\mathcal{N} = 2$ supersymmetric theories. More information about the theories can be obtained by putting them on different 3D backgrounds preserving rigid supersymmetry and evaluating partition functions. In [5] it was shown that one can construct rigid $\mathcal{N} = 2$ SUSY gauge theories on the ellipsoid S_b^3 with $U(1) \times U(1)$ isometry,

---

A citation of the form [V:x] refers to article number x in this volume.

K. Hosomichi (✉)
Department of Physics, National Taiwan University, Taipei 10617, Taiwan
e-mail: hosomiti@phys.ntu.edu.tw

© Springer International Publishing Switzerland 2016   307
J. Teschner (ed.), *New Dualities of Supersymmetric Gauge Theories*,
Mathematical Physics Studies, DOI 10.1007/978-3-319-18769-3_10

$$b^2(x_0^2 + x_1^2) + b^{-2}(x_2^2 + x_3^2) = 1, \tag{1.1}$$

with a suitable background vector and scalar fields. The additional fields which are required to make the ellipsoid supersymmetric have their origin in the off-shell supergravity [6], where the fully generalized form of Killing spinor equation appears as local SUSY transformation laws of fermions in the supergravity multiplet. The ellipsoid partition function was shown to depend on the squashing parameter b in a nontrivial manner. Another important background with rigid supersymmetry is $S^2 \times S^1$ which leads to the path integral definition of the 3D superconformal index [7, 8]. There are also results on more general 3D manifolds with a slightly different formalism based on topological twist [9, 10].

Another useful approach to find supersymmetric deformations of the round S^3 is the Scherk-Schwarz like reduction of $S^1 \times S^3$, which means that one includes finite rotation in the S^3 direction in the periodic identification of fields along S^1. This approach also makes an explicit connection between the 3D partition functions and 4D superconformal indices [11–14], and in particular the relation between nonzero angular momentum fugacity in 4D and the deformed geometry in 3D [15, 16]. As was shown in [17, 18], the dimensional reduction results in the familiar squashed S^3 with $SU(2) \times U(1)$ isometry, with some additional background fields turned on. However, there are two inequivalent reductions whose effect on the 3D physical quantities are totally different.

Meanwhile, the study of certain domain walls in 4D $\mathcal{N} = 2$ superconformal gauge theories in connection with AGT relation led to a conjecture that there is a precise agreement between quantities in 3D gauge theories on S^3 and the representation theory of Virasoro or W algebras [19, 20]. In general, compactification of a $(2, 0)$ theory on a Riemann surface Σ gives rise to several different (Lagrangian) descriptions that are related to one another by S-duality [21]. The S-duality domain walls are defined by gluing two mutually S-dual theories along an interface, and therefore have a natural connection to the elements of the mapping class group or Moore-Seiberg groupoid operation acting on conformal blocks. In this respect, it is important that the squashing parameter b corresponds to the Liouville or Toda coupling constant. Indeed, one of the building blocks of the ellipsoid partition function is the double-sine function $s_b(x)$, which in our context is most conveniently defined as the zeta-regularized infinite product [22]

$$s_b(x) = \prod_{m,n \in \mathbb{Z}_{\geq 0}} \frac{mb + nb^{-1} + \frac{Q}{2} - ix}{mb + nb^{-1} + \frac{Q}{2} + ix} \cdot \left(Q \equiv b + \frac{1}{b}\right) \tag{1.2}$$

The same function appears in the structure constants of Liouville or Toda CFTs with coupling b.

This review is organized as follows. In Sect. 2 we review the correspondence between a 3D gauge theory and 2D conformal field theories in the canonical example of the S-duality domain wall in $\mathcal{N} = 2^*$ $SU(2)$ super Yang-Mills theory. In Sect. 3 we review the localization computation for 3D gauge theories on the round S^3 and the ellipsoid S_b^3, and summarize the formulae for partition function as well as expectation values of loop observables. In Sect. 4 we review the path integral computation of 4D superconformal index, and see how the squashed S^3 background arises as a result of Scherk-Schwarz reduction.

2 3D AGT Relation

We review here the correspondence between 3D gauge theories and 2D conformal field theories in one typical example. The original idea was given in [19] which discussed the S-duality domain walls in 4D $\mathcal{N} = 2$ superconformal theories of *class S*, namely the compactification of $(2, 0)$-theories on punctured Riemann surfaces (see [V:2] for a review). It is important to recall here that, for this class of theories, there are different gauge theory descriptions corresponding to different pants decomposition σ of the surface Σ, and they are equivalent (S-dual) to one another. Also, if Lagrangian description is available, its gauge coupling q is determined from the complex structure of Σ which we regard to take values in Teichmüller space.

2.1 Janus and S-Duality Domain Walls

A Janus domain wall is a supersymmeric deformation of gauge theories which makes the complexified gauge coupling jump across the wall. Consider a theory of class S on S^4 with a Janus wall along the equator S^3 where the two (left and right) hemispheres with couplings q and q' meet. The 4D partition function in the presence of the wall should be given by

$$Z = \int dv(a) \, \mathcal{F}_{a,m}^{(\sigma)}(\bar{q}) \mathcal{F}_{a,m}^{(\sigma)}(q'), \tag{2.1}$$

as the product of the instanton partition functions $\mathcal{F}$ integrated over the real Coulomb branch parameters a with an appropriate measure. Here m denotes a collection of mass parameters, and σ labels a choice of pants decomposition. For generic complex structure q there is a natural pants decomposition which leads to a weakly coupled gauge theory description, and we choose σ to be the natural one at q.

As q' is varied away from q, the gauge theory on the right hemisphere becomes strongly coupled. To analytically continue the formula (2.1) in such a situation, one needs to S-dualize the right hemisphere and move to another pants decomposition σ' which gives a weakly coupled description at q'. We then have a system of two mutually S-dual theories meeting along the so-called S-duality domain wall. In the special case where q' is an image of q under the mapping class group, σ and σ' are equivalent so the theories on the two sides of the wall are the same. However, their degrees of freedom are connected across the wall via S-duality.

Under the AGT relation, the instanton partition functions correspond to Liouville or Toda conformal blocks labeled by a fusion channel σ and the internal and external momenta a, m. They should therefore transform under S-duality in the same way that the corresponding conformal blocks transform under the Moore-Seiberg groupoid operation g,

$$\mathcal{F}_{a,m}^{(\sigma)}(q') = \int d\nu(a') \, g_{a,a',m} \, \mathcal{F}_{a',m}^{(\sigma')}(q'). \tag{2.2}$$

By substituting (2.2) into (2.1) we obtain a formula for the S^4 partition function in the presence of an S-duality domain wall. Now the integration variables get doubled, as the Coulomb branch parameters on the two sides of the wall can vary independently. At this point, it is natural to expect that the integration kernel $g_{a,a',m}$ in (2.2) corresponds to the degrees of freedom localized on the S-duality wall between the two 4D theories in their vacua a, a'.

In general, the S-duality walls should be described by some local 3D worldvolume field theories coupled to the 4D bulk degrees of freedom. In the following we take the example of $\mathcal{N} = 2^*$ SYM theory, which is a deformation of $\mathcal{N} = 4$ SYM by a mass of the adjoint hypermultiplet. The S-duality transformations for this theory form the group $SL(2, \mathbb{Z})$ and we are interested in the wall corresponding to the "S-element". For $SU(N)$ gauge group, we expect the correspondence with the A_{N-1} Toda theory on a one-punctured torus. In the Liouville case $N = 2$, the kernel for the S-duality operation acting on torus 1-point conformal block is known explicitly [23],

$$g_{(p,p',p_E)} = \frac{2^{\frac{3}{2}}}{s_b(p_E)} \int_{\mathbb{R}} d\sigma \, \frac{s_b\left(p' + \sigma + \frac{1}{2}p_E + \frac{iQ}{4}\right) s_b\left(p' - \sigma + \frac{1}{2}p_E + \frac{iQ}{4}\right)}{s_b\left(p' + \sigma - \frac{1}{2}p_E - \frac{iQ}{4}\right) s_b\left(p' - \sigma - \frac{1}{2}p_E - \frac{iQ}{4}\right)} e^{4\pi i p \sigma}. \tag{2.3}$$

Here b is the Liouville coupling and $Q \equiv b + b^{-1}$. The Liouville momenta p, p', p_E are related to the conformal weight h labeling the Virasoro highest weight representations by the formula $h = p^2 + Q^2/4$. The double-sine function $s_b(x)$ is defined by (1.2), and will appear frequently later in this article.

2.2 Example: $\mathcal{N} = 2^*$ SYM

A classification of boundary conditions and domain walls for $\mathcal{N} = 4$ SYM theories with general gauge group G was given in [24, 25], and the action of S-duality on these objects was also studied. The 3D theory on the S-duality domain walls, called $T[G]$, plays a central role in this story. For $SU(N)$ gauge group, it was shown that the wall theory $T[SU(N)]$ is given by a 3D $\mathcal{N} = 4$ SUSY quiver gauge theory corresponding to the diagram in the left of Fig. 1. Here the circles and the square correspond respectively to the gauge symmetry $U(1) \times U(2) \times \cdots \times U(N - 1)$ and a global $U(N)$ symmetry, and the links correspond to hypermultiplets. The

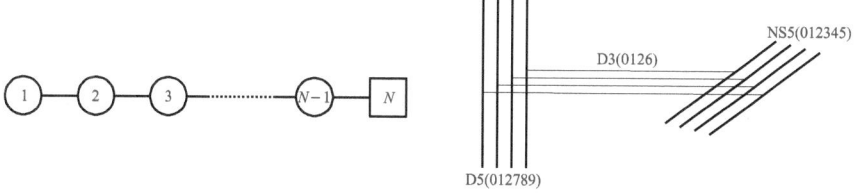

Fig. 1 The quiver diagram and a type IIB brane construction for the 3D gauge theory $T[SU(N)]$

Coulomb and Higgs branch moduli spaces both have an $SU(N)$ symmetry which can be coupled to the gauge fields in the bulk.

A simple type IIB brane construction can reproduce this fact. Consider N D3-branes stretched along the directions 0126 with $-L \leq x_6 \leq L$, ending on the D5-branes at $x_6 = \pm L$ extending in the directions 012789. Due to the boundary condition at D5-branes, the massless modes on D3-brane wordlvolume decompose into 3D $\mathcal{N} = 4$ vector and hypermultiplets. The vectormultiplet fields obey Dirichlet boundary condition, so for small L they are frozen to take vacuum configuration. As was explained in [24], to avoid D3-branes developing Nahm poles at the boundary, we need to introduce N D5-branes at each end so that each D5-brane has precisely one D3-brane ending on it. Nonzero (real) Coulomb branch parameter a can then be introduced by putting the ith D5-branes at, say, $(x_3, x_4, x_5) = (a_i, 0, 0)$ at each end.

Consider next the same brane configuration but now with an S-duality domain wall on the D3-brane worldvolume at $x_6 = 0$. It can be eliminated by applying the type IIB S-duality combined with the exchange of 345 and 789 directions to the right half space $x_6 \geq 0$, but then the N D5-branes at $x_6 = L$ turn into N NS5-branes (012345). The resulting brane configuration as shown on the right of Fig. 1 is what precisely gives rise to the above-mentioned quiver gauge theory. The D5-branes and NS5-branes are now free to move independently. The positions of NS5-branes a turn into $N - 1$ Fayet-Iliopoulos parameters, whereas those of D5-branes a' determine the masses of the $U(N - 1) \times U(N)$ bifundamental hypermultiplets.

Let us now focus on the simplest nontrivial case $N = 2$. In 3D $\mathcal{N} = 2$ terminology, the wall theory $T[SU(2)]$ is a $U(1)$ gauge theory with five chiral multiplets $\phi, q_1, q_2, \tilde{q}^1, \tilde{q}^2$. The neutral chiral field ϕ is a part of $\mathcal{N} = 4$ $U(1)$ vector multiplet and has R-charge 1. The two electrons q_1, q_2 and the two positrons $\tilde{q}^1, \tilde{q}^2$ have the R-charge $1/2$, and they form two flavors of hypermultiplets. $\mathcal{N} = 4$ supersymmetry requires a cubic superpotential of the form $\tilde{q}^i \phi q_i$.

As we have seen, the Coulomb branch parameter a appears in the wall theory as the $U(1)$ FI parameter, while a' is the mass for charged chiral fields which breaks the $SU(2)$ flavor symmetry to $U(1)$. In addition, the bulk $\mathcal{N} = 2^*$ mass parameter m should also show up in the wall theory in a way that preserves 3D $\mathcal{N} = 2$ supersymmetry as well as the $SU(2)$ isometries of the Coulomb and Higgs branches. It was argued in [20] that m is the mass for the chiral fields associated to the global symmetry under which $q_i, \tilde{q}^i$ have charge $+1$ and ϕ has charge -2.

It was observed in [20] that the exact partition function of this mass-deformed $T[SU(2)]$ theory on S^3 agrees precisely with the kernel of the S-duality transformation (2.3) for $b = 1$, under the identification

$$p = a, \quad p' = a', \quad p_E = m. \tag{2.4}$$

It was then shown in [5] that the formula (2.3) for general values of the coupling b can be reproduced by deforming the round S^3 into an ellipsoid S_b^3. The derivation of the formulae which are necessary to confirm this agreement will be reviewed in the next section.

2.3 A 3D Picture

As we have seen, Janus or S-duality domain walls correspond to smooth evolutions of the complex structure of a surface, and therefore have an interpretation as M5-branes wrapping three-manifolds. Let us explain this in the example of $\mathcal{N} = 2^*$ SYM.

Consider a Janus domain wall corresponding to a path in Teichmüller space between two points of extreme weak coupling that are S-dual image of each other. As one approaches towards one end from any point along the path, the torus Σ becomes thinner and thinner until it looks like the Moore-Seiberg graph Γ_1 for the torus one-point conformal blocks. In this process, two-dimensional part of the M5-brane worldvolume sweeps out a 3D solid torus B_1 with a codimension-2 defect Γ_1 left inside (Fig. 2 left). One of the two basis 1-cycles α, β of the torus, say α, shrinks to zero length inside B_1. Starting from the same point on the path and moving toward the other end, one obtains another solid torus B_2 with a defect Γ_2, inside which the cycle β shrinks to zero length. The two solid tori B_1 and B_2 glued together makes an S^3 with a defect Γ which is the union of the two graphs Γ_1, Γ_2 joined at the external

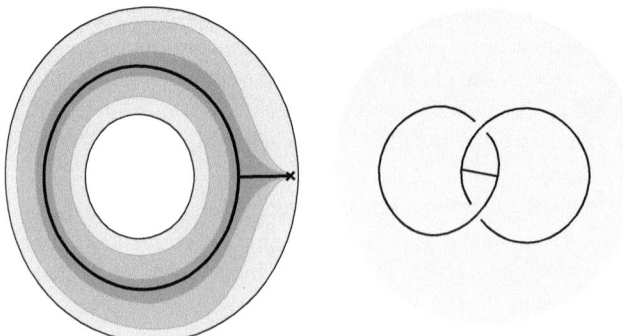

Fig. 2 (*left*) The process of a one-punctured torus degenerating into a Moore-Seiberg graph, thereby sweeping out a solid torus with a network of defect inside. (*right*) Two such solid tori glued together to make an S^3 with a network of defect

legs (Fig. 2 right). Γ therefore consists of two circle defects and a segment connecting them, and the three components are naturally labeled by the momenta p, p', p_E.

The 3D theories on domain walls or boundaries of 4D class S theories are now regarded as part of a much bigger class of theories which arise from M5-branes wrapping hyperbolic 3-manifolds. The relation between 3D SUSY gauge theories and hyperbolic 3-manifolds also gives rise to an AGT-like correspondence between 3D supersymmetric theories and Chern-Simons theories with non-compact gauge groups. For more details on this topic, see the review [V:11] in this volume.

3 3D Partition Function

In this section we review the construction of 3D $\mathcal{N} = 2$ supersymmetric gauge theories on a class of rigid SUSY backgrounds. Then we concentrate on the theories on the round sphere and the ellipsoids, and show how to compute partition function as well as the expectation values of Wilson and vortex loops using localization principle.

3.1 3D $\mathcal{N} = 2$ SUSY Theories

Let us begin by summarizing our convention for 3D spinor calculus. We use the standard Pauli's matrices for the Dirac matrices γ^a, and also $\gamma^{ab} = \frac{1}{2}(\gamma^a\gamma^b - \gamma^b\gamma^a)$. To define bilinear products of spinors, we use an anti-symmetric 2×2 matrix C with nonzero elements $C_{12} = -C_{21} = 1$. Writing the spinor indices explicitly, various bilinears are defined as follows.

$$\epsilon\psi \equiv \epsilon^\alpha C_{\alpha\beta}\psi^\beta, \quad \epsilon\gamma^a\psi \equiv \epsilon^\alpha C_{\alpha\beta}(\gamma^a)^\beta_{\ \gamma}\psi^\gamma, \quad \text{etc.} \tag{3.1}$$

In rigid SUSY theories on curved backgrounds, the parameters of SUSY transformation ϵ are no longer constants, but are solutions to the Killing spinor equation. For 3D $\mathcal{N} = 2$ supersymmetric theories, the SUSY is parametrized by two Killing spinors ϵ, $\bar{\epsilon}$ of R-charge $+1$, -1. The most general form of the Killing spinor equation can be found from off-shell supergravity [26] as the condition that gravitini are invariant under local SUSY for a suitable choice of parameters ϵ, $\bar{\epsilon}$.

$$D_m\epsilon = \left(\partial_m + \frac{1}{4}\omega_m^{ab}\gamma^{ab} - iV_m\right)\epsilon = iM\gamma_m\epsilon - iU_m\epsilon - \frac{1}{2}\varepsilon_{mnp}U^n\gamma^p\epsilon,$$

$$D_m\bar{\epsilon} = \left(\partial_m + \frac{1}{4}\omega_m^{ab}\gamma^{ab} + iV_m\right)\bar{\epsilon} = iM\gamma_m\bar{\epsilon} + iU_m\bar{\epsilon} + \frac{1}{2}\varepsilon_{mnp}U^n\gamma^p\bar{\epsilon}. \tag{3.2}$$

Here $\gamma_m \equiv e_m^a\gamma^a$ with e_m^a the vielbein, and throughout this article we regard ϵ, $\bar{\epsilon}$ as Grassmann even. Supersymmetric backgrounds are therefore characterized by the metric as well as the $U(1)_R$ gauge field V_m and other auxiliary fields M, U_m in the off-shell gravity multiplet. In this section we restrict our discussion to the backgrounds with

$$U_m = 0, \tag{3.3}$$

which include the round sphere and ellipsoids. More general supersymmetric backgrounds were studied systematically in three and four dimensions in [26–29]. For 3D $\mathcal{N} = 2$ systems it was shown that the existence of a Killing spinor implies that the background admits an almost contact metric structure.

The fields in 3D $\mathcal{N} = 2$ theories are grouped into two kinds of supermultiplets. A vector multiplet consists of a vector A_m, a real scalar σ, a pair of spinors $\lambda, \bar{\lambda}$ and an auxiliary scalar D which are all Lie algebra valued. They transform under supersymmetry as

$$\delta A_m = -\frac{i}{2}(\epsilon\gamma_m\bar{\lambda} + \bar{\epsilon}\gamma_m\lambda),$$

$$\delta\sigma = \frac{1}{2}(\epsilon\bar{\lambda} - \bar{\epsilon}\lambda),$$

$$\delta\lambda = \frac{1}{2}\gamma^{mn}\epsilon F_{mn} - \epsilon D - i\gamma^m\epsilon D_m\sigma,$$

$$\delta\bar{\lambda} = \frac{1}{2}\gamma^{mn}\bar{\epsilon} F_{mn} + \bar{\epsilon} D + i\gamma^m\bar{\epsilon} D_m\sigma,$$

$$\delta D = \frac{i}{2}\epsilon\left(\gamma^m D_m\bar{\lambda} + [\sigma, \bar{\lambda}] + iM\bar{\lambda}\right) - \frac{i}{2}\bar{\epsilon}\left(\gamma^m D_m\lambda - [\sigma, \lambda] + iM\lambda\right). \tag{3.4}$$

A chiral multiplet consists of a scalar ϕ, a spinor ψ and an auxiliary scalar F in an arbitrary representation R of the gauge group. Their conjugate fields $(\bar{\phi}, \bar{\psi}, \bar{F})$ are in the conjugate representation $\bar{R}$. If one assign the R-charge r to ϕ and $-r$ to $\bar{\phi}$, the R-charge of the remaining fields is determined from the supersymmetry as in Table 1. The transformation rule for these fields is given by

$$\delta\phi = \epsilon\psi, \quad \delta\psi = i\gamma^m\bar{\epsilon} D_m\phi + i\bar{\epsilon}\sigma\phi + \frac{2ri}{3}\gamma^m D_m\bar{\epsilon}\phi + \epsilon F,$$

$$\delta\bar{\phi} = \bar{\epsilon}\bar{\psi}, \quad \delta\bar{\psi} = i\gamma^m\epsilon D_m\bar{\phi} + i\epsilon\bar{\phi}\sigma + \frac{2ri}{3}\gamma^m D_m\epsilon\bar{\phi} + \bar{\epsilon}\bar{F},$$

$$\delta F = \bar{\epsilon}(i\gamma^m D_m\psi - i\sigma\psi - i\bar{\lambda}\phi) + \frac{i}{3}(2r-1)D_m\bar{\epsilon}\gamma^m\psi,$$

$$\delta\bar{F} = \epsilon(i\gamma^m D_m\bar{\psi} - i\bar{\psi}\sigma + i\bar{\phi}\lambda) + \frac{i}{3}(2r-1)D_m\epsilon\gamma^m\bar{\psi}. \tag{3.5}$$

Table 1 The scaling weight and the R-charge of the fields

Fields	ϵ	$\bar{\epsilon}$	A_a	σ	λ	$\bar{\lambda}$	$\tilde{D}$	ϕ	$\bar{\phi}$	ψ	$\bar{\psi}$	F	$\bar{F}$
Weight	$-\frac{1}{2}$	$-\frac{1}{2}$	1	1	$\frac{3}{2}$	$\frac{3}{2}$	2	r	r	$r+\frac{1}{2}$	$r+\frac{1}{2}$	$r+1$	$r+1$
R-charge	1	-1	0	0	1	-1	0	r	$-r$	$r-1$	$1-r$	$r-2$	$2-r$

Here the quantities in the representation R $(\bar{R})$ are regarded as the column vectors (resp. row vectors), so that the vector multiplet fields act on them from the left (right).

Supersymmetric Lagrangian consists of the following invariants. Those involving only vector multiplet fields are the Chern-Simons term (for which we write the action integral),

$$S_{CS} = \frac{ik}{4\pi} \int \mathrm{Tr}\left(AdA - \frac{2i}{3}A^3 - \sqrt{g}\mathrm{d}^3x\left(\bar{\lambda}\lambda + 2\sigma D + 4M\sigma^2\right)\right), \quad (3.6)$$

the Yang-Mills term and the Fayet-Iliopoulos term for abelian gauge symmetry.

$$\mathcal{L}_g = \mathrm{Tr}\left(\frac{1}{2}F_{mn}F^{mn} + D_m\sigma D^m\sigma + D^2 + i\bar{\lambda}\gamma^m D_m\lambda - i\bar{\lambda}[\sigma, \lambda] - M\bar{\lambda}\lambda\right),$$

$$\mathcal{L}_{FI} = -\frac{i\zeta}{\pi}(D + 4M\sigma). \quad (3.7)$$

The kinetic term for chiral matters is given by

$$\mathcal{L}_m = D_m\bar{\phi}D^m\phi + \bar{\phi}\sigma^2\phi + 4i(r-1)M\bar{\phi}\sigma\phi - 2r(2r-1)M^2\bar{\phi}\phi + \frac{rR}{4}\bar{\phi}\phi - i\bar{\phi}D\phi$$

$$+\bar{F}F - i\bar{\psi}\gamma^m D_m\psi + i\bar{\psi}\sigma\psi - (2r-1)M\bar{\psi}\psi + i\bar{\psi}\bar{\lambda}\phi - i\bar{\phi}\lambda\psi, \quad (3.8)$$

with R the scalar curvature of the background. The F-term of gauge invariant products of chiral multiplets with R-charge $r = 2$ is also invariant, but one can show that the result of localization computation does not depend on the F-term couplings. Note that, while the bosonic part of $\mathcal{L}_g$ is positive definite, that of $\mathcal{L}_m$ has positive definite real part only when the value of r is chosen appropriately. For example, for round sphere the positivity holds only when $0 < r < 2$.

The real mass for matters can be introduced by gauging the flavor symmetry by a background vector multiplet. The value of the background fields is chosen so as to preserve supersymmetry,

$$\sigma^{(bg)} = m \text{ (constant)}, \quad D^{(bg)} = A_m^{(bg)} = \lambda^{(bg)} = \bar{\lambda}^{(bg)} = 0. \quad (3.9)$$

3.2 SUSY Localization

To apply localization principle to supersymmetric path integrals, one first chooses an arbitrary supercharge δ, and then argue that the nonzero contribution to the path integral can be localized to the vicinity of saddle points, namely bosonic field configurations invariant under δ. This means that δ-transform of all the fermions must vanish on saddle points. For the theories of our interest, a useful observation is that both $\mathcal{L}_g$ and $\mathcal{L}_m$ are SUSY exact for any choice of δ, which follows from

$$\bar{\epsilon}\epsilon \cdot \mathcal{L}_{\mathrm{g}} = \delta_\epsilon \delta_{\bar{\epsilon}} \mathrm{Tr}\left(\bar{\lambda}\lambda + 4D\sigma + 8M\sigma^2\right),$$

$$\bar{\epsilon}\epsilon \cdot \mathcal{L}_{\mathrm{m}} = \delta_\epsilon \delta_{\bar{\epsilon}}\left(\bar{\psi}\psi - 2i\bar{\phi}\sigma\phi + 4M(r-1)\bar{\phi}\phi\right). \tag{3.10}$$

Namely, they can be written as δ-variation of some fermionic quantities, so they have to vanish at saddle points. A necessary condition for vector multiplet fields at saddle points follows from $\mathcal{L}_{\mathrm{g}} = 0$,

$$F_{mn} = D_m\sigma = D = 0. \tag{3.11}$$

This is actually sufficient for the saddle point condition $\delta\lambda = \delta\bar{\lambda} = 0$ to be satisfied. For theories on the round S^3 or its deformations, saddle points are thus labeled by constant scalar field σ and vanishing gauge field, up to gauge transformations. For non-simply connected manifolds such as lens spaces, one also has choices of Wilson lines along non-contractible loops [30–32]. For matter multiplets, an obvious solution to $\delta\psi = \delta\bar{\psi} = 0$ is

$$\phi = \bar{\phi} = F = \bar{F} = 0. \tag{3.12}$$

To show that this is the unique saddle point, the simplest way is to check that the kinetic operator for ϕ in $\mathcal{L}_{\mathrm{m}}$ has no zeromodes, so that $\mathcal{L}_{\mathrm{m}}$ vanishes only at (3.12). For theories on the round sphere, one can show by a full spectrum analysis that there are no zeromodes on all the saddle points as long as $0 < r < 2$. This allows us to assume that the spectrum remains free of zeromodes on the ellipsoids S_b^3 as long as b is reasonably close to 1. The exact partition function on S_b^3 turns out to be analytic in b, so it can be continued to arbitrary $b > 0$.

Since $\mathcal{L}_{\mathrm{g}}$ and $\mathcal{L}_{\mathrm{m}}$ are exact, the value of supersymmetric path integrals does not change if one adds them to the original Lagrangian with arbitrary coefficients t_{g}, t_{m}. By making those coefficients very large, one can bring the theory into extreme weak coupling. In this limit the path integral simplifies and can be performed in two steps. One first integrates over fluctuations around each saddle point, for which Gaussian approximation is exact. The result is then integrated over the space of saddle points labeled by constant σ.

3.3 Partition Function on the Round Sphere

As the simplest and yet the most important case, let us reproduce here the exact partition function of general $\mathcal{N} = 2$ SUSY theories on the unit round S^3.

We write the unit round metric as $ds^2 = e^a e^a$, and identify the dreibein $e^a = e_m^a dx^m$ with the left-invariant one-forms on the $SU(2)$ group manifold via

$$g^{-1}dg = ie^a\gamma^a, \quad g \in SU(2). \tag{3.13}$$

The isometry $SU(2)_\mathcal{L} \times SU(2)_\mathcal{R}$ acts on g from its left and right. Note that, under the above choice of the local Lorentz frame, $SU(2)_\mathcal{R}$ acts on fields as local Lorentz rotation as well as isometry rotation.

Let us summarize here the spectrum of free fields on the round sphere. We first notice that one can use the inverse dreibein e^{am} to define a triplet of vector fields $\mathcal{R}^a \equiv \frac{1}{2i} e^{am} \partial_m$ which generates $SU(2)_\mathcal{R}$. Using them, the kinetic terms for free complex scalars and spinors can be rewritten as

$$
\begin{aligned}
\bar{\phi}\, \Delta^{\text{scalar}}_{S^3} \phi &\equiv g^{mn} \partial_m \bar{\phi} \partial_n \phi = \bar{\phi} \cdot 4\mathcal{R}^a \mathcal{R}^a \phi, \\
-i\bar{\psi}\, \slashed{D}_{S^3} \psi &\equiv -i\bar{\psi}\gamma^m D_m \psi = \bar{\psi}\left(4S^a \mathcal{R}^a + \tfrac{3}{2}\right)\psi,
\end{aligned}
\tag{3.14}
$$

where $S^a = \frac{1}{2}\gamma^a$ is the generator of local Lorentz $SU(2)$ acting on spinors. Likewise, for a free Maxwell field $A = A^a e^a$ and its field strength $*dA = F^a e^a$, one finds

$$
F^a = 2i\varepsilon^{abc}\mathcal{R}^b A^c + 2A^a, \quad \text{or} \quad \vec{F} = (2 + 2\mathcal{R}^a T^a)\vec{A},
\tag{3.15}
$$

where T^a is the generator of local Lorentz $SU(2)$ in the triplet representation. The Maxwell kinetic operator for gauge field is given by $\Delta^{\text{vector}}_{S^3} \equiv (*d)^2$. The space of scalar, spinor and vector wave functions on S^3 thus form the following representation of $SU(2)_\mathcal{L} \times SU(2)_\mathcal{R}$.

$$
\begin{aligned}
\mathcal{H}_{\text{scalar}} &= \bigoplus_{n \geq 0} \left(\tfrac{n}{2}, \tfrac{n}{2}\right)_{n(n+2)}, \\
\mathcal{H}_{\text{spinor}} &= \bigoplus_{n \geq 0} \left\{\left(\tfrac{n}{2}, \tfrac{n+1}{2}\right)_{n+3/2} \oplus \left(\tfrac{n+1}{2}, \tfrac{n}{2}\right)_{-n-3/2}\right\}, \\
\mathcal{H}_{\text{vector}} &= \bigoplus_{n \geq 0} \left\{\left(\tfrac{n}{2}, \tfrac{n+2}{2}\right)_{(n+2)^2} \oplus \left(\tfrac{n+1}{2}, \tfrac{n+1}{2}\right)_0 \oplus \left(\tfrac{n+2}{2}, \tfrac{n}{2}\right)_{(n+2)^2}\right\}.
\end{aligned}
\tag{3.16}
$$

For convenience, we put the eigenvalue of $\Delta^{\text{scalar}}_{S^3}$, $-i\slashed{D}_{S^3}$ or $\Delta^{\text{vector}}_{S^3}$ for each irreducible representation as suffix. Note that the nonzero eigenmodes of $\Delta^{\text{vector}}_{S^3}$ are divergenceless vectors while the zero eigenmodes are total divergences.

On the unit round S^3, the simplest form of the Killing spinor equation

$$
\left(\partial_m + \tfrac{1}{4}\omega^{ab}_m \gamma^{ab}\right)\epsilon = iM\gamma_m\epsilon, \quad M = \pm\tfrac{1}{2}
\tag{3.17}
$$

has solutions. First, in the left-invariant local Lorentz frame, any constant spinor satisfies (3.17) with $M = +\frac{1}{2}$. The two independent solutions are left-invariant and transform as a doublet of $SU(2)_\mathcal{R}$. In addition, there are two independent solutions to (3.17) with $M = -\frac{1}{2}$ both of which are given by g^{-1} times a constant spinor. They are therefore right-invariant and form an $SU(2)_\mathcal{L}$ doublet. In this subsection, we choose the background $V_m = 0$, $M = \frac{1}{2}$.

Let us now turn to the computation of partition function using the localization principle. The supersymmetric saddle points are labeled by the constant value of the vector multiplet scalar $\sigma(x) = a$. The Chern-Simons or Fayet-Iliopoulos Lagrangians take nonzero value at the saddle point a according to the formula

$$e^{-S_{CS}} = e^{i\pi k \mathrm{Tr}(a^2)}, \quad e^{-S_{FI}} = e^{4\pi i \zeta a}, \tag{3.18}$$

In addition, we need the one-loop determinant which arise from integrating over all the fluctuation modes at the saddle point a under Gaussian (= one-loop) approximation.

We first study the vector multiplet for a non-abelian gauge symmetry G. Following the general prescription, we add to the original Lagrangian a SUSY exact regulator term $t_g \mathcal{L}_g$ and take $t_g \to \infty$. In this limit the regulator term dominates the path integral weight, and the Gaussian approximation becomes exact. The quadratic part of $\mathcal{L}_g$ in the Lorentz gauge $\partial^m A_m = 0$ is

$$\mathcal{L}_g = \mathrm{Tr}\left[\vec{A}\left(\Delta_{S^3}^{\mathrm{vector}} + a_{\mathrm{adj}}^2\right)\vec{A} + \hat{\sigma}\,\Delta_{S^3}^{\mathrm{scalar}}\hat{\sigma} + D^2 - \bar{\lambda}\left(-i\,\slashed{D}_{S^3} + \tfrac{1}{2} + i a_{\mathrm{adj}}\right)\lambda\right]. \tag{3.19}$$

Here we introduced the notation a_{adj} for a in the adjoint representation, namely $a_{\mathrm{adj}}\lambda \equiv [a, \lambda]$, and $\hat{\sigma}$ denotes the fluctuation of σ around its saddle point value a. To fix the gauge, we express the gauge field A as a sum of a divergenceless vector field $\hat{A}$ and a total derivative $d\varphi$, and insert the delta functional for φ. The Faddeev-Popov ghost determinant is trivial since gauge symmetry is just the shift of φ (up to terms irrelevant in the saddle-point approximation). But since $\mathrm{Tr}A_m A^m = \mathrm{Tr}(\hat{A}_m \hat{A}^m + \partial_m\varphi\partial^m\varphi)$, this change of integration variables gives rise to a Jacobian

$$\mathcal{D}A = \mathcal{D}\hat{A}\mathcal{D}'\varphi \cdot (\mathrm{Det}'\Delta_{S^3}^{\mathrm{scalar}})^{\frac{1}{2}\dim G}, \tag{3.20}$$

where the primes indicate that the constant modes are excluded. This Jacobian is canceled against the determinant arising from $\hat{\sigma}$-integration.

The integration over the remaining physical fields $\lambda, \bar{\lambda}$ and $\hat{A}$ gives rise to the following ratio of determinants,

$$\begin{aligned}
Z_{\mathrm{vec}}^{\text{1-loop}} &= \frac{\det_\lambda\left(-ia - \tfrac{1}{2} + i\,\slashed{D}_{S^3}\right)}{\det_{\hat{A}}(a^2 + \Delta_{S^3}^{\mathrm{vector}})^{\frac{1}{2}}} \\
&= \prod_{n\geq 0} \frac{[\det_{\mathrm{adj}}(-ia - n - 2)]^{(n+1)(n+2)} \cdot [\det_{\mathrm{adj}}(-ia + n + 1)]^{(n+1)(n+2)}}{[\det_{\mathrm{adj}}(a^2 + (n+2)^2)]^{(n+1)(n+3)}} \cdot \tag{3.21}
\end{aligned}$$

Let us take the Cartan-Weyl basis of G and assume that the saddle point parameter takes values in the Cargan subalgebra, namely $a = a_i H_i$ with H_i Cartan generators satisfying $\mathrm{Tr}(H_i H_j) = \delta_{ij}$. The above expression can then be rewritten further,

$$Z_{\text{vec}}^{\text{1-loop}} = \prod_{n \geq 1} n^{2\text{rk}G} \prod_{\alpha \in \Delta_+} (n^2 + (a \cdot \alpha)^2)^2 = (2\pi)^{\text{rk}G} \prod_{\alpha \in \Delta_+} \left(\frac{2 \sinh(\pi a \cdot \alpha)}{a \cdot \alpha} \right)^2,$$

$$(3.22)$$

where α runs over all the positive roots. The divergent infinite products were evaluated using zeta function regularization.

The constant value a of the scalar field can always be gauge-rotated into Cartan subalgebra. The domain of integration can therefore be reduced to Cartan subalgebra, but this in turn introduces a Vandermonde determinant in the measure which cancel nicely with the denominator of (3.22). The exact partition function for a theory with G vector multiplet is thus an integral over its Cartan subalgebra with the measure

$$\frac{1}{|\mathcal{W}|} \prod_i da_i \prod_{\alpha \in \Delta_+} \left(2 \sinh(\pi a \cdot \alpha) \right)^2. \qquad (3.23)$$

Here we modded out by the order of the Weyl group $\mathcal{W}$, which is the residual gauge symmetry after a has been gauge rotated into Cartan subalgebra.

Let us next turn to the matter fields. In the weak coupling limit, the action $\mathcal{L}_m$ for the matter fluctuations at the saddle point a is given by

$$\mathcal{L}_m = \bar{\phi}\{\Delta_{S^3}^{\text{scalar}} + a^2 + 2i(r-1)a + r(2-r)\}\phi + \bar{F}F + \bar{\psi}\left\{-i\slashed{D}_{S^3} + \frac{1}{2} + ia - r\right\}\psi.$$

$$(3.24)$$

Let us choose the basis vectors $\{|w\rangle\}$ of the matter representation R so as to diagonalize Cartan generators, i.e. $H_i|w\rangle = w_i|w\rangle$. Then the matter one-loop determinant becomes,

$$\begin{aligned} Z_{\text{matter}}^{\text{1-loop}} &= \frac{\det_\psi \left(\frac{1}{2} + ia - r - i\slashed{D}_{S^3} \right)}{\det_\phi (\Delta_{S^3}^{\text{scalar}} + 1 - (r - 1 - ia)^2)} \\ &= \prod_{n \geq 0} \frac{[\det_R(n + 2 + ia - r)]^{(n+1)(n+2)} \cdot [\det_R(-n - 1 + ia - r)]^{(n+1)(n+2)}}{[\det_R((n+1)^2 - (r - 1 - ia)^2)]^{(n+1)^2}} \\ &= \prod_{n=1}^\infty \prod_w \frac{(n + 1 - r + ia \cdot w)^n}{(n - 1 + r - ia \cdot w)^n} = \prod_w s_{b=1}(i(1 - r) - a \cdot w), \qquad (3.25) \end{aligned}$$

where w runs over all the weights of R.

We thus arrived at an integral formula for exact partition function of general 3D $\mathcal{N} = 2$ SUSY gauge theories on the unit round sphere. The basic building blocks for the integrand are the classical action evaluated at saddle points (3.18) and the matter one-loop determinant (3.25), and their product is integrated over the Cartan subalgebra of the gauge symmetry with the measure (3.23). For theories with matter mass, the mass parameter m of (3.9) enters into the one-loop determinant (3.25) in the same way as a, but we do not integrate over it.

3.4 Partition Function on Ellipsoids

Let us next consider the deformation from the round sphere to ellipsoids S_b^3 defined by (1.1). With a suitable polar coordinate system, the metric can be written as

$$ds^2 = \frac{1}{b^2} \cos^2 \theta d\varphi^2 + b^2 \sin^2 \theta d\chi^2 + f^2 d\theta^2,$$
$$f(\theta) = \sqrt{b^{-2} \sin^2 \theta + b^2 \cos^2 \theta}. \qquad (3.26)$$

A natural choice for the dreibein and the resulting spin connection are

$$e^1 = \frac{1}{b} \cos \theta d\varphi, \quad e^2 = b \sin \theta d\chi, \quad e^3 = f d\theta,$$
$$\omega^{12} = 0, \quad \omega^{13} = -\frac{1}{bf} \sin \theta d\varphi, \quad \omega^{23} = \frac{b}{f} \cos \theta d\chi. \qquad (3.27)$$

The ellipsoid can be made supersymmetric by turning on a suitable $U(1)_R$ gauge field in the background. This was found in [5] rather heuristically by taking a pair of Killing spinors on the (unit) round sphere with $V_m = U_m = 0$ and $M = \frac{1}{2}$,

$$\epsilon = \frac{1}{\sqrt{2}} \begin{pmatrix} -e^{\frac{i}{2}(\chi-\varphi+\theta)} \\ e^{\frac{i}{2}(\chi-\varphi-\theta)} \end{pmatrix}, \quad \bar{\epsilon} = \frac{1}{\sqrt{2}} \begin{pmatrix} e^{\frac{i}{2}(-\chi+\varphi+\theta)} \\ e^{\frac{i}{2}(-\chi+\varphi-\theta)} \end{pmatrix}, \qquad (3.28)$$

and studying the effect of squashing the metric. On the ellipsoid (3.26) they were found to satisfy the Killing spinor equation (3.2) with $U_m = 0$ and

$$V = -\frac{1}{2}\left(1 - \frac{1}{bf}\right)d\varphi + \frac{1}{2}\left(1 - \frac{b}{f}\right)d\chi, \quad M = \frac{1}{2f}. \qquad (3.29)$$

The supersymmetric observables on this background depend on the squashing parameter b in an nontrivial manner. Similar supersymmetric deformations from the round D-sphere into ellipsoids were studied for 4D $\mathcal{N} = 2$ theories by [33] and for 2D $\mathcal{N} = (2, 2)$ theories by [34].

Note that, in finding the dreibein, spin connection and background fields, the precise form of the function f is actually not needed as long as it is independent of φ and χ. It was pointed out in [35] that the above construction works for arbitrary smooth $f(\theta)$, with the only requirement coming from the smoothness at $\theta = 0$ and $\frac{\pi}{2}$,

$$f(\theta = 0) = b, \quad f\left(\theta = \frac{\pi}{2}\right) = \frac{1}{b}. \qquad (3.30)$$

More general supersymmetric backgrounds of sphere topology was studied in [29, 36], but it was also shown that supersymmetric observables depend on the background only through a single parameter b. See also [37, 38].

The partition function on the ellipsoid background can be computed again by applying the localization principle. First, the saddle points are given by the solutions to (3.11) and (3.12) as for the round sphere, and are therefore labeled by the constant value of the vector multiplet scalar σ. The value of the CS and FI actions S_{CS}, S_{FI} also remain the same as (3.18). However, the evaluation of the one-loop determinants on the ellipsoids (3.26) or other backgrounds with more general $f(\theta)$ becomes more complicated since one can no longer work out the full spectrum using spherical harmonics.

An alternative approach to compute the one-loop determinants is to study how the supersymmetry relates bosonic and fermionic eigenmodes of the Laplace or Dirac operators. Most of the eigenmodes are paired by the supersymmetry so that their net contribution to the one-loop determinant is trivial. It is therefore important to know the spectrum of the eigenmodes without superpartner.

Let us begin with a chiral multiplet in a representation R of the gauge group G. We first move to a new set of fields in terms of which the cancellation between bosonic and fermionic eigenvalues is most transparent. Let us introduce the Grassmann-odd scalar functions Ψ, $\bar{\Psi}$, Ψ', $\bar{\Psi}'$ and Grassmann-even scalars F', $\bar{F}'$ by

$$
\begin{aligned}
\psi = \epsilon \Psi' - \bar{\epsilon}\Psi, \quad & F = F' - i\bar{\epsilon}\gamma^m \epsilon D_m \phi, \\
\bar{\psi} = \bar{\epsilon}\bar{\Psi}' + \epsilon\bar{\Psi}, \quad & \bar{F} = \bar{F}' + i\epsilon\gamma^m \epsilon D_m \bar{\phi}.
\end{aligned}
\tag{3.31}
$$

They transform under supersymmetry as follows,

$$
\begin{aligned}
\delta\phi = \Psi, \quad & \delta\Psi = \mathcal{H}\phi, \quad & \delta\Psi' = F', \quad & \delta F' = \mathcal{H}\Psi', \\
\delta\bar{\phi} = \bar{\Psi}, \quad & \delta\bar{\Psi} = \mathcal{H}\bar{\phi}, \quad & \delta\bar{\Psi}' = \bar{F}', \quad & \delta\bar{F}' = \mathcal{H}\bar{\Psi}',
\end{aligned}
\tag{3.32}
$$

where $\mathcal{H}$ is the square of SUSY acting on scalar functions. To be more explicit, it acts on ϕ carrying the R-charge r as follows.

$$
\begin{aligned}
\mathcal{H}\phi &= i\bar{\epsilon}\gamma^m \epsilon D_m \phi - i\sigma\phi + \frac{r}{f}\phi \\
&= \left\{ -ib\partial_\varphi + ib^{-1}\partial_\chi - ia + \frac{Qr}{2} \right\}\phi. \quad (Q \equiv b + b^{-1})
\end{aligned}
\tag{3.33}
$$

Here the second equality holds up to non-linear terms which are irrelevant in the saddle point analysis.

To compute the one-loop determinant, we add a SUSY exact regulator $\mathcal{L}_{\text{reg}} = \delta V$ to the original Lagrangian with a large coefficient. We choose

$$
2V = (\bar{\phi}, \bar{F}') \begin{pmatrix} \mathcal{D} & 2\mathcal{D}_+ \\ 0 & 1 \end{pmatrix} \begin{pmatrix} \Psi \\ \Psi' \end{pmatrix} - (\bar{\Psi}, \bar{\Psi}') \begin{pmatrix} \mathcal{D} & 0 \\ 2\mathcal{D}_- & -1 \end{pmatrix} \begin{pmatrix} \phi \\ F' \end{pmatrix},
\tag{3.34}
$$

where

$$
\mathcal{D} \equiv \mathcal{H} + 2ia, \quad \mathcal{D}_+ \equiv -i\epsilon\gamma^m \epsilon D_m, \quad \mathcal{D}_- \equiv i\bar{\epsilon}\gamma^m \bar{\epsilon} D_m.
\tag{3.35}
$$

One can show that the operators $\mathcal{D}_\pm$ commutes with $\mathcal{H}$ by taking their R-charges ± 2 into account correctly. The regulator Lagrangian $\mathcal{L}_{\text{reg}}$ in the quadratic approximation consists of the following terms,

$$
\mathcal{L}_{\text{reg}}\big|_{\text{F}} = (\bar{\Psi}, \bar{\Psi}') \begin{pmatrix} \mathcal{D} & \mathcal{D}_+ \\ \mathcal{D}_- & -\mathcal{H} \end{pmatrix} \begin{pmatrix} \Psi \\ \Psi' \end{pmatrix},
$$

$$
\mathcal{L}_{\text{reg}}\big|_{\text{B}} = (\bar{\phi}, \bar{F}') \begin{pmatrix} \mathcal{D}\mathcal{H} & \mathcal{D}_+ \\ -\mathcal{D}_- & 1 \end{pmatrix} \begin{pmatrix} \phi \\ F' \end{pmatrix} = \bar{F}F + \bar{\phi}\Delta\phi,
$$

$$
\Delta = \mathcal{D}\mathcal{H} + \mathcal{D}_+\mathcal{D}_- = a^2 - (\bar{\epsilon}\gamma^m \epsilon D_m)^2 + \epsilon\gamma^m \epsilon D_m \cdot \bar{\epsilon}\gamma^n \bar{\epsilon} D_n. \tag{3.36}
$$

Note that the bosonic part is positive definite. Thus the one-loop determinant is given by the ratio of the determinants for the Dirac operator (the 2×2 matrix in the first line of (3.36)) and the Laplace operator Δ.

As was shown in [5], generically a scalar eigenmode of Δ and a pair of Dirac eigenmodes form a multiplet which yields no net contribution to the one-loop determinant. The modes which do not participate in this multiplet structure arise from ϕ in the kernel of $\mathcal{D}_-$ and Ψ' in the kernel of $\mathcal{D}_+$. It is easy to see from the matrix expression for $\mathcal{L}_{\text{reg}}$ that the one-loop determinant is given by the ratio of determinants of $\mathcal{H}$ evaluated on such modes,

$$
Z_{\text{mat}}^{\text{1-loop}} = \frac{\det_{\Psi'}(-\mathcal{H})\big|_{\text{Ker}\mathcal{D}_+}}{\det_\phi(\mathcal{H})\big|_{\text{Ker}\mathcal{D}_-}}. \tag{3.37}
$$

The spectrum of $\mathcal{H}$ which is relevant for the above one-loop determinant can be explicitly worked out. First, let us consider the spectrum of $\mathcal{H}$ on the scalar ϕ of R-charge r which is annihilated by $\mathcal{D}_-$. Assuming the form $\phi = \hat{\phi}(\theta)e^{im\varphi - in\chi}$, one finds

$$
e^{i(\chi-\varphi)}\bar{\epsilon}\gamma^n\bar{\epsilon}D_n\phi = \left\{ -\frac{b\sin\theta}{\cos\theta}(m - rV_\varphi) + \frac{\cos\theta}{b\sin\theta}(n + rV_\chi) - \frac{1}{f}\partial_\theta \right\}\phi = 0,
$$

$$
\mathcal{H}\phi = \left\{ mb + nb^{-1} - ia + \frac{Qr}{2} \right\}\phi. \tag{3.38}
$$

The first equation determines the form of $\hat{\phi}(\theta)$. In particular, from its behavior near the two ends $\theta = 0$ and $\frac{\pi}{2}$,

$$
\hat{\phi}(\theta) \sim \cos^m \theta \sin^n \theta, \tag{3.39}
$$

it follows that the eigenmode is normalizable only when $m, n \geq 0$. The same analysis can be repeated for the scalar Ψ' of R-charge $(r - 2)$ in the kernel of $\mathcal{D}_+$. We thus obtain the matter one-loop determinant

$$Z_{\text{mat}}^{\text{1-loop}} = \frac{\prod_{m,n\geq 0} \det_R \left(mb + nb^{-1} + ia - \frac{Q(r-2)}{2}\right)}{\prod_{m,n\geq 0} \det_R \left(mb + nb^{-1} - ia + \frac{Qr}{2}\right)}$$

$$= \prod_w s_b \left(\tfrac{iQ}{2}(1-r) - a \cdot w\right), \tag{3.40}$$

where w runs over all the weight vectors in the representation R. This generalizes the formula (3.25) on the round sphere.

The form of the matter one-loop determinant (3.37) shows that it can be computed from the index of the differential operators $\mathcal{D}_\pm$ which commute with $\mathcal{H}$. In [39] the relevant index was analyzed by regarding the ellipsoid as a Hopf fibration with the fiber direction $\partial_\varphi - \partial_\chi$. By decomposing the fields into Fourier modes carrying different KK momentum along the fiber, one can reduce the index to that of a differential operator on S^2 and apply the fixed point formula.

Let us next consider vector multiplet. Our starting point is the following formula for the one-loop determinant,

$$Z_{\text{vec}}^{\text{1-loop}} = \frac{\det_\lambda \left(-ia - \frac{1}{2f} + i\slashed{D}\right)}{\det_{\hat{A}}(a^2 + \Delta^{\text{vector}})^{1/2}} = \frac{\det_\lambda \left(-ia - \frac{1}{2f} + i\slashed{D}\right)}{\det_{\hat{A}}(-ia - *d)}, \tag{3.41}$$

which follows from the same gauge fixing procedure as for the round sphere (3.21). As before, the denominator is the determinant evaluated on the space of divergenceless vector wave functions. We evaluate this by finding out the maps between the spinor and vector eigenmodes,

$$v\lambda = -ia_{\text{adj}}\lambda + i\slashed{D}\lambda - \frac{1}{2f}\lambda, \tag{3.42}$$

$$v\hat{A}^m = -ia_{\text{adj}}\hat{A}^m - \varepsilon^{mnp}\partial_n \hat{A}_p, \quad D_m\hat{A}^m = 0. \tag{3.43}$$

We first notice the following identity holds for arbitrary vector field A_m.

$$\left(i\slashed{D} - \frac{1}{2f}\right)(\gamma^m \epsilon A_m) = i\epsilon \cdot D_m A^m - \gamma_m \epsilon \cdot \varepsilon^{mnp}\partial_n A_p. \tag{3.44}$$

It follows that, for each generic vector eigenmode $\hat{A}_m$, one can construct a spinor eigenmode λ of the same eigenvalue by the map $\lambda[\hat{A}] = \gamma^m \epsilon \hat{A}_m$. This map fails for the vector eigenmodes satisfying $\gamma^m \epsilon \hat{A}_m = 0$. Such modes can be expressed in terms of a scalar Y with R-charge -2 as,

$$\hat{A}_m = \epsilon \gamma_m \epsilon \cdot Y. \tag{3.45}$$

The divergence-free condition and the eigenmode equation (3.43) are translated into the following conditions on Y,

$$\mathcal{D}_+ Y = 0, \quad \mathcal{H}Y = \nu Y. \tag{3.46}$$

The normalizable solutions for Y are in one-to-one correspondence with the vector eigenmodes without spinor superpartners.

Next we notice that the following identity holds for arbitrary spinor λ,

$$\bar{\epsilon}\gamma_m\left(i\not{D}\lambda - \frac{1}{2f}\lambda\right)dx^m - id(\bar{\epsilon}\lambda) = - *d(\bar{\epsilon}\gamma_m\lambda dx^m). \tag{3.47}$$

It follows that, for each generic spinor eigenmode λ, one can construct the corresponding vector eigenmode $\hat{A}$ by the following map,

$$\hat{A}[\lambda] = (\nu + ia_{\mathrm{adj}})\bar{\epsilon}\gamma_m\lambda dx^m - id(\bar{\epsilon}\lambda). \tag{3.48}$$

To find the kernel of this map, let us introduce two scalar functions Λ_0, Λ_2 and denote $\lambda = \epsilon\Lambda_0 + \bar{\epsilon}\Lambda_2$. Then $\hat{A}[\lambda]$ vanishes when

$$\mathcal{D}_-\Lambda_0 = 0, \quad \mathcal{H}\Lambda_0 = \nu\Lambda_0, \quad \mathcal{D}_+\Lambda_0 = 2(\nu + ia_{\mathrm{adj}})\Lambda_2. \tag{3.49}$$

For any λ in the kernel, one can show by applying $*d$ onto (3.48) that the right hand side of (3.47) vanishes as long as $(\nu + ia_{\mathrm{adj}})$ is nonzero. Using this one can show that generic elements λ in the kernel automatically satisfies the eigenvalue equation (3.42). The only exceptional element in the kernel is $\lambda = \epsilon$ which does not satisfy (3.42), corresponding to $\Lambda_0 = \mathrm{const}$, $\Lambda_2 = 0$ and $\nu = -ia_{\mathrm{adg}}$. The normalizable solutions to (3.49) are thus in almost one-to-one correspondence with the spinor eigenodes without vector superpartners.

Thus the one-loop determinant for vector multiplet can be expressed again as the ratio of determinants of $\mathcal{H}$,

$$Z_{\mathrm{vec}}^{\text{1-loop}} = \frac{\det'_{\Lambda_0}(\mathcal{H})_{\mathrm{Ker}\mathcal{D}_-}}{\det_Y(\mathcal{H})_{\mathrm{Ker}\mathcal{D}_+}}, \tag{3.50}$$

where the prime in the enumerator indicates that the contribution from constant modes is excluded. Apart from this minor difference, it is just the inverse of the matter one-loop determinant for $r = 0$, $R = \mathrm{adj}$. Up to an a-independent overall constant, we obtain

$$Z_{\mathrm{vec}}^{\text{1-loop}} = \prod_{\alpha \in \Delta} \frac{s_b\left(a \cdot \alpha - \frac{iQ}{2}\right)}{(-ia \cdot \alpha)} = \prod_{\alpha \in \Delta_+} \frac{4\sinh(\pi ba \cdot \alpha)\sinh(\pi b^{-1}a \cdot \alpha)}{(a \cdot \alpha)^2}. \tag{3.51}$$

The general formula for the ellipsoid partition function can be summarized as follows. Vector multiplets yield the integration measure over Cartan subalgebra of the gauge symmetry algebra,

$$\frac{1}{|\mathcal{W}|} \prod_{i=1}^{r} da_i \prod_{\alpha \in \Delta_+} 4 \sinh(\pi ba \cdot \alpha) \sinh(\pi b^{-1} a \cdot \alpha), \tag{3.52}$$

chiral multiplets yields the determinants,

$$\prod_{w \in R} s_b \left(\frac{iQ}{2}(1 - r) - a \cdot w \right), \tag{3.53}$$

and the classical Lagrangians make the following contribution to the integrand.

$$e^{-S_{CS}} = e^{i\pi ka \cdot a}, \quad e^{-S_{FI}} = e^{4\pi i \zeta a}. \tag{3.54}$$

Let us compare the above formula with the known result in pure Chern-Simons theory [40]. Using the above formula together with Weyl denominator formula

$$\prod_{\alpha \in \Delta_+} 2 \sinh(\pi \alpha \cdot u) = \sum_{w \in \mathcal{W}} \epsilon(w) e^{2\pi w(\rho) \cdot u}, \tag{3.55}$$

one can express the partition function for pure SUSY Chern-Simons theory as a sum of simple Gaussian integrals. Assuming the level k to be positive, one finds

$$Z_{CS} = \frac{1}{|\mathcal{W}|} \int \prod_{i=1}^{r} da_i \prod_{\alpha \in \Delta_+} 4 \sinh(\pi ba \cdot \alpha) \sinh(\pi b^{-1} a \cdot \alpha) \cdot \exp(i\pi ka \cdot a)$$

$$= \exp \left(\frac{i\pi}{4} \dim G + \frac{i\pi}{12k}(b^2 + b^{-2}) y \dim G \right) \cdot k^{\frac{r}{2}} \prod_{\alpha \in \Delta_+} 2 \sin \left(\frac{\pi \alpha \cdot \rho}{k} \right). \tag{3.56}$$

Here y is the dual Coxeter number of G and ρ is the Weyl vector. We also used the formula

$$\rho^2 = \frac{1}{12} \dim G \, y. \tag{3.57}$$

Apart from some phase factors, we recover the the known answer for bosonic Chern-Simons theory at the level $k - y$. The mismatch in the level is because there is no finite renormalization of the Chern-Simons level for the case with $\mathcal{N} = 2$ supersymmetry [41].

3.5 Loop Observables

Here we introduce two kinds of supersymmetric Loop operators, the Wilson and vortex loops, and present the formulae for their expectation values. Similar loop operators in 4D $\mathcal{N} = 2$ theories are reviewed in [V:7] and play important role in understanding the AGT relation.

Supersymmetric Wilson loop operator is defined by

$$W_R(C) \equiv \mathrm{Tr}_R \mathrm{P} \exp \oint_C (iA + \sigma d\ell), \tag{3.58}$$

where C is a closed loop that winds along the direction of the Killing vector field $\bar{\epsilon} \gamma^m \epsilon$, and $d\ell$ denotes the length element along C. For theories on the unit round S^3 where the Killing vector is along the circle fiber of Hopf fibration, any C is a great circle of radius 2π. The expectation value of Wilson loops can be calculated in the same way as partition function, by just inserting into the integrand their classical value at the saddle point a,

$$W_R(C)\big|_{\mathrm{saddle}} = \mathrm{Tr}_R(e^{2\pi a}). \tag{3.59}$$

For theories on the ellipsoids with generic squashing parameter b (b^2 being irrational), the only supersymmetric closed loops are the ones at $\theta = 0$ and $\theta = \frac{\pi}{2}$ in the polar coordinate system (3.26), since no other curves along the Killing vector $\bar{\epsilon} \gamma^m \epsilon$ form closed loops. The two choices lead to different expectation values since they have radii b^{-1} and b, respectively.

$$W_R(\theta = 0)\big|_{\mathrm{saddle}} = \mathrm{Tr}_R(e^{2\pi a/b}), \quad W_R\left(\theta = \frac{\pi}{2}\right)\big|_{\mathrm{saddle}} = \mathrm{Tr}_R(e^{2\pi ab}). \tag{3.60}$$

There are additional supersymmetric loops for special values of the squashing parameter. When $b = \sqrt{p/q}$ with (p, q) coprime integers, torus knots winding p and q times along the φ and χ-directions at fixed $\theta \neq 0, \frac{\pi}{2}$ become supersymmetric [42].

The vortex loop is a one-dimensional defect along which the gauge field develops a singularity. For a vortex line lying along the z-axis of the flat Euclidean $\mathbb{R}^3(x, y, z)$, the gauge field strength has delta function singularity along the line,

$$F_{xy} = 2\pi H\delta(x)\delta(y) + \text{regular}, \tag{3.61}$$

where the flux H takes values in the Cartan subalgebra of the gauge symmarty algebra. In terms of the polar coordinate system on the xy-plane ($x + iy = re^{i\theta}$), the singular behavior of the gauge field near the vortex line is given by $A_\theta = H$. Also, it follows from (3.4) that we need to impose singular boundary condition on D as well,

$$D = 2\pi i H\delta(x)\delta(y) + \text{regular}, \tag{3.62}$$

in order to avoid the transformation rule of λ and $\bar{\lambda}$ becoming singular.

For a vortex loop to be supersymmetric, it has to lie along the direction of the Killing vector $\bar{\epsilon} \gamma^m \epsilon$. We orient the vortex loops so that the $+z$ direction always agrees with the direction of Killing vector and define the flux H accordingly. For generic ellipsoid backgrounds, supersymmetric vortex loops can only lie along the direction of $(-\varphi)$ at $\theta = 0$, or the direction of $(+\chi)$ at $\theta = \frac{\pi}{2}$. These two vortex loops are expressed by the flat gauge fields,

$$(\theta = 0)\ \ A = Hd\chi, \qquad \left(\theta = \frac{\pi}{2}\right)\ \ A = -Hd\varphi. \tag{3.63}$$

Let us hereafter restrict the discussion to the vortex loops in abelian gauge theory and evaluate their expectation value. First, notice that the introduction of a vortex loop with flux H in Chern-Simons theory at level k induces a Wilson loop with charge $-kH$. To see this, let us decompose the vector multipet fields in the presence of a vortex loop into the singular and regular parts, $A = A_{\text{sing}} + A_{\text{reg}}$. Then the SUSY Chern-Simons action integral for such A becomes

$$S_{\text{CS}}[A_{\text{sing}} + A_{\text{reg}}] = ikH \oint_C (A_{\text{reg}} - i\sigma_{\text{reg}}d\ell) + S_{\text{CS}}[A_{\text{reg}}]. \tag{3.64}$$

Therefore, the value of classical Chern-Simons action at the saddle point a gets shifted because of the vortex loop as

$$e^{i\pi k(a^2 + 2iab^{-1}H)} \quad \text{or} \quad e^{i\pi k(a^2 + 2iabH)}. \tag{3.65}$$

The value of the FI term $e^{-S_{\text{FI}}} = e^{4\pi i \zeta a}$ remains the same. Now one can go through the evaluation of the one-loop determinant again, where the only difference is that there is a nonzero flat gauge field in addition to a constant scalar a. Since it enters in the operator $\mathcal{H}$ as follows,

$$\mathcal{H}\phi = \left\{ -ib(\partial_\varphi - iA_\varphi) + ib^{-1}(\partial_\chi - iA_\chi) - ia + \frac{Qr}{2} \right\} \phi, \tag{3.66}$$

the effect of the vortex loop can be incorporated by shifting a in our previous formula by $-ibA_\varphi + ib^{-1}A_\chi$. Depending on whether the vortex loop is put at $\theta = 0$ or $\frac{\pi}{2}$, the saddle point parameter a is shifted by $ib^{-1}H$ or ibH.

Since the parameter a is to be integrated over, the shift of a by $ib^{\pm 1}H$ can be undone by shifting its integration contour. This also eliminates the shift of classical Chern-Simons action by a Wilson line. As a result, the effect of a vortex loop of flux H in abelian Chern-Simons theory at level k, FI coupling ζ just amounts to a multiplication of the factor

$$\exp\left(i\pi k b^{\mp 2} H^2 + 4\pi \zeta b^{\mp 1} H\right). \tag{3.67}$$

Our argument so far assumed that H is small. The computation of one-loop determinants on ellipsoids was based on the spectrum of normalizable eigenmodes of $\mathcal{H}$ in the kernel of the operators $\mathcal{D}_\pm$, but normalizability of the eigenmodes is affected by nonzero H. Also, the shift of a-integration contour may hit poles in the integrand. See [39, 43] for further discussions.

Vortex loops can also be introduced for flavor symmetry of matter chiral multiplets, by coupling the corresponding current to a singular background gauge field with nonzero flux H localized along a loop. Its effect is similar to that of real mass deformation, namely we have the appearance of $ib^{\mp 1}H$ in place of the real mass m in the matter one-loop determinants.

4   4D Superconformal Index

Superconformal index was introduced for 4D $\mathcal{N} = 1$ superconformal field theories by Römelsberger [11, 13] and for more general cases by Kinney et al. [12], as a quantity which encodes the spectrum of BPS operators. In superconformal theories, the spectrum of BPS operators is in correspondence with the spectrum of states in radial quantization. The index can therefore be formulated in terms of path integral on $S^1 \times S^3$, with an appropriate periodicity condition along the S^1. The periodicity can be twisted by various symmetries of the theory in such a way to preserve part of SUSY. The index is then a function of the fugacity variables that parametrize the twist.

The superconformal index is invariant under any SUSY-preserving continuous deformation of the theory and, in particular, independent of the gauge coupling. The indices of nontrivial theories at the RG fixed point can therefore be evaluated using the weak coupling description at high energy where saddle point approximation becomes exact.

Here we present the path integral derivation of the superconformal index for 4D $\mathcal{N} = 1$ SUSY theories. Our purpose here is to explain the connection between 3D partition functions on S^3 and 4D superconformal indices which was studied in [15–17]. Interestingly, some of the fugacity variables turn into parameters of supersymmetric deformations of the round S^3 upon dimensional reduction. As an important example, we reproduce two inequivalent SUSY backgrounds which are both based on the same squashed S^3 with $SU(2) \times U(1)$ isometry but characterized by different Killing spinor equations [5, 18].

The superconformal indices for 4D $\mathcal{N} = 2$ theories of class S are in correspondence with partition function of 2D q-deformed Yang-Mills theory, as reviewed in [V:9] in this volume.

4.1   4D $\mathcal{N} = 1$ SUSY Theories

We again begin by fixing the notations. In four dimensions there are two kinds of doublet spinors ψ_α and $\bar\psi^{\dot\alpha}$, corresponding to two copies of $SU(2)$ that form the 4D rotation symmetry. Their spinor indices are raised or lowered by antisymmetric ϵ

tensors with nonzero elements $\epsilon^{12} = -\epsilon_{12} = 1$. We introduce the 2×2 matrices,

$$\sigma_a = \bar{\sigma}_a = \text{Pauli matrix } (a = 1, 2, 3); \quad \sigma_4 = i, \quad \bar{\sigma}_4 = -i, \qquad (4.1)$$

with index structure $(\sigma_a)_{\alpha\dot{\beta}}$ and $(\bar{\sigma}_a)^{\dot{\alpha}\beta}$, satisfying standard algebra. We also use $\sigma_{ab} \equiv \frac{1}{2}(\sigma_a\bar{\sigma}_b - \sigma_b\bar{\sigma}_a)$ and $\bar{\sigma}_{ab} \equiv \frac{1}{2}(\bar{\sigma}_a\sigma_b - \bar{\sigma}_b\sigma_a)$.

Although 4D $\mathcal{N} = 1$ supersymmetric theories on general curved backgrounds and the equations for Killing spinors can be obtained from off-shell supergravity [6], here we take a heuristic approach. We consider the following Killing spinor equation,

$$D_m\epsilon = \sigma_m\bar{\kappa}, \quad D_m\bar{\epsilon} = \bar{\sigma}_m\kappa \quad \text{for some } \kappa, \bar{\kappa}. \qquad (4.2)$$

where the covariant derivative D_m contains the gauge field V_m for $U(1)_R$ under which $\epsilon, \bar{\epsilon}$ are charged $+1, -1$. Using these Killing spinors we set the transformation rule for $\mathcal{N} = 1$ vector multiplets,

$$\delta A_m = \frac{i}{2}(\epsilon\sigma_m\bar{\lambda} - \bar{\epsilon}\bar{\sigma}_m\lambda),$$

$$\delta\lambda = \frac{1}{2}\sigma^{mn}\epsilon F_{mn} - \epsilon D,$$

$$\delta\bar{\lambda} = \frac{1}{2}\bar{\sigma}^{mn}\bar{\epsilon} F_{mn} + \bar{\epsilon} D,$$

$$\delta D = -\frac{i}{2}\epsilon\sigma^m D_m\bar{\lambda} - \frac{i}{2}\bar{\epsilon}\bar{\sigma}^m D_m\lambda, \qquad (4.3)$$

and chiral multiplets,

$$\delta\phi = -\epsilon\psi, \quad \delta\psi = i\sigma^m\bar{\epsilon} D_m\phi + \frac{3ir}{4}\sigma^m D_m\bar{\epsilon}\phi + \epsilon F,$$

$$\delta\bar{\phi} = +\bar{\epsilon}\bar{\psi}, \quad \delta\bar{\psi} = i\bar{\sigma}^m\epsilon D_m\bar{\phi} + \frac{3ir}{4}\bar{\sigma}^m D_m\epsilon\bar{\phi} + \bar{\epsilon}\bar{F},$$

$$\delta F = i\bar{\epsilon}\bar{\sigma}^m D_m\psi + \frac{i(3r-2)}{4}D_m\bar{\epsilon}\bar{\sigma}^m\psi - i\bar{\epsilon}\bar{\lambda}\phi,$$

$$\delta\bar{F} = -i\epsilon\sigma^m D_m\bar{\psi} - \frac{i(3r-2)}{4}D_m\epsilon\sigma^m\bar{\psi} - i\epsilon\bar{\phi}\lambda. \qquad (4.4)$$

Here r is the R-charge of the field ϕ. The scaling weight and the R-charge of the fields are summarized in Table 2.

Table 2 The scaling weight and the R-charge of the fields

Fields	ϵ	$\bar{\epsilon}$	A_a	λ	$\bar{\lambda}$	D	ϕ	$\bar{\phi}$	ψ	$\bar{\psi}$	F	$\bar{F}$
Weight	$-\frac{1}{2}$	$-\frac{1}{2}$	1	$\frac{3}{2}$	$\frac{3}{2}$	2	$\frac{3r}{2}$	$\frac{3r}{2}$	$\frac{3r+1}{2}$	$\frac{3r+1}{2}$	$\frac{3r+2}{2}$	$\frac{3r+2}{2}$
R-charge	1	-1	0	1	-1	0	r	$-r$	$r-1$	$1-r$	$r-2$	$2-r$

Given a 3D background $\mathcal{M}$ with a pair of Killing spinors $\epsilon, \bar{\epsilon}$ satisfying (3.2) and (3.3), one can construct a 4D $\mathcal{N} = 1$ supersymmetric background $\mathcal{M} \times \mathbb{R}$ by choosing the metric and the $U(1)_R$ gauge field as follows.

$$ds^2_{(4D)} = e^a e^a = ds^2_{\mathcal{M}} + dt^2 \ (e^4 \equiv dt), \quad V_{(4D)} = V_{(3D)} - iM dt. \tag{4.5}$$

The 3D Killing spinors $\epsilon, \bar{\epsilon}$ are promoted to 4D Killing spinors satisfying

$$D_m \epsilon = -M\sigma_m \bar{\sigma}_4 \epsilon, \quad D_m \bar{\epsilon} = M\bar{\sigma}_m \sigma_4 \bar{\epsilon}. \tag{4.6}$$

The following supersymmetric Lagrangians on this background are relevant in the computation of the index.

$$\begin{aligned}
\mathcal{L}_g &= \mathrm{Tr}\left(\frac{1}{2} F_{mn} F^{mn} + D^2 + i\bar{\lambda}\bar{\sigma}^m D_m \lambda\right), \\
\mathcal{L}_m &= D_m \bar{\phi} D^m \phi + (3r-2)M(D_4\bar{\phi}\phi - \bar{\phi}D_4\phi) \\
&\quad + \left\{\frac{rR}{4} - 3r(3r-2)M^2\right\}\bar{\phi}\phi - i\bar{\phi}D\phi \\
&\quad - i\bar{\psi}\bar{\sigma}^m D_m \psi - i(3r-2)M\bar{\psi}\bar{\sigma}_4\psi + i\bar{\psi}\bar{\lambda}\phi + i\bar{\phi}\lambda\psi + \bar{F}F.
\end{aligned} \tag{4.7}$$

Here D_4 is the fourth component of $D_a \equiv e^m_a D_m$.

It is a useful observation that the above 4D transformation rules and Lagrangians can actually be obtained from the corresponding 3D quantities by the simple replacement $\sigma \to A_t + i\partial_t$.

4.2 Path Integral Formulation of the Index

Let us choose $\mathcal{M}$ to be the unit round sphere and set $M = \frac{1}{2}, V = -\frac{i}{2}dt$. The Killing spinor equation (4.6) on this background has two independent solutions for each of ϵ and $\bar{\epsilon}$, which are all constant spinors in the left-invariant frame. Besides these four solutions, there are four solutions to (4.6) with the right hand side sign-flipped. These eight solutions correspond to the eight supercharges in the 4D $\mathcal{N} = 1$ superconformal algebra, but the Lagrangians in (4.7) with $M = \frac{1}{2}$ are invariant only under the first four.

From the four Killing spinors satisfying (4.6), let us pick up the two characterized by $\gamma_3 \epsilon = -\epsilon$ and $\gamma_3 \bar{\epsilon} = \bar{\epsilon}$, and denote the corresponding supercharges by $\mathbf{S}$ and $\mathbf{Q}$. The R-charges and $SU(2)_R$ spins of $\mathbf{S}, \mathbf{Q}$ are opposite to those of the corresponding Killing spinors, so $\mathbf{S}$ has $\mathbf{R} = -1, \mathbf{J}^3_R = +\frac{1}{2}$ while $\mathbf{Q}$ has $\mathbf{R} = 1, \mathbf{J}^3_R = -\frac{1}{2}$. The anticommutator of $\mathbf{S}$ and $\mathbf{Q}$ can be found from the algebra of the corresponding SUSY transformations acting on fields. With a suitable normalization of $\epsilon, \bar{\epsilon}$ one finds

$$\{\mathbf{S}, \mathbf{Q}\} = -\partial_t + iA_t - 2\mathbf{J}^3_R - \mathbf{R}, \tag{4.8}$$

where A_t is the component of dynamical gauge field and $\mathbf{J}_{\mathcal{R}}^a$ is the sum of isometry rotation of S^3 and local Lorentz rotation. Note that, since we have turned on the background $U(1)_R$ gauge field so that the Killing spinors corresponding to $\mathbf{S}, \mathbf{Q}$ are time independent, the time derivative $-\partial_t + iA_t$ should not be simply related to the dilation $\mathbf{D}$. Rather it should be identified with $\mathbf{D} - \frac{1}{2}\mathbf{R}$ which commutes with the supercharges $\mathbf{S}$ and $\mathbf{Q}$. Thus we have reproduced an important subalgebra of the 4D $\mathcal{N} = 1$ superconformal algebra,

$$\{\mathbf{S}, \mathbf{Q}\} = \mathbf{D} - 2\mathbf{J}_{\mathcal{R}}^3 - \frac{3}{2}\mathbf{R} \equiv \mathbf{H}. \tag{4.9}$$

Now let us compactify the time direction $t \sim t + \beta$. The path integral on the resulting background $S^3 \times S^1$ defines the superconformal index. In the simplest example where all the fields obey periodic boundary condition, one obtains

$$I = \mathrm{Tr}\left[(-1)^{\mathbf{F}} q^{\mathbf{D} - \frac{1}{2}\mathbf{R}}\right]. \quad (q \equiv e^{-\beta}) \tag{4.10}$$

This form can be generalized by twisting the periodicity of fields by various symmetries which commute with the supercharges $\mathbf{S}, \mathbf{Q}$. Some of such symmetries are in the superconformal algebra. The Cartan subalgebra of its bosonic part is generated by the dilation $\mathbf{D}$, the $U(1)$ R-charge $\mathbf{R}$ and the two rotation generators $\mathbf{J}_{\mathcal{L}}^3, \mathbf{J}_{\mathcal{R}}^3$, of which three linear combinations commute with $\mathbf{S}$ and $\mathbf{Q}$. Also, in theories with additional global symmetry, one can use any of its elements $\mathbf{m}$ to modify the periodicity. The fully generalized index is then given by

$$I = \mathrm{Tr}\left[(-1)^{\mathbf{F}} q^{\mathbf{D} - \frac{1}{2}\mathbf{R}} x^{2\mathbf{J}_{\mathcal{R}}^3 + \mathbf{R}} y^{2\mathbf{J}_{\mathcal{L}}^3} e^{i\mathbf{m}\beta}\right], \quad q = e^{-\beta}, x = e^{i\beta\xi}, y = e^{i\beta\eta} \tag{4.11}$$

and is a function of the fugacity parameters ξ, η and $\mathbf{m}$ as well as β. An important remark here is that the only states which contribute to the index are those annihilated by the supercharges $\mathbf{Q}, \mathbf{S}$ and also by their anticommutator $\mathbf{H}$. The index therefore depends on q and x only through their product $qx = e^{-\beta + i\beta\xi}$.

The index (4.11) is given by a path integral over fields obeying twisted periodicity condition. By a suitable field redefinition, it can be rewritten into a path integral over ordinary periodic fields but with a deformed Lagrangian. In this process, the twists by R- or flavor symmetries turn into a constant background gauge fields along the t direction. On the other hand, the twist by rotational symmetries means Scherk-Schwarz like compactification,

$$(t, g) \sim (t + \beta, e^{-i\beta\eta\gamma^3} g e^{i\beta\xi\gamma^3}). \tag{4.12}$$

This can be brought into a system with ordinary time periodicity by a suitable change of coordinates, but then the metric written in the new coordinates squires off-diagonal components

$$ds^2 = dt^2 + g_{mn}^{(S^3)}(dx^m + u^m dt)(dx^n + u^n dt), \quad u \equiv 2i\xi \mathcal{R}^3 + 2i\eta \mathcal{L}^3. \quad (4.13)$$

Here the vector fields $\mathcal{L}^a$, $\mathcal{R}^a$ are properly normalized generators of $SU(2)_{\mathcal{L},\mathcal{R}}$. In fact, the effect of this deformation of the metric on field theory is simply to modify the time derivative ∂_t by the rotation generator. For example, the kinetic term for a free scalar becomes

$$\frac{1}{2}(\partial_t \phi - u^m \partial_m \phi)^2 + \frac{1}{2}g_{(S^3)}^{mn} \partial_m \phi \partial_n \phi. \quad (4.14)$$

A little more work shows that, for spinor fields, the time derivative is modified by a combination of $u^m \partial_m$ and a local Lorentz transformation which makes precisely the action of the rotation symmetry. Summarizing, the general index (4.11) can be computed by path integral over periodic fields on $S^1 \times S^3$, with the following replacement in the Lagrangian (4.7)

$$i\partial_t \longmapsto i\hat{\partial}_t \equiv i\partial_t + \xi(2\mathbf{J}_{\mathcal{R}}^3 + \mathbf{R}) + 2\eta \mathbf{J}_{\mathcal{L}}^3 + \mathbf{m}. \quad (4.15)$$

4.3 Evaluation of the Index

Let us turn to the evaluation of the index. Since the index is invariant under deformations preserving the algebra of $\mathbf{S}, \mathbf{Q}, \mathbf{H}$, we introduce the sum of $\mathcal{L}_g$ and $\mathcal{L}_m$ in (4.7) with a large overall coefficient into the path integral weight so that the argument of exact saddle point analysis apply. This time, the saddle points are labeled by the constant value of gauge field along time direction $A_t = a$.

Let us evaluate the one-loop determinant, first for the vector multiplet with gauge group G. It is most convenient to work in the temporal gauge $A_t = a$, for which we need to introduce ghosts with kinetic term $\mathrm{Tr}(\bar{c}D_t c)$. The Gaussian integral over fluctuations gives

$$\frac{\mathrm{Det}_\lambda \left(\hat{\partial}_t - ia - \frac{1}{2} + i\not{D}_{S^3}\right) \mathrm{Det}_c'(\hat{\partial}_t - ia)}{\mathrm{Det}_A(-(\hat{\partial}_t - ia)^2 + \Delta_{S^3}^{\mathrm{vector}})^{\frac{1}{2}}}. \quad (4.16)$$

Here the prime indicates that the constant modes of the ghosts are excluded, and $\hat{\partial}_t$ is defined in (4.15). Expanding the fields into spherical harmonics which diagonalizes the Laplace or Dirac operators on S^3, the above determinant can be rewritten into an infinite product of 1D Dirac determinants on the circle of circumference β,

$$\det(\partial_t - ix) = \prod_{k \in \mathbb{Z}}(2\pi ik\beta^{-1} - ix) = -2i \sin \frac{\beta x}{2}. \quad (4.17)$$

The integral over the ghost modes with $SU(2)_{\mathcal{L}} \times SU(2)_{\mathcal{R}}$ spin $(0, 0)$ yields

$$\det'_{\text{adj}}(\partial_t - ia) = \beta^{\text{rk}G} \prod_{\alpha \in \Delta_+} \left(\frac{2\sin(\beta \alpha \cdot a/2)}{\alpha \cdot a} \right)^2, \tag{4.18}$$

where we assumed a to take values in Cartan torus. Combined with the Vandermonde determinant, this gives an appropriate measure factor for the integration over Cartan torus.

$$d\mu(a) = \frac{1}{|\mathcal{W}|} \prod_{i=1}^{r} \frac{d\hat{a}_i}{2\pi} \prod_{\alpha \in \Delta_+} 4\sin^2 \frac{\alpha \cdot \hat{a}}{2}. \quad (\hat{a} \equiv \beta a) \tag{4.19}$$

The integral over the remaining modes of all the fields gives, after an enormous cancellation between bosonic and fermionic contributions, the following.

$$I_{\text{vec}} = \prod_{n \geq 1} \det_{\text{adj}}\left(\partial_t - ia + n(1 - i\xi + i\eta)\right)\det_{\text{adj}}\left(\partial_t - ia + n(1 - i\xi - i\eta)\right).$$

$$= I_0(q_1, q_2)^{\text{rk}G} \cdot \prod_{\alpha \in \Delta, n \geq 1} (1 - q_1^n e^{i\alpha \cdot \hat{a}})(1 - q_2^n e^{i\alpha \cdot \hat{a}}), \tag{4.20}$$

where

$$I_0(q_1, q_2) \equiv \prod_{n \geq 1}(1 - q_1^n)(1 - q_2^n),$$

$$q_1 \equiv qxy = e^{-\beta(1 - i\xi - i\eta)}, \quad q_2 \equiv qx/y = e^{-\beta(1 - i\xi + i\eta)}. \tag{4.21}$$

The first line in (4.20) can be regarded as a refinement of the 3D result (3.22) corresponding to the addition of one more dimension with periodicity β and twists ξ, η. Note that, in going to the second line, an infinite zero-point energy has been regularized so that the result agree with what we would obtain from canonical quantization.

To compute the index from canonical formalism, we decompose the vector multiplet fields on $S^1 \times S^3$ using spherical harmonics and reduce the free super-Yang-Mills theory to a quantum mechanics of infinitely many bosonic and fermionic harmonic oscillators. The oscillator modes all carry definite eigenvalues of $\mathbf{R}$, $\mathbf{J}_{\mathcal{L}}^3$, $\mathbf{J}_{\mathcal{R}}^3$, and their frequency determines the eigenvalue of $\mathbf{D} - \frac{1}{2}\mathbf{R}$. In computing the index as a trace over the Fock space, it is convenient to first consider the trace over one-particle states called the letter index. For a vector multiplet for gauge group G it is given by

$$i_{\text{vec}} \equiv \text{Tr}_{(1\text{p})}\left[(-1)^{\mathbf{F}} q^{\mathbf{D} - \frac{1}{2}\mathbf{R}} x^{2\mathbf{J}_{\mathcal{R}}^3 + \mathbf{R}} y^{2\mathbf{J}_{\mathcal{L}}^3} e^{i\hat{a}}\right]$$

$$= \text{tr}_{\text{adj}} U \cdot \sum_{n \geq 0} \left(q^{n+2} \chi_{\frac{n+2}{2}, \frac{n}{2}} + q^{n+2} \chi_{\frac{n}{2}, \frac{n+2}{2}} - xq^{n+1} \chi_{\frac{n+1}{2}, \frac{n}{2}} - x^{-1} q^{n+2} \chi_{\frac{n}{2}, \frac{n+1}{2}} \right),$$

$$U \equiv e^{i\hat{a}} \in G, \quad \chi_{j,\bar{j}} \equiv \text{tr}_{(j,\bar{j})}[x^{2\mathbf{J}_{\mathcal{R}}^3} y^{2\mathbf{J}_{\mathcal{L}}^3}]. \tag{4.22}$$

In fact, all the oscillators not saturating the bound $\mathbf{H} \geq 0$ form pairs and do not contribute to the letter index. Indeed, the above letter index can be simplified as follows,

$$i_{\text{vec}} = -\left(\frac{q_1}{1-q_1} + \frac{q_2}{1-q_2}\right) \text{tr}_{\text{adj}}(U). \tag{4.23}$$

The full index is then obtained as its plethystic exponential,

$$I_{\text{vec}} = \text{PE}\left[i_{\text{vec}}(q_1, q_2, U)\right] \equiv \exp\left(\sum_{n\geq 1} \frac{1}{n} i_{\text{vec}}(q_1^n, q_2^n, U^n)\right), \tag{4.24}$$

integrated over U in the Cartan torus with the invariant measure (4.19).

Let us next consider the chiral multiplet of R-charge r in the representation R of the gauge group. Its one-loop determinant is

$$
\begin{aligned}
I_{\text{mat}} &= \frac{\text{Det}_\psi\left(\hat{\partial}_t - ia + r - \frac{1}{2} + i\not{D}_{S^3}\right)}{\text{Det}_\phi(-(\hat{\partial}_t - ia + r - 1)^2 + \Delta_{S^3}^{\text{scalar}} + 1)} \\
&= \prod_{m,n\geq 0} \frac{\det_R(-\partial_t + ia + (1 - i\xi)(m + n + 2 - r) - i\eta(m - n))}{\det_R(\partial_t - ia + (1 - i\xi)(m + n + r) - i\eta(m - n))}. \tag{4.25}
\end{aligned}
$$

This can again be regarded as a refinement of the one-loop determinant (3.25) for 3D chiral multiplet. With an appropriate regularization of the zero-point energy, one can rewrite this further as a product over the weights of the representation R,

$$I_{\text{mat}} = \prod_w \Gamma(e^{iw\cdot\hat{a}}(q_1 q_2)^{\frac{r}{2}}; q_1, q_2), \tag{4.26}$$

where $\Gamma(z; q_1, q_2)$ is the elliptic Gamma function

$$\Gamma(z; q_1, q_2) = \prod_{m,n\geq 0} \frac{1 - z^{-1} q_1^{m+1} q_2^{n+1}}{1 - z q_1^m q_2^n}. \tag{4.27}$$

This result can also be obtained from canonical formalism, as the plethystic exponential of the letter index,

$$
\begin{aligned}
i_{\text{mat}} &= \text{tr}_R(U) \cdot \sum_{n\geq 0}\left(x^r q^{n+r} \chi_{\frac{n}{2},\frac{n}{2}} - x^{r-1} q^{n+r+1} \chi_{\frac{n+1}{2},\frac{n}{2}}\right) \\
&\quad + \text{tr}_{\bar{R}}(U) \cdot \sum_{n\geq 0}\left(x^{-r} q^{n+2-r} \chi_{\frac{n}{2},\frac{n}{2}} - x^{1-r} q^{n+r+1} \chi_{\frac{n}{2},\frac{n+1}{2}}\right) \\
&= \frac{(q_1 q_2)^{\frac{r}{2}} \text{tr}_R(U) - (q_1 q_2)^{1-\frac{r}{2}} \text{tr}_R(U^{-1})}{(1 - q_1)(1 - q_2)}. \tag{4.28}
\end{aligned}
$$

4.4 Squashed S^3 from Twisted Compactifications

In the limit $\beta \to 0$ where one can neglect the KK modes, the 4D superconformal index reduces to 3D partition function, but with a new dependence on additional parameters ξ, η. They enter into the 3D partition function through the squashing parameter b,

$$b^2 = \frac{1 - i\xi + i\eta}{1 - i\xi - i\eta}. \tag{4.29}$$

Recall that we have chosen the background $S^3 \times S^1$ with $M = \frac{1}{2}$ at the beginning of Sect. 4.2, and that our computation was preserving a pair of left-invariant supercharges $\mathbf{Q}, \mathbf{S}$. In this case, the above relation shows that the twist by $\mathbf{J}_\mathcal{R}^3$ (accompanied by an appropriate R-twist) has a rather trivial effect on the partition function, but the twist by $\mathbf{J}_\mathcal{L}^3$ does change the partition function in a non-trivial manner. So the different Scherk-Schwarz twists lead to qualitatively different 3D backgrounds after dimensional reduction.

To understand the effect of two different twists upon 3D geometry, let us consider instead the twisted compactification with $\xi \neq 0, \eta = 0$ and try different choices of unbroken supersymmetry. After moving to the coordinate system with ordinary time periodicity, the metric is given by

$$ds^2 = E^a E^a = e^1 e^1 + e^2 e^2 + (e^3 + \xi dt)^2 + dt^2. \tag{4.30}$$

On this space, one can either preserve left-invariant or right-invariant supercharges by choosing the background $U(1)_R$ gauge field appropriately to make the corresponding Killing spinors t-independent. For $V_t = -\frac{i}{2} + \xi$, the Killing spinor equation (4.6) with $M = \frac{1}{2}$ has a pair of time-independent solutions satisfying $\gamma_3\epsilon = -\epsilon$ and $\gamma_3\bar\epsilon = +\bar\epsilon$, which we identified with the left-invariant supercharges $\mathbf{S}, \mathbf{Q}$. The solutions corresponding to the other pair of left-invariant supercharges become time-independent when $V_t = -\frac{i}{2} - \xi$. For $V_t = \frac{i}{2}$, the Killing spinor equation (4.6) with $M = -\frac{1}{2}$ has solutions corresponding to the four right-invariant supercharges.

To do the dimensional reduction along S^1, we rewrite the metric (4.30) into the form

$$ds^2 = \hat{E}^a \hat{E}^a = e^1 e^1 + e^2 e^2 + u^2 e^3 e^3 + u^{-2}(dt + u^2\xi \, e^3)^2, \tag{4.31}$$

where $u \equiv (1 + \xi^2)^{-1/2}$. Since this can be regarded as a local Lorentz transformation, the Killing spinors on the new local Lorentz frame satisfy

$$D_m\epsilon = -M\sigma_m(\bar\sigma_a h^a)\epsilon, \quad D_m\bar\epsilon = M\bar\sigma_m(\sigma_a h^a)\bar\epsilon, \tag{4.32}$$

where $M = \frac{1}{2}$ or $-\frac{1}{2}$ for the left- or right-invariant Killing spinors, and

$$h^a = (0, 0, -u\xi, u). \tag{4.33}$$

By dropping the last term on the right hand side of (4.31) we obtain the 3D metric of the familiar squashed S^3 with $SU(2)_{\mathcal{L}} \times U(1)_{\mathcal{R}}$ isometry. But the nature of the dimensionally reduced theory depends also on which supersymmetries have been preserved in the reduction.

If we set $M = \frac{1}{2}$ and $V_t = -\frac{i}{2} + \xi$ upon dimensional reduction, the supersymmetry of the resulting 3D theory is characterized by the Killing spinor equation

$$\left(\partial_m + \frac{1}{4}\omega_m^{ab}\gamma^{ab} + iu\xi^2 V_m\right)\epsilon = \frac{iu}{2}\gamma_m\epsilon,$$

$$\left(\partial_m + \frac{1}{4}\omega_m^{ab}\gamma^{ab} - iu\xi^2 V_m\right)\bar{\epsilon} = \frac{iu}{2}\gamma_m\bar{\epsilon}, \qquad (4.34)$$

where $V_m \equiv \hat{E}_m^3 = ue_m^3$. The above Killing spinor equation takes the form of (3.2) with $U_m = 0$, and $1/4$ of the supersymmetry on the round S^3 remains unbroken after squashing due to the background $U(1)_R$ gauge field $-u\xi^2 V_m$. It was shown in [5] that the exact partition function on this squashed S^3 background is essentially the same as that on the round S^3, in consistency with the discussion in the previous subsection. For the case $M = \frac{1}{2}$ and $V_t = -\frac{i}{2} + \xi$, the 3D Killing spinor equation takes the same form as above but the $U(1)_R$ gauge field appears with the opposite sign.

If we set $M = -\frac{1}{2}$ and $V_t = \frac{i}{2}$, the Killing spinor equation of the 3D theory is

$$\left(\partial_m + \frac{1}{4}\omega_m^{ab}\gamma^{ab}\right)\epsilon = -\frac{iu}{2}\gamma_m\epsilon - u\xi V^n\gamma_{mn}\epsilon,$$

$$\left(\partial_m + \frac{1}{4}\omega_m^{ab}\gamma^{ab}\right)\bar{\epsilon} = -\frac{iu}{2}\gamma_m\bar{\epsilon} + u\xi V^n\gamma_{mn}\bar{\epsilon}, \qquad (4.35)$$

again with $V_m \equiv \hat{E}_m^3 = ue_m^3$. This case preserves $1/2$ of the Killing spinors on the round S^3. The above Killing spinor equation can be identified with (3.2) with $U_m \neq 0$. It was shown in [18] that the partition function on this background depends nontrivially on ξ through the squashing parameter

$$b = u(1 - i\xi). \qquad (4.36)$$

For a real ξ, the squashing parameter b is a complex phase.

Acknowledgments The author thanks Naofumi Hama, Sungjay Lee and Jaemo Park for collaboration on the materials discussed in this article. The author also thanks the string theory group at Yukawa Institute for Theoretical Physics, Kyoto University where the major part of this article was written.

References

[1]   Kapustin, A., Willett, B., Yaakov, I.: Exact results for Wilson loops in superconformal Chern-Simons theories with matter. JHEP **1003**, 089 (2010). arXiv:0909.4559 [hep-th]

[2]   Jafferis, D.L.: The exact superconformal R-symmetry extremizes Z. JHEP **1205**, 159 (2012). arXiv:1012.3210 [hep-th]

[3]   Hama, N., Hosomichi, K., Lee, S.: Notes on SUSY gauge theories on three-sphere. JHEP **1103**, 127 (2011). arXiv:1012.3512 [hep-th]

[4]   Pestun, V.: Localization of gauge theory on a four-sphere and supersymmetric Wilson loops. Commun. Math. Phys. **313**, 71–129 (2012). arXiv:0712.2824 [hep-th]

[5]   Hama, N., Hosomichi, K., Lee, S.: SUSY gauge theories on squashed three-spheres. JHEP **1105**, 014 (2011). arXiv:1102.4716 [hep-th]

[6]   Festuccia, G., Seiberg, N.: Rigid supersymmetric theories in curved superspace. JHEP **1106**, 114 (2011). arXiv:1105.0689 [hep-th]

[7]   Kim, S.: The Complete superconformal index for $N = 6$ Chern-Simons theory. Nucl. Phys. **B821**, 241–284 (2009). arXiv:0903.4172 [hep-th]

[8]   Imamura, Y., Yokoyama, S.: Index for three dimensional superconformal field theories with general R-charge assignments. JHEP **1104**, 007 (2011). arXiv:1101.0557 [hep-th]

[9]   Kallen, J.: Cohomological localization of Chern-Simons theory. JHEP **1108**, 008 (2011). arXiv:1104.5353 [hep-th]

[10]   Ohta, K., Yoshida, Y.: Non-abelian localization for supersymmetric Yang-Mills-Chern-Simons theories on Seifert manifold. Phys. Rev. **D86**, 105018 (2012). arXiv:1205.0046 [hep-th]

[11]   Romelsberger, C.: Counting chiral primaries in $N = 1$, $d = 4$ superconformal field theories. Nucl. Phys. **B747**, 329–353 (2006). arXiv:hep-th/0510060 [hep-th]

[12]   Kinney, J., Maldacena, J.M., Minwalla, S., Raju, S.: An index for 4 dimensional super conformal theories. Commun. Math. Phys. **275**, 209–254 (2007). arXiv:hep-th/0510251 [hep-th]

[13]   Romelsberger, C.: Calculating the superconformal index and seiberg duality. arXiv:0707.3702 [hep-th]

[14]   Dolan, F., Osborn, H.: Applications of the superconformal index for protected operators and q-hypergeometric identities to $N = 1$ dual theories. Nucl. Phys. **B818**, 137–178 (2009). arXiv:0801.4947 [hep-th]

[15]   Gadde, A., Yan, W.: Reducing the 4d index to the S^3 partition function. JHEP **1212**, 003 (2012). arXiv:1104.2592 [hep-th]

[16]   Dolan, F., Spiridonov, V., Vartanov, G.: From 4d superconformal indices to 3d partition functions. Phys. Lett. **B704**, 234–241 (2011). arXiv:1104.1787 [hep-th]

[17]   Imamura, Y.: Relation between the 4d superconformal index and the S^3 partition function. JHEP **1109**, 133 (2011). arXiv:1104.4482 [hep-th]

[18]   Imamura, Y., Yokoyama, D.: N = 2 supersymmetric theories on squashed three-sphere. Phys. Rev. **D85**, 025015 (2012). arXiv:1109.4734 [hep-th]

[19]   Drukker, N., Gaiotto, D., Gomis, J.: The virtue of defects in 4D gauge theories and 2D CFTs. JHEP **1106**, 025 (2011). arXiv:1003.1112 [hep-th]

[20]   Hosomichi, K., Lee, S., Park, J.: AGT on the S-duality wall. JHEP **1012**, 079 (2010). arXiv:1009.0340 [hep-th]

[21]   Gaiotto, D.: $N = 2$ dualities. JHEP **1208**, 034 (2012). arXiv:0904.2715 [hep-th]

[22]   Quine, J., Heydari, S., Song, R.: Zeta regularized products. Trans. Am. Math. Soc. **338**, 213 (1993)

[23]   Teschner, J.: From Liouville theory to the quantum geometry of Riemann surfaces. arXiv:hep-th/0308031 [hep-th]

[24]   Gaiotto, D., Witten, E.: Supersymmetric boundary conditions in $N = 4$ super Yang-Mills theory. J. Stat. Phys. **135**, 789–855 (2009). arXiv:0804.2902 [hep-th]

[25]   Gaiotto, D., Witten, E.: S-duality of boundary conditions in $N = 4$ super Yang-Mills theory, Adv. Theor. Math. Phys. **13** (2009). arXiv:0807.3720 [hep-th]

[26]   Closset, C.T., Dumitrescu, T., Festuccia, G., Komargodski, Z.: Supersymmetric field theo-
       ries on three-manifolds. arXiv:1212.3388 [hep-th]
[27]   Klare, C., Tomasiello, A., Zaffaroni, A.: Supersymmetry on curved spaces and holography.
       JHEP **1208**, 061 (2012). arXiv:1205.1062 [hep-th]
[28]   Dumitrescu, T.T., Festuccia, G., Seiberg, N.: Exploring curved superspace. JHEP **1208**, 141
       (2012). arXiv:1205.1115 [hep-th]
[29]   Alday, L.F., Martelli, D., Richmond, P., Sparks, J.: Localization on three-manifolds.
       arXiv:1307.6848 [hep-th]
[30]   Gang, D.: Chern-simons theory on L(p, q) lens spaces and localization. arXiv:0912.4664
       [hep-th]
[31]   Benini, F., Nishioka, T., Yamazaki, M.: 4d Index to 3d Index and 2d TQFT. Phys. Rev. **D86**,
       065015 (2012). arXiv:1109.0283 [hep-th]
[32]   Imamura, Y., Yokoyama, D.: $S^3/\mathbb{Z}_n$ partition function and dualities. JHEP **1211**, 122 (2012).
       arXiv:1208.1404 [hep-th]
[33]   Hama, N., Hosomichi, K.: Seiberg-Witten theories on ellipsoids. JHEP **1209**, 033 (2012).
       arXiv:1206.6359 [hep-th]
[34]   Gomis, J., Lee, S.: Exact Kahler potential from gauge theory and mirror symmetry.
       arXiv:1210.6022 [hep-th]
[35]   Martelli, D., Passias, A., Sparks, J.: The gravity dual of supersymmetric gauge theories on
       a squashed three-sphere. Nucl. Phys. **B864**, 840–868 (2012). arXiv:1110.6400 [hep-th]
[36]   Closset, C., Dumitrescu, T.T., Festuccia, G., Komargodski, Z.: The geometry of supersym-
       metric partition functions. JHEP **1401**, 124 (2014). arXiv:1309.5876 [hep-th]
[37]   Nian, J.: Localization of supersymmetric Chern-Simons-Matter theory on a squashed
       S^3 with $SU(2) \times U(1)$ isometry. arXiv:1309.3266 [hep-th]
[38]   Tanaka, A.: Localization on round sphere revisited. JHEP **1311**, 103 (2013).
       arXiv:1309.4992 [hep-th]
[39]   Drukker, N., Okuda, T., Passerini, F.: Exact results for vortex loop operators in 3d super-
       symmetric theories. arXiv:1211.3409 [hep-th]
[40]   Witten, E.: Quantum field theory and the Jones polynomial. Commun. Math. Phys. **121**,
       351 (1989)
[41]   Kao, H.-C., Lee, K.-M., Lee, T.: The Chern-Simons coefficient in supersymmetric Yang-
       Mills Chern-Simons theories, Phys. Lett. **B373**, 94–99 (1996). arXiv:hep-th/9506170
       [hep-th]
[42]   Tanaka, A.: Comments on knotted 1/2 BPS Wilson loops. JHEP **1207**, 097 (2012).
       arXiv:1204.5975 [hep-th]
[43]   Kapustin, A., Willett, B., Yaakov, I.: Exact results for supersymmetric abelian vortex loops
       in 2+1 dimensions. arXiv:1211.2861 [hep-th]

3d Superconformal Theories
from Three-Manifolds

Tudor Dimofte

We review here some aspects of the 3d $\mathcal{N} = 2$ SCFT's that arise from the compactification of M5 branes on 3-manifolds. The program to systematically describe these theories and their properties began in a series of papers [1–3], inspired by earlier physical studies [4–6], and has since been extended and clarified in [7–12], among other works.

Part of the "3d-3d correspondence" includes an analogue of the AGT relation [13–16] and the index-TQFT relation of [17–19], discussed in much of the rest of this volume. Recall that for theories of class $\mathcal{S}$, i.e. 4d $\mathcal{N} = 2$ theories $T_K[C]$ obtained by wrapping K M5 branes on C, one expects

Partition function of $T_K[C]$ on __		Partition function of __ on C	
S_b^4	$=$	Liouville theory	
$\mathbb{R}_\epsilon^4 \simeq D_b^4$	$=$	Liouville theory (conformal block)	(1)
$S^3 \times_q S^1$	$=$	q-Yang-Mills or generalizations	

The basic logic is that one takes the 6d geometry supporting the $(2,0)$ theory on the worldvolume of the M5 branes to be of the form $X \times C$, where X is one of the geometries in the left column of (1); then compactifying first on C leads to $T_K[C]$ on X, whereas compactifying first on X should lead to some other theory (the right

---

A citation of the form [V:x] refers to article number x in this volume.

T. Dimofte (✉)
Institute for Advanced Study, Einstein Dr., Princeton, NJ 08540, USA
e-mail: dd@ias.edu

T. Dimofte
Trinity College, CB2 1TQ, Cambridge, UK

© Springer International Publishing Switzerland 2016
J. Teschner (ed.), *New Dualities of Supersymmetric Gauge Theories*,
Mathematical Physics Studies, DOI 10.1007/978-3-319-18769-3_11

column of (1)) on C. Similarly, if we denote by $T_K[M]$ the effective 3d field theory obtained from wrapping K M5 branes on M, we expect[1]

Part'n function of $T_K[C]$ on		Part'n function of ___ on M
S_b^3	$=$	$SL(K, \mathbb{C})$ Chern-Simons at level $k = 1$ [1, 11]
$S^2 \times_q S^1$	$=$	$SL(K, \mathbb{C})$ Chern-Simons at level $k = 0$ [3, 12]
$\mathbb{R}^2 \times_q S^1$	$=$	holomorphic sector of $SL(K, \mathbb{C})CS$ [20, 4, 21]
SUSY vacua on $\mathbb{R}^2 \times S^1$	$=$	flat $SL(K, \mathbb{C})$ connections on M [20, 4].

$$(2)$$

The 3d-3d and 2d-4d correspondences fit very nicely together when M has a boundary. We will describe in Sect. 1 that when ∂M is nontrivial, the theory $T_K[M]$ is best interpreted as a boundary condition or domain wall for the 4d $\mathcal{N} = 2$ theory $T_K[\partial M]$ [1, 8, 22]. This has some natural implications for partition functions. For example, if M has two distinct boundaries of the same type, $\partial M = C \sqcup C$, then $T_K[M]$ describes a domain wall in the 4d theory $T_K[C]$. In turn, the 3d partition functions of $T_K[M]$ on a space Y (from the right column of (2)) should *act* on the 4d partition functions of $T_K[C]$ on a half-space X with $\partial X = Y$. Examples of this type and others have been explored in [5, 6, 23–25], and we will elaborate a bit further on them in Sect. 2.2. Similar ideas about domain walls also constituted a major ingredient in the recent 4d-2d correspondence of [26].

The current successes of the "3d-3d" program include a systematic prescription for associating theories $\widetilde{T}_K[M]$ to a wide class of 3-manifolds M with boundary [1, 2, 7], which we discuss in Sects. 3–4. Sometimes the theories $\widetilde{T}_K[M]$ only contain a subsector[2] of the full theory $T_K[M]$ of K M5 branes on M; though in special cases one does recover the full $T_K[M]$. In particular, one recovers the full $T_K[M]$ when M is a 3-manifold encoding a duality domain wall in a 4d $\mathcal{N} = 2$ $T_K[C]$, as long as $\chi(C) < 0$. We will revisit this subtlety in Sect. 4.1; in the following we drop the tilde on $\widetilde{T}_K[M]$ to simplify notation.

The main technique of [1, 2, 7] is to triangulate the manifold, cutting it up into tetrahedra, and then to "glue" $T_K[M]$ together from elementary 3d theories T_Δ associated to the tetrahedron pieces. One obtains this way an abelian Chern-Simons-matter theory—a theory of "class $\mathcal{R}$"—that flows to the desired SCFT $T_K[M]$ in the infrared. Quite beautifully, different triangulations of M lead to different UV Chern-Simons-matter theories that flow to the *same* $T_K[M]$. In other words, the UV theories are

---

[1] Here S_b^3 denotes a "squashed" 3-sphere with ellipsoidal metric. It is also useful to note that complex $SL(K, \mathbb{C})$ Chern-Simons theory has *two* coupling constants or levels (k, σ), one quantized and the other continuous, cf. Sects. 2.1 and 2.2. It is only the quantized level that is being fixed in (2). The general pattern following from work of [11] is that $T_K[C]$ on a squashed Lens space $L(k, 1)_b$ is equivalent to $SL(K, \mathbb{C})$ Chern-Simons at level k.

[2] To be precise: after compactification on S^1, the subsectors only contain SUSY vacua corresponding to irreducible $SL(K, \mathbb{C})$ flat connections on M, with given boundary conditions, rather than *all* flat connections as prescribed by (2). The relation between these subsectors and the "full" $T_K[M]$ began to be analyzed in [27].

related by a generalized 3d mirror symmetry. The 3d-3d program therefore leads to the *geometric* classification of a huge subset of abelian 3d mirror symmetries.

Mathematically, the study of 3-manifold theories based on triangulations has led to the new concepts of "framed" 3-manifolds and moduli spaces of "framed" flat connections on them [7, 8]. They generalize the framework of [28] for studying higher Teichmüller theory on 2d surfaces—which in turn played a central role in the 2d-4d explorations of Gaiotto, Moore, and Neitzke, cf. [29, 30].

Despite many exciting achievements, there is still much to develop in the 3d-3d program. One interesting direction of study would be to find *nonabelian* UV descriptions for theories $T_K[M]$, dual to the abelian ones that come from triangulations.[3] This may come from cutting manifolds into simpler pieces along smooth surfaces (rather than sharp tetrahedron boundaries, which have edges and corners), much as was done for cutting 2d surfaces in [32]. Such smooth cutting and gluing should provide the construction of $T_K[M]$ for general closed 3-manifolds as well, and may circumvent the difficulties with irreducible flat connections and subsectors (cf. Footnote 2) encountered so far. Finally, while computations of sphere partition functions and indices of $T_K[M]$ are easy and accessible, it would be extremely interesting to analyze the actual Q-cohomology of the space of BPS states of a theory $T_K[M]$ on (say) $S^2 \times \mathbb{R}$. This would have immediate applications to the categorification of quantum 3-manifold invariants, along the lines of [20, 33].

1 The 6d Setup

Before discussing methods to construct $T_K[M]$, let's first try to understand exactly what it *means* to associate a 3d $\mathcal{N} = 2$ theory to an oriented 3-manifold M, and what properties the theory should have.

One way to think about this is to start in 11-dimensional M-theory, wrapping K M5 branes on $M \times \mathbb{R}^3$. If we want to preserve supersymmetry we must make sure that M is a supersymmetric cycle. Taking the ambient 11-dimensional geometry to be a cotangent bundle $T^*M \times \mathbb{R}^5$ (with M its zero-section), we can preserve at least four supercharges.[4] If we subsequently decouple gravity, taking a field-theory limit on the M5 branes, and flow to low energy so that fluctuations along M can be neglected, we expect to obtain a 3-dimensional $\mathcal{N} = 2$ theory on $\mathbb{R}^3$. In the far infrared, the theory generically hits a superconformal fixed point, which we might call $T_K[M]$.

---

[3] In a few examples, nonabelian duals are already known: the basic tetrahedron theory has an $SU(2)$ dual discussed in [31]; and the theory for the basic S-duality wall in 4d $\mathcal{N} = 2$ $SU(2)$ theory with $N_f = 4$ (associated to the manifold in Fig. 4b) has an $SU(2)$ dual found in [24]. Some basic ideas about smooth gluing were also discussed in [5].

[4] The counting goes as follows. First, the cotangent bundle T^*M is a noncompact Calabi-Yau manifold. M-theory on a generic Calabi-Yau background preserves eight supercharges (cf. [34, 35]). An M5 brane wrapping a special Lagrangian cycle in the Calabi-Yau (such as the zero-section M in T^*M) is half-BPS, and preserves four of the eight supercharges.

In this brane construction, the starting metric on M might be chosen arbitrarily. All the details of the metric enter (as couplings) into an effective field theory on $\mathbb{R}^3$. However, in the process of flowing to the infrared the metric is expected to "uniformize," acquiring constant curvature.[5] Correspondingly, renormalization flow washes away most of the coupling dependence in the effective theory on $\mathbb{R}^3$. Most topological 3-manifolds admit a metric with constant negative curvature [38], i.e. a hyperbolic metric, and they are the ones we'll be interested in.[6] Moreover, if M is closed, the hyperbolic metric is unique [39]. In this case, $T_K[M]$ is indeed expected to be a superconformal theory, which depends only on the *topology* of M, has no flavor symmetry, and admits no (obvious) marginal deformations. Just as the hyperbolic structure on M is rigid, we might say that $T_K[M]$ is rigid.

We may also understand $T_K[M]$ directly in field theory. The 6d theory on K M5 branes is the $(2, 0)$ SCFT with Lie algebra A_{K-1}. It must be topologically twisted along M in order to preserve supersymmetry. (In general, the required topological twist is prescribed by the normal geometry of the supersymmetric cycle $M \subset T^*M$ [40]; but in this case the choice is unique.) In particular, the $SO(3)_E$ part of the Lorentz group corresponding to M is twisted by an $SO(3)_R$ subgroup of the $SO(5)_R$ R-symmetry group (cf. [4, 20]). The unbroken R-symmetry is the commutant of $SO(3)_R \subset SO(5)_R$, namely $SO(2)_R \simeq U(1)_R$, as appropriate for an $\mathcal{N} = 2$ theory in 3d. We again are welcome to choose any metric on M that we want. In the UV, the effective field theory on $\mathbb{R}^3$ will depend on the metric, but after flowing to the IR one hopes to obtain an SCFT that does not.

This is all entirely analogous to compactification of K M5 branes, or the A_{K-1} (2,0) theory, on 2d surfaces C. In that case, the IR theory $T_K[C]$ (a theory of "class S") depends on the conformal class of a metric on C, which is equivalent to a choice of hyperbolic metric. In contrast to 3d, the hyperbolic metric on a closed surface allows continuous deformations, and the 4d $\mathcal{N} = 2$ theory $T_K[C]$ has corresponding exactly marginal gauge couplings [V:2].

The story becomes much more interesting, and in many ways much more manageable, if we allow M to have defects and boundaries.

Codimension-two defects placed along knots in M add flavor symmetry to $T_K[M]$. In the 6d A_{K-1} (2,0) theory, there are different types of "regular" defects, labelled by partitions of K, and carrying various subgroups of $SU(K)$ as their flavor symmetry [32].[7] In M-theory, each regular defect along a knot $\mathcal{K} \subset M$ comes from a stack of K or fewer "probe" M5 branes that wrap the noncompact supersymmetric 3-cycle $N^*\mathcal{K} \subset T^*M$ (the conormal bundle of $\mathcal{K}$) as well as $\mathbb{R}^3$. The flavor symmetry can

---

[5]See, e.g., the supergravity solutions of [36] involving special Lagrangian 3-cycles. For the analogous compactifications on 2d surfaces, the flow of the metric to constant curvature was analyzed in [37].

[6]Notable exceptions include spheres, tori, lens spaces, and more general Seifert-fibered manifolds, which have the structure of an S^1 fibration over a surface. The 3d theories resulting from compactification on such manifolds are qualitatively different from the hyperbolic case. For example, compactification on a 3-torus yields $\mathcal{N} = 8$ SYM in 3d, while compactification on the 3-sphere yields a gapped theory that breaks SUSY.

[7]Also described in Sect. 3.1–3.2 of *Families of* $\mathcal{N} = 2$ *field theories* by D. Gaiotto.

be understood as arising from the symmetry group of the probes. In the presence of a defect, the hyperbolic metric on M acquires a cusp-like singularity, cf. [41].

In order to add boundaries to M, we must be somewhat more creative, since M5 branes cannot end. Alternatively, the $(2, 0)$ theory does not admit ordinary supersymmetric boundary conditions because it is chiral. We create boundaries for M at "infinity" by allowing asymptotic regions that look like $\mathbb{R}_+ \times C$ for some surface C (Fig. 1). Then M is no longer compact. Wrapping M5 branes on M leads not to an isolated 3d theory but to a half-BPS superconformal boundary condition (preserving 3d $\mathcal{N} = 2$ SUSY) for the 4d theory $T_K[C]$. We might call this boundary condition $T_K[M]$. If M has multiple asymptotic regions with cross-sections C_i, then $T_K[M]$ is a common boundary condition for a product of theories $T_K[C_i]$, which do not interact with each other in the 4d bulk; equivalently, $T_K[M]$ can be thought of as a half-BPS domain wall between one subset of 4d theories $\prod_{i<I} T_K[C_i]$ and its complement $\prod_{i\geq I} T_K[C_i]$ (Fig. 2). Note that defects in M (orange lines in the figures) may enter asymptotic regions, where they look like punctures in the surfaces C_i.

In the presence of asymptotic boundaries C_i, the hyperbolic metric on M is no longer rigid. It depends (at least) on a choice of hyperbolic structure for each surface C_i, i.e. on a choice of boundary conditions. This choice, of course, parametrizes the bulk couplings of $\prod_i T_K[C_i]$.

We can try to transform the boundary condition $T_K[M]$ into a stand-alone 3d $\mathcal{N} = 2$ theory by decoupling the 4d bulk theories $\prod_i T_K[C_i]$. However, there is no unique way to do this. Suppose, for example, that there's just a single boundary C. Working with a nonabelian SCFT $T_K[C]$, we attain a (non-canonical) weak-coupling limit by adjusting the hyperbolic metric on C so as to stretch it into pairs of pants

Fig. 1 Compactifying M5's on a 3-manifold with asymptotic boundaries to obtain a boundary condition $T[M]$

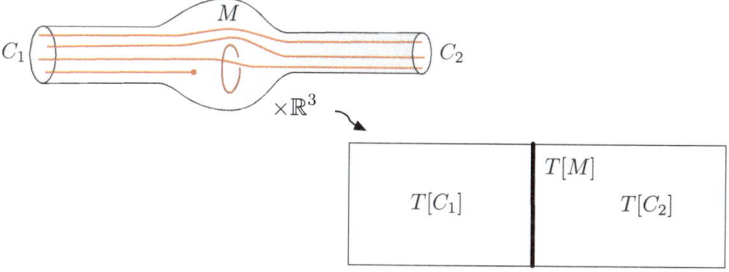

Fig. 2 Re-interpreting the boundary condition of Fig. 1 as a domain wall

Fig. 3 Shrinking a pants
decomposition of a
one-punctured torus into a
network of defects

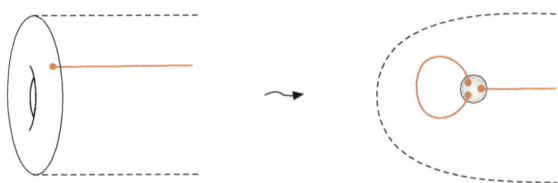

connected by long, thin tubes [32]. There is a weakly coupled $SU(K)$ gauge group
in $T_K[C]$ associated to each tube. In the limit of infinite stretching, we may hope to
leave behind a 3d theory $T_K[M, \mathbf{p}]$, labelled by the chosen pants decomposition $\mathbf{p}$
of C. $T_K[M, \mathbf{p}]$ should have a residual $SU(K)$ flavor symmetry for every stretched
tube (which would get gauged in re-coupling to a 4d bulk).

We may represent this 4d–3d decoupling geometrically by "shrinking" C to a trivalent network of maximal codimension-two defects, as dictated by the pants decomposition $\mathbf{p}$ (Fig. 3).[8] This effectively compactifies M. The trivalent junctures of defects
survive as asymptotic regions of M with the cross-section of a 3-punctured sphere.
Thus, the theory $T_K[M, \mathbf{p}]$ is still potentially coupled to a collection of 4d "trinion"
theories, and this coupling takes a little extra work to undo. For example, in the case
$K = 2$, the trinion theory just consists of four free hypermultiplets, coupled to the 3d
boundary theory by superpotentials. One can adjust bulk parameters to make some
of these hypermultiplets very massive. This is discussed in [1] and especially [8].

Alternatively, we can move onto the 4d Coulomb branch of $T_K[C]$ and flow to
the IR. Then $T_K[C]$ is a Seiberg-Witten theory, with some abelian gauge symmetry
$U(1)^d$. The electric-magnetic duality group is $Sp(2d, \mathbb{Z})$. We decouple the Seiberg-
Witten theory by choosing an electric-magnetic duality frame Π, and adjusting parameters and moduli so that all the electric gauge couplings in that frame become
weak. Again, a little more is needed to decouple BPS hypermultiplets. In the end, we
obtain a purely 3d theory $T_K[M, \Pi]$ with $U(1)^d$ flavor symmetry left over from the
bulk gauge group. We will usually represent the manifold giving rise to $T_K[M, \Pi]$
as simply having its asymptotic region $C \times \mathbb{R}_+$ cut off at finite distance.

1.1 Duality Walls

A very simple application of the above constructions is to represent duality walls
for 4d $\mathcal{N} = 2$ theories of class $\mathcal{S}$ by 3d geometries [5, 6, 8]. To this end, we take
$M = \mathbb{R} \times C$ for some punctured surface C. In other words, M has two asymptotic
boundaries C. The punctures of C just become defects running the entire "length" of

---

[8]This "shrinking" procedure turns parts of M that look like $S^1 \times \mathbb{R} \times \mathbb{R}_+$ (i.e. the neighborhoods
of tubes) into defects. An identical setup was used to create defects in [32].

M. Naively, $T_K[M]$ just becomes a trivial domain wall between two copies of $T_K[C]$. However, we can make it look non-trivial by taking different decoupling limits on the two ends.

For example, if we work with $T_K[C]$ as a UV SCFT, we can take two different weak-coupling limits corresponding to pants decompositions $\mathbf{p}, \mathbf{p}'$. The 3d theory $T_K[M, \mathbf{p}, \mathbf{p}']$ that is left behind is the theory of an S-duality domain wall. For example, if C is a punctured torus (with a minimal puncture), then $T_K[C]$ is 4d $\mathcal{N} = 2^*$ theory with gauge group $SU(K)$. Letting $\mathbf{p}$ and $\mathbf{p}'$ shrink the A and B-cycles of the punctured torus, respectively, we should obtain the S-duality wall whose 3d theory is usually called (mass-deformed) $T[SU(K)]$ [42].

As discussed above, we can represent decoupling limits geometrically by shrinking appropriate legs/tubes of C to defects at the two ends of M, so that we obtain a compact manifold $M_{\mathbf{p},\mathbf{p}'}$ with a trivalent network of defects. In the case of S-duality for a one-punctured torus, the resulting manifold is a "Hopf network" of defects in S^3, shown in Fig. 4a. By using the methods of Sect. 4, its 3d theory was shown in [8] to be equivalent to $T[SU(2)]$ (for $K = 2$). Similarly, if we take C to be a four-punctured sphere (with appropriate minimal/maximal punctures) and set $\mathbf{p}, \mathbf{p}'$ to correspond to its "s and t channel" decompositions, we get the basic S-duality for $\mathcal{N} = 2$ SQCD with $N_f = 2K = 2N_c$. The 3d geometry for the duality wall is shown in Fig. 4b; its associated 3d theory appeared in [8, 24].

We can also put theories $T_K[C]$ on their Coulomb branch, and choose decoupling limits Π, Π' at the two ends of M that are appropriate for Seiberg-Witten theory. The theory $T_K[M, \Pi, \Pi']$ becomes a "Seiberg-Witten duality wall" that implements abelian IR dualities. The simplest such walls (involving duality for gauge multiplets alone) were discussed from a field-theory perspective in [43]. In general, one can also act on hypermultiplets, as discussed in [1].

Finally, decoupling one end of M in the UV and one in the IR (on the Coulomb branch), we can obtain the 3d theory $T_K[M, \mathbf{p}, \Pi]$ for an "RG wall" [8]. It has the property that operators hitting the wall on the UV side are decomposed into a basis of IR operators on the other side, cf. [44]. For supersymmetric line operators, such UV-IR maps have been discussed (e.g.) in [45, 46], and RG walls give them a novel physical interpretation. The 3d geometry representing an RG wall for $\mathcal{N} = 2^*$ theory is shown in Fig. 4c. (Note how two of these geometries can be glued along their outer boundaries to form the UV S-duality manifold of Fig. 4a.)

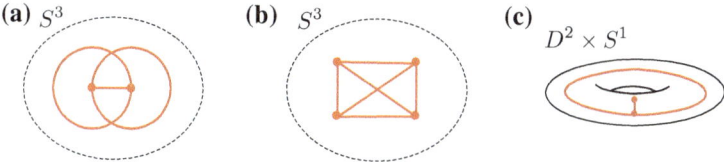

Fig. 4 Geometries representing various S-duality and RG walls: **a** a Hopf network of defects in S^3; **b** a tetrahedral network of defects in S^3; and **c** a network of defects in a solid torus, corresponding to a particular pants decomposition and connecting to a puncture on the boundary

2 3d Theories, $SL(K)$ Connections, and Chern-Simons

One of the most interesting geometric properties of a 3-manifold theory $T_K[M]$ is the relation between its vacua and flat $SL(K, \mathbb{C})$ connections on M. The other AGT-like correspondences between partition functions of $T_K[M]$ and Chern-Simons theory on M in (2) can be understood as quantizations of this basic semi-classical relation. Strictly speaking, the relation to flat connections holds when $T_K[M]$ is compactified on a circle S^1 of finite radius. So let us do this, assuming that the full 6d geometry is now $M \times \mathbb{R}^2 \times S^1$.

The 6d (2,0) theory on a circle gives rise to 5d maximally supersymmetric Yang-Mills on $M \times \mathbb{R}^2$, with gauge group $SU(K)$,[9] and with a partial topological twist along M. We may explicitly write down the 5d BPS equations. The partial twist transforms three real scalars in the gauge multiplet into an adjoint-valued 1-form φ on M. The BPS equations on M then take the form of "Hitchin equations" generalized to three dimensions[10]:

$$[D_i, D_j] = 0 \quad (i, j = 1, 2, 3), \quad \sum_{ij} g^{ij}[D_i, D_j^\dagger] = 0, \quad (3)$$

where D_i is the gauge-covariant derivative with respect to a *complexified* gauge field $A_i + i\varphi_i$, and g_{ij} is a chosen background metric on M, cf. [4, 20, 47]. The equations are invariant under real $SU(K)$ gauge transformations. The set of solutions to (3), modulo real gauge transformations, is equivalent (up to a lower-dimensional subset) to the solutions of the equations $[D_i, D_j] = 0$ alone, modulo *complex* $SL(K, \mathbb{C})$ gauge transformations. But this means that the solutions are complex $SL(K, \mathbb{C})$ flat connections on M. Let us denote this moduli space as

$$\widetilde{\mathcal{L}}_K(M) = \{ \text{flat } SL(K, \mathbb{C}) \text{ connections on } M \}. \quad (4)$$

We expect it to correspond to the space of vacua of $T_K[M]$ on $S^1 \times \mathbb{R}^2$.

In terms of branes, the M5 branes wrapping $M \times S^1 \times \mathbb{R}^2$ become D4 branes wrapping $M \times \mathbb{R}^2$. The worldvolume theory of the D4's is 5d SYM. The three adjoint scalar fields that were promoted to a 1-form φ are the translation modes of the D4's in the fibers of the cotangent bundle T^*M. In the infrared, one expects the stack of K D4 branes to *separate* in T^*M, becoming a single multiply-wrapped brane, and forming a spectral cover $\widetilde{M}$ of M. The pattern of separation then is encoded in the eigenvalues of the 1-form φ.

---

[9]It is also possible to arrive at a theory where the center of $SU(K)$, or subgroups of the center, are not gauged. Then instead of getting a relation to $SL(K, \mathbb{C})$ connections, we find a relation to $PSL(K, \mathbb{C})$ connections, or similar. The details are subtle (see [8]), but the correct relation can ultimately be derived by examining the charges of fundamental line operators in $T_K[M]$.

[10]The structure of Hitchin equations in two dimensions and their relation to 4d $\mathcal{N} = 2$ theory on a circle is reviewed in [V:3].

We might remark that starting from a flat complex connection $\mathcal{A}$ and obtaining the spectral 1-form φ is *not* an easy task. To do so, one must find the right complex gauge transformation h so that the transformed $\mathcal{A}_h$ satisfies the real equation $g^{ij}[D_i, D_j^\dagger] = 0$, in addition to the complex flatness equations. Then the imaginary part of this particular $\mathcal{A}_h$ is φ. Therefore, φ and the spectral cover it encodes depend on the choice of metric g_{ij} for M—even though the notion of a flat complex connection does not.

It is also useful to observe that after splitting equations (3) into real and imaginary parts they reduce to $F_A = dA + A^2 = \varphi^2$, along with $d_A \varphi = d_A * \varphi = 0$. The latter equations say that φ is a covariantly harmonic one-form on M. The eigenvalues of φ give rise (roughly) to a harmonic one-form on the spectral cover, which plays the role of a Seiberg-Witten form for $T_K[M]$ [2].

If M is compact and hyperbolic, the flat $SL(K, \mathbb{C})$ connections on M typically turn out to be rigid. We note, however, that mathematically it is still unknown precisely when rigidity holds.[11] If the flat connections are indeed rigid, then $\widetilde{\mathcal{L}}_K(M)$ consists of a discrete collection of points, and $T_K[M]$ will have isolated vacua (no moduli space) on $\mathbb{R}^2 \times S^1$.

A more interesting situation arises when M has asymptotic boundary C, so that $T_K[M]$ is a boundary condition for the 4d theory $T_K[C]$. Suppose that we move onto the Coulomb branch of $T_K[C]$. After compactification on a circle S^1, the theory $T_K[C]$ can be described in the IR as a 3d sigma model whose target is the moduli space of flat $SL(K, \mathbb{C})$ connections on C [48]

$$\mathcal{P}_K(C) = \{\text{flat } SL(K, \mathbb{C}) \text{ connections on } C\}. \tag{5}$$

This space arises physically from a standard 2d version of Hitchen's equations (3). It is actually a hyperkähler space, as appropriate for 4d $\mathcal{N} = 2$ supersymmetry. However, we will only consider it in a single complex structure—the complex structure associated to the 3d $\mathcal{N} = 2$ subalgebra that the boundary condition $T_K[M]$ preserves. Then, for us, $\mathcal{P}_K[C]$ is simply a complex symplectic space. Its holomorphic symplectic form is given by the Atiyah-Bott formula

$$\Omega = \int_C \text{Tr}\left[\delta\mathcal{A} \wedge \delta\mathcal{A}\right], \tag{6}$$

where $\delta\mathcal{A}$ is the deformation of a complex connection. The holomorphic coordinates on $\mathcal{P}_K[C]$ are eigenvalues or traces of $SL(K, \mathbb{C})$ holonomies (or some more elementary cross-ratio coordinates, *à la* [28], from which holonomies can be constructed, see

---

[11]One can attempt to use algebraic Mostow rigidity [39] to analyze the problem. This requires knowing that the representation $\rho : \pi_1(M) \to SL(K, \mathbb{C})$ defined by the holonomies of a flat connection $\mathcal{A}$ is a lattice. That is, $\rho(\pi_1(M)) \subset SL(K, \mathbb{C})$ is a discrete subgroup, with no accumulation points, such that $SL(K, \mathbb{C})/\rho(\pi_1(M))$ has finite volume. This is true if M is hyperbolic and $\mathcal{A}$ is the flat connection related to the hyperbolic metric; but is unknown in general.

Sect. 4.1). In $T_K[C]$ these coordinates are the vevs of supersymmetric line operators that wrap S^1 [45, 49].[12]

Now, the moduli space $\widetilde{\mathcal{L}}_K(M)$ of flat connections on M generically projects to a Lagrangian submanifold $\mathcal{L}_K(M) \subset \mathcal{P}_K(C)$, which parameterizes the flat connections on the boundary C that extend to M:

$$\mathcal{L}_K(M) = \{\text{flat } SL(K, \mathbb{C}) \text{ connections on } \partial M \text{ that extend to } M\}. \qquad (7)$$

The expectation that this is Lagrangian follows from the fact that flatness equations are elliptic; at a basic level, only *half* of the classical parameters on the boundary are needed to specify a flat connection in the bulk. Moreover, both $\mathcal{P}_K(C)$ and $\mathcal{L}_K(M)$ are algebraic. The equations that cut out $\mathcal{L}_K(M)$ can be interpreted as Ward identities for line operators in $T_K[C]$ in the presence of the boundary condition $T_K[M]$. In the effective 3d sigma-model to $\mathcal{P}_K(C)$, $\mathcal{L}_K(M)$ is quite literally a Lagrangian brane boundary condition [1].

If we decouple the 4d bulk theory $T_K[C]$ to leave behind a 3d theory $T_K[M, \mathbf{p}]$ or $T_K[M, \Pi]$, the Lagrangian $\mathcal{L}_K(M)$ acquires a more intrinsic interpretation. Let us consider the Seiberg-Witten description of $T_K[C]$ for simplicity. Then the choice of duality frame Π needed for the decoupling maps precisely to a choice of *polarization* for $\mathcal{P}_K(C)$. This is a local splitting of coordinates into "positions" x (corresponding to IR Wilson lines of $T_K[C]$) and "momenta" p (corresponding to IR 't Hooft lines).

The decoupled theory $T_K[M, \Pi]$ has $U(1)^d$ flavor symmetry, where $2d = \dim_{\mathbb{C}} \mathcal{P}_K(C)$. The positions x are twisted masses[13] for each $U(1)$ symmetry, complexified by $U(1)$ Wilson lines around S^1. The momenta p can be thought of as effective FI parameters for the flavor symmetries; or equivalently as the vevs of complexified moment map operators for each $U(1)$. The Lagrangian $\mathcal{L}_K(M)$ then describes the subset of twisted masses and effective FI parameters that allow supersymmetric vacua to exist on $\mathbb{R}^2 \times S^1$—it is the "supersymmetric parameter space" of $T_K[M, \Pi]$.

More concretely, by compactifying $T_K[M, \Pi]$ on a circle we obtain a 2-dimensional $\mathcal{N} = (2, 2)$ theory, whose IR behavior is governed by an effective twisted superpotential $\widetilde{W}$. After extremizing $\widetilde{W}$ with respect to dynamical fields, it retains a dependence on complexified masses x. The supersymmetric parameter space is then defined by [6][14]

$$\mathcal{L}_K(M): \quad \exp\left(x_i \partial_{x_i} \widetilde{W}(x)\right) = p_i, \qquad i = 1, \ldots, d. \qquad (8)$$

---

[12] See Sect. 2 of *Hitchin systems in $\mathcal{N} = 2$ field theory* by A. Neitzke.

[13] Explicitly, if we re-introduce the radius β of the compactification circle, these dimensionless coordinates arise as $x = \exp\left(\beta m_{3d} + i \oint_{S^1} A\right)$, where A is the background gauge field for a 3d flavor symmetry, and m_{3d} is its real mass. A factor of β also enters (8) to keep $\widetilde{W}$ dimensionless.

[14] This Lagrangian and its quantization also plays a role in the study of surface operators in 4d $\mathcal{N} = 2$ theories, and their lifts to 3d defects in 5d theories—see Sect. 2.4 of [V:8].

The description of $\mathcal{L}_K(M)$ and $\mathcal{P}_K(\partial M)$ can be generalized to geometries M that include codimension-two defects. It is necessary to impose boundary conditions for flat connections at the defects. These effectively increase the dimension of $\mathcal{P}_K(\partial M)$, basically as if all defects had been regularized to small tubular pieces of boundary. This is natural, since defects enlarge the flavor symmetry group of $T_K[M]$. Mathematically, $\mathcal{P}_K(\partial M)$ and $\mathcal{L}_K(M)$ most accurately take the form of moduli spaces of "framed" flat connections, which we discuss in Sect. 4.1.

2.1 Quantization and 3d-3d Relations

Having understood the fundamental relation between flat connections and the parameters/observables of $T_K[M]$, one can further deform the $\mathbb{R}^2 \times S^1$ geometry to quantize the pair $\mathcal{L}_K(M) \subset \mathcal{P}_K(\partial M)$. The basic idea is that adding angular momentum, so that $\mathbb{R}^2 \simeq \mathbb{C}$ fibers over S^1 with twist $z \to qz$, leads to a non-commutative algebra of Wilson and 't Hooft line operators that satisfy $\hat{p}\hat{x} = q\,\hat{x}\hat{p}$ [45, 50], ([V:3], Sect. 3). The algebraic equations for $\mathcal{L}_K(M)$ are promoted to operators that annihilate partition functions of $T_K[M, \Pi]$ (or $T_K[M, \mathbf{p}]$), enforcing Ward identities in the twisted geometry.

The quantization of the pair $\mathcal{L}_K(M) \subset \mathcal{P}_K(\partial M)$ also has a natural interpretation on the "geometric" side of the 3d-3d correspondence. It is useful to recall that flat $SL(K, \mathbb{C})$ connections on a 3-manifold are the classical solutions of quantum $SL(K, \mathbb{C})$ Chern-Simons theory. The space $\mathcal{P}_K(\partial M)$ is just the semi-classical phase space that Chern-Simons theory associates to a boundary of M, and its quantization produces the algebra of operators acting on a quantum Chern-Simons Hilbert space $\mathcal{H}_K(\partial M)$ [51–53]. Similarly, the Lagrangian $\mathcal{L}_K(M)$ is just a semi-classical wavefunction, and its quantization produces a distinguished element of the operator algebra that annihilates the Chern-Simons wavefunction on M, an element of $\mathcal{H}_K(\partial M)$.

One expects, therefore, that partition functions of $T_K[M, *]$ on spacetimes with angular momentum are equivalent to wavefunctions in complex Chern-Simons theory, leading to the correspondences of (2). A precise choice of spacetime is required to fully specify how the Chern-Simons Hilbert space should be quantized—in particular to specify the level of the Chern-Simons theory. However, the structure of the quantum line-operator algebra (the algebra of operators in CS theory) remains essentially independent of this choice. Here are some options that have been studied:

- On spinning $\mathbb{R}^2 \times_q S^1$ as above, the partition function of $T_K[M, \Pi]$ depends on a discrete choice α of boundary condition (basically a massive vacuum) at infinity on $\mathbb{R}^2$, in addition to q and the complex masses x. Geometrically, α is a choice of flat connection on M given boundary conditions x. The resulting partition functions $B_\alpha(x; q)$ [21, 54], which count BPS states of $T_K[M, *]$, correspond to partition functions in analytically continued $SU(K)$ Chern-Simons theory on M, with exotic choices of integration contour labelled by α, much as in [53, 55, 56].

- The partition function of $T_K[M, *]$ on a spinning $S^2 \times_q S^1$ geometry computes a supersymmetric index [57–59]. It was conjectured in [3] and derived in [12] that the index corresponds to a wavefunction of $SL(K, \mathbb{C})$ Chern-Simons theory at level $k = 0$. This is not a trivial theory! To be more precise, we must recall that complex Chern-Simons theory has two levels (k, σ), one quantized and one continuous. Here only the quantized level is set to zero; the continuous σ is related to the spin in the index geometry as $q \sim e^{2\pi/\sigma}$.
- The partition function of $T_K[M, *]$ on an ellipsoid S_b^3, computed via methods of [60, 61],[15] was conjectured in [1, 5, 6] to correspond to an $SL(K, \mathbb{R})$-like Chern-Simons wavefunction. A careful supergravity calculation in [11, 62] then derived a direct relation to $SL(K, \mathbb{C})$ Chern-Simons theory at level $k = 1$. The Hilbert spaces of these two Chern-Simons theories are very similar—see Sect. 2.2.
- It was conjectured in [21] that the index and ellipsoid partition functions can both be written as sums of products of "holomorphic blocks" $B_\alpha(x; q)$, providing a direct relation between the three types of partition functions above. This is essentially holomorphic-antiholomorphic factorization in complex Chern-Simons theory, and involves a 3d analogue of topological/anti-topological fusion [63, 64] for $T_K[M, *]$.
- Extending the results of [11], one expects that the partition function of $T_K[M, *]$ on a squashed lens space $L(k, 1)_b$ (which can be computed via methods of [65]) agrees with a wavefunction of $SL(K, \mathbb{C})$ Chern-Simons theory at general level k.

The relation between S_b^3 partition functions of $T_2[M, *]$ and complex Chern-Simons theory provided some of the first concrete tests of 3d-3d duality. For 3-manifolds with boundary, the relevant Chern-Simons partition functions could be computed using methods of [66–68] (and are now understood to capture $SL(2, \mathbb{C})$ Chern-Simons at level $k = 1$). In the case of $S^2 \times_q S^1$, however, techniques for computing the index of $T_K[M, *]$ led to a *new* algorithm for computing $SL(K, \mathbb{C})$ Chern-Simons wavefunctions at level $k = 0$, which has since been formalized mathematically [69, 70]. Repeating this exercise for squashed lens spaces should prove equally interesting.

2.2 Connection to AGT

As anticipated in the introduction, the fact that 3d theories $T_K[M, *]$ naturally define boundary conditions for 4d theories $T_K[\partial M]$ of class $\mathcal{S}$ leads to a close interplay between the partition functions involved in 3d-3d and 2d-4d relations.

The basic physical idea is that if X is a 4-manifold with boundary allowing supersymmetric compactification of $\mathcal{N} = 2$ theories, the partition function $\mathcal{Z}_X(T_K[\partial M], \mathbf{p})$ should depend on supersymmetric boundary conditions, and can be interpreted as a wavefunction in some Hilbert space $\mathcal{H}_K[\partial M, \mathbf{p}]$. Here we write

---

[15]See also *A review on SUSY gauge theories on S^3* by K. Hosomichi.

$\mathcal{Z}_X\big(T_K[\partial M], \mathbf{p}\big)$ to emphasize that the way one prescribes boundary conditions may depend on a choice of weak-coupling duality frame for $T_K[\partial M]$, given (say) by a pants decomposition $\mathbf{p}$ for ∂M. For example, if $X = D_b^4$ is half of the squashed 4-sphere S_b^4 (equivalently, for computational purposes, to the omega-background $X = \mathbb{R}_\epsilon^4$), then $\mathcal{Z}_X\big(T_2[\partial M], \mathbf{p}\big)$ is an instanton partition function of $T_2[\partial M]$. The instanton partition function depends on Coulomb moduli a_i for each gauge group that is manifest in the duality frame $\mathbf{p}$. Via the AGT correspondence, it is natural to identify the instanton partition function with a wavefunction in the Hilbert space of Liouville conformal blocks $\mathcal{H}_2[\partial M, \mathbf{p}]$.

Here we should emphasize a technical point. In this interpretation, $\mathcal{H}_K[\partial M, \mathbf{p}]$ is not the (enormous) full physical Hilbert space of $T_K[\partial M]$ on ∂X. Rather, $\mathcal{H}_K[\partial M, \mathbf{p}]$ is a "BPS" subsector of the full Hilbert space, whose elements are supersymmetric ground states of $T_K[M]$ on ∂X. The supersymmetric partition functions that we describe belong to this subsector, which has finite functional dimension.

Now if M is any 3-manifold with boundary ∂M, then the partition function of $T_K[M, \mathbf{p}]$ on ∂X should *also* be a wavefunction in the Hilbert space $\mathcal{H}_K[\partial M, \mathbf{p}]$. In order to calculate the partition function of $T_K[\partial M]$ on X, coupled to the theory $T_K[\partial M, \mathbf{p}]$ on ∂X, we simply take an inner product

$$\big\langle \mathcal{Z}_X\big(T_K[\partial M], \mathbf{p}\big) \,\big|\, \mathcal{Z}_{\partial X}\big(T_K[M, \mathbf{p}]\big) \big\rangle. \tag{9}$$

For example, if $X = D_b^4$, then $\partial X = S_b^3$, and $\mathcal{Z}_{\partial X}\big(T_2[M, \mathbf{p}]\big)$ is simply the ellipsoid partition function of the 3d theory $T_2[M, \mathbf{p}]$. Note that the 3d theory $T_2[M, \mathbf{p}]$ has flavor symmetries with complexified twisted masses a_i for every gauge symmetry of the bulk theory $T_2[\partial M]$ (in duality frame $\mathbf{p}$); thus both the right and left sides of (9) depend on the same parameters a_i, and taking an inner product just means integrating them out with the right measure.

By using the doubling trick of Fig. 2, these constructions can easily be extended to domain walls. For example, one might insert an S-duality domain wall carrying theory $T_2[M; \mathbf{p}, \mathbf{p}']$ on the equator $S_b^3 \subset S_b^4$. Here $\partial M = \overline{C} \sqcup C$ for some surface C, so the ellipsoid partition function belongs to a product of Hilbert spaces $\mathcal{Z}_{S_b^3}\big(T_2[M; \mathbf{p}, \mathbf{p}']\big) \in \mathcal{H}_2[C]^* \otimes \mathcal{H}_2[C]$. The partition function on the whole S_b^4 with the domain wall becomes

$$\big\langle \mathcal{Z}_{D_b^4}\big(T_2[C], \mathbf{p}\big) \,\big|\, \mathcal{Z}_{S_b^3}\big(T_2[M; \mathbf{p}, \mathbf{p}']\big) \,\big|\, \mathcal{Z}_{D_b^4}\big(T_2[C], \mathbf{p}'\big) \big\rangle. \tag{10}$$

Such configurations with S-duality domain walls in S_b^4 have been studied at length, e.g. in [5, 6, 23, 24] (see [V:10]). The 3d partition function $\mathcal{Z}_{S_b^3}\big(T_2[M, \mathbf{p}, \mathbf{p}']\big)$ can be identified with a Moore-Seiberg kernel in Liouville theory—it acts naturally on $\mathcal{H}_2[C]$, changing the basis from one labelled by $\mathbf{p}$ to one labelled by $\mathbf{p}'$. In this case, the Lagrangian $\mathcal{L}_2(M)$ and its quantization describes the transformation of line operators from one side of the wall to the other. An analogous setup involving domain walls on the equator of the index geometry $S^2 \times_q S^1 \subset S^3 \times_q S^1$ was considered in [3, 25, 71].

We remark that while physically it is clear that all wavefunctions appearing in formulas such as (9) must belong to the same Hilbert space—namely, the space describing basic supersymmetric boundary conditions on ∂X—this is sometimes a little less clear on the "geometric" side of the 3d-3d and 2d-4d correspondences. There remain a few interesting details to be worked out here. For example, the 2d-4d correspondence says that $T_2[\partial M]$ belongs to a space of Liouville conformal blocks on ∂M, while the 3d-3d correspondence says that $T_2[M, *]$ belongs to the Hilbert space of $SL(2, \mathbb{C})$ Chern-Simons theory, at level $k = 1$, on ∂M. These are not obviously equivalent. A promising observation is that the Liouville Hilbert space is a boundary Hilbert space for $SL(2, \mathbb{R})$ Chern-Simons [72].[16] In turn, the quantization of a model phase space $(\mathbb{R})^2$ in $SL(2, \mathbb{R})$ theory yields $\mathcal{H} = L^2(\mathbb{R})$; while the quantization of a model $(\mathbb{C}^*)^2$ in $SL(2, \mathbb{C})$ theory at level k yields $\mathcal{H} = L^2(\mathbb{R}) \otimes V_k$, where $\dim V_k = k$ [3]; these model descriptions agree when $k = 1$.

3 Top-Down Construction

Currently there exist two closely related approaches for producing 3d $\mathcal{N} = 2$ Lagrangian gauge theories that flow in the IR to 3-manifold theories $T_K[M]$. Both approaches lead to abelian Chern-Simons-matter theories of class $\mathcal{R}$, whose superpotentials may contain nonperturbative monopole operators. Going in reverse chronological order, we will first introduce the more intuitive "top-down" construction of [2] here, and then discuss the more concrete but also more technical "bottom-up" construction of [1, 7] in Sect. 4.

It is important to keep in mind that many different UV Lagrangian theories can have the same IR fixed point $T_K[M]$. We will say that such UV theories are "mirror symmetric," after the first dualities of this type found in [73–76]. The phenomenon is entirely analogous to Seiberg duality for 4d $\mathcal{N} = 1$ theories (and sometimes even arises from reducing 4d dualities [77]). For now we note that the abelian Lagrangians described here could easily have non-abelian mirrors.

The basic idea of [2] is to derive the BPS particle content and interactions for a UV description of $T_K[M]$ from the geometry of a K-fold spectral cover $\tilde{M}$ of M, and then to use them (optimistically) to reconstruct an entire 3d Lagrangian. For example, in M-theory, the K coincident M5 branes wrapping M are expected to deform at low energy[17] in the fiber directions of T^*M, recombining into a single brane that wraps the cover $\tilde{M}$. The BPS states and their interactions then arise from M2 branes that end on this M5.

---

[16]Quantization of $SL(2, \mathbb{R})$ flat connections on a surface is reviewed in this volume in *Supersymmetric gauge theories, quantization of $\mathcal{M}_{\text{flat}}$, and conformal field theory* by J. Teschner.

[17]Here we mean low energy from the point of view of M-theory dynamics, which is still UV for 3d field theories on $\mathbb{R}^3$. See related comments below about being able to choose arbitrary metric for M.

We can understand the appearance of a spectral cover $\widetilde{M}$, governed by a multi-valued harmonic one-form λ on M (or a single-valued harmonic one-form λ on $\widetilde{M}$), directly in M theory. In order to preserve supersymmetry, an M5 brane must wrap a special Lagrangian 3-cycle in T^*M. The zero-section M is one such cycle, but it can be deformed. Small deformations preserving the special Lagrangian condition are precisely parametrized by real harmonic 1-forms λ on M. We should emphasize again that λ depends on a choice of metric for M, which is entirely up to us—we are *not* working in the ultra low-energy limit where only the hyperbolic metric is relevant.

Alternatively, we could obtain the spectral cover in field theory by starting with the nonabelian Hitchin-like construction of Sect. 2, and sending the compactification radius to infinity. This radius β implicitly entered the definition of the complexified connection $A + i\beta\varphi$ in (3); as $\beta \to \infty$, a rescaled Higgs field φ survives. So long as the three components of φ are simultaneously diagonalizable, we saw that their eigenvalues define a multi-valued harmonic 1-form. A more direct 6d construction, along the lines of [32, 78], would extract λ from certain operators of the 6d $(2, 0)$ theory.

From $\widetilde{M}$ and λ, one can attempt to read off the content of a UV Lagrangian description of $T_K[M]$, which we'll call $\widetilde{T}_K[M]$. First, the integral of λ around any 1-cycle $\gamma \subset \widetilde{M}$ produces a real scalar σ in a 3d $\mathcal{N} = 2$ vector multiplet. The integral of the (abelian) M5-brane B-field on the same cycle leads to the actual 3d abelian gauge field A_μ, the superpartner of σ. Thus, to a first approximation, the number of gauge multiplets in $\widetilde{T}_K[M]$ is the first Betti number $b_1(M)$. In fact, if there is any torsion in $H_1(\widetilde{M}, \mathbb{Z})$, it indicates the presence of additional gauge multiplets that are killed (dynamically) by nonzero Chern-Simons terms. The full claim is that if

$$H_1(\widetilde{M}, \mathbb{Z}) \simeq \mathbb{Z}\langle\gamma_1, \ldots, \gamma_d\rangle\big/\big(\Sigma_j k_{ij}\gamma_j = 0\big), \tag{11}$$

then $\widetilde{T}_K[M]$ has d abelian gauge multiplets coupled with a Chern-Simons level matrix k_{ij}.

If M has defects, they lift to defects in the spectral cover $\widetilde{M}$. Then, much as in the setting of compactification on 2d surfaces with punctures, the non-trivial 1-cycles in $\widetilde{M}$ that link the defects give rise to non-dynamical gauge fields and flavor symmetries in $\widetilde{T}_K[M]$. Note that defects impose boundary conditions on λ that forbid a trivial solution $\lambda \equiv 0$.

Similarly, if M has an asymptotic boundary of the form $C \times \mathbb{R}_+$, the spectral cover $\widetilde{M}$ will have asymptotic regions of the form $\Sigma \times \mathbb{R}_+$, where Σ is a K-fold cover of C. It is the Seiberg-Witten curve for the 4d theory $T_K[C]$. If we pass to a weak-coupling limit Π of $T_K[C]$ to obtain a pure 3d theory $T_K[M, \Pi]$, half of the cycles in the Seiberg-Witten curve will get pinched off. The remaining cycles contribute to $H_1(\widetilde{M})$, and lead to non-dynamical $U(1)$ gauge multiplets in $\widetilde{T}_K[M, \Pi]$, corresponding to the expected $U(1)$ flavor symmetries.

Most interestingly, M2 branes ending on the M5 wrapping the spectral cover lead to BPS particles and superpotential interactions in $T_K[M]$. The basic case is a non-

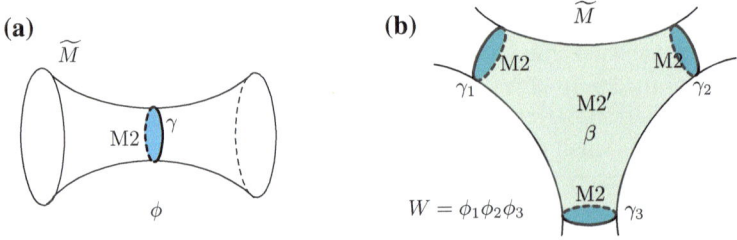

Fig. 5 Producing BPS chirals and superpotentials from M2 branes wrapped on $\widetilde{M}$

contractible cycle $\gamma \subset \widetilde{M}$ of *minimum volume* that bounds a disc $D_\gamma \subset T^*M$. An M2 brane wrapping $D \times \mathbb{R} \subset T^*M \times \mathbb{R}^3$ gives rise to a BPS particle of charge γ, hence a chiral multiplet ϕ in $\widetilde{T}_K[M]$ (Fig. 5a). If the M2 brane instead ends on a 2-cycle $\beta \subset \widetilde{M}$ (filling in a ball in T^*M), then it looks like an instanton in $\mathbb{R}^3$, which can generate a superpotential involving a monopole operator. It is the monopole for the gauge field associated to the 1-cycle γ dual to β. Finally, suppose we have a collection of M2 branes wrapping some discs D_i, with $\partial D_i = \gamma_i$, giving rise to chirals ϕ_i. Then an additional M2' brane might wrap a ball in T^*M whose boundary is a union of the discs D_i and an open 2-cycle in $\widetilde{M}$ connecting their boundaries γ_i (Fig. 5b). This latter M2' brane also looks like an instanton in $\mathbb{R}^3$, and generates a superpotential interaction among the chirals, $W = \prod_i \phi_i$.

Altogether, the vector multiplets and their Chern-Simons interactions, and the chiral multiplets and their superpotential interactions, all obtained geometrically from $\widetilde{M}$, could specify the abelian Chern-Simons matter theory $\widetilde{T}_K[M]$ (or $\widetilde{T}_K[M, \Pi]$, etc.). Unfortunately, the prescription can be extremely difficult to implement in general. The problem is that, given an arbitrary background metric on M, one cannot easily solve for the harmonic form λ and the minimum-volume cycles on $\widetilde{M}$.

One way to circumvent this problem is to deform the metric on M so that the cover $\widetilde{M}$ becomes "especially nice," making it easy to read off the particle content of $\widetilde{T}_K[M]$. We will explain this further in the next sections. Often there are multiple "especially nice" limits, which lead to different mirror-symmetric theories $\widetilde{T}_K[M]$.

3.1 Seiberg-Witten Domain Walls

A basic scenario that can allow a simple description of the spectral cover $\widetilde{M}$ is for M representing a Seiberg-Witten domain wall, as discussed in Sect. 1.1. Such manifolds were the focus of study in [2, 9].

We take $M = \mathbb{R} \times C$, where C is a punctured surface. The punctures become defects running the entire length of M. At the two asymptotic ends of M, we consider the theory $T_K[C]$ on its Coulomb branch. Globally, we picture the spectral cover

$\widetilde{M}$ as a fibration over the infinite direction $\mathbb{R}$, whose fiber over a point $x^3 \in \mathbb{R}$ is a Seiberg-Witten curve Σ_{x^3} for $T_K[C]$. The Seiberg-Witten curve comes with a holomorphic Seiberg-Witten differential $\lambda^{SW}(x^3)$. As x^3 varies from $-\infty$ to ∞, we want to smoothly vary the UV gauge couplings (i.e. the metric on C), as well as mass parameters coming from the defects and Coulomb moduli in such a way that the theory $T_K[C]$ decouples at $x = \pm\infty$ according to some chosen polarizations Π, Π'.

In order to preserve 3-dimensional $\mathcal{N} = 2$ supersymmetry, the variation we choose cannot be completely arbitrary. Geometrically, we need the real part of the varying Seiberg-Witten differential $\lambda^{SW}(x^3)$ to form two of the three components of a harmonic 1-form λ on $\widetilde{M}$. Alternatively, in field-theory terms, we recall that the 3d $\mathcal{N} = 2$ central charges are the real parts[18] of 4d $\mathcal{N} = 2$ central charges (just as the scalar in a 3d gauge multiplet is the real part of the scalar in a 4d gauge multiplet). A necessary condition for unbroken 3d SUSY is

$$\partial_3 \mathrm{Re}[a(x_3)] = \partial_3 \mathrm{Re}[a_D(x_3)] = \partial_3 \mathrm{Re}[m(x_3)] = 0, \tag{12}$$

i.e. the real parts of all 4d central charges, coming from periods of λ^{SW}, are fixed. A 4d theory $T_K[C]$ whose parameters vary[19] in the x^3 direction subject to (12) can be called a generalized *Janus configuration*, cf. [80]. The condition (12) ensures that $\partial_3 \mathrm{Re}\, \lambda^{SW}$ is an exact 2-form on Σ, i.e. $\partial_3 \mathrm{Re}\, \lambda^{SW} = d_\Sigma f$, where d_Σ is the exterior derivative along Σ. Then $\mathrm{Re}\, \lambda^{SW} - f\, dx^3$ is a closed real 1-form on $\widetilde{M}$, which can be further corrected[20] to produce the harmonic 1-form λ.

The fundamental example of a Seiberg-Witten domain wall involves the Seiberg-Witten curve

$$\Sigma_\Delta : \quad z^2 = -w^2 + m, \qquad \lambda^{SW} = z\, dw, \tag{13}$$

where m is a complex mass parameter. Note that the curve is a double cover of the complex w-plane, which we identify as C, with branch points at $w = \pm\sqrt{m}$, and that the only nontrivial period comes from the cycle γ connecting the branch points:

$$\frac{1}{\pi} \oint_\gamma \lambda^{SW} = \frac{2}{\pi} \int_{-\sqrt{m}}^{\sqrt{m}} \lambda^{SW} = m. \tag{14}$$

Indeed, the Seiberg-Witten theory corresponding to the curve (13) has a single BPS hypermultiplet of central charge m (and mass $|m|$). More generally, the curve (13)

---

[18]More generally, we have $Z_{3d} = \mathrm{Re}[\zeta^{-1} Z_{4d}]$, where the phase ζ characterizes the $4d \to 3d$ supersymmetry breaking. The 4d R-symmetry group $SU(2)_R \times U(1)_r$ is broken to $U(1)_R$ (a Cartan of $SU(2)_R$), and this ζ is rotated by the broken $U(1)_r$. This same phase also happens to select the complex structure that one should use for the hyperkähler moduli spaces of flat connections [29, 48], as discussed in Sect. 2.

[19]Similar half-BPS configurations in 3d $\mathcal{N} = 2$ theories were discussed in [79].

[20]The correction requires solving the potential problem $\nabla^2 \sigma = \partial_3 f$. Then $\lambda = \mathrm{Re}\, \lambda^{SW} - f\, dx^3 + d\sigma$.

can also be thought of as a local model for any Seiberg-Witten fibration $\Sigma \to C$ where two branch points are coming close together.

To build a domain wall from (13), we vary the imaginary part of m while keeping the real part fixed, say $m = m_0 + ix_3$. The two branch points of $\sigma \to C$ sweep out branch lines of a 3d fibration $\tilde{M} \to M$. As $x_3 \to \pm\infty$, the branch lines move very far apart, the mass $|m|$ of the 4d BPS state grows infinitely, and the 4d theory $T_K[C]$ decouples. At $x_3 = 0$, the branch lines are minimally separated, and an M2 brane wrapping the cycle γ between them produces a "trapped" 3d BPS chiral ϕ. Its 3d real mass is m_0. We find that $T_K[M, \Pi, \Pi'] =: T_\Delta$ (which will eventually be called the "tetrahedron theory") contains a single free chiral transforming under the $U(1)$ flavor symmetry coming from the cycle γ. If we want a true SCFT, we should set $m_0 = 0$; otherwise the 3d theory is mass-deformed.

In field-theory terms, the full domain wall $T_K[M]$, can be understood roughly as follows. Let us denote by $T_K[C^-]$ and $T_K[C^+]$ the 4d Seiberg-Witten theories on the left and right half-spaces $\mathbb{R}^3 \times \mathbb{R}_\pm$. Each of these theories has a BPS hypermultiplet Φ^- and Φ^+, which we rewrite as a pair of 3d $\mathcal{N} = 2$ chirals (X^-, Y^-) and (X^+, Y^+). Here X and Y have opposite flavor charge. On both the left and the right, we give $X^\pm$ Dirichlet boundary conditions and $Y^\pm$ Neumann boundary conditions. Then, at $x_3 = 0$, we couple the (free) boundary values of $Y^\pm$ to our 3d chiral ϕ via a superpotential [1]

$$W = Y^-\phi - \phi Y^+ \big|_{x_3=0}. \tag{15}$$

These couplings modify the Dirichlet b.c. for the X's to $X^-|_{x^3=0} = \phi = X^+|_{x^3=0}$, via a mechanism studied in [81, 82].

In the far infrared, we can simply use (15) to integrate out ϕ, obtaining $Y^+ = Y^-$ and $X^+ = X^-$. Thus we recover a single 4d theory $T_K[C]$ on all of $\mathbb{R}^4$. This is not unexpected: in the deep IR, all Seiberg-Witten "duality" walls are basically trivial! However, if we first send Im $m \to \infty$ on the left and right sides of the wall to freeze out the 4d hypers, we are left with the decoupled 3d theory T_Δ containing a nontrivial chiral ϕ.

Note that the choices Π and Π' that we made to decouple the two sides in this example had nothing to do with dynamical electric/magnetic gauge fields. They simply selected which halves of the hypers $(X^\pm, Y^\pm)$ got Neumann versus Dirichlet boundary conditions.[21] More generally, one may augment couplings to 3d chirals as in (15) with true changes of polarization, which are implemented by pure 3d $\mathcal{N} = 2$ Chern-Simons theories living on the domain wall [43] (see also Sect. 4.2).

---

[21] It may seem like $\Pi = \Pi'$ in this example. This is not the case, due to the relative orientation on the two halves. The setup corresponding to $\Pi = \Pi'$ involves X getting Dirichlet b.c. on one side and Y getting Dirichlet b.c. on the other, with the remaining (Neumann) halves coupled directly by a superpotential $W = Y^- X^+$ at $x^3 = 0$. This flows immediately to $T_K[C]$ on all of $\mathbb{R}^4$.

4 Bottom-Up Construction: Symplectic Gluing

In the last section, we mentioned that a judicious choice of metric on M can lead to an especially simple spectral cover $\widetilde{M}$, so that the full abelian Chern-Simons Lagrangian of a theory $\widetilde{T}_K[M]$ can be read off. What we had in mind was a cover branched along a set of lines, so that the branch lines are well separated almost everywhere. In a few isolated regions, the branch lines pass close by one another, and each such region might be modeled on the example (13) of Sect. 3.1. Graphically, each region of closest-approach may be represented as a tetrahedron Δ in a 3d triangulation of M. Then we can attempt to associate a canonical "tetrahedron theory" T_Δ to each tetrahedron—basically the theory of a free 3d chiral multiplet—and then to glue them together properly. This is what was done in [1] for $K = 2$, and generalized to arbitrary $K \geq 3$ in [7].

The idea of [1] was to develop a complete, consistent set of gluing rules for tetrahedron theories, working from the ground up. Physically, the gluing rules amount to introducing superpotential couplings for internal edges in a triangulated manifold, and possibly gauging $U(1)$ flavor symmetries. The rules are very precise, and make many properties of $T_K[M]$ manifest—such as the presence of various marginal and relevant operators, and the existence of an unbroken $U(1)_R$ symmetry in the infrared. On the other hand, one always obtains abelian Chern-Simons matter Lagrangians with abelian flavor symmetries, and it can be quite nontrivial to see that some of the flavor symmetries have expected nonabelian enhancements, e.g. to $SU(K)$. More seriously, as mentioned in the introduction, the theories obtained from triangulations sometimes capture only a sub-sector of the full $T_K[M]$; we will explain why in Sect. 4.1.1.

Geometrically, the approach of [1] mimics a construction of classical and quantum flat $SL(K)$ connections on 3-manifolds via "symplectic gluing." The method of symplectic gluing for quantized connections on triangulated manifolds was developed in [68], generalizing classical observations of Neumann and Zagier [83] and Thurston [84] in hyperbolic geometry. The basic idea, going back to work of Atiyah and A. Weinstein, is that when gluing $M = M_1 \cup_\Sigma M_2$ along some boundary Σ, the standard notion of "taking an inner product of wavefunctions in boundary Hilbert spaces" can be replaced by a formally equivalent procedure of quantum symplectic reduction. The latter procedure is easy to implement even when only partial pieces of boundary are glued.

Since the gluing rules for theories $T_K[M]$ are built to match the gluing of quantum connections, many of the relations between sphere partition functions of $T_K[M]$ and Chern-Simons wavefunctions on M that were summarized in Sect. 2 can be proven combinatorially. More interestingly, one realizes that for a manifold M with boundary, the theory $T_K[M, \Pi]$ should *itself* be viewed as a sort of wavefunction—with its flavor symmetries playing the role of "position variables" that the wavefunction depends on.

We proceed to summarize some of the results of [1, 7, 68], starting with symplectic gluing in geometry and then extending the gluing to 3d gauge theory.

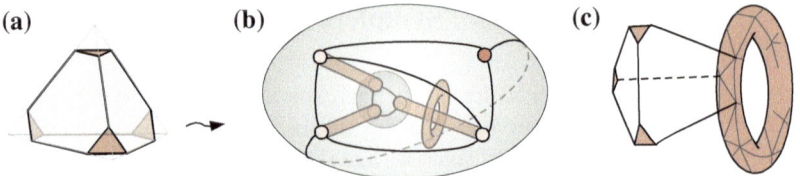

Fig. 6 Truncated tetrahedra (**a**), which can be glued together to form a framed 3-manifold M (**b**). The small vertex-triangles of tetrahedra tile the small tubular boundaries of M (**c**)

4.1 Framed 3-Manifolds and Framed Flat Connections

It is useful to introduce a topological class of *framed 3-manifolds* [7, 8], which represent the 3-manifolds with asymptotic boundaries and networks of defects from Sect. 1 that were used to compactify the 6d $(2, 0)$ theory. A framed 3-manifold[22] is a 3-manifold M with non-empty boundary ∂M, along with a separation of ∂M into "big" and "small" pieces:

- The big boundary consists of surfaces C of arbitrary genus g and $h \geq 1$ holes, such that $-\chi(C) = 2g - 2 + h > 0$. (In particular, these surfaces admit 2d hyperbolic metrics.)
- The small boundary consists of discs, annuli, or tori. The S^1 boundaries of small discs and annuli connect to the holes on the big boundary.

Each of the big boundaries C is meant to represent an asymptotic boundary of a compactification manifold—or rather an asymptotic boundary that has been "cut off" to isolate a 3d theory. Each of the small boundaries represents a codimension-two defect that has been regularized to a long, thin tube.

An oriented framed 3-manifold can be glued together from oriented, truncated tetrahedra (Fig. 6), which are themselves framed 3-manifolds. The big boundary of a tetrahedron is a 4-holed sphere, tiled by four big hexagons. The small boundary consists topologically of four small discs, the triangular vertex neighborhoods. In order to form any more complicated framed 3-manifolds, the big hexagons on tetrahedron faces are glued together in pairs—so some parts of the big boundary may remain unglued—while the small boundary is *never* glued.

Notice that a 3d triangulation of a framed 3-manifold induces a 2d "ideal triangulation" of its big boundary, i.e. a triangulation where all edges begin and end at the holes/punctures. Having fixed the big-boundary triangulation, all possible 3d triangulations of the interior are related by performing sequences of 2–3 moves, shown below in Fig. 9.

Geometrically, on a framed 3-manifold M we can study *framed flat connections*. This is a precise mathematical object that ultimately reproduces (an algebraically open subset of) the correct supersymmetric parameter space of a theory $T_K[M, \Pi]$

---

[22] Such manifolds were called "admissible" in [7].

on a circle, refining (4)–(7). (Framed flat connections in two dimensions played a prominent role in [28, 30].)

A framed flat $PSL(K, \mathbb{C})$ connection on M is a standard flat $PSL(K, \mathbb{C})$ connection together with a choice of invariant flag on each small boundary component. It might be useful to recall that a flag is a set of nested subspaces

$$\{0\} \subset F_1 \subset \cdots \subset F_K = \mathbb{C}^K, \qquad \dim F_K = K. \tag{16}$$

For example, a flag in $\mathbb{C}^2$ is just a complex line in $\mathbb{C}^2$, *a.k.a.* a point in $\mathbb{CP}^1$. What we require for the framing of a flat connection is a choice of flat section of an associated flag bundle on ∂M that's invariant under the $PSL(K, \mathbb{C})$ holonomy around each small boundary. Then we set

$$\mathcal{P}_K(\partial M) = \{\text{framed flat } PSL(K, \mathbb{C}) \text{ connections on } \partial M \setminus (\text{all small discs})\}, \tag{17}$$
$$\mathcal{L}_K(M) = \{\text{ connections in } \mathcal{P}_K(\partial M) \text{ that extend to framed flat connections on } M\}.$$

As discussed in Footnote 9, one sometimes needs to lift these spaces to $SL(K)$ rather than $PSL(K)$, depending on the precise theory of interest. Here we will use $PSL(K)$ for concreteness.

The choice of framing for a flat connection is usually unique, or almost so. For example, a $PSL(K)$ holonomy matrix with distinct eigenvalues has a unique set of K eigenvectors. Choosing an ordering of the eigenvectors, one can then construct an invariant flag. On the other hand, if eigenvalues coincide there may be a continuous choice of invariant flag. This choice resolves singularities in the naive moduli spaces $\mathcal{P}_K(M)$, $\mathcal{L}_K(M)$. An analogous physical resolution of moduli spaces is well known to exist in the presence of defects on surfaces, cf. [29, 50, 55].

The fundamental example of a framed pair (17) is for a truncated tetrahedron Δ, with $K = 2$. On the boundary $\partial \Delta$, viewed as a sphere with four holes, we consider framed flat connections with unipotent holonomy around the holes. (It is necessary to ask for unipotent holonomy, i.e. unit eigenvalues, in order for flat connections to potentially extend to the interior.) At each hole, we choose a complex line in $\mathbb{C}^2$ that's an eigenline of the holonomy there. If the holonomy is parabolic, of the form $\begin{pmatrix} 1 & a \\ 0 & 1 \end{pmatrix}$ with $a \neq 0$, the eigenline is unique. On the other hand, if the holonomy becomes trivial $\begin{pmatrix} 1 & 0 \\ 0 & 1 \end{pmatrix}$, the eigenline is completely undetermined. This extra choice in the latter scenario blows up a singularity in the unframed moduli space.

We can parametrize a generic framed flat $PSL(2)$ connection on $\partial \Delta$ with "cross-ratio coordinates" of Fock and Goncharov [28], as follows.[23] Every edge E in the natural triangulation of $\partial \Delta$ is contained in a unique (truncated) quadrilateral. We parallel-transport the eigenlines at the four vertices of this quadrilateral to any common point p_E inside the quadrilateral, and take their cross-ratio[24] to define a co-

---

[23]These coordinates generalize Thurston's classic shear coordinates in Teichmüller theory, later studied by Penner, Fock, and others.

[24]Recall that lines in $\mathbb{C}^2$ are just points in $\mathbb{CP}^1$, so an $SL(2)$-invariant cross-ratio can be formed.

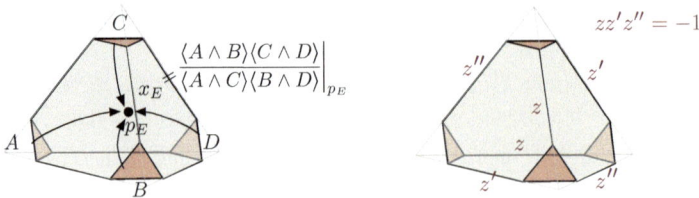

Fig. 7 Defining six edge-coordinates for a tetrahedron by parallel-transporting lines A, B, C, D to common points p_E, then taking cross-ratios

ordinate x_E. The product of these cross-ratio coordinates around any tetrahedron vertex is -1 (due to the unipotent holonomy), which also implies that coordinates on opposite edges are equal. Relabeling the edge-coordinates z, z', z'' as on the left of Fig. 7, we find that

$$\mathcal{P}_2(\partial \Delta) \approx \left\{ z, z', z'' \in \mathbb{C}^* \,\middle|\, zz'z'' = -1 \right\} =: \mathcal{P}_{\partial \Delta}, \qquad (18)$$

with expected complex dimension 2. The complex symplectic structure on $\mathcal{P}_{\partial \Delta}$ induces Poisson brackets $\{\log z, \log z'\} = \{\log z', \log z''\} = \{\log z'', \log z\} = 1$.

Similarly, we may consider framed flat connections in the bulk of Δ. But now, since Δ is contractible, any flat connection is gauge-equivalent to a trivial one. Nevertheless, the choice of four eigenlines at the vertices (modulo the overall action of $PSL(2)$) remains, and is parametrized by the Lagrangian submanifold

$$\mathcal{L}_\Delta = \{z'' + z^{-1} - 1 = 0\} \subset \mathcal{P}_{\partial \Delta}. \qquad (19)$$

The relation $z'' + z^{-1} - 1 = 0$ (which could equivalently be written as $z + z'^{-1} - 1 = 0$ or $z' + z''^{-1} - 1 = 0$) is simply a standard Plücker relation among the cross-ratio coordinates, reflecting the fact that after the tetrahedron is filled in we may parallel-transport all eigenlines to a common point in the interior of Δ and simultaneously calculate all cross-ratios there.

For a general framed 3-manifold M, we may choose a 2d triangulation of the big boundary and again construct cross-ratio coordinates x_E there. Their Poisson bracket is such that

$$\{\log x_E, \log x_{E'}\} = \text{oriented \# of faces shared by } E, E'. \qquad (20)$$

These are supplemented by holonomy eigenvalues around A- and B-cycles of small torus boundaries, and by a combination of holonomy eigenvalues and canonically conjugate "twist" coordinates for each small annulus, altogether forming a system of coordinates for an algebraically open patch of $\mathcal{P}_2(\partial M)$ that's isomorphic to a complex torus $(\mathbb{C}^*)^{2d}$. The fundamental result is that if M is cut into N truncated tetrahedra (in any manner that's consistent with the chosen boundary triangulation) then this patch of $\mathcal{P}_2(\partial M)$ is a symplectic quotient

$$\mathcal{P}_2(\partial M) = \left(\prod_{i=1}^{N} \mathcal{P}_{\partial \Delta_i}\right) /\!\!/ (\mathbb{C}^*)^{N-d}. \tag{21}$$

The $N - d$ moment maps μ_I in the symplectic reduction are simply the products of tetrahedron edge-coordinates z_i, z_i', z_i'' around every internal edge E_I created in the gluing. Fixing $\mu_I = 1$ ensures that a classical flat connection is smooth at that edge. In addition, every $\mathbb{C}^*$ coordinate in $\mathcal{P}_2(\partial M)$ is expressed as a Laurent monomial in tetrahedron edge-coordinates (well defined up to multiplication by the μ_I). For example, every x_E on the big boundary of M is a product of the tetrahedron edge-coordinates incident to the edge E.

The Lagrangian $\mathcal{L}_2(M) \subset \mathcal{P}_2(\partial M)$ can also be obtained[25] by "pulling" a canonical product Lagrangian $\prod_i \mathcal{L}_{\Delta_i} \subset \prod_i \mathcal{P}_{\partial \Delta_i}$ through the symplectic reduction (21). This means projecting $\prod_i \mathcal{L}_{\Delta_i}$ along the $(\mathbb{C}^*)^{N-d}$ flows of the moment maps μ_I, and intersecting with the locus $\mu_I = 1$. This gives a very hands-on algebraic construction of a moduli space that otherwise may appear extremely complicated.

It is known how to generalize the symplectic-gluing construction of $\mathcal{L}_K(M) \subset \mathcal{P}_K(\partial M)$ to arbitrary K. Moreover, it is straightforward to quantize the entire construction [68]. Combinatorially, quantization requires taking logarithms of all cross-ratio coordinates, and consistently keeping track of their imaginary parts. This corresponds physically to keeping track of the $U(1)_R$ symmetry of $T_K[M]$ on curved backgrounds.

4.1.1 Limitations

We have noted in passing that when we construct Lagrangian $\mathcal{L}_K[M]$ from tetrahedra by symplectic gluing, we may only recover an algebraically open patch of the full moduli space of framed flat connections on M. The basic limitation is that all cross-ratio coordinates z, z', z'' for tetrahedra in a triangulation of M must be non-degenerate: not equal to 0, 1, or ∞. Equivalently, the four framing flags at the vertices of any tetrahedron must be distinct after parallel transport to the center. This restriction can sometimes cause the glued Lagrangian $\mathcal{L}_K[M]$ to miss entire families of flat connections. Then, if we use an analogous gluing construction to build a 3d $\mathcal{N} = 2$ theory, as in the next section, we may only recover a subsector of the full $T_K[M]$, whose vacua on S^1 correspond only to *some* of the flat connections on M. This was recently emphasized in [27].

---

[25]Strictly speaking, this is true only for a sufficiently generic or refined triangulation of M. In particular, one must make sure that the $(\mathbb{C}^*)^{N-d}$ action in the quotient is transverse to the product Lagrangian $\prod_i \mathcal{L}_{\Delta_i}$.

To illustrate what we mean in terms of flat connections, suppose that M is a knot complement, i.e. S^3 with a knotted defect inside, which has been regularized to a small torus boundary. A flat $SL(2, \mathbb{C})$ connection on M induces (via its holonomies) a representation $\rho : \pi_1(M) \to SL(2, \mathbb{C})$, and can be classified by the "reducibility" of this representation, i.e. the subgroup of $SL(2, \mathbb{C})$ that commutes with the image $\rho(\pi_1(M))$. For example, only the identity element commutes with a fully irreducible representation, while a full $GL(1) \subset SL(2, \mathbb{C})$ commutes with an "abelian" representation (whose holonomies can all be simultaneously diagonalized). Typically both types of representations exist: there is always an abelian representation, while for hyperbolic knot complements the holonomy of the hyperbolic metric is always irreducible. If we now choose a triangulation for M and choose a framing line on $\partial M = T^2$, we find that all vertices of all tetrahedra share the same framing line (since all vertices land on the same T^2), and the only way to get non-degenerate cross-ratios is to have non-trivial parallel transport inside the tetrahedra. However, the parallel transport of an abelian flat connection acts trivially on the framing lines—and tetrahedron cross-ratios for an abelian flat connection are always degenerate. Therefore, only non-abelian representations are captured by symplectic gluing of tetrahedra.

This is not a serious problem when $K = 2$ and all components of ∂M have genus > 1, such as for manifolds encoding duality domain walls in theories $T_2[C]$ of class $\mathcal{S}$, when C has negative Euler character. In this case, generic choices of boundary conditions (eigenvalues of boundary holonomies) completely forbid reducible flat connections on M. For example, the manifold in Fig. 4a, encoding the S-duality wall for $\mathcal{N} = 2^*$ theory, has a total boundary of genus 2. Then triangulation methods readily reconstruct $T_2[M] \simeq T[SU(2)]$, without missing any branches of vacua [8].

In higher rank ($K \geq 3$) the issue is more severe. Non-degeneracy of cross-ratios requires all the defects in a manifold M to be of "maximal" type, carrying maximal $SU(K)$ flavor symmetry (so that all eigenvalues of boundary holonomies can be distinct). Subsequently, only fully irreducible flat connections are captured by the standard symplectic gluing of [7].

The precise physical significance of the subsector of $T_K[M]$ coming from gluing tetrahedra is still being elucidated. Thinking of $T_K[M]$ as the theory of K M5 branes wrapping $M \times \mathbb{R}^3 \subset T^*M \times \mathbb{R}^3 \times \mathbb{R}^2$, as in Sect. 1, a plausible conjecture is that the subsector obtained by gluing tetrahedra only captures the physics of configurations where the K M5's reconnect into a *single* M5 wrapping a spectral cover of M. Thus the subsector is missing configurations where the K M5's reconnect into multiple components (or remain fully disconnected), and are thus able to separate in the $\mathbb{R}^2$ direction. Such configurations would correspond to the missing branches of vacua. This conjecture is in line with findings of [27], where it was argued in examples that the full $T_K[M]$ contains an additional $U(1)_t$ flavor symmetry, involving rotations of $\mathbb{R}^2$.

4.2 The Tetrahedron Theories

Just as framed 3-manifolds are glued together from tetrahedra, the 3-manifold theories $T_K[M]$ (or more precisely $T_K[M, \Pi]$ or $T_K[M, \mathbf{p}]$) are glued together from tetrahedron theories. For simplicity, we will review how this works in the case $K = 2$.

The first step is to identify the theory of a single truncated tetrahedron. As we first tried to motivate physically in Sect. 1, however, there should be no *unique* tetrahedron theory. Rather, there is an infinite family of 3d theories $T_2[\Delta, \Pi]$ labelled by choices of polarization Π on the boundary of the tetrahedron—*a.k.a.* ways of decoupling an abelian 4d bulk gauge theory from a 3d boundary condition. Now we can understand the polarization in a purely geometric setting: Π is a choice of "electric" $\mathbb{C}^*$ position coordinate and canonically conjugate "magnetic" $\mathbb{C}^*$ momentum coordinate for $\mathcal{P}_{\partial\Delta}$. Choosing

$$\Pi = \Pi_z := \begin{pmatrix} \text{position} = z \\ \text{momentum} = z'' \end{pmatrix}, \tag{22}$$

with canonical Poisson bracket $\{\log z'', \log z\} = 1$, the tetrahedron theory was conjectured in [1] to be

$$T_\Delta := T_2[\Delta, \Pi_z] = \begin{cases} \text{free chiral } \phi_z \text{ with } U(1)_z \text{ flavor symmetry};\\ \text{background CS level } -1/2 \text{ for } U(1)_z. \end{cases} \tag{23}$$

This agrees beautifully[26] with the theory intuited from an analysis of the tetrahedron's spectral cover in Sect. 3.1.

The symplectic group $Sp(2, \mathbb{Z})$ acts both on a formal polarization vector such as (22) and on a 3d SCFT with a $U(1)$ flavor symmetry, as described in [43]. The provides a concrete way to change the polarization of a theory; for example, we expect

$$T_2[\Delta, g \circ \Pi_z] = g \circ T_2[\Delta, \Pi_z], \qquad g \in Sp(2, \mathbb{Z}). \tag{24}$$

Concretely, the generator $T = \begin{pmatrix} 1 & 0 \\ 1 & 1 \end{pmatrix}$ acts on a theory by adding $+1$ to the background Chern-Simons level for the flavor symmetry. The generator $S = \begin{pmatrix} 0 & -1 \\ 1 & 0 \end{pmatrix}$ gauges the flavor $U(1)$, after which there appears a new "topological" flavor symmetry $U(1)_J$. These actions can be understood as the effect of electric-magnetic duality on the 3d boundary of a 4d abelian gauge theory.

Although we can choose any polarization we want for the tetrahedron theory, three of them are special: the polarizations in which one of the edge-coordinates themselves (i.e. z, z', or z'' rather than an arbitrary Laurent monomial like $z^3 z'^{-1}$)

---

[26]Note that the half-integer background Chern-Simons term is corrected by the standard parity anomaly of a 3d $\mathcal{N} = 2$ theory (cf. [74]) to be an integer in the IR, given any nonzero real mass for ϕ_z.

is a position. We can call these Π_z, $\Pi_{z'}$, and $\Pi_{z''}$. In fact, since the cyclic rotation symmetry of the tetrahedron permutes $z \to z' \to z'' \to z$, we might even expect that the resulting theories are all equivalent:

$$T_2[\Delta, \Pi_z] \simeq T_2[\Delta, \Pi_{z'}] \simeq T_2[\Delta, \Pi_{z''}]. \tag{25}$$

This is indeed true. For example, to pass from Π_z to $\Pi_{z'}$, we act with $ST \in Sp(2, \mathbb{Z})$,

$$\Pi_{z'} = \begin{pmatrix} z' \\ z \end{pmatrix} = \begin{pmatrix} -\frac{1}{zz''} \\ z \end{pmatrix} = \begin{pmatrix} -1 & -1 \\ 1 & 0 \end{pmatrix} \cdot \begin{pmatrix} z \\ z'' \end{pmatrix} = ST \circ \Pi_z, \tag{26}$$

where the linear transformation acts *multiplicatively* (i.e. $\begin{pmatrix} a & b \\ c & d \end{pmatrix} \cdot \begin{pmatrix} z \\ w \end{pmatrix} = \begin{pmatrix} z^a w^b \\ z^c w^d \end{pmatrix}$), and we are ignoring signs[27] such as $(-1)\frac{1}{zz''}$. Correspondingly, we find

$$T_2[\Delta, \Pi_{z'}] = ST \circ T_2[\Delta, \Pi_z] = \begin{cases} U(1) \text{ gauge theory with chiral } \phi_{z'} \text{ of charge } +1 \, ; \\ \text{CS level } +1/2 \text{ for the dynamical } U(1) \, ; \\ \text{topological } U(1)_{z'} \text{ flavor symemtry.} \end{cases} \tag{27}$$

In the infrared, this theory flows to the *same* SCFT T_Δ as in (23). The monopole operator of (27) (which creates free vortices) matches the free chiral of (23) [1, 74]. This match is strong evidence that the tetrahedron theory has been properly identified.

Yet another piece of evidence that (23) is correct comes from compactifying the theory on a circle S^1 and calculating its supersymmetric parameter space (8). A straightforward summation of Kaluza-Klein modes (cf. [85]) leads to the twisted superpotential $\widetilde{W}(z) = \text{Li}_2(z^{-1})$, where $\log z$ is the complexified mass associated to the $U(1)_z$ flavor symmetry. Then the definition of the effective FI parameter

$$\exp \frac{\partial \widetilde{W}(z)}{\partial z} = z'' \quad \Rightarrow \quad z'' + z^{-1} - 1 = 0 \tag{28}$$

reproduces the tetrahedron Lagrangian $\mathcal{L}_\Delta$ from (19), as desired.

4.3 Gluing Together Theories

Now suppose that a framed 3-manifold M is glued together from N tetrahedra. In order to define an isolated 3d theory $T_2[M, \Pi]$, we need to choose a polarization Π

---

[27] The signs, and indeed the full lift to logarithms of the edge-coordinates, becomes relevant when keeping track of a choice of $U(1)_R$ symmetry for a theory. Then symplectic $Sp(2N, \mathbb{Z})$ actions are promoted to affine-symplectic $ISp(2N, \mathbb{Z})$ actions.

for the big boundary of M,[28] or rather for the part of $\mathcal{P}_2(\partial M)$ corresponding to the big boundary. For any small tori in ∂M, we also choose A- and B-cycles. For small annuli, though, the choice of non-contractible "A-cycles" (and so the polarization) is canonical.

We build $T_2[M, \Pi]$ by first taking a "tensor product" of tetrahedron theories

$$T_\times = T_{\Delta_1} \times \cdots \times T_{\Delta_N}, \tag{29}$$

which is basically a collection of N free chirals ϕ_{z_i} with flavor symmetry $\prod_i U(1)_{z_i} \simeq U(1)^N$. This product theory corresponds to a product polarization $\Pi_\times = $ (positions z_i; momenta z_i") on the product phase space $\prod_i \mathcal{P}_{\partial \Delta_i}$.

Now the symplectic group $Sp(2N, \mathbb{Z})$ acts to change the polarization of $T_\times$. This is a natural extension of the $Sp(2, \mathbb{Z})$ action on theories with a single $U(1)$ symmetry: the action of an element $g \in Sp(2N, \mathbb{Z})$ just modifies various CS levels, gauges some of the $U(1)$'s in $U(1)^N$, and/or permutes the $U(1)$ factors in $U(1)^N$.

We then choose a *new* polarization $\widetilde{\Pi}_\times = g \circ \Pi_\times$ for $T_\times$, determined by the following algebraic properties:

1. all the position and momentum coordinates of Π (as monomial functions on $\prod_i \mathcal{P}_{\partial \Delta_i}$) are positions and momenta, respectively, in $\widetilde{\Pi}_\times$; and
2. all the moment maps μ_I (products of tetrahedron edge-coordinates around internal edges in M) are positions in $\widetilde{\Pi}_\times$.

The first requirement simply makes $\widetilde{\Pi}_\times$ compatible with our desired final polarization Π. The second requirement, however, is absolutely crucial for the gluing: it guarantees[29] that the transformed product theory $g \circ T_\times$ will contain *chiral operators* $\mathcal{O}_I$ associated to each internal edge E_I of M. Each of these operators $\mathcal{O}_I$ will transform under a flavor symmetry associated to the internal-edge coordinate μ_I.

The final step in the gluing is to add the $N - d$ internal-edge operators $\mathcal{O}_I$ to the superpotential of $g \circ T_\times$. This breaks $N - d$ $U(1)$ flavor symmetries, and implements the symplectic reduction (21) on the gauge-theory level. The result is a UV abelian Chern-Simons-matter theory with manifest $U(1)^d$ flavor symmetry, which flows in the IR to $T_2[M, \Pi]$.

---

[28] In Sect. 1, we also talked about isolating 3d theories $T_K[M, \mathbf{p}]$ based on a pants decomposition $\mathbf{p}$ of the topological boundary of M. This was meant to correspond to decoupling a nonabelian 4d gauge theory in some duality frame. Such a choice is already *built in* to the definition of a *framed* manifold M: a pants decomposition for a boundary component $\mathcal{C}$ corresponds to a splitting of that boundary into a network of small annuli connected by big 3-punctured spheres when selecting a framing.

[29] Just like in the gluing of classical Lagrangian submanifolds, some extra regularity conditions need to be imposed on a 3d triangulation to truly guarantee the existence of the gluing operators $\mathcal{O}_I$. See Sect. 4.1 of [1] or the Appendix A of [7].

5 Examples

We finish with a brief look at two simple framed 3-manifolds M and their effective theories at $K = 2$. We'll mainly follow the bottom-up approach of symplectic gluing from tetrahedra; though both examples are amenable to top-down analyses as well.

The first example, introduced in [1], is a triangular bipyramid (Fig. 8, left). Like a truncated tetrahedron, it only has disc-like small boundaries (at the five truncated vertices), and a big boundary consisting of a five-holed sphere. The bipyramid can be assembled from gluing either two or three tetrahedra together. The IR equivalence of the glued theories that result (containing either two or three chirals) provides the local proof of triangulation independence for general glued theories $T_2[M, \Pi]$ (in fact also for $K > 2$).

The second example is a 3-manifold with topology $M = C \times I$, where C is a cylinder and $I = \{0 \le t \le 1\}$ an interval. We picture M as a solid cylinder with a core drilled out (Fig. 8, right). To specify M as a *framed* 3-manifold, we take the boundary C_0 at $t = 0$ (the core in the solid-cylinder picture) to be a small annulus. The remainder of ∂M is split into a big annulus C_1, glued to two big punctured discs (the ends of the solid cylinder, $\partial C \times I$), with two additional small discs sandwiched inbetween (drawn as tiny triangular regions in Fig. 8). Thus, topologically, total full big boundary of M is a 4-holed sphere. This manifold turns out to be the basic building block of RG domain walls, as well as more general UV S-duality walls, as discussed in Sect. 1.1 (and in great detail in [8]). Geometrically, M represents the local shrinking of an annular region on any surface to a long, thin tube, and ultimately to a defect. We will see that the theory $T_2[M, \Pi]$ has $SU(2) \times U(1)$ flavor symmetry, allowing a coupling to a nonabelian 4d gauge group on one side, and an abelian gauge group on the other.

5.1 2–3 Move and Mirror Symmetry

Let M be the triangular bipyramid. Let's first observe that M has a boundary phase space $\mathcal{P}_2(\partial M) \simeq (\mathbb{C}^*)^4$. It is easy to see this: one can construct cross-ratio coordinates x_E for each of the nine edges on the boundary, while each of the five vertices imposes a relation that the product of edge-coordinates around that vertex equals ± 1 (for unipotent holonomy). Thus $\dim_\mathbb{C} \mathcal{P}_2(\partial M) = 9 - 5 = 4$. We will choose a

Fig. 8 The bipyramid (*left*) and the thickened annulus representing the RG manifold (*right*)

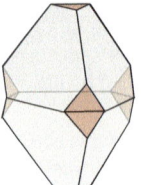
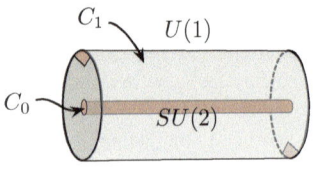

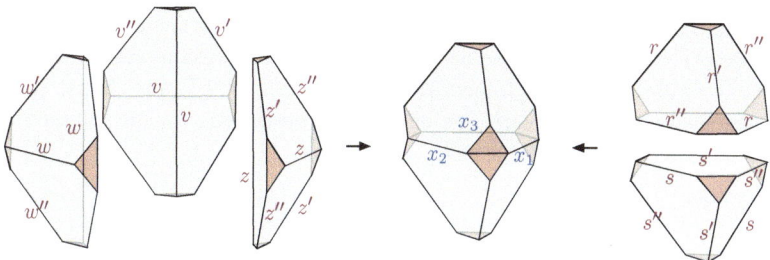

Fig. 9 Gluing together the bipyramid from two or three tetrahedra

polarization Π_{eq} for $\mathcal{P}_2(\partial M)$ such that two of the three equatorial edges of the bipyramid (x_1, x_2) carry electric/position coordinates, as in the center of Fig. 9. Since the product of all equatorial edges is one, this implies that the third edge $x_3 = x_1^{-1}x_2^{-1}$ is electric or "mutually local" as well. Note that specifying the position (but not momentum) coordinates in a polarization is sufficient to define a theory $T_2[M, \Pi_{eq}]$ up to background Chern-Simons levels.

Now, suppose that we glue together a bipyramid M from three tetrahedra, as on the LHS of Fig. 9. We must polarize the tetrahedra, and we choose standard polarizations (22), in such a way that the unprimed position coordinates z, w, v all lie along the internal edge of M. Now the three equatorial edges on the boundary of the bipyramid also get coordinates z, w, v (from opposite edges of the three tetrahedra). So no change of polarization is needed to make the product polarization $\Pi_z \times \Pi_w \times \Pi_v$ on the tetrahedra compatible with our final desired Π_{eq}. The bipyramid theory $T_2[M, \Pi_{eq}]$ is then easy to write down: it is just the product $T_{\Delta z} \times T_{\Delta w} \times T_{\Delta v}$ containing three chirals ϕ_z, ϕ_w, ϕ_v, in which the $U(1)^3$ flavor symmetry is broken to $U(1)^2$ by a cubic superpotential

$$\mathcal{O}_I = \phi_z \phi_w \phi_v \tag{30}$$

corresponding to the internal edge. This theory is usually called the "XYZ model." Note how the individual operators ϕ_z, ϕ_w, ϕ_v are each associated to one of the electric edges on ∂M.

Let us also explain the symplectic reduction geometrically. We can explicitly write the boundary phase space as

$$\mathcal{P}_2(\partial M) = \left(\mathcal{P}_{\partial \Delta z} \times \mathcal{P}_{\partial \Delta z} \times \mathcal{P}_{\partial \Delta z}\right) /\!/ \mathbb{C}^*$$
$$\simeq \{z, z'', w, w'', v, v'' \in \mathbb{C}^*\} / (z'', w'', v'') \sim (tz'', tw'', tv'')\big|_{zwv = 1}, \tag{31}$$

where we have quotiented with respect to the flows of the moment map $\mu_I = zwv$, and intersected with the locus $\mu_I = 1$. Notice that all products of tetrahedron coordinates on external edges (such as z, w, v, or $z'w''$, $w'v''$, etc.), *commute* with μ_I, and so form good coordinates x_E on the quotient. (For a computation of the Lagrangian submanifold $\mathcal{L}_2(M)$ and its quantization, see [68] or [1].)

Alternatively, if we form the bipyramid from two tetrahedra, there are no internal edges created, but a nontrivial change of polarization is required. Let us assign triples of coordinates to the tetrahedra as on the RHS of Fig. 9, and choose standard polarizations Π_r, Π_s for them. The equatorial coordinates for the bipyramid are related to tetrahedron coordinates as

$$x_1 = rs'', \quad x_2 = r''s, \quad \left(x_3 = r's' = (rr''ss'')^{-1} = (x_1 x_2)^{-1}\right), \tag{32}$$

and so involve both tetrahedron positions (r, s) and momenta (r'', s''). The $Sp(4, \mathbb{Z})$ change of polarization that relates $\Pi_r \times \Pi_s$ to Π_{eq} acts on the theory $T_{\Delta r} \times T_{\Delta s}$ by gauging[30] the anti-diagonal subgroup of the flavor symmetry group $U(1)_r \times U(1)_s$. The resulting theory is just 3d $\mathcal{N} = 2$ SQED, which is mirror symmetric to the XYZ model [74]. It has an axial $U(1)_{ax}$ and a topological $U(1)_J$ flavor symmetry, matching the $U(1)^2$ flavor symmetry of the XYZ model. Moreover, it has monopole and anti-monopole operators $\eta_\pm$ in addition to the gauge-invariant meson $\varphi = \phi_r \phi_s$, which together match the three chiral operators ϕ_z, ϕ_w, ϕ_v of the XYZ model, and label the equatorial edges of the bipyramid.

5.2 The Basic RG Wall

Now let M be the RG-wall manifold. Just like the bipyramid, it also has a 4-complex dimensional phase space. Independent coordinates on $\mathcal{P}_2(M)$ are now given by cross-ratios (x_m, x_d) on two edges of the big annulus C_1 (compare Figs. 8 and 10) together with an eigenvalue λ of the $PSL(2)$ holonomy[31] around the girth of the small annulus C_0 and its canonical conjugate, a twist coordinate τ:

$$\mathcal{P}_2(M) \simeq \{x_m, x_d, \lambda, \tau\} \simeq (\mathbb{C}^*)^4, \tag{33}$$

$$\{\log x_d, \log x_m\} = 2, \quad \{\log \tau, \log \lambda\} = 1, \quad \text{other brackets vanishing.}$$

We will choose a polarization Π_e with position coordinates λ and $x_e = (x_m x_d)^{-1/2}$.

We can build M from two truncated tetrahedra, as shown in Fig. 10. There are no internal edges, so no superpotentials will be needed. We give the tetrahedra edge-coordinates z, z', z'' and w, w', w'' and standard polarizations Π_z, Π_w. Then we find

$$\lambda = \sqrt{\frac{z}{w}}, \quad x_e = \sqrt{zw} \tag{34}$$

---

[30]The precise $Sp(4, \mathbb{Z})$ action first removes the background Chern-Simons coupling for the anti-diagonal subgroup of $U(1)_r \times U(1)_s$, and then gauges it. It is a nice exercise to demonstrate this.

[31]Two technical clarifications here: first, the choice of eigenvalue λ versus λ^{-1} depends on the choice of framing for the flat connection at the small annulus; second, to get a well defined sign for λ one actually needs to lift to $SL(2)$ rather than $PSL(2)$ holonomies around the small annulus.

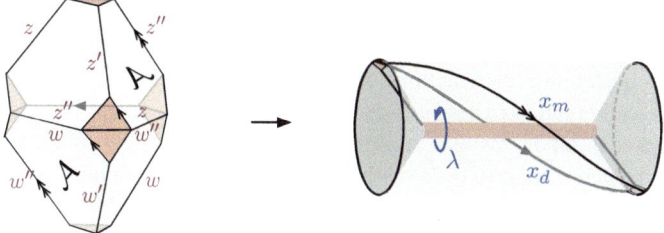

Fig. 10 Forming the RG-wall manifold M by identifying two faces of the bipyramid, as indicated by labels '$\mathcal{A}$' on the *left*. On the *right*, we show the triangulation on the big boundary of M

(as well as $x_m = z''w''$, $x_d = z'w'$, $\tau = \lambda z''/w''$). Since λ and x_e are just made from tetrahedron positions z, w, the change of polarization $\Pi_z \times \Pi_w \rightarrow \Pi_e$ involves no gauging, just a redefinition of flavor symmetries. We find that $T_2[M, \Pi_e]$ is a theory of two free chirals ϕ_z, ϕ_w transforming with charges $(+1, -1)$ and $(+1, +1)$, respectively, under $U(1)_\lambda$ and $U(1)_e$ flavor symmetries associated to λ and x_e. Of course the vector $U(1)_\lambda$ symmetry is actually enhanced to $SU(2)_\lambda$. As promised, the extremely simple theory $T_2[M, \Pi_e]$ can couple both to $SU(2)$ and $U(1)$ 4d gauge groups.

Alternatively, had we chosen a polarization Π_m with λ and x_m as positions, we would instead have described $T_2[M, \Pi_m]$ as a theory of two chirals ϕ_z, ϕ_w whose axial $U(1)_e$ symmetry is gauged at Chern-Simons level -1, and replaced by a topological $U(1)_m$. This is roughly the UV GLSM description of a 3d $\mathbb{CP}^1$ sigma model. Now the theory has a monopole operator $\mathcal{O}_m$ associated to the external "electric" edge with coordinate x_m. Similarly, we could have chosen a polarization Π_d to obtain a theory $T_2[M, \Pi_d]$ whose axial $U(1)_e$ is gauged at Chern-Simons level $+1$.

The claim of [8], a full review of which is beyond our scope, is that the theories $T_2[M, *]$ are effective theories for an RG domain wall in pure $SU(2)$ Seiberg-Witten theory. In the respective polarizations Π_e, Π_m, Π_d, the 3d theories couple to the abelian 4d theory on its Coulomb branch—in 4d duality frames so that the electric, magnetic, or dyonic gauge fields are fundamental. In all these polarizations, the 3d theory couples on the other side of the wall to the nonabelian UV gauge group $SU(2)_\lambda$.

One way to create an RG wall in pure $SU(2)$ theory is by engineering a Janus configuration (cf. (12)) where the UV cutoff Λ varies (relative to a fixed observation scale) as a function of the space coordinate x^3. To the left of the wall, Λ can be arbitrarily close to zero, effectively putting the 4d theory in the UV; while to the right of the wall Λ can be sent close to infinity. We observe the theory at an intermediate energy scale throughout. This traps 3d degrees of freedom on the wall. We can even make an educated guess at what they should be.

Passing through the wall from left to right, the imaginary part of $a(x^3)$ is forced to infinity (relative to our observation scale), breaking $SU(2) \rightarrow U(1)$ and Higgsing the 4d theory. However, close to the (left of the) wall, the $SU(2)$ gauge fields are effectively non-dynamical, since the gauge coupling is infinitesimally small. Thus Goldstone bosons cannot be eaten up by W-bosons, and parametrize a $\mathbb{CP}^1 \simeq SU(2)/U(1)$ -worth of degrees of freedom at the wall. This beautifully matches the bottom-up constructions of $T_2[M, *]$.

The RG walls (and nonabelian S-duality walls) of more complicated 4d theories always involve components that look like the theories $T_2[M, *]$. Indeed, whenever one has a framed 3-manifold $\widehat{M}$ with a network of small annuli connecting big boundaries, the neighborhood of every small annulus can be made to look exactly like our RG-manifold M. This proves, among other things, that in a bottom-up construction of $T_2[\widehat{M}]$, all the $U(1)$ symmetries associated to small annuli will be enhanced to $SU(2)$'s—as must be the case if the small annuli are to represent defects in a 6d compactification.

Finally, let us see what information is contained in the Lagrangian submanifold $\mathcal{L}_2(M)$ of the RG-wall manifold. By rewriting the tetrahedron Lagrangians $z'' + z^{-1} - 1 = w'' + w^{-1} - 1 = 0$ in terms of x_m, x_d, λ, τ and $x_e = 1/\sqrt{x_m x_d}$, we find

$$\left(\text{Wilson}_{\frac{1}{2}}\right) \qquad \lambda + \lambda^{-1} = x_e + x_e^{-1} - x_e x_m \qquad (35a)$$

$$\left(\text{'t Hooft }_{\frac{1}{2}}\right) \qquad \frac{(\tau\lambda)^{\frac{1}{2}} - (\tau\lambda)^{-\frac{1}{2}}}{\lambda - \lambda^{-1}} = \frac{1}{\sqrt{x_m}} \qquad (35b)$$

$$\left(\text{'t Hooft-Wilson}_{\frac{1}{2}}\right) \qquad \frac{(\tau/\lambda)^{\frac{1}{2}} - (\tau/\lambda)^{-\frac{1}{2}}}{\lambda - \lambda^{-1}} = \sqrt{x_d}. \qquad (35c)$$

The first equation relates the spin-1/2 UV Wilson line of pure $SU(2)$ Seiberg-Witten theory to IR line operators of abelian electric and magnetic charge [6, 45]. The second and third equations (which are not independent) relate the spin-1/2 UV 't Hooft lines and mixed 't Hooft-Wilson lines to the IR magnetic and dyonic line operators. The honest $SU(2)$ theory should only contain magnetic UV operators of spin-one, corresponding (roughly) to squaring equations (35b-35c), which then gets rid of the square roots. The quantization of relations (35) turns out to match operator equations known from quantum Teichmüller theory on the annulus [86, 87], giving a beautiful geometric interpretation of the latter.

Acknowledgments It is a pleasure to thank Christopher Beem, Clay Córdova, Davide Gaiotto, and Sergei Gukov for discussions and advice during the writing of this review, and especially Andrew Neitzke and Jeorg Teschner for careful readings and comments.

References

[1]   Dimofte, T., Gaiotto, D., Gukov, S.: Gauge theories labelled by three-manifolds. arXiv:1108.4389

[2]   Cecotti, S., Cordova, C., Vafa, C.: Braids, walls, and mirrors. arXiv:1110.2115

[3]   Dimofte, T., Gaiotto, D., Gukov, S.: 3-manifolds and 3d indices. arXiv:1112.5179

[4]   Dimofte, T., Gukov, S., Hollands, L.: Vortex counting and Lagrangian 3-manifolds. arXiv:1006.0977

[5]   Terashima, Y., Yamazaki, M.: SL(2, R) Chern-Simons Liouville, and gauge theory on duality walls. JHEP **1108**, 135 (2011). arXiv:1103.5748

[6]   Dimofte, T., Gukov, S.: Chern-Simons theory and S-duality. arXiv:1106.4550

[7]   Dimofte, T., Gabella, M., Goncharov, A.B.: K-decompositions and 3d gauge theories. arXiv:1301.0192

[8]   Dimofte, T., Gaiotto, D., van der Veen, R.: RG domain walls and hybrid triangulations. arXiv:1304.6721

[9]   Cordova, C., Espahbodi, S., Haghighat, B., Rastogi, A., Vafa, C.: Tangles, generalized reidemeister moves, and three-dimensional mirror symmetry. arXiv:1211.3730

[10] Fuji, H., Gukov, S., Stosic, M., Sułkowski, P.: 3d analogs of Argyres-Douglas theories and knot homologies. arXiv:1209.1416

[11] Cordova, C., Jafferis, D.L.: Complex Chern-Simons from M5-branes on the squashed three-sphere. arXiv:1305.2891

[12] Lee, S., Yamazaki, M.: 3d Chern-Simons theory from M5-branes. arXiv:1305.2429

[13] Alday, L.F., Gaiotto, D., Tachikawa, Y.: Liouville correlation functions from four-dimensional gauge theories. Lett. Math. Phys. **91**(2), 167–197 (2010). arXiv:0906.3219

[14] Alday, L.F., Gaiotto, D., Gukov, S., Tachikawa, Y., Verlinde, H.: Loop and surface operators in N = 2 gauge theory and Liouville modular geometry. JHEP **1001**, 113 (2010). arXiv:0909.0945

[15] Drukker, N., Gomis, J., Okuda, T., Teschner, J.: Gauge theory loop operators and Liouville theory. JHEP **1002**, 057 (2010). arXiv:0909.1105

[16] Hama, N., Hosomichi, K.: Seiberg-Witten theories on ellipsoids. arXiv:1206.6359

[17] Gadde, A., Pomoni, E., Rastelli, L., Razamat, S.S.: S-duality and 2d topological QFT. JHEP **1003**, 032 (2010). arXiv:0910.2225

[18] Gadde, A., Rastelli, L., Razamat, S.S., Yan, W.: The 4d superconformal index from q-deformed 2d Yang-Mills. Phys. Rev. Lett. **106**, 241602 (2011). arXiv:1104.3850

[19] Gaiotto, D., Rastelli, L., Razamat, S.S.: Bootstrapping the superconformal index with surface defects. arXiv:1207.3577

[20] Witten, E.: Fivebranes and knots. Quantum Topol. **3**(1), 1–137 (2012). arXiv:1101.3216

[21] Beem, C., Dimofte,T., Pasquetti, S.: Holomorphic blocks in three dimensions. arXiv:1211.1986

[22] Drukker, N., Gaiotto, D., Gomis, J.: The virtue of defects in 4D gauge theories and 2D CFTs. arXiv:1003.1112

[23] Hosomichi, K., Lee, S., Park, J.: AGT on the S-duality wall. JHEP **1012**, 079 (2010). arXiv:1009.0340

[24] Teschner, J., Vartanov, G.S.: 6j symbols for themodular double, quantum hyperbolic geometry, and supersymmetric gauge theories. arXiv:1202.4698

[25] Gang, D., Koh, E., Lee, S., Park, J.: Superconformal index and 3d-3d correspondence for mapping cylinder/torus. arXiv:1305.0937

[26] Gadde, A., Gukov, S., Putrov, P.: Fivebranes and 4-manifolds. arXiv:1306.4320

[27] Chung, H.-J., Dimofte, T., Gukov, S., Sułkowski, P.: 3d-3d correspondence revisited. arXiv:1405.3663

[28] Fock, V.V., Goncharov, A.B.: Moduli spaces of local systems and higher Teichmuller theory. Publ. Math. Inst. Hautes Etudes Sci. **103**, 1–211 (2006). arXiv:math/0311149v4

[29] Gaiotto, D., Moore, G.W., Neitzke, A.: Wall-crossing, Hitchin Systems, and the WKB approximation. Adv. Math. **234**, 239–403 (2013). arXiv:0907.3987

[30] Gaiotto, D., Moore, G.W., Neitzke, A.: Spectral networks and snakes. arXiv:1209.0866

[31] Jafferis, D., Yin, X.: A duality appetizer. arXiv:1103.5700

[32] Gaiotto, D.: N = 2 dualities. JHEP **1208**, 034 (2012). arXiv:0904.2715

[33] Gukov, S., Iqbal, A., Kozcaz, C., Vafa, C.: Link homologies and the refined topological vertex. arXiv:0705.1368

[34] Cadavid, A.C., Ceresole, A., D'Auria, R., Ferrara, S.: 11-dimensional supergravity compactified on Calabi-Yau threefolds. Phys. Lett. **B357**, 76–80 (1995). arXiv:hep-th/9506144v1

[35] Papadopoulos, G., Townsend, P.K.: Compactification of D = 11 supergravity on spaces of exceptional holonomy. Phys. Lett. **B357**, 300–306 (1995). arXiv:hep-th/9506150v2

[36] Gauntlett, J.P., Kim, N., Waldram, D.: M-fivebranes wrapped on supersymmetric cycles. Phys. Rev. **D63**, 126001 (2001). arXiv:hep-th/0012195v2

[37] Anderson, M.T., Beem, C., Bobev, N., Rastelli, L.: Holographic uniformization. arXiv:1109.3724

[38] Thurston, W.P.: Three dimensional manifolds, Kleinian groups, and hyperbolic geometry. Bull. AMS **6**(3), 357–381 (1982)

[39] Mostow, G.: Strong rigidity of locally symmetric spaces (Ann. Math. Stud.), vol. 78. Princeton University Press, Princeton, NJ (1973)

[40] Bershadsky, M., Sadov, V., Vafa, C.: D-branes and topological field theories. Nucl. Phys. **B463**, 420–434 (1996). arXiv:hep-th/9511222v1

[41] Gaiotto, D., Maldacena, J.: The gravity duals of N = 2 superconformal field theories. arXiv:0904.4466

[42] Gaiotto, D., Witten, E.: S-duality of boundary conditions in N = 4 super Yang-Mills theory. Adv. Theor. Math. Phys. **13**(2), 721–896 (2009). arXiv:0807.3720

[43] Witten, E.: SL(2, Z) action on three-dimensional conformal field theories with Abelian symmetry. hep-th/0307041v3

[44] Gaiotto, D.: Domain walls for two-dimensional renormalization group flows. arXiv:1201.0767

[45] Gaiotto, D., Moore, G.W., Neitzke, A.: Framed BPS states. arXiv:1006.0146

[46] Cordova C., Neitzke, A.: Line defects, tropicalization, and multi-centered quiver quantum mechanics. arXiv:1308.6829

[47] Gaiotto, D., Witten, E.: Knot invariants from four-dimensional gauge theory. arXiv:1106.4789

[48] Gaiotto, D., Moore, G.W., Neitzke, A.: Four-dimensional wall-crossing via three-dimensional field theory. Commun. Math. Phys. **299**(1), 163–224 (2010). arXiv:0807.4723

[49] Nekrasov, N., Rosly, A., Shatashvili, S.: Darboux coordinates, Yang-Yang functional, and gauge theory. Nucl. Phys. Proc. Suppl. **216**, 69–93 (2011). arXiv:1103.3919

[50] Gukov, S., Witten, E.: Gauge theory, ramification, and the geometric Langlands program. Curr. Dev. Math. **2006**, 35–180 (2008). hep-th/0612073v2

[51] Axelrod, S., Pietra, S.D., Witten, E.: Geometric quantization of Chern-Simons gauge theory. J. Differ. Geom. **33**(3), 787–902 (1991)

[52] Witten, E.: Quantization of Chern-Simons gauge theory with complex gauge group. Commun. Math. Phys. **137**, 29–66 (1991)

[53] Gukov, S.: Three-dimensional quantum gravity, Chern-Simons theory, and the A-polynomial. Commun. Math. Phys. **255**(3), 577–627 (2005). arXiv:hep-th/0306165v1

[54] Pasquetti, S.: Factorisation of N = 2 theories on the squashed 3-sphere. arXiv:1111.6905

[55] Witten, E.: Analytic continuation of Chern-Simons theory. arXiv:1001.2933

[56] Witten, E.: A new look at the path integral of quantum mechanics. arXiv:1009.6032

[57] Kim, S.: The complete superconformal index for N = 6 Chern-Simons theory. Nucl. Phys. **B821**, 241–284 (2009). arXiv:0903.4172

[58] Imamura, Y., Yokoyama, S.: Index for three dimensional superconformal field theories with general R-charge assignments. arXiv:1101.0557

[59] Kapustin, A., Willett, B.: Generalized superconformal index for three dimensional field theories. arXiv:1106.2484

[60] Kapustin, A., Willett, B., Yaakov, I.: Exact results for Wilson loops in superconformal Chern-Simons theories with matter. JHEP **1003**, 089 32 pp. (2010). arXiv:0909.4559

[61] Hama, N., Hosomichi, K., Lee, S.: SUSY gauge theories on squashed three-spheres. arXiv:1102.4716

[62] Cordova, C., Jafferis, D.L.: Five-dimensional maximally supersymmetric Yang-Mills in supergravity backgrounds. arXiv:1305.2886

[63] Cecotti, S., Vafa, C.: Topological-anti-topological fusion. Nucl. Phys. **B367**(2), 359–461 (1991)

[64]   Cecotti, S., Gaiotto, D., Vafa, C.: tt* geometry in 3 and 4 dimensions. arXiv:1312.1008

[65]   Benini, F., Nishioka, T., Yamazaki, M.: 4d Index to 3d Index and 2d TQFT. arXiv:1109.0283

[66]   Hikami, K.: Generalized volume conjecture and the A-polynomials—the Neumann-Zagier potential function as a classical limit of quantum invariant. J. Geom. Phys. **57**(9), 1895–1940 (2007). arXiv:math/0604094v1

[67]   Dimofte, T., Gukov, S., Lenells, J., Zagier, D.: Exact results for perturbative Chern-Simons theory with complex gauge group. Commun. Number Theory Phys. **3**(2), 363–443 (2009). arXiv:0903.2472

[68]   Dimofte, T.: Quantum Riemann surfaces in Chern-Simons theory. arXiv:1102.4847

[69]   Garoufalidis, S.: The 3D index of an ideal triangulation and angle structures. arXiv:1208.1663

[70]   Garoufalidis, S., Hodgson, C.D., Rubinstein, J.H., Segerman, H.: 1-efficient triangulations and the index of a cusped hyperbolic 3-manifold. arXiv:1303.5278

[71]   Gang, D., Koh, E., Lee, K.: Superconformal index with duality domain wall. JHEP **10**, 187 (2012). arXiv:1205.0069

[72]   Verlinde, H.: Conformal field theory, two-dimensional quantum gravity and quantization of Teichmüller space. Nucl. Phys. **B337**(3), 652–680 (1990)

[73]   Intriligator, K., Seiberg, N.: Mirror symmetry in three dimensional gauge theories. Phys. Lett. **B387**, 513–519 (1996). arXiv:hep-th/9607207v1

[74]   Aharony, O., Hanany, A., Intriligator, K., Seiberg, N., Strassler, M.J.: Aspects of N = 2 supersymmetric gauge theories in three dimensions. Nucl. Phys. **B499**(1–2), 67–99 (1997). arXiv:hep-th/9703110v1

[75]   de Boer, J., Hori, K., Ooguri, H., Oz, Y., Yin, Z.: Mirror symmetry in three-dimensional gauge theories, SL(2, Z) and D-brane moduli spaces. Nucl. Phys. **B493**, 148–176 (1996). arXiv:hep-th/9612131v1

[76]   de Boer, J., Hori, K., Oz, Y.: Dynamics of N = 2 supersymmetric gauge theories in three dimensions. Nucl. Phys. **B500**, 163–191 (1997). arXiv:hep-th/9703100v3

[77]   Aharony, O., Razamat, S.S., Seiberg, N., Willett, B.: 3d dualities from 4d dualities. arXiv:1305.3924

[78]   Witten, E.: Solutions of four-dimensional field theories via M theory. Nucl. Phys. **B500**, 3 (1997). arXiv:hep-th/9703166v1

[79]   Gadde, A., Gukov, S., Putrov, P.: Walls, lines, and spectral dualities in 3d gauge theories. arXiv:1302.0015

[80]   Gaiotto, D., Witten, E.: Janus configurations, Chern-Simons couplings, and the theta-angle in N = 4 super Yang-Mills theory. JHEP **1006**, 097 (2010). arXiv:0804.2907

[81]   Gaiotto, D., Witten, E.: Supersymmetric boundary conditions in N = 4 super Yang-Mills theory. J. Stat. Phys. **135**, 789–855 (2009). arXiv:0804.2902

[82]   DeWolfe, O., Freedman, D.Z., Ooguri, H.: Holography and defect conformal field theories. Phys. Rev. **D66**, 025009 (2002). arXiv:hep-th/0111135v3

[83]   Neumann, W.D., Zagier, D.: Volumes of hyperbolic three-manifolds. Topology **24**(3), 307–332 (1985)

[84]   Thurston, W.: The geometry and topology of three-manifolds. Lecture Notes at Princeton University, Princeton University Press, Princeton (1980)

[85]   Nekrasov, N.A., Shatashvili, S.L.: Supersymmetric vacua and Bethe Ansatz. Nucl. Phys. B, Proc. Suppl. **192–193**, 91–112 (2009). arXiv:0901.4744

[86]   Kashaev, R.: The quantum dilogarithm and Dehn twists in quantum Teichmüller theory, Integrable structures of exactly solvable two-dimensional models of quantum field theory (Kiev, 2000). NATO Sci. Ser. II Math. Phys. Chem. **35**(2001), 211–221 (2000)

[87]   Teschner, J.: An analog of a modular functor from quantized Teichmuller theory. Handbook of Teichmuller Theory, vol. I, pp. 685–760 (2007). arXiv:math/0510174v4

Supersymmetric Gauge Theories, Quantization of $\mathcal{M}_{\text{flat}}$, and Conformal Field Theory

Jörg Teschner

Abstract We review the relations between $\mathcal{N} = 2$-supersymmetric gauge theories, Liouville theory and the quantization of moduli spaces of flat connections on Riemann surfaces.

1 Introduction

Alday et al. [AGT] discovered remarkable relations between the instanton parti-tion functions of certain four-dimensional $\mathcal{N} = 2$-supersymmetric gauge theo-ries and the conformal field theory called Liouville theory. These relations will be referred to as the AGT-correspondence. We will discuss an explanation for the AGT-correspondence based on the observation that both instanton partition func-tions and Liouville conformal blocks are naturally related to certain wave-functions in the quantum theory obtained by quantising the moduli spaces of flat $\mathrm{PSL}(2, \mathbb{R})$-connections on certain Riemann surfaces C. We will be considering a class of gauge theories referred to as class $\mathcal{S}$, see ([V:8], [GMN2]) or the contribution [V:2] in this volume. The gauge theories $\mathcal{G}_{C,\mathfrak{g}}$ have elements labelled by the choice of a Riemann surface C and a Lie-algebra $\mathfrak{g}$ of type A, D or E. In the following we will restrict attention to the case where $\mathfrak{g} = A_1$, and denote the corresponding gauge theories as $\mathcal{G}_C$. However, the reader will notice that many of the arguments below generalise easily to more general theories of class $\mathcal{S}$.

The root for the relations between the gauge theories and moduli spaces of flat connections will be found in the identification of the algebra generated by the super-symmetric Wilson- and 't Hooft loop operators with the algebra of trace-functions which represent natural coordinates for the moduli spaces of flat connections. This algebra may become non-commutative if the gauge theories are defined on curved spaces, or deformed by supersymmetry-preserving deformations like the Omega-

---

A citation of the form [V:x] refers to article number x in this volume.

J. Teschner (✉)
DESY Theory, Notkestr. 85, 22603 Hamburg, Germany
e-mail: teschner@mail.desy.de

© Springer International Publishing Switzerland 2016
J. Teschner (ed.), *New Dualities of Supersymmetric Gauge Theories*,
Mathematical Physics Studies, DOI 10.1007/978-3-319-18769-3_12

deformation [N]. It turns out that the resulting non-commutativity is the same as the one resulting from the quantisation of the relevant moduli spaces of flat connections.

Concerning the other side of the coin we are going to review the definition of the conformal blocks of Liouville theory. Formulated in the right way, part of the relation to the quantization of moduli spaces of flat connections becomes obvious. There furthermore exists a natural representation of the quantized algebra of trace functions on the spaces of conformal blocks.

We are going to explain how the AGT-correspondence follows from the relation between supersymmetric loop operators and trace functions, combined with certain consequences of unbroken supersymmetry. Knowing precisely which algebra is generated by the supersymmetric loop operators, one may reconstruct expectation values of loop operators on backgrounds like the four-ellipsoid. From these data one may in particular recover the low-energy effective actions of the considered gauge theories. This approach relates the AGT-correspondence to some of the work of Gaiotto et al. [GMN2, GMN3]. It is in some respects similar to the one used by Nekrasov et al. [NRS] to study the case with Omega-deformation preserving two-dimensional $\mathcal{N} = 2$ super-Poincaré invariance.

2 Theories of Class $\mathcal{S}$

2.1 A_1 Theories of Class $\mathcal{S}$

To a Riemann surface C of genus g and n punctures one may associate ([G09, GMN2], [V:2]) a four-dimensional gauge theory $\mathcal{G}_C$ with $\mathcal{N} = 2$ supersymmetry, gauge group $(SU(2))^h$, $h := 3g - 3 + n$ and flavor symmetry $(SU(2))^n$. The theories in this class are UV-finite, and therefore characterised by a collection of gauge coupling *constants* $g_1, \ldots, g_h$. In the cases where $(g, n) = (0, 4)$ and $(g, n) = (1, 1)$ one would get the supersymmetric gauge theories commonly referred to as $N_f = 4$ and $\mathcal{N} = 2^*$-theory, respectively. The correspondence between data associated to the surface C and the gauge theory $\mathcal{G}_C$ is summarised in the table below.

Riemann surface C	Gauge theory $\mathcal{G}_C$
Pants decomposition $\mathcal{C}$ + trivalent graph Γ on C, $\sigma = (\mathcal{C}, \Gamma)$	Lagrangian description with action functional S_τ^σ
Gluing parameters $q_r = e^{2\pi i \tau_r}$, $r = 1, \ldots, 3g - 3 + n$	UV-couplings $\tau = (\tau_1, \ldots, \tau_h)$, $\tau_r = \dfrac{4\pi i}{g_r^2} + \dfrac{\theta_r}{2\pi}$
rth tube	rth vector multiplet $(A_{r,\mu}, \phi_r, \ldots)$
n boundaries	n hypermultiplets
Change of pants decomposition	S-duality

More details can be found in [V:2] and references therein. To the kth boundary there corresponds a flavor group $SU(2)_k$ with mass parameter M_k. The hypermultiplet masses are linear combinations of the parameters m_k, $k = 1, \ldots, n$ as explained in more detail in [G09, AGT]. The relevant definitions and results from Riemann surface theory are collected in Appendix 1. It is necessary to refine the pants decomposition by introducing the trivalent graph Γ in order to have data that distinguish action functionals with theta angles θ_r differing by multiples of 2π. This will be done such that

$$S^{\sigma}_{\tau + e_r} = S^{\delta_r . \sigma}_{\tau}, \tag{2.1}$$

where e_r is the unit vector with rth component equal to one, and $\delta_r . \sigma$ denotes the action of the Dehn twist along the rth tube on $\sigma = (\mathcal{C}, \Gamma)$, which will map the graph Γ on C to another one.

2.2 Realisation of S-Duality

Different Lagrangian descriptions of the theories $\mathcal{G}_C$ are related by S-duality. Two actions $S^{\sigma_1}_{\tau_1}$ and $S^{\sigma_2}_{\tau_2}$ describe different perturbative expansions for one and the same theory. The respective perturbative expansions will be valid in the regimes where all coupling constants $g_{1,r}$ and $g_{2,r}$ are small. To formulate the meaning of S-duality more precisely let us assume that there exists a non-perturbative definition of $\mathcal{G}_C$ allowing us to define normalised expectation values of observables $\mathcal{O}$ like $\langle\!\langle \mathcal{O} \rangle\!\rangle_{\mathcal{G}_{C_\tau}}$ non-perturbatively as functions of τ, a set of parameters for the complex structure on C. S-duality holds if for each observable $\mathcal{O}$ there exist functionals $\mathcal{F}^{\sigma_i}_{\mathcal{O}}$ constructed using the fields in actions $S^{\sigma_i}_{\tau_i}$ together with choices of coupling constants $\tau_i = \tau_i(\tau)$, $i = 1, 2$, such that

$$\langle\!\langle \mathcal{O} \rangle\!\rangle_{\mathcal{G}_{C_\tau}} \asymp \langle\!\langle \mathcal{F}^{\sigma_1}_{\mathcal{O}} \rangle\!\rangle_{S^{\sigma_1}_{\tau_1}} \quad \text{and} \quad \langle\!\langle \mathcal{O} \rangle\!\rangle_{\mathcal{G}_{C_\tau}} \asymp \langle\!\langle \mathcal{F}^{\sigma_2}_{\mathcal{O}} \rangle\!\rangle_{S^{\sigma_2}_{\tau_2}}, \tag{2.2}$$

in the sense of equality of asymptotic expansions.

The passage from one Lagrangian description S^{σ}_{τ} to another may be decomposed into the elementary S-duality transformations corresponding to the cases where one of the coupling constants g_r gets large, while all others g_s, $s \neq r$ stay small. The arguments given in [G09] suggest that S-duality is realized in the following way: In the regime where $q_r = e^{2\pi i \tau_r} \to 1$ one may use the Lagrangian description with action $S^{\sigma_r}_{\tau'}$ associated to the data $\sigma_{;r} = (\mathcal{C}_{;r}, \Gamma_{;r})$ obtained from $\sigma = (\mathcal{C}, \Gamma)$ by a local modification which is defined as follows: There is a unique subsurface $C_r \hookrightarrow C$ isomorphic to either $C_{0,4}$ or $C_{1,1}$ that contains γ_r in the interior of C_r. $\sigma_{;r} = (\mathcal{C}_{;r}, \Gamma_{;r})$ is defined by local substitutions within C_r depicted in Figs. 1 and 2 for the two cases, respectively. If $C_r = C_{0,4}$ there is another strongly coupled regime which can be described in terms of a dual action. It corresponds to $q_r \to \infty$, and the dual action $S^{\sigma_r}_{\tau'}$ is associated to the data $\sigma_{;r}$ obtained from σ by the composition of the B-move depicted in Fig. 3 with an F-move.

Fig. 1 The F-move

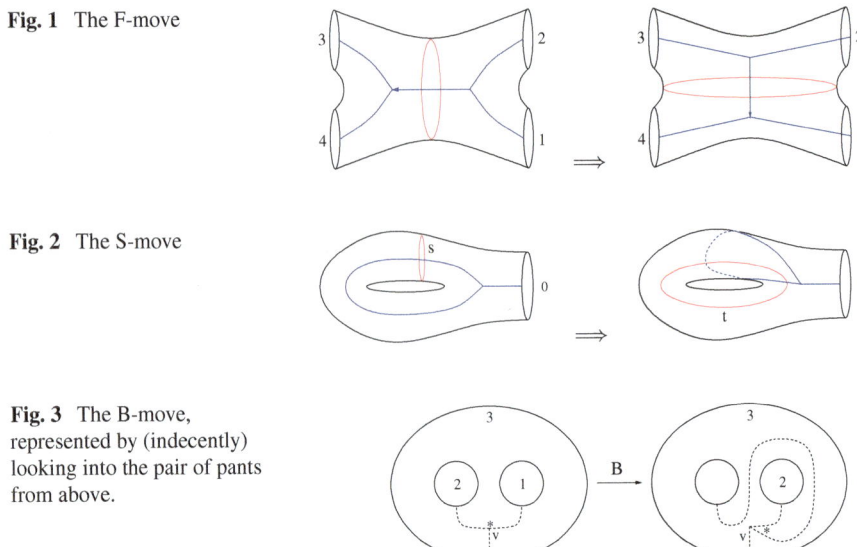

Fig. 2 The S-move

Fig. 3 The B-move,
represented by (indecently)
looking into the pair of pants
from above.

An important feature of the mapping between the respective sets of observables is
that the Wilson- and 't Hooft loops defined using S_τ^σ will correspond to the 't Hooft
and Wilson loops defined using $S_{\tau'}^{\sigma,r}$, respectively. This is the main feature we shall
use in the following.

Any transition between two pants decompositions σ_1 and σ_2 can be decomposed
into the elementary F-, S-, and B-moves. It follows that the groupoid of S-duality
transformations coincides with the Moore-Seiberg groupoid for the gauge theories
of class $\mathcal{S}$, see Appendix 1, subsection "The Moore-Seiberg Groupoid".

2.3 Gauge Theories $\mathcal{G}_C$ on Ellipsoids

It may be extremely useful to study quantum field theories on compact Euclidean
space-times or on compact spaces rather than $\mathbb{R}^4$. Physical quantities get finite size
corrections which encode deep information on the quantum field theory we study.
The zero modes of the fields become dynamical, and have to be treated quantum-
mechanically.

In the case of supersymmetric quantum field theories there are not many compact
background space-times that allow us to preserve part of the supersymmetry. A particularly interesting family of examples was studied in [HH], extending the seminal
work of Pestun [Pe]. A review can be found in the Article [V:6] in this volume.

Let us consider gauge theories $\mathcal{G}_C$ on the four-dimensional ellipsoid

$$E_{\epsilon_1,\epsilon_2}^4 := \{(x_0, \ldots, x_4) \,|\, x_0^2 + \epsilon_1^2(x_1^2 + x_2^2) + \epsilon_2^2(x_3^2 + x_4^2) = 1\}. \qquad (2.3)$$

It was shown in [Pe, HH], see also [V:6], for some examples of gauge theories $\mathcal{G}_C$ that one of the supersymmetries Q is preserved on $E^4_{\epsilon_1,\epsilon_2}$. It should be possible to generalize the proof of existence of an unbroken supersymmetry Q to all four-dimensional $\mathcal{N} = 2$ supersymmetric field theories with a Lagrangian description.

Interesting physical quantities include the partition function $\mathcal{Z}_{\mathcal{G}_C}$, or more generally expectation values of supersymmetric loop operators $\mathcal{L}_\gamma$ such as the Wilson- and 't Hooft loops. Such quantities are formally defined by the path integral over all fields on $E^4_{\epsilon_1,\epsilon_2}$. It was shown in a few examples for gauge theories from class $\mathcal{S}$ in [Pe, HH], reviewed in [V:6], how to evaluate this path integral by means of the localization technique. A variant of the localization argument was used to show that the integral over all fields actually reduces to an integral over the locus in field space where the scalars ϕ_r take constant *real* values $\phi_r = \text{diag}(a_r, -a_r) = \text{const}$, and all other fields vanish. This immediately implies that the path integral reduces to an ordinary integral over the variables a_r. It seems clear that this argument can be generalized to all theories of class $\mathcal{S}$ with a Lagrangian.

For some theories $\mathcal{G}_C$ it was found in [Pe] that the result of the localization calculation of the partition function takes the form

$$Z^{\mathcal{G}_C}_{E^4_{\epsilon_1,\epsilon_2}} (m, \tau; \epsilon_1, \epsilon_2) = \int d\mu(a) \, |\mathcal{Z}^{\text{inst}}(a, m, \tau; \epsilon_1, \epsilon_2)|^2 . \tag{2.4}$$

The main ingredients are the instanton partition function $\mathcal{Z}^{\text{inst}}(a, m, \tau; \epsilon_1, \epsilon_2)$ which depends on the zero modes $a = (a_1, \ldots, a_h)$ of the scalar fields, hypermultiplet mass parameters $m = (m_1, \ldots, m_n)$, UV gauge coupling constants $\tau = (\tau_1, \ldots, \tau_h)$, and two parameters ϵ_1, ϵ_2. The instanton partition functions can be defined as the partition function of the Omega-deformation of $\mathcal{G}_C$ on $\mathbb{R}^4$ [N], and may be calculated by means of the instanton calculus [LNS, MNS1, MNS2, NS04], as reviewed in [V:4] in this volume.

It is expected that the form (2.4) will hold for arbitrary theories $\mathcal{G}_C$, but the instanton partition function $\mathcal{Z}^{\text{inst}}(a, m, \tau; \epsilon_1, \epsilon_2)$ can only be calculated for the cases where C has genus 0 or 1, and the pants decomposition is of linear or circular quiver type, respectively.

2.4 Supersymmetric Loop Operators

Supersymmetric Wilson loops can be defined as path-ordered exponentials of the general form

$$W_{r,i} := \text{Tr} \, \mathcal{P} \exp \left[\oint_C ds \, (i\dot{x}^\mu A^r_\mu + |\dot{x}|\phi^r) \right]. \tag{2.5a}$$

The choice of contour C is severely constrained by the requirement that the resulting observable is supersymmetric. Two possible choices for the four-manifold M^4 of interest are $M^4 = \mathbb{R}^3 \times S^1$ and the four-ellipsoid. In the first case one may take a contour $\mathcal{C}$ that wraps the S^1. For the case $M^4 = E^4_{\epsilon_1 \epsilon_2}$ it was shown in [Pe, HH, GOP] that these observables are left invariant by the supersymmetry Q preserved on $E^4_{\epsilon_1, \epsilon_2}$ if $\mathcal{C}$ is one of the contours $\mathcal{C}_i$, $i = 1, 2$, with $\mathcal{C}_1$ and $\mathcal{C}_2$ being the circles with constant $(x_0, x_3, x_4) = 0$ and $(x_0, x_1, x_2) = 0$, respectively. Throughout this section we will assume that $\mathcal{C}$ is identified with one of the two $\mathcal{C}_i$.

The 't Hooft loop observables $T_{r,i}$, $i = 1, 2$, can be defined semiclassically for vanishing theta-angles $\theta = 0$ by the boundary condition

$$F_r \sim \frac{B_r}{4} \epsilon_{klm} \frac{x^k}{|\vec{x}|^3} dx^m \wedge dx^l , \tag{2.6}$$

near the contour $\mathcal{C}$. The coordinates x^k, $k = 1, 2, 3$, are local coordinates for the space transverse to $\mathcal{C}_i$, and B_r is an element of the Cartan subalgebra of $SU(2)$. In order to get supersymmetric observables one needs to have a corresponding singularity at S^1_i for the scalar fields ϕ_r. For the details of the definition and the generalization to $\theta \neq 0$ we refer to [GOP].

Application of the localisation technique to the calculation of Wilson loop operators [Pe, HH], see [V:6, V:7] for reviews, leads to results of the form

$$\left\langle W_{r,i} \right\rangle_{E^4_{\epsilon_1 \epsilon_2}} = \int d\mu(a) \, |\mathcal{Z}^{\text{inst}}(a, m, \tau; \epsilon_1, \epsilon_2)|^2 \, 2 \cosh(2\pi a_r/\epsilon_i) , \tag{2.7}$$

where $i = 1, 2$. A rather nontrivial extension of the method from [Pe] allows one to treat the case of 't Hooft loops [GOP] as well, see [V:7] for a review. The result is of the following form:

$$\left\langle T_{r,i} \right\rangle_{E^4_{\epsilon_1, \epsilon_2}} = \int d\mu(a) \, (\mathcal{Z}_{\text{inst}}(a, m, \tau; \epsilon_1, \epsilon_2))^* \, \mathcal{D}_{r,i} \, \mathcal{Z}_{\text{inst}}(a, m, \tau; \epsilon_1, \epsilon_2) , \tag{2.8}$$

with $\mathcal{D}_{r,i}$ being a difference operator acting only on the variable a_r of $\mathcal{Z}_{\text{inst}}(a, m, \tau; \epsilon_1, \epsilon_2)$, which has coefficients that depend on a, m and ϵ_i, in general.

2.5 Relation to Quantum Liouville Theory

The authors of [AGT] observed in some examples of theories from class $\mathcal{S}$ that one has (up to inessential factors $\mathcal{Z}^{\text{spur}}(m, \tau; \epsilon_1, \epsilon_2)$) an equality between the instanton partition functions and the conformal blocks $\mathcal{Z}^{\text{Liou}}(\beta, \alpha, \tau; b)$ of Liouville theory,

$$\mathcal{Z}^{\text{inst}}(a, m, \tau; \epsilon_1, \epsilon_2) = \mathcal{Z}^{\text{spur}}(m, \tau; \epsilon_1, \epsilon_2) \, \mathcal{Z}^{\text{Liou}}(\beta, \alpha, q; b) , \tag{2.9}$$

assuming a suitable dictionary between the variables involved. The "spurious" factor $\mathcal{Z}^{\text{spur}}(m, \tau; \epsilon_1, \epsilon_2)$ will turn out to be inessential, dropping out of normalised expectation values

$$\langle\!\langle \mathcal{L}_\gamma \rangle\!\rangle_{E^4_{\epsilon_1,\epsilon_2}} := \left(\langle\!\langle 1 \rangle\!\rangle_{E^4_{\epsilon_1,\epsilon_2}}\right)^{-1} \langle\!\langle \mathcal{L}_\gamma \rangle\!\rangle_{E^4_{\epsilon_1,\epsilon_2}}, \tag{2.10}$$

as follows easily from the general form of the results for the expectation values quoted in (3.7), and is therefore called "spurious".

We'll now briefly review the definition of the right hand side of (2.9) for the cases of Riemann surfaces C of genus zero with n punctures. The definition for Riemann surfaces C of arbitrary genus is discussed in [TV13].

The Virasoro algebra Vir_c has generators L_n, $n \in \mathbb{Z}$, and relations

$$[L_n, L_m] = (n - m)L_{n+m} + \frac{c}{12}n(n^2 - 1)\delta_{n+m,0}. \tag{2.11}$$

The relevant conformal blocks can be constructed using chiral vertex operators. Let us use the notation $\Delta_\alpha := \alpha(Q - \alpha)$, with Q being a variable parameterising the value c of the central element in (2.11) as $c = 1 + 6Q^2$. We will denote the highest weight representation with weight Δ_β by $\mathcal{V}_\beta$. A chiral vertex operator is an operator $V^\alpha_{\beta_2\beta_1}(z) : \mathcal{V}_{\beta_1} \to \mathcal{V}_{\beta_2}$ that satisfies the crucial intertwining property

$$[L_n, V^\alpha_{\beta_2\beta_1}(z)] = z^n(z\partial_z + (n + 1)\Delta_\alpha)V^\alpha_{\beta_2\beta_1}(z). \tag{2.12}$$

The property (2.12) defines the operator $V^\alpha_{\beta_2\beta_1}(z)$ as a formal power series in z^k uniquely up to multiplication with a complex number. The normalization freedom can be parameterized by the number $N^\alpha_{\beta_2\beta_1}$ defined by

$$V^\alpha_{\beta_2\beta_1}(z)\, e_{\beta_1} = z^{\Delta_{\beta_2} - \Delta_{\beta_1} - \Delta_\alpha}\left[N^\alpha_{\beta_2\beta_1}e_{\beta_2} + \mathcal{O}(z)\right], \tag{2.13}$$

where e_β is the highest weight vector of the representation $\mathcal{V}_\beta$. A particularly useful choice for the normalization factor $N^\alpha_{\beta_2\beta_1}$ will be

$$N^\alpha_{\beta_2\beta_1} = \sqrt{C(\bar{\alpha}_3, \alpha_2, \alpha_1)}, \tag{2.14}$$

where $\bar{\alpha}_3 = Q - \alpha_3$, and $C(\alpha_3, \alpha_2, \alpha_1)$ is the three-point function in Liouville theory. An explicit formula for $C(\alpha_3, \alpha_2, \alpha_1)$ was conjectured in [DO, ZZ], and a derivation was subsequently presented in [T01].

Using the invariant bilinear form $\langle ., . \rangle_\beta : \mathcal{V}_\beta \otimes \mathcal{V}_\beta \to \mathbb{C}$ one may then construct conformal blocks as matrix elements of products of chiral vertex operators such as

$$\mathcal{Z}^{\text{Liou}}_s(\beta, \alpha, q; b) := \left\langle e_{\alpha_n}, V^{\alpha_{n-1}}_{\alpha_n,\beta_{n-3}}(z_{n-1}) V^{\alpha_{n-2}}_{\beta_{n-3},\beta_{n-4}}(z_{n-2}) \cdots V^{\alpha_2}_{\beta_1\alpha_1}(z_2)\, e_{\alpha_1} \right\rangle_{\alpha_n}. \tag{2.15}$$

The parameters $q = (q_1, \ldots, q_{n-3})$ are given by the ratios $q_r = z_{r+1}/z_{r+2}$, with $r = 1, \ldots, n - 3$. Equation (2.15) defines conformal blocks associated to particular pants decompositions of $C_{0,n}$. In the case of $n = 4$, for example, one gets the conformal blocks associated to the pants decomposition depicted on the left of Fig. 1.

We may now state the dictionary between the variables appearing in the relation (2.9) between Liouville conformal blocks and the instanton partition functions of the corresponding gauge theories:

$$q_r = \frac{z_{r+1}}{z_{r+2}} = e^{2\pi i \tau_r}, \qquad \beta_r = \frac{Q}{2} + i \frac{a_r}{\hbar}, \qquad r = 1, \ldots, n - 3, \qquad (2.16a)$$

$$\alpha_k = \frac{Q}{2} + i \frac{M_k}{\hbar}, \quad k = 1, \ldots, n, \qquad \hbar^2 = \epsilon_1 \epsilon_2. \qquad (2.16b)$$

In order to construct conformal blocks associated to general pants decompositions of surfaces $C_{0,n}$ of genus zero let us introduce the descendants of a chiral vertex operator $V^{\alpha}_{\beta_2 \beta_1}(z)$. The descendants may be defined as the family of operators $V^{\alpha}_{\beta_2 \beta_1}[v](z)$: $\mathcal{V}_{\beta_1} \to \mathcal{V}_{\beta_2}$ that satisfy

$$V^{\alpha}_{\beta_2 \beta_1}[L_{-2}v](z) =: T(z) V^{\alpha}_{\beta_2 \beta_1}[v](z) :, \qquad \qquad V^{\alpha}_{\beta_2 \beta_1}[e_\alpha](z) = V^{\alpha}_{\beta_2 \beta_1}(z), \qquad (2.17)$$

$$V^{\alpha}_{\beta_2 \beta_1}[L_{-1}v](z) = \partial_z V^{\alpha}_{\beta_2 \beta_1}[v](z),$$

where $: T(z) V^{\alpha}_{\beta_2 \beta_1}[v]z :$ is defined as

$$: T(z) V^{\alpha}_{\beta_2 \beta_1}[v](z) := \sum_{n \leq -2} z^{-n-2} L_n V^{\alpha}_{\beta_2 \beta_1}[v]z + V^{\alpha}_{\beta_2 \beta_1}[v](z) \sum_{n \geq -1} z^{-n-2} L_n. \quad (2.18)$$

With the help of the descendants one has a new way to compose chiral vertex operators, allowing us, for example, to construct conformal blocks on $C = \mathbb{P}^1 \setminus \{0, z_2, z_3, \infty\}$ as

$$\mathcal{Z}_t^{\text{Liou}}(\beta, \alpha, z; b) := \left\langle e_{\alpha_4}, V^{\beta}_{\alpha_4 \alpha_1} \left[V^{\alpha_3}_{\beta \alpha_2}(z_3 - z_2) e_{\alpha_2} \right](z_2) e_{\alpha_1} \right\rangle_{\alpha_4}. \qquad (2.19)$$

This conformal block is associated to the pants decomposition on the right of Fig. 1. By considering arbitrary compositions of chiral vertex operators one may construct conformal blocks associated to arbitrary pants decompositions of a surface C with genus zero and n boundaries.

The relations (2.9) have fully been proven [AFLT] in the cases where the relevant conformal blocks are of the from (2.15) corresponding to the so-called linear quiver gauge theories. It is not straightforward to generalise this proof to more general pants decompositions like those corresponding to conformal blocks of the form (2.19). The technical difficulties encountered for more general pants decompositions are considerable and not yet resolved in general, see [HKS] for partial results in this direction.

3 Reduction to Quantum Mechanics

3.1 Localization as Reduction to Zero Mode Quantum Mechanics

We may assign to the expectation values $\langle \mathcal{L} \rangle$ of a loop observable $\mathcal{L}$ an interpretation in terms of expectation values of operators $\mathsf{L}_{\mathcal{L}}$ which act on the Hilbert space obtained by canonical quantization of the gauge theory $\mathcal{G}_C$ on the space-time $\mathbb{R} \times E^3_{\epsilon_1, \epsilon_2}$, where $E^3_{\epsilon_1, \epsilon_2}$ is the three-dimensional ellipsoid defined as

$$E^3_{\epsilon_1, \epsilon_2} := \{(x_1, \ldots, x_4) \,|\, \epsilon_1^2(x_1^2 + x_2^2) + \epsilon_2^2(x_3^2 + x_4^2) = 1\}. \qquad (3.1)$$

This is done by interpreting the coordinate x_0 for $E^4_{\epsilon_1, \epsilon_2}$ as Euclidean time. Noting that $E^4_{\epsilon_1, \epsilon_2}$ looks near $x_0 = 0$ as $\mathbb{R} \times E^3_{\epsilon_1, \epsilon_2}$, we expect to be able to represent partition functions $\mathcal{Z}_{\mathcal{G}_C}(E^4_{\epsilon_1, \epsilon_2})$ or expectation values $\langle \mathcal{L} \rangle_{\mathcal{G}_C(E^4_{\epsilon_1, \epsilon_2})}$ as matrix elements of states in the Hilbert space $\mathcal{H}_{\mathcal{G}_C}$ defined by canonical quantization of $\mathcal{G}_C$ on $\mathbb{R} \times E^3_{\epsilon_1, \epsilon_2}$. More precisely

$$\mathcal{Z}_{\mathcal{G}_C}(E^4_{\epsilon_1, \epsilon_2}) = \langle \tau | \tau \rangle, \qquad \langle \mathcal{L} \rangle_{E^4_{\epsilon_1, \epsilon_2}} = \langle \tau | \mathsf{L}_{\mathcal{L}} | \tau \rangle, \qquad (3.2)$$

where $\langle \tau |$ and $| \tau \rangle$ are the states created by performing the path integral over the upper/lower half-ellipsoid

$$E^{4,\pm}_{\epsilon_1, \epsilon_2} := \{(x_0, \ldots, x_4) \,|\, x_0^2 + \epsilon_1^2(x_1^2 + x_2^2) + \epsilon_2^2(x_3^2 + x_4^2) = 1 \,,\, \pm x_0 > 0\}, \quad (3.3)$$

respectively, and $\mathsf{L}_{\mathcal{L}}$ is the operator that represents the observable $\mathcal{L}$ within $\mathcal{H}_{\mathcal{G}_C}$.

The form (2.7), (2.8) of the loop operator expectation values is naturally interpreted in the Hamiltonian framework as follows. In the functional Schroedinger picture one would represent the expectation values $\langle \mathcal{L} \rangle_{E^4_{\epsilon_1, \epsilon_2}}$ schematically in the following form

$$\langle \mathcal{L} \rangle_{E^4_{\epsilon_1, \epsilon_2}} = \int [\mathcal{D}\Phi] \, (\Psi[\Phi])^* \, \mathsf{L}_{\mathcal{L}} \Psi[\Phi], \qquad (3.4)$$

the integral being extended over all field configuration on the three-ellipsoid $E^3_{\epsilon_1, \epsilon_2}$ at $x_0 = 0$. The wave-functional $\Psi[\Phi]$ is defined by means of the path integral over the lower half-ellipsoid $E^{4,-}_{\epsilon_1, \epsilon_2}$ with Dirichlet-type boundary conditions defined by the field configuration Φ.

The fact that the path integral localizes to the locus Loc_C defined by constant values $\phi_r = \mathrm{diag}(a_r, -a_r) = \mathrm{const.}$ of the scalars and zero values for all other fields ([Pe, HH], [V:6]) implies that the path integral in (3.4) can be reduced to an ordinary integral of the form

$$\langle \mathcal{L} \rangle_{E^4_{\epsilon_1, \epsilon_2}} = \int da \, (\Psi_\tau(a))^* \, \pi_0(\mathsf{L}_{\mathcal{L}}) \Psi_\tau(a), \qquad (3.5)$$

with $\Psi_\tau(a)$ defined by means of the path integral over the lower half-ellipsoid $E^{4,-}_{\epsilon_1,\epsilon_2}$ with Dirichlet boundary conditions $\Phi \in \mathsf{Loc}_C$, $\phi_r = \mathrm{diag}(a_r, -a_r)$, $r = 1, \ldots, h$. The Dirichlet boundary condition $\Phi \in \mathsf{Loc}_C$, $\phi_r = a_r$ is naturally interpreted as defining a Hilbert subspace $\mathcal{H}_0$ within $\mathcal{H}_{\mathcal{G}_C}$. States in $\mathcal{H}_0$ can, by definition, be represented by wave-functions $\Psi(a)$, $a = (a_1, \ldots, a_h)$. $\pi_0(\mathsf{L}_c)$ is the projection of L_c to $\mathcal{H}_0$.

Note that the boundary condition $\Phi \in \mathsf{Loc}_C$ preserves the supercharge Q used in the localization calculations of ([Pe, HH], [V:6])—that's just what defined the locus Loc_C in the first place. We may therefore use the arguments from [Pe, HH] to identify the wave-functions $\Psi_\tau(a)$ in (3.5) with the instanton partition functions,

$$\Psi_\tau(a) = \mathcal{Z}^{\mathrm{inst}}(a, m, \tau; \epsilon_1, \epsilon_2) . \tag{3.6}$$

The form of the results for expectation values of loop observables quoted in (2.7), (2.8) is thereby naturally explained.

3.2 S-Duality of Expectation Values

In each Lagrangian description with action S^σ_τ one will be able to express loop operator expectation values in the form

$$\langle \mathcal{L} \rangle^{S^\sigma_\tau}_{E^4_{\epsilon_1,\epsilon_2}} = \int da \ (\Psi^\sigma_\tau(a))^* \mathcal{D}^\sigma_\mathcal{L} \Psi^\sigma_\tau(a) , \tag{3.7}$$

defining representations of the algebra $\mathcal{A}_{\epsilon_1\epsilon_2}$ in terms of operators $\mathcal{D}^\sigma_\mathcal{L}$. The Wilson loops $W_{r,1}$ and $W_{r,2}$ act diagonally as operators of multiplication by $2\cosh(2\pi a_r/\epsilon_1)$ and $2\cosh(2\pi a_r/\epsilon_2)$, respectively. The 't Hooft loops $T_{r,i}$ will be represented by difference operators denoted as $\mathcal{D}^\sigma_{r,i}$.

In order for S-duality to hold, we need that the representations of the algebra of loop operators associated to any two pants decompositions σ_1 and σ_2 are unitarily equivalent. This means in particular that the eigenfunctions $\Psi^{\sigma_1}_\tau(a)$ and $\Psi^{\sigma_2}_\tau(a)$ must be related by an integral transformations of the form[1]

$$\Psi^{\sigma_2}_\tau(a_2) = \int da_1 \ K_{\sigma_2\sigma_1}(a_2, a_1) \ \Psi^{\sigma_1}_\tau(a_1) . \tag{3.8}$$

If S^σ_τ and $S^{\sigma'}_{\tau'}$ are two actions with τ and τ' differing only by shifts of the theta-angle $\theta_r \to \theta_r + 2\pi k_r$, it follows from (2.1) that we must have

$$\Psi^\sigma_{\tau'}(a) = \Psi^{\sigma'}_\tau(a) , \tag{3.9}$$

---

[1] Considering theories $\mathcal{G}_C$ associated to Riemann surfaces with genus $g > 1$ one has to allow for an additional factor on the right hand side of the relation (3.8). This is discussed in [TV13].

with $\tau' = \tau + k_r e_r$. By using the transformations (3.8) one finds that we must have

$$\Psi^\sigma_{\mu,\tau}(a) \;=\; \Psi^{\mu,\sigma}_\tau(a)\,, \tag{3.10}$$

for any Dehn twist $\mu \in \mathrm{MCG}(C)$. The notation $\Psi^\sigma_{\mu,\tau}(a)$ on the left hand side denotes the analytic continuation of $\Psi^\sigma_\tau(a)$ with respect to τ defined by the element $\mu \in \mathrm{MCG}(C)$.

By combining (3.8) and (3.10) we get

$$\Psi^\sigma_{\mu,\tau}(a_2) \;=\; \int da_1 \; K_{\mu,\sigma,\sigma}(a_2, a_1) \, \Psi^\sigma_\tau(a_1)\,. \tag{3.11}$$

Assuming that we know the kernels $K_{\mu,\sigma,\sigma}(a_2, a_1)$, we would thereby get a Riemann-Hilbert type problem[2] for the wave-functions $\Psi^\sigma_\tau(a_1)$. Equation (3.11) describes the effect of a monodromy in the gauge theory parameter space in terms of an integral transformation with kernel $K_{\mu,\sigma,\sigma}(a_2, a_1)$.

Let's note, however, that the kernels $K_{\mu,\sigma,\sigma}(a_2, a_1)$ are by no means arbitrary: They are strongly constrained by the fact that (3.8) must intertwine the representations of the algebra $\mathcal{A}_{\epsilon_1 \epsilon_2}$ defined by the actions S^{σ_1} and S^{σ_2}, respectively. Concretely, we must have, in particular,

$$\overrightarrow{\mathcal{D}^{\sigma_2}_{r,i}} \cdot K_{\sigma_2 \sigma_1}(a_2, a_1) \;=\; K_{\sigma_2 \sigma_1}(a_2, a_1) \, 2\cosh(2\pi a_{1,r}/\epsilon_i)\,,$$
$$K_{\sigma_2 \sigma_1}(a_2, a_1) \cdot \overleftarrow{\mathcal{D}^{\sigma_1}_{r,i}} \;=\; 2\cosh(2\pi a_{2,r}/\epsilon_i)\, K_{\sigma_2 \sigma_1}(a_2, a_1)\,, \tag{3.12}$$

expressing the fact that S-duality exchanges Wilson and 't Hooft loops. The equations represent a system of difference equations that turns out to determine $K_{\sigma_2 \sigma_1}(a_2, a_1)$ uniquely up to normalization. This means that the kernels are essentially determined by the representation theory of the algebra $\mathcal{A}_{\epsilon_1 \epsilon_2}$.

We will in the following describe how to identify the algebra $\mathcal{A}_{\epsilon_1 \epsilon_2}$. This information may then be used [TV13] to determine the kernels $K_{\mu,\sigma,\sigma}(a_2, a_1)$ defining the Riemann-Hilbert problem (3.11). Fixing the τ-asymptotics by means of perturbative information one gets a Riemann-Hilbert problem which has an essentially unique solution, thereby characterizing the wave-functions $\Psi^\sigma_\tau(a)$ completely.

Keeping in mind (3.6) we conclude that the instanton partition functions $\mathcal{Z}^{\text{inst}}$ can be characterized using the representation theory of $\mathcal{A}_{\epsilon_1 \epsilon_2}$. Note that the prepotential $\mathcal{F}$ giving the low-energy effective action of $\mathcal{G}_C$ is recovered from $\mathcal{Z}^{\text{inst}}$ via $\mathcal{F} = \lim_{\epsilon_1, \epsilon_2 \to 0} \epsilon_1 \epsilon_2 \log \mathcal{Z}^{\text{inst}}$. This means that the low-energy effective action is

---

[2]The Riemann-Hilbert problem is often formulated as the problem to find vectors of multivalued analytic functions on a punctured Riemann surface C with given monodromy, a representation of $\pi_1(C)$ in $SL(N, \mathbb{C})$. Our Eq. (3.11) generalises the Riemann-Hilbert problem in two ways: The Riemann surface C is replaced by the moduli space $\mathcal{M}(C)$ of complex structures on the surface C, and the monodromy takes values in the group of unitary transformations of an infinite-dimensional Hilbert-space rather than $SL(N, \mathbb{C})$.

encoded abstractly within the algebra $\mathcal{A}_{\epsilon_1 \epsilon_2}$. These observations motivate why this algebra was called "non-perturbative skeleton" of $\mathcal{G}_C$ in [TV13].

4 The Algebra of Loop Operators

In order to realise the program outlined at the end of the previous section it will be essential to know the algebra $\mathcal{A}_{\epsilon_1 \epsilon_2}$ precisely. We are now going to explain how $\mathcal{A}_{\epsilon_1 \epsilon_2}$ is related to the non-commutative algebra obtained by quantising the space of functions on the moduli space of flat PSL(2)-connections. This section is meant to give a guide to the literature on the known relations between the algebra generated by the supersymmetric Wilson- and 't Hooft loop operators on the one hand, and the (quantised) algebra of functions on the moduli space of flat PSL(2)-connections on the other hand.

4.1 The Algebra of Supersymmetric Loop Operators

The algebra of gauge theory observables contains the supersymmetric Wilson- and 't Hooft loop observables. The product of such loop operators will generate further loop observables supported at the same loop $\mathcal{C}$. The generalizations of Wilson- and 't Hooft loop operators $\mathcal{L}_\gamma$ that are generated in this way describe the effect of inserting heavy "dyonic" probe particles, and can therefore be labelled by pairs $\gamma = (r, s)$ of electric and magnetic charge vectors, see [DMO], and the article [V:7] for a review. We will be interested in the algebra $\mathcal{A}$ generated by polynomial functions of the loop operators.

One should note that the labelling of loop operators by charges is based on a given Lagrangian description of the theory. A particularly simple example for the dependence of the underlying Lagrangian is provided by the Witten-effect: Two actions S_1 and S_2 which differ only by a shift of the theta-angle θ_r by 2π will define the same expectation values after proper identification of the loop operators: A loop observable with charge γ_1 defined by S_1 gets identified with the loop observable with charge γ_2 defined by S_2 iff the magnetic charges coincide and the electric charges of γ_1 and γ_2 differ by certain multiples of the magnetic charges. A precise statement for the A_1 theories of class $\mathcal{S}$ of interest here can be found in [DMO], see also [V:7].

It should also be remarked that the precise specification of a gauge theory of class $\mathcal{S}$ depends on certain discrete topological data [AST, Ta13] defining in particular the set of allowed charges for the line operators. This phenomenon is related to interesting subtleties showing up when the gauge theory $\mathcal{G}_C$ is studied on more general four-manifolds, but it is not relevant for what is discussed in this article as we are exclusively dealing with four manifolds having the topology of the four-sphere.

4.2 UV Versus IR Loop Operators

It will be instructive to consider the four-ellipsoid $E^4_{\epsilon_1\epsilon_2}$ in the limit where $\epsilon_1 = 0$. In this case the four-ellipsoid $E^4_{\epsilon_1,\epsilon_2}$ degenerates into $E^2_{\epsilon_2} \times \mathbb{R}^2$, where

$$E^2_{\epsilon_2} := \{(x_0, \ldots, x_2) \mid x_0^2 + \epsilon_2^2(x_3^2 + x_4^2) = 1\}. \tag{4.1}$$

This implies that only the Wilson- and 't Hooft loops wrapped on the remaining circle C_2 will remain. We may still relate expectation values to matrix elements by choosing x_0 as (Euclidean) time coordinate. Near the "equator" $x_0 = 0$, the two-ellipsoid $E^2_{\epsilon_2}$ looks like $\mathbb{R} \times S^1$. One may expect that studying the gauge theory $\mathcal{G}_C$ on $\mathbb{R}^2 \times E^2_{\epsilon_1}$ will allow us make contact with the work of Gaiotto et al. [GMN1, GMN2, GMN3], who have studied the gauge theories $\mathcal{G}_C$ on the circle compactification $\mathbb{R}^3 \times S^1$. Aspects relevant for us are reviewed in [V:3].

Considering the theory $\mathcal{G}_C$ on $\mathbb{R}^3 \times S^1$ at low energies, it was argued in [GMN1] that $\mathcal{G}_C$ becomes effectively represented by a three-dimensional sigma model with hyperkähler target space $\mathcal{M}(C)$. This means in particular that the hyperkähler space $\mathcal{M}(C)$ represents the moduli space of vacua of $\mathcal{G}_C$ on $\mathbb{R}^3 \times S^1$.

The supersymmetric Wilson- and 't Hooft loops supported on S^1 are called UV line operators in ([GMN3], [V:3]). Vacuum expectation values of these line operators[3]

$$L_\gamma(m) := \langle \mathcal{L}_\gamma \rangle_m, \qquad m \in \mathcal{M}(C), \tag{4.2}$$

represent coordinate functions on the moduli space of vacua of $\mathcal{G}_C$ on $\mathbb{R}^3 \times S^1$. We see that the algebra $\mathcal{A}$ of UV line operators must coincide with a (sub-)algebra of the algebra of functions on the moduli space of vacua $\mathcal{M}(C)$.

Other useful sets of coordinate functions for $\mathcal{M}(C)$ have been defined in [GMN1] using the effective low-energy description of $\mathcal{G}_C$: They are denoted as $\mathcal{X}_\eta(m)$, are labelled by vectors η in the charge lattice Γ, and represent Darboux coordinates for the holomorphic symplectic structure Ω on $\mathcal{M}(C)$. The functions $\mathcal{X}_\eta(m)$ have been interpreted in [GMN3] as expectation values of IR line operators describing the effect of the insertion of a heavy dyonic source of charge η into the low-energy effective field theory.

It has been argued in [GMN3, CN], see also [V:3], that the expectation values $L_\gamma(m)$ can be alternatively computed using the effective IR description of $\mathcal{G}_C$ on $\mathbb{R}^3 \times S^1$, leading to a relation between UV and IR line operators of the following form:

$$L_\gamma(m) = \sum_{\eta\in\Gamma} \overline{\Omega}_{\gamma,\eta}\, \mathcal{X}_\eta(m). \tag{4.3}$$

---

[3]Comparing with [V:3] let us note that on $\mathbb{R}^3 \times S^1$ one may consider families of line operators preserving different supersymmetries, parameterised by a parameter ζ in [V:3]. We here focus on the case $\zeta = 1$ corresponding to the line operators studied on $E^4_{\epsilon_1\epsilon_2}$. Let us furthermore note that the label γ used for UV line operators here is used for IR line operators in [V:3].

The positive-integer coefficients $\overline{\Omega}_{\gamma,\eta}$ have an interesting physical interpretation as an index counting certain BPS states that exist in the presence of line defects [GMN3].

4.3 Relation with Moduli Spaces of Flat Connections

The following table summarizes known connections between the moduli spaces $\mathcal{M}_{flat}(C)$ of flat $SL(2)$-connections[4] on the surfaces C and the moduli space of vacua $\mathcal{M}(C)$ on $\mathbb{R}^3 \times S^1$:

Riemann surface C	Gauge theory $\mathcal{G}_C$
Moduli space of flat connections $\mathcal{M}_{flat}(C)$	Moduli space of vacua $\mathcal{M}(C)$ on $\mathbb{R}^3 \times S^1$
Trace functions L_γ on $\mathcal{M}_{flat}(C)$	UV line operators $\mathcal{L}\gamma$
Fock-Goncharov coordinates	IR line operators $\mathcal{X}_\eta$

We have gathered the relevant definitions and results concerning $\mathcal{M}_{flat}(C)$ in Appendix 2. The mapping between trace functions L_γ and UV line operators $\mathcal{L}_\gamma$ is defined by identifying the Dehn-Thurston parameters classifying closed loops on C (see Appendix 2, subsection "Topological Classification of Closed Loops" for a short summary) with the charge labels $\gamma = (r, s)$ of the line operators ([DMO], [V:7]).[5] The definition of the Fock-Goncharov coordinates for $\mathcal{M}_{flat}(C)$ is briefly reviewed in Appendix 2, subsection "Fock–Goncharov Coordinates", and the relations to IR line operators are discussed in [GMN2, GMN3].

An argument in favor of the identification between $\mathcal{M}_{flat}(C)$ and $\mathcal{M}(C)$ starts by considering the six-dimensional $(2, 0)$ theory on $S^1 \times \mathbb{R}^3 \times C$. Compactifying first on C and then on S^1 gives the three-dimensional sigma model with target space $\mathcal{M}(C)$, as mentioned above. It may alternatively be obtained from the six-dimensional theory by first compactifying on S^1 followed by compactification on C. After compactifying on S^1 one would then find the maximally supersymmetric five-dimensional super-Yang-Mills theory on $\mathbb{R} \times \mathbb{R}^2 \times C$. Further compactification on C yields a nonlinear sigma-model with target being $\mathcal{M}_{Hit}(C)$, the moduli space of solutions to Hitchin's self-duality equations using a variant of the argument presented in [BJSV]. More details and references can be found in ([GMN2], Sect. 3.1). $\mathcal{M}_{Hit}(C)$ is a hyperkähler space naturally related to $\mathcal{M}_{flat}(C)$ in one of its hyperkähler structures ([Hi], [V:3]).

---

[4]This may be $SL(2, \mathbb{C})$- or $SL(2, \mathbb{R})$-connections depending on the context, as will be discussed later.

[5]The set of allowed charges $\gamma = (r, s)$ in a theory $\mathcal{G}_C$ is generically smaller than the set of allowed Dehn-Thurston parameters [AST, Ta13]. This subtlety does not affect our discussions: For each allowed Dehn-Thurston parameter there *exists* a choice of the extra discrete data specifying gauge theories $\mathcal{G}_C$ such that the corresponding UV line operator $\mathcal{L}_\gamma$ can be defined within $\mathcal{G}_C$. Having determined the set of allowed charges in the duality frame corresponding to a particular pants decomposition, one may figure out the allowed charges in any other duality frame by some simple rules.

A way to find the identification between particular coordinate functions on $\mathcal{M}_{\text{flat}}(C)$ and the UV line operators summarised in the table above was described in ([GMN3], Sect. 7).

4.4 Quantization

An interesting generalization of the set-up considered in Sect. 4.2 (compactification on S^1) is obtained by imposing certain twisted boundary conditions with parameter b along S^1 [GMN3, IOT]. The resulting deformation, denoted $\mathbb{R}^3 \times_b S^1$ of the background $\mathbb{R}^3 \times S^1$ is related to the Omega-deformation, and it can be used to model the residual effect of the curvature in the vicinity of the circles $\mathcal{C}_i$ on $E^4_{\epsilon_1,\epsilon_2}$ which represent the support of the loop operators [V:7].

It has been argued in [GMN3, IOT], see also [V:7], that the effect of the twisted boundary conditions is to deform the algebra $\mathcal{A}$ into a non-commutative algebra $\mathcal{A}_b$. In the case of the A_1 theories of class $\mathcal{S}$ it was argued in [GMN3] that the resulting algebra is nothing but the quantized algebra of functions on $\mathcal{M}_{\text{flat}}(C)$, denoted $\text{Fun}_\hbar(\mathcal{M}_{\text{flat}}(C))$, here with $\hbar = b^2$. There should in particular exist a deformed version of the relation (4.3) between UV and IR line operators. The left hand side of this relation, the deformed UV line operator, should be independent of the choice of coordinates that appear on the right hand side. As different sets of coordinates $\mathcal{X}_\eta$ are related by (quantized-) cluster transformations, it will suffice to figure out the quantum analog of (4.3) for particular triangulations. This is what was done in ([GMN3], Sect. 11) for the A_1-case, leading to the conclusion that the algebra generated by the deformed UV line operators is the quantisation of the the the algebra of trace functions on $\mathcal{M}_{\text{flat}}(C)$ that will be described in more detail in the following section.

Highly nontrivial support for this proposal has been given by explicit calculations for some theories of class $\mathcal{S}$ ([IOT], [V:7]). A rather different line of arguments leading to the same conclusion was proposed by Nekrasov and Witten [NW].

4.5 Back to the Ellipsoid

As mentioned above, one may expect that the twisted boundary conditions defining $\mathbb{R}^3 \times_b S^1$ would model the residual effect of the curvature in the vicinity of the curves $\mathcal{C}_i$ on $E^4_{\epsilon_1,\epsilon_2}$ ([IOT], [V:7]), at least as far as the algebraic properties of loop operators are concerned. The comparison of the results of localisation calculations on the two spaces ([GOP] for S^4, and [IOT] for $\mathbb{R}^3 \times_b S^1$) provides highly nontrivial quantitative evidence for this claim. In the case of the four ellipsoid $E^4_{\epsilon_1,\epsilon_2}$ one thereby expects to get a (twisted) product of two copies of $\text{Fun}_\hbar(\mathcal{M}_{\text{flat}}(C))$ associated to the two circles $\mathcal{C}_i$ supporting supersymmetric loop observables.

However, there is a crucial difference between the cases of $\mathbb{R}^3 \times_b S^1$ and $E^4_{\epsilon_1,\epsilon_2}$. In the case of $\mathbb{R}^3 \times_b S^1$ one will generically get complex values for expectation values of loop observables which are functions of the scalar expectation values at infinity, the holonomy of the gauge field around S^1 and the complexified gauge coupling constants τ. The precise relation was given in [IOT].

In the case of $E^4_{\epsilon_1,\epsilon_2}$, on the contrary, one gets only real numbers larger than 2 for the expectation values of Wilson loops from the localisation calculations of [Pe, HH]. By S-duality this will imply that the 't Hooft loops will define positive self-adjoint operators on $\mathcal{H}_0$ with the same spectrum. This means that the relevant moduli spaces to consider in this case will not be the moduli spaces $\mathcal{M}^{\mathbb{C}}_{\mathrm{flat}}(C)$ of flat $PSL(2, \mathbb{C})$-connections, but rather its real slice $\mathcal{M}^{\mathbb{R}}_{\mathrm{flat}}(C)$ defined by having real values bounded below by 2 for all trace coordinates.

It is known that $\mathcal{M}^{\mathbb{R}}_{\mathrm{flat}}(C)$ breaks up into finitely many disconnected components $\mathcal{M}^{\mathbb{R},d}_{\mathrm{flat}}(C)$, $|d| = 0, \ldots, 2g - 2 + n$, and there exists a distinguished component $\mathcal{M}^{\mathbb{R},0}_{\mathrm{flat}}(C)$ which has the necessary properties. This component is isomorphic to the Teichmüller spaces of Riemann surfaces [Go88, Hi] and therefore referred to as the Teichmüller component, see Appendix 2, subsection "The Teichmüller Component".

The resulting situation is summarised in the table below.

Riemann surface C	Gauge theory $\mathcal{G}_C$
Quantised algebras of functions	Algebra $\mathcal{A}_b$ generated by
$\mathrm{Fun}_{b^2}(\mathcal{M}_{\mathrm{flat}}(C))$	Wilson- and 't Hooft loops on $\mathbb{R}^3 \times_b S^1$
Quantized algebras of functions	Algebra $\mathcal{A}_{\epsilon_1\epsilon_2}$ generated by
$\mathrm{Fun}_{b^2}(\mathcal{M}^{\mathbb{R},0}_{\mathrm{flat}}(C)) \,\tilde{\times}\, \mathrm{Fun}_{b^{-2}}(\mathcal{M}^{\mathbb{R},0}_{\mathrm{flat}}(C))$	Wilson- and 't Hooft loops on $E^4_{\epsilon_1,\epsilon_2}$

The notation $\tilde{\times}$ indicates that the representatives of the factors commute only up to a sign, in general.

5 Quantization of Moduli Spaces of Flat Connections

We now have the input we need to develop the program outlined in Sect. 3.2—the reconstruction of instanton partition functions from the algebra of loop operators. In the rest of this section we shall briefly describe the quantization of $\mathcal{M}^{\mathbb{R},0}_{\mathrm{flat}}(C)$.

5.1 Quantization of the Fock-Goncharov Coordinates

The simplicity of the Poisson brackets of the Fock-Goncharov coordinates makes part of the quantization quite simple. To each edge e of a triangulation t of a Riemann surface $C_{g,n}$ associate a quantum operator X^t_e corresponding to the classical phase space function $\mathcal{X}_e^t$. Canonical quantization of the Poisson brackets (6.59) yields an algebra $\mathcal{B}_t$ with generators X_e^t and relations

$$X_e^t, X_{e'}^t = e^{2\pi i b^2 n_{ee'}} X_{e'}^t X_e^t, \tag{5.1}$$

where $n_{ee'}$ is the number of intersections of e with e', counted with a sign.

Note furthermore that the variables $\mathcal{X}_e$ are positive for the Teichmüller component. The scalar product of the quantum theory should realize the phase space functions $\mathcal{X}_e$ as *positive* self-adjoint operators X_e^t. By choosing a polarization one may define a Schrödinger type representations π_t in terms of multiplication and finite shift operators. It can be realized on suitable dense subspaces of the Hilbert space $\mathcal{H}_t \simeq L^2(\mathbb{R}^{3g-3+n})$.

There exists a family of automorphisms which describe the relation between the quantized variables associated to different triangulations [F97, Ka1, CF1]. If triangulation t_e is obtained from t by changing only the diagonal in the quadrangle containing e, we have

$$X_{e'}^{t_e} = \begin{cases} X_{e'}^t \displaystyle\prod_{a=1}^{|n_{e'e}|} \left(1 + e^{\pi i (2a-1)b^2} (X_e^t)^{-\text{sgn}(n_{e'e})}\right)^{-\text{sgn}(n_{e'e})} & \text{if } e' \neq e, \\ (X_e^t)^{-1} & \text{if } e' = e. \end{cases} \tag{5.2}$$

It follows that the quantum theory of $\mathcal{M}_{\text{flat}}^{\mathbb{R},0}(C)$ has the structure of a quantum cluster algebra [FG2].

It is possible to construct [Ka1] unitary operators T_{t_1,t_2} that represent the quantum cluster transformations (5.2) in the sense that

$$X_e^{t_2} = T_{t_1 t_2}^{-1} \cdot X_e^{t_1} \cdot T_{t_1 t_2}. \tag{5.3}$$

The operators $T_{t_2 t_1}$ describe the change of representation when passing from the quantum theory associated to triangulation t_1 to the one associated to t_2. It follows that the resulting quantum theory does not depend on the choice of a triangulation in an essential way.

As indicated in Sect. 4.4, one may intepret the coordinates $\mathcal{X}_e^\tau$ as expectation values of IR line operators. The formula (5.1) describes the quantum deformation induced by the twisted boundary condition on $\mathbb{R}^3 \times_b S^1$, and (5.2) describes the behavior of the IR line operators under (quantum-) wall-crossing ([GMN1, GMN3], [V:3]).

5.2 Quantization of the Trace Functions

There is a simple algorithm (reviewed in Appendix 2, subsection "Trace Functions in Terms of Fock-Goncharov Coordinates") for calculating the trace functions in terms of the variables $\mathcal{X}_e^t$ leading to Laurent polynomials in the variables $\mathcal{X}_e$ of the form

$$\mathcal{L}_\gamma = \sum_{\nu \in \mathbb{F}} C_\gamma^t(\nu) \prod_e (\mathcal{X}_e^t)^{\frac{1}{2}\nu_e}, \tag{5.4}$$

where the summation is taken over a finite set $\mathbb{F}$ of vectors $\nu \in \mathbb{Z}^{3g-3+2n}$ with components ν_e.

According to ([GMN3], [V:3]) one may interpret the trace functions as UV line operators. Formula (5.4) thereby becomes identified with (4.3).

For curves γ having $C_\gamma^t(\nu) \in \{0, 1\}$ for all $\nu \in \mathbb{F}$ it has turned out to be sufficient to replace $(\mathcal{X}_e^t)^{\nu_e}$ in (5.4) by $\exp(\sum_e \nu_e \log \mathsf{X}_e^t)$ in order to define the quantum operator L_γ^t associated to a classical trace function $\mathcal{L}_\gamma$. For other triangulations one may define $\mathsf{L}_\gamma^{t'}$ using

$$\mathsf{L}_\gamma^{t'} = \mathsf{T}_{tt'}^{-1} \cdot \mathsf{L}_\gamma^t \cdot \mathsf{T}_{tt'}. \tag{5.5}$$

It turns out that this is sufficient to define the operators L_γ^t in general [T05]. It follows from (5.5) that we may regard the algebras of quantised trace functions generated by the operators L_γ^t as different representations π_t of an abstract algebra $\mathcal{A}_b$ which does not depend on the choice of a triangulation, $\mathsf{L}_\gamma^t \equiv \pi_t(L_\gamma)$ for $L_\gamma \in \mathcal{A}_b$.

The operators L_γ^t are positive self-adjoint with spectrum bounded from below by 2, as follows from the result of [Ka4]. Two operators $\mathsf{L}_{\gamma_1}^t$ and $\mathsf{L}_{\gamma_2}^t$ commute if the intersection of γ_1 and γ_2 is empty. It is therefore possible to diagonalise simultaneously the quantised trace functions associated to a maximal set of non-intersecting closed curves defining a pants decomposition [T05, TV13].

5.3 Representations Associated to Pants Decompositions

Mutual commutativity of the quantized trace-functions $\mathsf{L}_{\gamma_r}^t$ ensures existence of operators $\mathsf{R}_{\sigma|t}$ which map the operators $\mathsf{L}_{\gamma_r}^t$, $r = 1, \ldots, h$ associated to the curves $\mathcal{C} = (\gamma_1, \ldots, \gamma_h)$ defining a pants decomposition to the operators of multiplication by $2\cosh(l_r/2)$. The states in the image $\mathcal{H}_\sigma$ of $\mathsf{R}_{\sigma|t}$ can be represented by functions $\psi(l)$, $l = (l_1, \ldots, l_h)$ depending on variables $l_r \in \mathbb{R}^+$ which parameterise the eigenvalues of $\mathsf{L}_{\gamma_r}^t$. The operators $\mathsf{R}_{\sigma|t}$ define a new family of representations π_σ of $\mathcal{A}_b$ via

$$\pi_\sigma(L_\gamma) := \mathsf{R}_{\sigma|t} \cdot \pi_t(L_\gamma) \cdot (\mathsf{R}_{\sigma|t})^{-1}. \tag{5.6}$$

The representations are naturally labelled by the data $\sigma = (\mathcal{C}, \Gamma)$ we had encountered before. The unitary operators $\mathsf{R}_{\sigma|t} : \mathcal{H}_t \to \mathcal{H}_\sigma$ were constructed explicitly in [T05, TV13].

5.3.1 Transitions Between Representation

The passage between the representations π_{σ_1} and π_{σ_2} associated to two different pants decompositions is then described by

$$\mathsf{U}_{\sigma_2\sigma_1} := \mathsf{R}_{\sigma_2|\mathrm{t}} \cdot (\mathsf{R}_{\sigma_1|\mathrm{t}})^{-1} \,.$$

The unitary operators $\mathsf{U}_{\sigma_2\sigma_1}$ intertwine the representations π_{σ_1} and π_{σ_2},

$$\pi_{\sigma_2}(\mathcal{L}_\gamma) \cdot \mathsf{U}_{\sigma_2\sigma_1} = \mathsf{U}_{\sigma_2\sigma_1} \cdot \pi_{\sigma_1}(\mathcal{L}_\gamma) \,. \tag{5.7}$$

Explicit representations for the operators $\mathsf{U}_{\sigma_2\sigma_1}$ have been calculated in [NT, TV13] for pairs $[\sigma_2, \sigma_1]$ related by the generators of the Moore-Seiberg groupoid. The B-move is represented as

$$(\mathsf{B}\psi_s)(\beta) = B_{l_2l_1}^{l_3}\, \psi_s(\beta)\,, \qquad B_{l_2l_1}^{l_3} = e^{\pi i (\Delta_{l_3} - \Delta_{l_2} - \Delta_{l_1})}\,, \tag{5.8}$$

where $\Delta_l = (1 + b^2)/4b + (l/4\pi b)^2$. The F-move is represented in terms of an integral transformation of the form

$$\psi_s(l_s) \equiv (\mathsf{F}\psi_t)(l_s) = \int_{\mathbb{R}^+} dl_t\; F_{l_sl_t}\begin{bmatrix} l_3 & l_2 \\ l_4 & l_1 \end{bmatrix} \psi_t(l_t)\,. \tag{5.9}$$

A similar formula exists for the S-move. The explicit expressions can be found in [TV13].

The operators $\mathsf{U}_{\sigma_2\sigma_1}$ define a unitary projective representation of the Moore-Seiberg groupoid,

$$\mathsf{U}_{\sigma_3\sigma_2} \cdot \mathsf{U}_{\sigma_2\sigma_1} = \zeta_{\sigma_3\sigma_2\sigma_1} \mathsf{U}_{\sigma_3\sigma_1}\,, \tag{5.10}$$

where $\zeta_{\sigma_3\sigma_2\sigma_1} \in \mathbb{C}$, $|\zeta_{\sigma_3\sigma_2\sigma_1}| = 1$. The explicit formulae for the relations of the Moore-Seiberg groupoid in the quantisation of $\mathcal{M}_{\mathrm{flat}}^0(C)$ are listed in [TV13].

Having a representation of the Moore-Seiberg groupoid automatically produces a representation of the mapping class group. An element of the mapping class group μ represents a diffeomorphism of the surface C, and therefore maps any MS graph σ to another one denoted $\mu.\sigma$. Note that the Hilbert spaces $\mathcal{H}_\sigma$ and $\mathcal{H}_{\mu.\sigma}$ are canonically isomorphic. Indeed, the Hilbert spaces H_σ depend only on the combinatorics of the graphs σ, but not on their embedding into C. We may therefore define an operator $\mathsf{M}_\sigma(\mu) : \mathcal{H}_\sigma \to \mathcal{H}_\sigma$ as

$$\mathsf{M}_\sigma(\mu) := \mathsf{U}_{\mu.\sigma,\sigma}\,. \tag{5.11}$$

It is automatic that the operators $\mathsf{M}(\mu)$ define a projective unitary representation of the mapping class group $\mathrm{MCG}(C)$ on $\mathcal{H}_\sigma$.

The kernels of the operators $\mathsf{U}_{\sigma_2\sigma_1}$, $\mathsf{T}_{t_2t_1}$ *and* $\mathsf{R}_{\sigma|t}$ *are related to the partition functions of* $d = 3$ *gauge theories on duality walls, see* ([DGV], [V:11]) *and references therein. The relations are summarised in the following table:*

Riemann surface C	Gauge theory $\mathcal{G}_C$
Kernels representing operators $\mathsf{U}_{\sigma_2\sigma_1}$	UV duality walls $T_2[M, \mathbf{p}, \mathbf{p}']$
Kernels representing operators $\mathsf{R}_{\sigma|t}$	RG domain walls $T_2[M, \mathbf{p}, \Pi]$
Kernels representing operators $\mathsf{T}_{t_2t_1}$	IR duality walls $T_2[M, \Pi, \Pi']$

5.3.2 Representations

The representations $\pi_\sigma(\mathsf{L}_\gamma)$ were calculated explicitly for the generators of $\mathcal{A}_b$ in [TV13].

As a prototypical example let us consider the case where σ corresponds to the pants decomposition of $C_{0,4}$ depicted on the left of Fig. 1. We may associate generators L_s, L_t and L_u of $\mathcal{A}_b$ to the simple closed curves γ_s, γ_t, and γ_u introduced in Appendix 2, subsection "Generators and Relations", respectively. The generators $\mathcal{L}_r$ $r = 1, \ldots, 4$ are associated to the boundary components of $C \simeq C_{0,4}$. The representation of $\mathcal{A}_b$ will be generated from the operators L_s, L_t and L_u defined as follows:

$$\mathsf{L}_s := 2\cosh(\mathsf{l}/2). \tag{5.12a}$$

$$\mathsf{L}_t := \frac{1}{2(\cosh \mathsf{l}_s - \cos 2\pi b^2)}\left(2\cos \pi b^2(L_2 L_3 + L_1 L_4) + \mathsf{L}_s(L_1 L_3 + L_2 L_4)\right) \tag{5.12b}$$

$$+ \sum_{\epsilon=\pm 1} \frac{1}{\sqrt{2\sinh(\mathsf{l}_s/2)}} e^{\epsilon\mathsf{k}/2} \frac{\sqrt{c_{12}(\mathsf{L}_s)c_{34}(\mathsf{L}_s)}}{2\sinh(\mathsf{l}_s/2)} e^{\epsilon\mathsf{k}/2} \frac{1}{\sqrt{2\sinh(\mathsf{l}_s/2)}}$$

where

$$\mathsf{l}\,\psi_\sigma(l) = l\psi_\sigma(l), \qquad \mathsf{k}\,\psi_\sigma(l) = -4\pi i b^2\, \partial_l\psi_\sigma(l),$$

and $c_{ij}(L_s)$ is defined as

$$c_{ij}(L_s) = L_s^2 + L_i^2 + L_j^2 + L_s L_i L_j - 4. \tag{5.12c}$$

L_u is given by a similar expression [TV13]. The operators l and k are quantum counterparts of the Fenchel-Nielsen coordinates, see Appendix 2, subsection "Fenchel-Nielsen Coordinates for $\mathcal{M}_{\mathrm{flat}}^{\mathbb{R},0}(C)$" for a definition.

As indicated above, one may interpret the trace functions L_s, L_t, L_u *as UV line operators, here for the* $N_f = 4$ *theory associated to* $C_{0,4}$. L_s, L_t *and* L_u *correspond to the Wilson loop, 't Hooft loop, and simplest dyonic loop, respectively. The formulae* *above are directly related to the expectation values of these line operators on $\mathbb{R}^3 \times_b S^1$ calculated in [IOT].*

5.3.3 The Algebra of Trace Functions

Using the explicit representations for the generators of $\mathcal{A}_b$ obtained in [TV13] it becomes straightforward to calculate the relations that they satisfy. As a prototypical example, let us again consider the case $C = C_{0,4}$. There are two main relations: Quadratic relation:

$$Q(\mathcal{L}_s, \mathcal{L}_t, \mathcal{L}_u) := e^{\pi i b^2} \mathcal{L}_s \mathcal{L}_t - e^{-\pi i b^2} \mathcal{L}_t \mathcal{L}_s \tag{5.13}$$
$$- (e^{2\pi i b^2} - e^{-2\pi i b^2}) \mathcal{L}_u - (e^{\pi i b^2} - e^{-\pi i b^2})(\mathcal{L}_1 \mathcal{L}_3 + \mathcal{L}_2 \mathcal{L}_4).$$

Cubic relation:

$$P(\mathcal{L}_s, \mathcal{L}_t, \mathcal{L}_u) = -e^{\pi i b^2} \mathcal{L}_s \mathcal{L}_t \mathcal{L}_u \tag{5.14}$$
$$+ e^{2\pi i b^2} \mathcal{L}_s^2 + e^{-2\pi i b^2} \mathcal{L}_t^2 + e^{2\pi i b^2} \mathcal{L}_u^2$$
$$+ e^{\pi i b^2} \mathcal{L}_s (\mathcal{L}_3 \mathcal{L}_4 + \mathcal{L}_1 \mathcal{L}_2) + e^{-\pi i b^2} \mathcal{L}_t (\mathcal{L}_2 \mathcal{L}_3 + \mathcal{L}_1 \mathcal{L}_4)$$
$$+ e^{\pi i b^2} \mathcal{L}_u (\mathcal{L}_1 \mathcal{L}_3 + \mathcal{L}_2 \mathcal{L}_4)$$
$$+ \mathcal{L}_1^2 + \mathcal{L}_2^2 + \mathcal{L}_3^2 + \mathcal{L}_4^2 + \mathcal{L}_1 \mathcal{L}_2 \mathcal{L}_3 \mathcal{L}_4 - (2 \cos \pi b^2)^2.$$

The generators L_k, $k = 1, \ldots, 4$ are central elements in $\mathcal{A}_b(C_{0,4})$, associated to the boundary components. The quadratic relations represent the deformation of the Poisson bracket (6.57), while the cubic relation is a deformation of the relation (6.50a).

6 Relation to Liouville Theory

Having worked out the quantization of $\mathcal{M}_{\text{flat}}^{\mathbb{R},0}(C)$, we have determined the monodromy data we need to define the Riemann-Hilbert type problem discussed in Sect. 3.2. In order to derive the AGT-correspondence along these lines it remains to observe that the Liouville conformal blocks provide solutions to this Riemann-Hilbert problem.

Our goal in this section is to explain why Liouville conformal blocks are the wave-functions solving the Riemann-Hilbert problem (3.11). To this aim we are going to explain that

> Liouville theory is just another way to represent the quantum theory of $\mathcal{M}_{\text{flat}}^{\mathbb{R},0}(C)$ defined in Sect. 5.

The identification between conformal blocks and wave-functions in the quantum theory of moduli spaces of flat connections will follow naturally.

6.1 Complex-Analytic Darboux Coordinates for $\mathcal{M}_{\text{flat}}^0(C)$

Our explanations will be based on the fact that $\mathcal{M}_{\text{flat}}^{\mathbb{R},0}(C)$ is isomorphic to the Teichmüller space $T(C)$ (see Appendix 2, subsection "The Teichmüller Component"). This implies that there exists an alternative quantisation scheme using *holomorphic* coordinates for $T(C)$. We are going to explain that the quantum theory in the resulting quantisation scheme is naturally related to conformal field theory.

For simplicity, we will here restrict attention to $C = C_{0,4} = \mathbb{P}^1 \setminus \{z_1, z_2, z_3, z_4\}$. We do not lose generality when we assume that $z_1 = 0$, $z_3 = 1$, $z_4 = \infty$. The value of $q := z_2$ defines a complex-analytic coordinate for the moduli space $\mathcal{M}(C)$ of complex structures on C. The Fuchsian group corresponding to the complex structure parameterized by a value of q defines a flat $PSL(2, \mathbb{R})$-connection. We may therefore regard q as a local coordinate for $\mathcal{M}_{\text{flat}}^0(C)$ which is related to the Fenchel-Nielsen coordinates (k, l) in a very complicated way. The relation becomes reasonably simple only in the limit $|q| \to 0 \Leftrightarrow l \to 0$, where one has

$$\frac{l}{2\pi} \simeq \frac{\pi}{\log(1/|q|)}, \qquad 2\pi k \simeq \arg(q), \tag{6.1}$$

where the notation $\simeq$ indicates equality to leading order in this limit.

The complicated nature of the dependence of q on the Darboux coordinates (k, l) is reflected in the fact that the Poisson structure on $\mathcal{M}_{\text{flat}}^{\mathbb{R},0}(C)$ is represented in terms of q in a much more complicated way. A useful way to describe the Poisson structure using the coordinate q is to find a function $h = h(q, \bar{q})$ that is canonically conjugate to q in the sense that

$$\{q, h(q, \bar{q})\} = -i . \tag{6.2}$$

Such a function can be found from the metric $ds^2 = e^{2\varphi} dy d\bar{y}$ of constant negative curvature associated to q by writing the function $t(y) = -(\partial_y \varphi)^2 + \partial_y^2 \varphi$ in the form

$$t(y) = \frac{\delta_3}{(y-1)^2} + \frac{\delta_1}{y^2} + \frac{\delta_2}{(y-q)^2} + \frac{v}{y(y-1)} + \frac{q(q-1)}{y(y-1)} \frac{h}{y-q} . \tag{6.3}$$

The residue $h = h(q, \bar{q})$ in (6.3) is indeed the sought-for conjugate variable to q, as follows from the beautiful results [TZ87a, CMS, TZ03] that the classical Liouville action $S_{\text{cl}}[\varphi]$ is the Kähler potential for the symplectic form on $T(C) \simeq \mathcal{M}_{\text{flat}}^0(C)$ corresponding to the Poisson-structure we consider, and that

$$h(q, \bar{q}) = -\frac{\partial}{\partial q} S_{\text{cl}}[\varphi] . \tag{6.4}$$

The function $h(q, \bar{q})$ is called the accessory parameter. Having real monodromy (subgroup of $PSL(2, \mathbb{R})$) of the differential operator $\partial_y^2 + t$ clearly requires fine-tuning of the residue h in (6.3) in a way that depends on the complex structure q.

6.2 Quantization of Complex-Analytic Darboux Coordinates for $\mathcal{M}_{\text{flat}}^0(C)$

One may then consider an alternative representation for the quantum theory of $\mathcal{M}_{\text{flat}}^0(C)$ which is such that the operator representing the complex-analytic coordinate q is realized as a multiplication operator q,

$$\mathsf{q}\psi(q) = q\psi(q). \tag{6.5}$$

The quantization of the observable h should then give an operator h that satisfies

$$[\mathsf{h}, \mathsf{q}] = b^2, \tag{6.6}$$

and can therefore be represented as

$$\mathsf{h}\psi(q) = b^2 \frac{\partial}{\partial q}\psi(q). \tag{6.7}$$

In order for such a representation to be equivalent to the representation we had previously defined using the Darboux coordinates (k, l) we should consider wave-functions $\phi(q)$ that are holomorphic in q. Such a representation can be seen as an analog of the coherent state representation of quantum mechanics.

It will be useful for us to think of the wave-functions $\psi(q)$ in such a representation as overlaps $\langle q | \psi \rangle$ of the abstract state $| \psi \rangle$ with an eigenstate $\langle q |$ of the operator q.

6.3 Geometric Definition of the Conformal Blocks

In order to see how the quantisation of $\mathcal{T}(C)$ is related to conformal field theory, let us present a more geometric approach to the definition of the conformal blocks going back to [BPZ].

Let C be the Riemann surface $C = \mathbb{P}^1 \setminus \{z_1, \ldots, z_n\}$ of genus 0 with n marked points $z_1, \ldots, z_n$. At each of the marked points z_r, $r = 1, \ldots, n$, let us choose the local coordinates $w_r = y - z_r$. We associate highest weight representations $\mathcal{V}_r$, of Vir_c to P_r, $r = 1, \ldots, n$. The representations $\mathcal{V}_r$ are generated from highest weight vectors e_r with weights Δ_r.

The conformal blocks are then defined to be the linear functionals $\mathcal{F} : \mathcal{V}_{[n]} \equiv \otimes_{r=1}^n \mathcal{V}_r \to \mathbb{C}$ that satisfy the invariance property

$$\mathcal{F}(T[\chi] \cdot v) = 0 \quad \forall v \in \mathcal{V}_{[n]}, \quad \forall \chi \in \mathfrak{V}_{\text{out}}, \tag{6.8}$$

where $\mathfrak{V}_{\text{out}}$ is the Lie algebra of meromorphic differential operators on C which may have poles only at $z_1, \ldots, z_n$. The action of $T[\chi]$ on $\otimes_{r=1}^n \mathcal{R}_r \to \mathbb{C}$ is defined as

$$T[\chi] = \sum_{r=1}^n \mathrm{id} \otimes \cdots \otimes \underset{(\text{rth})}{L[\chi^{(r)}]} \otimes \cdots \otimes \mathrm{id}, \quad L[\chi^{(r)}] := \sum_{k \in \mathbb{Z}} L_k \chi_k^{(r)} \in \mathrm{Vir}_c,$$
$$\tag{6.9}$$

where $\chi_k^{(r)}$ are the coefficients of the Laurent expansions of χ at the points $P_1, \ldots P_n$,

$$\chi(z_r) = \sum_{k \in \mathbb{Z}} \chi_k^{(r)} w_r^{k+1} \partial_{w_r} \in \mathbb{C}((w_r)) \partial_{w_r}, \tag{6.10}$$

with $\mathbb{C}((t))$ being the space of Laurent series in the variable t.

The vector space of conformal blocks associated to the Riemann surface C with representations $\mathcal{V}_r$ associated to the marked points $P_r, r = 1, \ldots, n$ will be denoted as $\mathsf{CB}(\mathcal{V}_{[n]}, C)$. It is the space of solutions to the defining invariance conditions (6.8).

The space $\mathsf{CB}(\mathcal{V}_{[n]}, C)$ is infinite-dimensional in general. Considering the case $n = 4$, for example, one may see this more explicitly by noting that the defining invariance property allows us to express the values $\mathcal{F}(v_4 \otimes v_3 \otimes v_2 \otimes v_1)$ in terms of the complex numbers

$$\mathcal{Z}^{(k)}(\mathcal{F}, C) := \mathcal{F}(e_4 \otimes e_3 \otimes L_{-1}^k e_2 \otimes e_1), \quad k \in \mathbb{Z}^{>0}, \tag{6.11}$$

were e_i are the highest weight vectors of $\mathcal{V}_i, i = 1, 2, 3, 4$. We note that $\mathcal{F}$ is completely defined by the values $\mathcal{Z}^{(k)}(\mathcal{F}, C)$. The space of conformal blocks $\mathsf{CB}(\mathcal{V}_{[4]}, C)$ is therefore isomorphic as a vector space to the space of *formal* power series in one variable.

This definition of conformal blocks is closely related, but not quite identical to the one introduced previously in Sect. 2.5. To indicate the relation let us note that matrix elements like

$$\mathcal{F}_\beta(v_4 \otimes v_3 \otimes v_2 \otimes v_1) := \left\langle e_{\alpha_4}, V_{\alpha_4, \beta}^{\alpha_3}[V_3](z_3) V_{\beta, \alpha_1}^{\alpha_2}[V_2](z_2) e_{\alpha_1} \right\rangle_{\alpha_4}, \tag{6.12}$$

will represent particular examples for conformal blocks as defined in this section. Validity of the defining invariance property (6.8) follows from the covariance properties of the chiral vertex operators.

6.4 Deformations of the Complex Structure of C

A key point that needs to be discussed about the spaces of conformal blocks is the dependence on the complex structure of C, here specified by the positions $z_1, \ldots z_n$ of the marked points. There is a natural way to represent infinitesimal variations of the complex structure of C on the spaces of conformal blocks. By combining the definition of conformal blocks with the so-called "Virasoro uniformization" of the moduli space $\mathcal{M}_{0,n}$ of complex structures on $C = C_{0,n}$ one may construct a natural representation of infinitesimal motions on $\mathcal{M}_{0,n}$ on the space of conformal blocks.

The "Virasoro uniformization" of the moduli space $\mathcal{M}_{0,n}$ may be formulated as the statement that the tangent space $T\mathcal{M}_{0,n}$ to $\mathcal{M}_{0,n}$ at C can be identified with the double quotient

$$T\mathcal{M}_{0,n} = \Gamma(C \setminus \{z_1, \ldots, z_n\}, \Theta_C) \Big\backslash \bigoplus_{k=1}^{n} \mathbb{C}((w_k))\partial_k \Big/ \bigoplus_{k=1}^{n} w_k \mathbb{C}[[w_k]]\partial_k, \quad (6.13)$$

where $\mathbb{C}[[w_k]]$ are the spaces of Taylor series in the local coordinates w_k for $k = 1, \ldots, n$, respectively, and $\Gamma(C \setminus \{z_1, \ldots, z_n\}, \Theta_C)$ is the space of vector fields that are holomorphic on $C \setminus \{z_1, \ldots, z_n\}$, embedded into $\bigoplus_{k=1}^{n} \mathbb{C}((w_k))\partial_k$ via (6.10).

Given a tangent vector $\vartheta \in T\mathcal{M}_{0,n}$, it follows from the Virasoro uniformization (6.13) that we may find elements η_ϑ of $\bigoplus_{k=1}^{n} \mathbb{C}((t_k))\partial_k$, which represent ϑ via (6.13). Let us then consider $\mathcal{F}(T[\eta_\vartheta]v)$ with $T[\eta]$ being defined in (6.9) in the case that v is the product of highest weight vectors, $v = e_n \otimes \cdots \otimes e_1$. Equation (6.13) allows us to define the derivative $\delta_\vartheta \mathcal{F}(v)$ of $\mathcal{F}(v)$ in the direction of $\vartheta \in T\mathcal{M}_{0,n}$ as

$$\delta_\vartheta \mathcal{F}(v) := \mathcal{F}(T[\eta_\vartheta]v), \quad (6.14)$$

Dropping the condition that v is a product of highest weight vectors, one may use (6.14) to define $\delta_\vartheta \mathcal{F}$ in general. And indeed, it is well-known that (6.14) leads to the definition of a canonical flat connection on the space $\mathsf{CB}(\mathcal{V}_{[n]}, C)$ of conformal blocks [BF].

6.5 Conformal Blocks Versus Function on $\mathcal{T}_{0,n}$

In the case $n = 4$ it is easy to see that (6.14) can be reduced simply to

$$\partial_z \mathcal{F}(v_4 \otimes v_3 \otimes v_2 \otimes v_1) = \mathcal{F}(v_4 \otimes v_3 \otimes L_{-1}v_2 \otimes v_1). \quad (6.15)$$

Let us introduce the notation

$$\mathcal{Z}^{\text{Liou}}(\mathcal{F}, C) = \mathcal{F}(e_1 \otimes \cdots \otimes e_n), \quad (6.16)$$

for the value of $\mathcal{F}$ on the product of highest weight vectors. Equation (6.15) allows us to identify the values $\mathcal{Z}^{(k)}(\mathcal{F}, C)$ defined in (6.11) as the kth derivatives of the partition functions $\mathcal{Z}^{\text{Liou}}(\mathcal{F}, C)$. We had seen above that the collection of the numbers $\mathcal{Z}^{(k)}(\mathcal{F}, C)$ characterizes the conformal blocks $\mathcal{F} \in \mathsf{CB}(\mathcal{V}_{[4]}, C)$ completely.

One may define the parallel transport of conformal blocks over $\mathcal{M}_{0,4}$ via

$$\mathcal{Z}^{\text{Liou}}(\mathcal{F}, C_w) = \sum_{k=0}^{\infty} \frac{1}{n!} (w - z)^n \mathcal{Z}^{(k)}(\mathcal{F}, C_z). \tag{6.17}$$

We see that there is a one-to-one correspondence between functions $\mathcal{Y}(w)$ defined on some open, simply connected neighborhood $\mathcal{U}_z$ of a point z in $\mathcal{T}_{0,4}$ and conformal blocks $\mathcal{F}$ for which the series (6.17) converges for all $w \in \mathcal{U}_z$: The Taylor expansion coefficients $\mathcal{Y}_k$ of $\mathcal{Y}(w)$ can be used to define a conformal block $\mathcal{F}_{\mathcal{Y}} \in \mathsf{CB}(\mathcal{V}_{[4]}, C)$ such that $\mathcal{Z}^{(k)}(\mathcal{F}_{\mathcal{Y}}, C) = \mathcal{Y}_k$. Conversely, for "well-behaved" conformal blocks $\mathcal{F} \in \mathsf{CB}(\mathcal{V}_{[4]}, C_z)$ one may use (6.17) to define a family of conformal blocks in a neighborhood $\mathcal{U}_z(\mathcal{F})$.

We are ultimately not interrested in the most crazy conformal blocks, but rather in those whose partition functions can be analytically continued over all of $\mathcal{T}(C)$, and which have reasonably mild singular behaviour at the boundaries of $\mathcal{T}(C)$. Such a subspace will be denoted $\mathsf{CB}^{\text{reg}}(\mathcal{V}_{[4]}, C)$. It was proposed in [TV13] that the conformal blocks defined previously with the help of chiral vertex operators generate a basis for $\mathsf{CB}^{\text{reg}}(\mathcal{V}_{[4]}, C)$ in a suitable sense. This proposal is based on the highly nontrivial results of [T01, T03a] that the partition functions $\mathcal{Z}^{\text{Liou}}$ can be analytically continued over all of $\mathcal{T}(C)$, and that the bases associated to different pants decompositions are linearly related.

6.6   Verlinde Loop Operators

The construction of conformal blocks using chiral vertex operators, or more generally by gluing conformal blocks associated to three-punctured spheres gives another way to define a natural family of operators acting on spaces of conformal blocks. The resulting operators will be identified with quantized trace functions. We will describe the construction in the case of genus 0 in terms of chiral vertex operators.

Let us consider chiral vertex operators $V^{\alpha}_{\beta_2\beta_1}(z)$ in the special case where $\alpha = -b/2$, assuming that Q is represented as $Q = b + b^{-1}$. If furthermore β_2 and β_1 are related as $\beta_2 = \beta_1 \mp b/2$, the vertex operators $\psi_s(y) \equiv \psi_{\beta_1,s}(y) := V^{-b/2}_{\beta_1-sb/2,\beta_1}(y)$, $s \in \{1, -1\}$, are well-known to satisfy a differential equation of the form

$$\partial_y^2 \psi_{\beta_1,s}(y) + b^2 : T(y)\psi_{\beta_1,s}(y) := 0, \tag{6.18}$$

with normal ordering defined in (2.18). The chiral vertex operators $\psi_{\beta_1,s}(y)$ are called degenerate fields. It follows from (6.18) that matrix elements such as

$$\mathcal{F}_{ss'}(\alpha; \beta \,|\, z \,|\, y_0 \,|\, y) := \langle \alpha_4 \,|\, \psi_s(y_0)\psi_{s'}(y) \,|\, \Theta_{s+s'} \rangle, \tag{6.19}$$

$$|\, \Theta_{s+s'} \rangle := V^{\alpha_3}_{\alpha_4+(s+s')\frac{b}{2}, \beta}(z_3) V^{\alpha_2}_{\beta\alpha_1}(z_2) V^{\alpha_1}_{\alpha_1,0}(z_1) |\, 0 \rangle,$$

will satisfy the partial differential equation $\mathcal{D}_{\text{BPZ}}\mathcal{F} = 0$, with

$$\mathcal{D}_{\text{BPZ}} := \frac{1}{b^2} \frac{\partial^2}{\partial y^2} + \frac{\Delta_{-\frac{b}{2}}}{(y-y_0)^2} + \frac{1}{y-y_0} \frac{\partial}{\partial y_0} + \sum_{k=1}^{3} \left(\frac{\Delta_{\alpha_k}}{(y-z_k)^2} + \frac{1}{y-z_k} \frac{\partial}{\partial z_k} \right). \tag{6.20}$$

As explained previously, we may regard the matrix elements (6.19) as the partition functions of conformal blocks in $\mathsf{CB}(\mathcal{V}'_{[6]}, C_{0,6})$, where now $\mathcal{V}'_{[6]} = \mathcal{V}_{[4]} \otimes \mathcal{V}^{\otimes 2}_{-b/2}$.

Using these ingredients it is straightforward to show that the analytic continuation of the matrix elements $\mathcal{F}_{ss'}(\alpha; \beta \,|\, z \,|\, y_0 \,|\, y)$, $s, s' \in \{1, -1\}$ with respect to y along closed paths γ on $C_{0,5}$ can be expressed as a linear combination of the matrix elements $\mathcal{F}_{ss''}(\alpha; \beta' \,|\, z \,|\, y_0 \,|\, y)$ having parameters β' that differ from β by integer multiples of the parameter b,

$$\mathcal{F}_{ss'}(\alpha; \beta \,|\, z \,|\, y_0 \,|\, y) = \sum_{s''=\pm} \mathsf{M}_\gamma(\beta, \mathsf{T}_\beta)^{s''}_{ss'} \cdot \mathcal{F}_{ss''}(\alpha; \beta \,|\, z \,|\, y_0 \,|\, y) \tag{6.21}$$

where T is the operator which shifts the argument β of $\mathcal{F}_{ss'}$ by the amount b. The matrices M_γ define representations of the fundamental group $\pi_1(C_{0,5})$ on the space $\mathsf{CB}(\mathcal{V}'_{[6]}, C_{0,6})$.

The definition of the Verlinde loop operators is based on the simple fact that $\mathsf{CB}(\mathcal{V}_{[4]} \otimes \mathcal{V}_0, C_{0,5})$ is canonically isomorphic to $\mathsf{CB}(\mathcal{V}_{[4]}, C_{0,4})$ if $\mathcal{V}_0$ is the vacuum representation. One may furthermore note that there exists a linear combination $\sum_s K_s \psi_s(y_0)\psi_{-s}(y)$ which is a descendant of the chiral vertex operator $V^0_{\beta,\beta}[\psi_+(y_0 - y)e_{-\frac{b}{2}}](y)$ associated to the vacuum representation. These observations allow us to define both an embedding $\imath$ and a projection $\wp$,

$$\imath : \mathsf{CB}^{\text{reg}}(\mathcal{V}_{[4]}, C_{0,4}) \hookrightarrow \mathsf{CB}^{\text{reg}}(\mathcal{V}'_{[6]}, C_{0,6}),$$
$$\wp : \mathsf{CB}^{\text{reg}}(\mathcal{V}'_{[6]}, C_{0,6}) \rightarrow \mathsf{CB}^{\text{reg}}(\mathcal{V}_{[4]}, C_{0,4}), \tag{6.22}$$

in a natural way. The Verlinde loop operators can then be defined as the composition

$$\mathsf{V}_\gamma := \wp \circ \mathsf{M}_\gamma \circ \imath. \tag{6.23}$$

Concretely this boils down to taking a certain linear combination of the matrix elements $\mathsf{M}_\gamma(\beta, \mathsf{T}_\beta)^{s''}_{ss'}$ representing the monodromy along γ on $\mathsf{CB}^{\text{reg}}(\mathcal{V}_{[6]}, C_{0,6})$. The explicit calculations of the operator V_γ in [AGGTV, DGOT] shows that the Verlinde loop operators define a representation of $\text{Fun}_\hbar(\mathcal{M}_{\text{flat}}(C))$ on the space of conformal blocks $\mathsf{CB}^{\text{reg}}(\mathcal{V}_{[4]}, C_{0,n})$ which is equivalent to the one defined in Sect. 5.3.2 if we identify variables

$$\beta = \frac{Q}{2} + i\frac{l}{4\pi b}, \qquad \alpha_k = \frac{Q}{2} + i\frac{l_k}{4\pi b}, \quad k = 1, \ldots, 4. \qquad (6.24)$$

Let us summarize the observations made in this section in the following table:

Quantisation of ...	... is realised in CFT via
Darboux coordinates (q, h)	conformal Ward indentities
Fenchel-Nielsen coordinates (k, l)	Verlinde loop operators

The degenerate fields have a beautiful interpretation in the gauge theories $\mathcal{G}_C$ in terms of a family of observables called surface operators [AGGTV], explained also in the article [V:8] in this collection.

6.7 Liouville Conformal Blocks as Solutions to the Riemann-Hilbert Problem

We claim that the solution to the Riemann-Hilbert type problem defined in Sect. 3.2 is given by the Liouville conformal blocks in the following sense

$$\mathcal{Z}_\sigma^{\text{inst}}(a, m; \tau; \epsilon_1, \epsilon_2) = \mathcal{Z}_\sigma^{\text{spur}}(\alpha; \tau; b) \, \mathcal{Z}_\sigma^{\text{Liou}}(\beta, \alpha; q; b). \qquad (6.25)$$

The solution of the Riemann-Hilbert problem defined in Sect. 3.2 is unique up to multiplication with meromorphic functions which may have poles only at the boundary of $\mathcal{M}(C_{0,4})$. The resulting freedom can be absorbed into $\mathcal{Z}_\sigma^{\text{spur}}(\alpha, \tau; b)$.

In order to verify (6.25) we need to show that the representation of the mapping class group on spaces of Liouville conformal blocks is the same as the one coming from the quantum theory of $\mathcal{M}_{\text{flat}}^{\mathbb{R},0}(C)$ as described in Sect. 5. This boils down to the comparison of the respective realizations of B and F-moves. The coincidence of B-moves is trivial to verify. The realization of the F-move on Liouville conformal blocks was calculated in [T01], where a relation of the form

$$\mathcal{Z}_s^{\text{Liou}}(\beta_1, q) = \int_{\mathbb{S}} d\beta_2 \, F_{\beta_1\beta_2} \begin{bmatrix} \alpha_3 & \alpha_2 \\ \alpha_4 & \alpha_1 \end{bmatrix} \mathcal{Z}_t^{\text{Liou}}(\beta_2, q), \quad \mathbb{S} \equiv \frac{Q}{2} + i\mathbb{R}^+, \qquad (6.26)$$

was found. For the normalization defined in (2.14) we find the *same* kernel $F_{\beta_1\beta_2} \begin{bmatrix} \alpha_3 & \alpha_2 \\ \alpha_4 & \alpha_1 \end{bmatrix}$ in the relation (6.26), as was found within the quantum theory of $\mathcal{M}_{\text{flat}}^{\mathbb{R},0}(C)$ described in Sect. 5.

This is good enough to conclude that (6.25) must hold. To round off the picture, let us exhibit the meaning of the partition functions $\mathcal{Z}_\sigma^{\text{Liou}}(\beta, \alpha; q; b)$ within the quantisation of $\mathcal{M}_{\text{flat}}^{\mathbb{R},0}(C)$.

In this section we have described two different representations for the quantisation of one and the same Poisson-manifold, obtained by quantisation of the coordinates (q, h) and (l, k), respectively. One may expect that these two representations should be unitarily equivalent. The eigenstates $|l\rangle$ of the operator l are complete in $\mathcal{H}_{\sigma_s}$. It should therefore be possible to relate the wave-function $\psi(q) \equiv \langle q|\psi\rangle$ representing a state $|\psi\rangle$ in the holomorphic representation to the wave-function $\Psi(l) \equiv \langle l|\psi\rangle$ representing the same state in the length representation as

$$\psi(q) \equiv \langle q \,|\, \psi \rangle \;=\; \int dl \, \langle q \,|\, l \rangle \langle l \,|\, \psi \rangle . \tag{6.27}$$

The kernel $\langle q \,|\, l \rangle$ is the complex conjugate of the wave-function $\langle l \,|\, q \rangle$ of the "coherent" state $|q\rangle$ in the length representation.

One may use essentially the same arguments as presented in Sect. 3.2 to conclude that $\langle q \,|\, l \rangle$ must solve the same Riemann-Hilbert problem as discussed above. Combined with a discussion of the asymptotics at the boundary of $\mathcal{T}(C)$ we may thereby conclude [T03b, TV13] that

$$\langle q \,|\, l \rangle \;=\; \mathcal{Z}_\sigma^{\text{Liou}}(\beta, \alpha; q; b) . \tag{6.28}$$

We have thereby identified more precisely which wave-functions the conformal blocks are: They describe the change of representation between the two natural representations for the quantum theory of $\mathcal{M}_{flat}^{\mathbb{R},0}(C)$ discussed in this section.

6.8 The Nekrasov-Shatashvili Limit

The results reviewed in this article are related to the work [NRS] in an interesting way. In order to explain the relations to [NRS] let us consider the limit $\epsilon_2 \to 0$ corresponding to the classical limit for the quantum theory discussed in the previous sections. This will also provide further insight into the meaning of $\langle q \,|\, l \rangle = \mathcal{Z}^{\text{Liou}}$. It can be shown [T10] that the conformal blocks behave as

$$\log \mathcal{Z}_\sigma^{\text{Liou}}(\beta, \alpha; q; b) \sim -\frac{1}{\epsilon_2} \mathcal{Y}(l, m; q; \epsilon_1) , \tag{6.29}$$

assuming that the variables are related by (2.16). $\mathcal{Y}(l, m; q; \epsilon_1)$ is defined as follows: For given values of l and q let us consider differential operators of the form $\epsilon_1^2(\partial_y^2 + t(y))$, with $t(y)$ of the form (6.3). It can be argued that there is a unique choice $h = h(l, q)$ for the residue h such that $\epsilon_1^2(\partial_y^2 + t(y))$ has monodromy with trace equal to $2\cosh(l/2)$. $\mathcal{Y}(l, m; q; \epsilon_1)$ is defined up to a constant by the condition that

$$\frac{\partial}{\partial q} \mathcal{Y}(l, m; q; \epsilon_1) \;=\; -h(l, q) . \tag{6.30}$$

The constant can be fixed by demanding that the constant term $\mathcal{Y}_0(l, m; \epsilon_1)$ in

$$\mathcal{Y}(l, m; q; \epsilon_1) \underset{q \to 0}{\sim} -(\delta - \delta_1 - \delta_2) \log q + \mathcal{Y}_0(l, m; \epsilon_1) + \mathcal{O}(q), \qquad (6.31)$$

is one half of the sum of the Liouville actions on the three-punctured spheres into which $C_{0,4}$ splits when $q \to 0$. We furthermore have

$$k(l, q) = 4\pi i \frac{\partial}{\partial l} \mathcal{Y}(l, m; q; \epsilon_1). \qquad (6.32)$$

To verify (6.32) note, on the one hand, that the Verlinde loop operators reduce to the trace functions in the limit $\epsilon_2 \to 0$. Recall that the trace functions may be parameterised by the (complexified) Fenchel-Nielsen coordinates (l, k). The resulting expression may be compared to one following from (5.12) and (6.29) in this limit, giving (6.32).

The pairs of coordinates $(l, k(l, q))$ describe a Lagrangian sub-manifold denoted $\mathrm{Op}_{\mathfrak{sl}_2}(C)$ within $\mathcal{M}_{\mathrm{flat}}(C)$ sometimes called the "brane of opers". It follows from (6.32), (6.30) that $\mathcal{Y}(l, m; q; \epsilon_1)$ is the generating function of this sub-manifold. We thereby arrive at the description for the ϵ_2-limit of the instanton partition functions that was proposed in [NRS]. One may therefore view the results of [TV13] reviewed in this article as the generalisation of the results from [NRS] to nonzero ϵ_2.

6.9 Quantization of Seiberg-Witten Theory

It will furthermore be instructive to consider the limit where both $\epsilon_1, \epsilon_2 \to 0$, in which $E^4_{\epsilon_1, \epsilon_2}$ turns into $\mathbb{R}^4$, and we can make contact with Seiberg Witten theory.

To begin with, let us note that $\epsilon_1^2 \partial_y^2 + t(y)$ turns into the quadratic differential $-\vartheta(y)$ when $\epsilon_1 \to 0$. Using $\vartheta(y)$ we define the Seiberg-Witten curve Σ as usual by

$$\Sigma = \{ (v, u) \mid v^2 = \vartheta(u) \}. \qquad (6.33)$$

It follows by WKB analysis of the differential equation $(\epsilon_1^2 \partial_y^2 + t(y))\chi = 0$ that the coordinates l_e have asymptotics that can be expressed in terms of the Seiberg-Witten differential Λ on Σ defined such that $\Lambda^2 = \vartheta(u)(du)^2$. We find

$$l \sim \frac{4\pi}{\epsilon_1} a, \qquad k \sim \frac{4\pi}{\epsilon_1} a^{\mathrm{D}}, \qquad (6.34)$$

where a and a^{D} are periods of the Seiberg-Witten differential Λ defined as

$$a := \int_\alpha \Lambda, \qquad a^{\mathrm{D}} := \int_\beta \Lambda, \qquad (6.35)$$

with α and β being lifts of γ_s and γ_t to cycles on Σ that project to zero, respectively. The prepotential $\mathcal{F}$ is obtained in the limit $\epsilon_1, \epsilon_2 \to 0$ as follows:

$$
\begin{aligned}
\mathcal{F}(a, m, q) :&= -\lim_{\epsilon_1, \epsilon_2 \to 0} \epsilon_1 \epsilon_2 \log \mathcal{Z}_\sigma^{\text{inst}}(a, m, q; \epsilon_1, \epsilon_2) \\
&= \lim_{\epsilon_1 \to 0} \epsilon_1 \mathcal{Y}(a, m; q; \epsilon_1).
\end{aligned}
\tag{6.36}
$$

$\mathcal{F}(a, m; q)$ satisfies the relations

$$
a^{\mathrm{D}} = \frac{\partial}{\partial a} \mathcal{F}(a, q), \qquad h = -\frac{\partial}{\partial q} \mathcal{F}(a, q).
\tag{6.37}
$$

A proof of the relations (6.37) that is valid for all A_1-theories of class $\mathcal{S}$ was given in ([GT], Sect. 7.3.2). The relations (6.37) are equivalent to the statement that both the coordinates (a, a^{D}) describing the special geometry underlying Seiberg-Witten theory, and the coordinates (q, h) introduced above can be seen as systems of Darboux coordinates for the same space $T^* \mathcal{T}(C)$. The prepotential $\mathcal{F}(a, m, q)$ is nothing but the generating function of the change of variables between (a, a^{D}) and (q, h).

These observations show that the relations between the quantum theory on $\mathcal{M}_{\text{flat}}^0(C)$ and the gauge theories $\mathcal{G}_C$ discussed in this article can be seen as the quantization of the special geometry used in Seiberg-Witten theory. The dual zero modes a and a^{D} turn into the Darboux coordinates l and k upon partial compactification to $S^1 \times \mathbb{R}^3$ or $E_{\epsilon_1}^2 \times \mathbb{R}^2$. Further compactification to a four-ellipsoid leads to the quantization of these zero modes.

It is intriguing to observe that very similar ideas have been discussed in the context of topological string theory, where it has been proposed that the partition function of the topological string has an interpretation as a wave-function arising from the quantization of special geometry. The geometric engineering of gauge theories within string theory leads to relations between topological string and instanton partition functions, see the articles [V:13, V:14] in this volume for a review. One may hope that the relations with the quantization of moduli spaces of vacua discussed in this article may help us to get a more unified picture.

Acknowledgments The author would like to thank T. Dimofte, M. Gabella, A. Neitzke and T. Okuda for very useful remarks on a preliminary version of this article.

Appendix 1: Riemann Surfaces: Basic Definitions and Results

This appendix introduces basic definitions and results concerning Riemann surfaces C that will be used throughout the paper. A Riemann surface C is a two-dimensional topological surface S together with a choice of complex structure on S. We will denote by $\mathcal{M}(S)$ the moduli space of complex structures on a two-dimensional surface S, and by $\mathcal{T}(C)$ the Teichmüller space of deformations of complex structures on the Riemann surface C.

Complex Analytic Gluing Construction

A convenient family of particular coordinates on the Teichmüller spaces $\mathcal{T}(C)$ is defined by means of the complex-analytic gluing construction of Riemann surfaces C from three punctured spheres [Ma, HV]. Let us briefly review this construction.

Let C be a (possibly disconnected) Riemann surface. Fix a complex number q with $|q| < 1$, and pick two points Q_1 and Q_2 on C together with coordinates $z_i(P)$ in a neighborhood of Q_i, $i = 1, 2$, such that $z_i(Q_i) = 0$, and such that the discs D_i,

$$D_i := \{ P_i \in C; \, |z_i(P_i)| < |q|^{-\frac{1}{2}} \}, \qquad i = 1, 2,$$

do not intersect. One may define the annuli A_i,

$$A_i := \{ P_i \in C; \, |q|^{\frac{1}{2}} < |z_i(P_i)| < |q|^{-\frac{1}{2}} \}, \qquad i = 1, 2.$$

To glue A_1 to A_2 let us identify two points P_1 and P_2 on A_1 and A_2, respectively, iff the coordinates of these two points satisfy the equation

$$z_1(P_1)z_2(P_2) = q. \tag{6.38}$$

If C is connected one creates an additional handle, and if $C = C_1 \sqcup C_2$ has two connected components one gets a single connected component after performing the gluing operation. In the limiting case where $q = 0$ one gets a nodal surface which represents a component of the boundary $\partial \mathcal{M}(S)$ defined by the Deligne-Mumford compactification $\overline{\mathcal{M}}(S)$.

By iterating the gluing operation one may build any Riemann surface C of genus g with n punctures from three-punctured spheres $C_{0,3}$. Embedded into C we naturally get a collection of annuli $A_1, \ldots, A_h$, where

$$h := 3g - 3 + n. \tag{6.39}$$

The construction above can be used to define a $3g - 3 + n$-parametric family of Riemann surfaces, parameterised by a collection $q = (q_1, \ldots, q_h)$ of complex parameters. These parameters can be taken as coordinates for a neighbourhood of a component in the boundary $\partial \mathcal{M}(S)$ which are complex-analytic with respect to its natural complex structure [Ma].

Conversely, assume given a Riemann surface C and a cut system, a collection $\mathcal{C} = \{\gamma_1, \ldots, \gamma_h\}$ of homotopy classes of non-intersecting simple closed curves on

C. Cutting along all the curves in $\mathcal{C}$ produces a pants decompostion, $C \setminus \mathcal{C} \simeq \bigsqcup_v C_{0,3}^v$, where the $C_{0,3}^v$ are three-holed spheres.

Having glued C from three-punctured spheres defines a distinguished cut system, defined by a collection of simple closed curves $\mathcal{C} = \{\gamma_1, \ldots, \gamma_h\}$ such that γ_r can be embedded into the annulus A_r for $r = 1, \ldots, h$.

An important deformation of the complex structure of C is the Dehn-twist: It corresponds to rotating one end of an annulus A_r by 2π before regluing, and can be described by a change of the local coordinates used in the gluing construction. The coordinate q_r can not distinguish complex structures related by a Dehn twist in A_r. It is often useful to replace the coordinates q_r by logarithmic coordinates τ_r such that $q_r = e^{2\pi i \tau_r}$. This corresponds to replacing the gluing identification (6.38) by its logarithm. In order to define the logarithms of the coordinates z_i used in (6.38), one needs to introduce branch cuts on the three-punctured spheres, an example being depicted in Fig. 4.

By imposing the requirement that the branch cuts chosen on each three-punctured sphere glue to a connected three-valent graph Γ on C, one gets an unambiguous definition of the coordinates τ_r. We see that the logarithmic versions of the gluing construction that define the coordinates τ_r are parameterized by the pair of data $\sigma = (\mathcal{C}_\sigma, \Gamma_\sigma)$, where $\mathcal{C}_\sigma$ is the cut system defined by the gluing construction, and Γ_σ is the three-valent graph specifying the choices of branch cuts. In order to have a handy terminology we will call the pair of data $\sigma = (\mathcal{C}_\sigma, \Gamma_\sigma)$ a *pants decomposition*, and the three-valent graph Γ_σ will be called the Moore-Seiberg graph, or MS-graph associated to a pants decomposition σ. The construction outlined above gives a set of coordinates for the neighbourhood $\mathcal{U}_\sigma$ of the boundary component of $\mathcal{T}(C)$ corresponding to σ.

The gluing construction depends on the choices of coordinates around the punctures Q_i. There exists an ample supply of choices for the coordinates z_i such that the union of the neighbourhoods $\mathcal{U}_\sigma$ produces a cover of $\mathcal{M}(C)$ [HV]. For a fixed choice of these coordinates one produces families of Riemann surfaces fibred over the multi-discs $\mathcal{U}_\sigma$ with coordinates q. Changing the coordinates z_i around Q_i produces a family of Riemann surfaces which is locally biholomorphic to the initial one [RS].

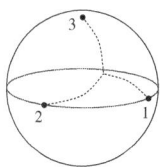

Fig. 4 A sphere with three punctures, and a choice of branch cuts for the definition of the logarithms of local coordinates around the punctures.

The Moore-Seiberg Groupoid

Let us note [MS, BK] that any two different pants decompositions σ_2, σ_1 can be connected by a sequence of elementary moves localized in subsurfaces of $C_{g,n}$ of type $C_{0,3}$, $C_{0,4}$ and $C_{1,1}$. The elementary moves are called the B, F, Z and S-moves, respectively. Graphical representations for the elementary moves F, S and B are given in Figs. 1, 2 and 3, respectively. The Z-move is just the change of distinguished boundary component in a three-punctured sphere.

One may formalize the resulting structure by introducing a two-dimensional CW complex $\mathcal{M}(C)$ with set of vertices $\mathcal{M}_0(C)$ given by the pants decompositions σ, and a set of edges $\mathcal{M}_1(C)$ associated to the elementary moves. The Moore-Seiberg groupoid is defined to be the path groupoid of $\mathcal{M}(C)$. It can be described in terms of generators and relations, the generators being associated with the edges of $\mathcal{M}(C)$, and the relations associated with the faces of $\mathcal{M}(C)$. The classification of the relations was first presented in [MS], and rigorous mathematical proofs have been presented in [FG1, BK]. The relations are all represented by sequences of elementary moves localized in subsurfaces $C_{g,n}$ with genus $g = 0$ and $n = 3, 4, 5$ punctures, as well as $g = 1$, $n = 1, 2$. Graphical representations of the relations can be found in [MS, FG1, BK].

Uniformization

The classical uniformization theorem ensures existence and uniqueness of a hyperbolic metric, a metric of constant negative curvature, on a Riemann surface C. In a local chart with complex analytic coordinates y one may represent this metric in the form $ds^2 = e^{2\varphi}dyd\bar{y}$, with φ being a solution to the Liouville equation $\partial\bar{\partial}\varphi = \mu e^{2\varphi}dyd\bar{y}$.

The solutions to the Liouville equation may be parameterized by a function $t(y)$ related to φ as

$$t := -(\partial_y\varphi)^2 + \partial_y^2\varphi. \tag{6.40}$$

$t(y)$ is holomorphic as a consequence of the Liouville equation. The solution to the Liouville equation can be reconstructed from $t(y)$ by first finding the solutions to

$$(\partial_y^2 + t(y))\chi = 0. \tag{6.41}$$

Picking two linearly independent solutions $\chi_\pm$ of (6.41) with $\chi'_+\chi_- - \chi'_-\chi_+ = 1$ allows us to represent $e^{2\varphi}$ as $e^{2\varphi} = -(\chi_+\bar{\chi}_- - \chi_-\bar{\chi}_+)^{-2}$. The hyperbolic metric $ds^2 = e^{2\varphi}dyd\bar{y}$ may then be written in terms of the quotient $A(y) := \chi_+/\chi_-$ as

$$ds^2 = e^{2\varphi}dyd\bar{y} = \frac{\partial A\bar{\partial}\bar{A}}{(\text{Im}(A))^2}. \tag{6.42}$$

It follows that $A(y)$ represents a conformal mapping from C to a domain Ω in the upper half plane $\mathbb{U}$ with its standard constant curvature metric. The monodromies of the solution χ are represented on $A(y)$ by Moebius transformations. These Moebius transformations describe the identifications of the boundaries of the simply-connected domain Ω in $\mathbb{U}$ which represents the image of C under A. C is therefore conformal to $\mathbb{U}/\Gamma$, where the Fuchsian group Γ is the monodromy group of the differential operator $\partial_y^2 + t(y)$.

Appendix 2: Moduli Spaces of Flat Connections

In this appendix we shall review some of the basic definitions and results concerning the moduli spaces $\mathcal{M}_{flat}(C)$.

Moduli of Flat Connections and Character Variety

We will consider flat $\mathrm{PSL}(2,\mathbb{C})$-connections $\nabla = d - A$ on Riemann surfaces C. Let $\mathcal{M}_{flat}(C)$ be the moduli space of all such connections modulo gauge transformations.

Given a flat $\mathrm{PSL}(2,\mathbb{C})$-connection $\nabla = d - A$, one may define its holonomy $\rho(\gamma)$ along a closed loop γ as $\rho(\gamma) = \mathcal{P}\exp(\int_\gamma A)$. The assignment $\gamma \mapsto \rho(\gamma)$ defines a representation of $\pi_1(C)$ in $\mathrm{PSL}(2,\mathbb{C})$. As any flat connection is locally gauge-equivalent to the trivial connection, one may characterize gauge-equivalence classes of flat connections by the corresponding representations $\rho : \pi_1(C) \to \mathrm{PSL}(2,\mathbb{C})$. This allows us to identify the moduli space $\mathcal{M}_{flat}(C)$ of flat $\mathrm{PSL}(2,\mathbb{C})$-connections on C with the so-called character variety

$$\mathcal{M}_{char}(C) := \mathrm{Hom}(\pi_1(C), \mathrm{PSL}(2,\mathbb{C}))/\mathrm{PSL}(2,\mathbb{C}). \tag{6.43}$$

The moduli space $\mathcal{M}_{flat}(C)$ has a natural real slice, the moduli space $\mathcal{M}_{flat}^{\mathbb{R}}(C)$ of flat $\mathrm{PSL}(2,\mathbb{R})$-connections.

The Teichmüller Component

There is a well-known relation between the Teichmüller space $\mathcal{T}(C)$ and a connected component of the moduli space $\mathcal{M}_{flat}^{\mathbb{R}}(C)$ of flat $\mathrm{PSL}(2,\mathbb{R})$-connections on C. This component is called the Teichmüller component and will be denoted as $\mathcal{M}_{flat}^{\mathbb{R},0}(C)$. The relation between $\mathcal{T}(C)$ and $\mathcal{M}_{flat}^0(C)$ may be described as follows. To a hyperbolic metric $ds^2 = e^{2\varphi}dyd\bar{y}$ let us associate the connection $\nabla = \nabla' + \nabla''$,

$$\nabla'' = \bar{\partial}, \qquad \nabla' = \partial + M(y)dy, \qquad M(y) = \begin{pmatrix} 0 & -t \\ 1 & 0 \end{pmatrix}, \qquad (6.44)$$

with t constructed from $\varphi(y, \bar{y})$ as in (6.40). This connection is flat since $\partial_y \bar{\partial}_{\bar{y}} \varphi = \mu e^{2\varphi}$ implies $\bar{\partial} t = 0$. The Fuchsian group Γ characterizing the uniformization of C is nothing but the holonomy ρ of the connection ∇ defined in (6.44).

The Fuchsian groups Γ fill out the connected component $\mathcal{M}_{\mathrm{char}}^{\mathbb{R},0}(C) \simeq \mathcal{T}(C)$ in $\mathcal{M}_{\mathrm{flat}}^{\mathbb{R}}(C)$ called the Teichmüller component.

Fock–Goncharov Coordinates

Let τ be a triangulation of the surface C such that all vertices coincide with marked points on C. An edge e of τ separates two triangles defining a quadrilateral Q_e with corners being the marked points $P_1, \ldots, P_4$. For a given local system $(\mathcal{E}, \nabla)$, let us choose four sections s_i, $i = 1, 2, 3, 4$ that obey the condition $\nabla s_i = 0$, and are eigenvectors of the monodromy around P_i. Out of the sections s_i form [FG1, GMN2]

$$\mathcal{X}_e^\tau := -\frac{(s_1 \wedge s_2)(s_3 \wedge s_4)}{(s_2 \wedge s_3)(s_4 \wedge s_1)}, \qquad (6.45)$$

where all sections are evaluated at a common point $P \in Q_e$. It is not hard to see that $\mathcal{X}_e^\tau$ does not depend on the choice of P.

There exists a simple description of the relations between the coordinates associated to different triangulations. If triangulation τ_e is obtained from τ by changing only the diagonal in the quadrangle containing e, we have

$$\mathcal{X}_{e'}^{\tau_e} = \begin{cases} \mathcal{X}_{e'}^\tau \left(1 + (\mathcal{X}_e^\tau)^{-\mathrm{sgn}(n_{e'e})} \right)^{-n_{e'e}} & \text{if } e' \neq e, \\ (\mathcal{X}_e^\tau)^{-1} & \text{if } e' = e. \end{cases} \qquad (6.46)$$

This reflects part of the structure of a cluster algebra that $\mathcal{M}_{\mathrm{flat}}(C)$ has.

Trace Functions

The trace functions

$$L_\gamma := \nu_\gamma \mathrm{tr}(\rho(\gamma)), \qquad (6.47)$$

represent useful coordinate functions for $\mathcal{M}_{\mathrm{flat}}^{\mathbb{C}}(C)$. The signs ν_γ will be chosen such that the restriction to L_γ to the Teichmüller component $\mathcal{M}_{\mathrm{char}}^{\mathbb{R},0}(C)$ satisfies $L_\gamma = 2\cosh(l_\gamma/2) > 2$, where l_γ is the length of the hyperbolic geodesic on $\mathbb{U}/\Gamma$ isotopic to γ.

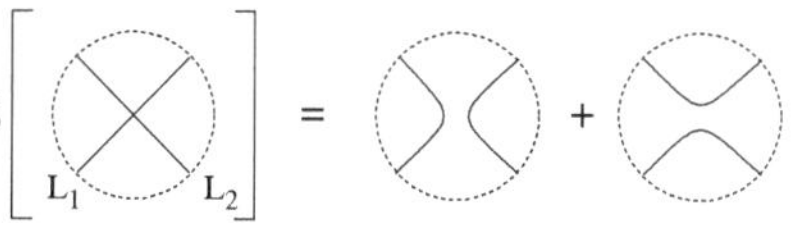

Fig. 5 The symmetric
smoothing operation

$$S\left[\begin{array}{c}\text{⊗}\\ L_1 \qquad L_2\end{array}\right] = \text{)(} + \text{⊗}$$

The coordinate functions L_γ generate the commutative algebra $\mathcal{A}(C) \simeq$ $\text{Fun}^{\text{alg}}(\mathcal{M}_{\text{flat}}(C))$ of functions on $\mathcal{M}_{\text{flat}}(C)$. The well-known relation $\text{tr}(g)\text{tr}(h) = \text{tr}(gh) + \text{tr}(gh^{-1})$ valid for any pair of $SL(2)$-matrices g, h implies that the geodesic length functions satisfy the so-called skein relations,

$$L_{\gamma_1} L_{\gamma_2} = L_{S(\gamma_1, \gamma_2)}, \tag{6.48}$$

where $S(\gamma_1, \gamma_2)$ is the loop obtained from γ_1, γ_2 by means of the smoothing operation, defined as follows. The application of S to a single intersection point of γ_1, γ_2 is depicted in Fig. 5. The general result is obtained by applying this rule at each intersection point, and summing the results.

Topological Classification of Closed Loops

With the help of pants decompositions one may conveniently classify all non-selfintersecting closed loops on C up to homotopy. To a loop γ let us associate the collection of integers (r_e, s_e) associated to all edges e of Γ_σ which are defined as follows. Recall that there is a unique curve $\gamma_e \in \mathcal{C}_\sigma$ that intersects a given edge e on Γ_σ exactly once, and which does not intersect any other edge. The integer r_e is defined as the number of intersections between γ and the curve γ_e. Having chosen an orientation for the edge e_r we will define s_e to be the intersection index between e and γ.

Dehn's theorem (see [DMO] for a nice discussion) ensures that the curve γ is up to homotopy uniquely classified by the collection of integers (r, s), subject to the restrictions

(i) $r_e \geq 0$,

(ii)   if $r_e = 0 \Rightarrow s_e \geq 0$,   (6.49)

(iii) $r_{e_1} + r_{e_2} + r_{e_3} \in 2\mathbb{Z}$ whenever $\gamma_{e_1}, \gamma_{e_2}, \gamma_{e_3}$ bound the same trinion.

We will use the notation $\gamma_{(r,s)}$ for the geodesic which has parameters $(r, s) : e \mapsto (r_e, s_e)$.

Generators and Relations

The pants decompositions allow us to describe $\mathcal{A}(C)$ in terms of generators and relations. As set of generators for $\mathcal{A}(C)$ one may take the functions $L_{(r,s)} \equiv L_{\gamma_{(r,s)}}$. The skein relations imply various relations among the $L_{(r,s)}$. It is not hard to see that these relations allow one to express arbitrary $L_{(r,s)}$ in terms of a finite subset of the set of $L_{(r,s)}$.

Let us temporarily restrict attention to surfaces with genus zero and $n = 4$ boundaries. The Moore-Seiberg graph Γ_σ will then have only one internal edge, allowing us to drop the index e labelling the edges. Let us introduce the geodesics $\gamma_s = \gamma_{(0,1)}$, $\gamma_t = \gamma_{(2,0)}$ and $\gamma_u = \gamma_{(2,1)}$. The geodesics γ_s and γ_t are depicted as red curves on the left and right half of Fig. 1. We will denote $L_k \equiv L_{\gamma_k}$, where $k \in \{s, t, u\}$. The trace functions L_s, L_t and L_u generate $\mathcal{A}(C)$.

These coordinates are not independent, though. Further relations follow from the relations in $\pi_1(C)$. It can be shown (see e.g. [Go09] for a review) that the coordinate functions L_s, L_t and L_u satisfy an algebraic relation of the form

$$P(L_s, L_t, L_u) = 0. \tag{6.50a}$$

The polynomial P in (6.50) is explicitly given as[6]

$$
\begin{aligned}
P(L_s, L_t, L_u) := & - L_s L_t L_u + L_s^2 + L_t^2 + L_u^2 \\
& + L_s(L_3 L_4 + L_1 L_2) + L_t(L_2 L_3 + L_1 L_4) + L_u(L_1 L_3 + L_2 L_4) \\
& - 4 + L_1^2 + L_2^2 + L_3^2 + L_4^2 + L_1 L_2 L_3 L_4.
\end{aligned}
\tag{6.50b}
$$

In the expressions above we have denoted $L_i := L_{\gamma_i}$, where γ_i, $i = 1, 2, 3, 4$ represent the boundary components of $C_{0,4}$, labelled according to the convention defined in Fig. 1.

Trace Functions in Terms of Fock-Goncharov Coordinates

Assume given a path ϖ_γ on the fat graph homotopic to a simple closed curve γ on $C_{g,n}$. Let the edges be labelled e_i, $i = 1, \ldots, r$ according to the order in which they appear on ϖ_γ, and define σ_i to be 1 if the path turns left at the vertex that connects edges e_i and e_{i+1}, and to be equal to -1 otherwise. Consider the following matrix,

$$X_\gamma = V^{\sigma_r} E(z_{e_r}) \cdots V^{\sigma_1} E(z_{e_1}), \tag{6.51}$$

where $z_e = \log X_e$, and the matrices $E(z)$ and V are defined respectively by

---

[6]Comparing to [Go09] note that some signs were absorbed by a suitable choice of the signs ν_γ in (6.47).

$$E(z) = \begin{pmatrix} 0 & +e^{+\frac{z}{2}} \\ -e^{-\frac{z}{2}} & 0 \end{pmatrix}, \quad V = \begin{pmatrix} 1 & 1 \\ -1 & 0 \end{pmatrix}. \tag{6.52}$$

Taking the trace of X_γ one gets the hyperbolic length of the closed geodesic isotopic to γ via [F97]

$$L_\gamma \equiv 2\cosh(l_\gamma/2) = |\text{tr}(X_\gamma)|. \tag{6.53}$$

We may observe that the classical expression for $L_\gamma \equiv 2\cosh\frac{1}{2}l_\gamma$ as given by formula 6.53 is a linear combination of monomials in the variables $u_e^{\pm 1} \equiv e^{\pm\frac{z_e}{2}}$ of the very particular form (5.4).

Fenchel-Nielsen Coordinates for $\mathcal{M}_{\text{flat}}^{\mathbb{R},0}(C)$

One may express L_s, L_t and L_u in terms of the Fenchel-Nielsen coordinates l and k [Ok, Go09]. Explicit expressions are for $C_{0,4}$,

$$L_s = 2\cosh(l/2), \tag{6.54a}$$

$$\begin{aligned} L_t\big((L_s)^2 - 4\big) = {} & 2(L_2 L_3 + L_1 L_4) + L_s(L_1 L_3 + L_2 L_4) \\ & + 2\cosh(k)\sqrt{c_{12}(L_s)c_{34}(L_s)}, \end{aligned} \tag{6.54b}$$

$$\begin{aligned} L_u\big((L_s)^2 - 4\big) = {} & L_s(L_2 L_3 + L_1 L_4) + 2(L_1 L_3 + L_2 L_4) \\ & + 2\cosh((2k-l)/2)\sqrt{c_{12}(L_s)c_{34}(L_s)}, \end{aligned} \tag{6.54c}$$

where $L_i = 2\cosh\frac{l_i}{2}$, and $c_{ij}(L_s)$ is defined as

$$c_{ij}(L_s) = L_s^2 + L_i^2 + L_j^2 + L_s L_i L_j - 4. \tag{6.55}$$

These expressions ensure that the algebraic relations $P_e(L_s, L_t, L_u) = 0$ are satisfied. By complexifying (l, k) one gets (local) coordinates for $\mathcal{M}_{\text{flat}}^{\mathbb{C}}(C)$ [NRS].

Poisson Structure

There is also a natural Poisson bracket on $\mathcal{A}(C)$ [Go86], defined such that

$$\{L_{\gamma_1}, L_{\gamma_2}\} = L_{A(\gamma_1, \gamma_2)}, \tag{6.56}$$

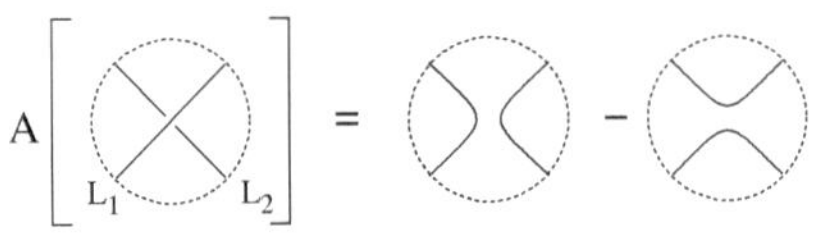

Fig. 6 The anti-symmetric smoothing operation

where $A(\gamma_1, \gamma_2)$ is the loop obtained from γ_1, γ_2 by means of the anti-symmetric smoothing operation, defined as above, but replacing the rule depicted in Fig. 5 by the one depicted in Fig. 6. This Poisson structure coincides with the Poisson structure coming from the natural symplectic structure on $\mathcal{M}_{\text{flat}}(C)$ which was introduced by Atiyah and Bott.

The resulting expression for the Poisson bracket $\{ L_s, L_t \}$ can be written elegantly in the form

$$\{ L_s, L_t \} = \frac{\partial}{\partial L_u} P(L_s, L_t, L_u). \tag{6.57}$$

It is remarkable that the same polynomial appears both in (6.50) and in (6.57), which indicates that the symplectic structure on $\mathcal{M}_{\text{flat}}$ is compatible with its structure as algebraic variety.

The Fenchel-Nielsen coordinates are known to be Darboux-coordinates for $\mathcal{M}_{\text{flat}}(C)$, having the Poisson bracket

$$\{ l, k \} = 2. \tag{6.58}$$

The Poisson structure is also rather simple in terms of the Fock-Goncharov coordinates,

$$\{\mathcal{X}_e^\tau, \mathcal{X}_{e'}^\tau\} = n_{e,e'} \, \mathcal{X}_{e'}^\tau \, \mathcal{X}_e^\tau, \tag{6.59}$$

where $n_{e,e'}$ is the number of faces e and e' have in common, counted with a sign.

References

[AST]   Aharony, O., Seiberg, N., Tachikawa, Y.: Reading between the lines of four-dimensional gauge theories. JHEP **1308**, 115 (2013)
[AFLT]   Alba, V.A., Fateev, V.A., Litvinov, A.V., Tarnopolsky, G.M.: On combinatorial expansion of the conformal blocks arising from AGT conjecture. Lett. Math. Phys. **98**, 33–64 (2011)
[AGT]   Alday, L.F., Gaiotto, D., Tachikawa, Y.: Liouville correlation functions from four-dimensional gauge theories. Lett. Math. Phys. **91**, 167–197 (2010)
[AGGTV]   Alday, L.F., Gaiotto, D., Gukov, S., Tachikawa, Y., Verlinde, H.: Loop and surface operators in $\mathcal{N} = 2$ gauge theory and Liouville modular geometry. J. High Energy Phys. **1001**, 113 (2010)
[BPZ]   Belavin, A.A., Polyakov, A.M., Zamolodchikov, A.B.: Infinite conformal symmetry in two-dimensional quantum field theory. Nucl. Phys. **B241**, 333–380 (1984)

[BF]   Frenkel, E., Ben-Zvi, D.: Vertex algebras and algebraic curves. Mathematical Surveys
            and Monographs, vol. 88, 2nd edn. American Mathematical Society, Providence (2004)
[BJSV]   Bershadsky, M., Johansen, A., Sadov, V., Vafa, C.: Topological reduction of 4D SYM
            to 2D sigma models. Nucl. Phys. **B448**, 166–186 (1995)
[BK]   Bakalov, B., Kirillov Jr, A.: On the Lego-Teichmüller game. Transform. Groups **5**(3),
            207–244 (2000)
[CF1]   Chekhov, L.O., Fock, V.: A quantum Teichmüller space. Theor. Math. Phys. **120**, 1245–
            1259 (1999)
[CMS]   Cantini, L., Menotti, P., Seminara, D.: Proof of Polyakov conjecture for general elliptic
            singularities. Phys. Lett. **B517**, 203–209 (2001)
[CN]   Cordova, C., Neitzke, A.: Line defects, tropicalization, and multi-centered quiver quan-
            tum mechanics. JHEP **09**, 099 (2014)
[DGV]   Dimofte, T., Gaiotto, D., van der Veen, R.: RG Domain Walls and Hybrid Triangulations.
            arXiv:1304.6721 [hep-th]
[DMO]   Drukker, N., Morrison, D.R., Okuda, T.: Loop operators and S-duality from curves on
            Riemann surfaces. JHEP **0909**, 031 (2009)
[DO]   Dorn, H., Otto, H.-J.: Two and three-point functions in Liouville theory. Nucl. Phys.
            B429, 375–388 (1994)
[DGOT]   Drukker, N., Gomis, J., Okuda, T., Teschner, J.: Gauge theory loop operators and
            Liouville theory. J. High Energy Phys. **1002**, 057 (2010)
[F97]   Fock, V.: Dual Teichmüller spaces. arXiv:dg-ga/9702018
[FG1]   Fock, V.V., Goncharov, A.: Moduli spaces of local systems and higher Teichmller
            theory. Publ. Math. Inst. Hautes Études Sci. **103**, 1–211 (2006)
[FG2]   Fock, V.V., Goncharov, A.B.: The quantum dilogarithm and representations of quantum
            cluster varieties. Invent. Math. **175**, 223–286 (2009)
[G09]   Gaiotto, D.: $N = 2$ dualities. JHEP **1208**, 034 (2012)
[GMN1]   Gaiotto, D., Moore, G., Neitzke, A.: Four-dimensional wall-crossing via three-
            dimensional field theory. Commun. Math. Phys. **299**, 163–224 (2010)
[GMN2]   Gaiotto, D., Moore, G., Neitzke, A.: Wall-crossing, Hitchin systems, and the WKB
            approximation. Adv. Math. **234**, 239–403 (2013)
[GMN3]   Gaiotto, D., Moore, G., Neitzke, A.: Framed BPS states. Adv. Theor. Math. Phys. **17**,
            241–397 (2013)
[Go86]   Goldman, W.: Invariant functions on Lie groups and Hamiltonian flows of surface group
            representations. Invent. Math. **85**, 263–302 (1986)
[Go88]   Goldman, W.: Topological components of spaces of representations. Invent. Math. **93**,
            557–607 (1988)
[Go09]   Goldman, W.: Trace coordinates on fricke spaces of some simple hyperbolic surfaces,
            Handbook of Teichmüller theory, vol. II, pp. 611–684, IRMA Lect. Math. Theor. Phys.,
            13, Eur. Math. Soc., Zürich (2009)
[GOP]   Gomis, J., Okuda, T., Pestun, V.: Exact results for 't Hooft loops in Gauge theories on
            S^4. JHEP **1205**, 141 (2012)
[GT]   Gaiotto, D., Teschner, J.: Irregular singularities in Liouville theory and Argyres-
            Douglas type gauge theories. JHEP **1212**, 050 (2012)
[Hi]   Hitchin, N.: The self-duality equations on a Riemann surface. Proc. Lond. Math. Soc.
            55(3), 59–126 (1987)
[HH]   Hama, N., Hosomichi, K.: Seiberg-witten theories on ellipsoids. JHEP **1209**, 033
            (2012). (Addendum-ibid, vol. 1210, p. 051)
[HKS]   Hollands, L., Keller, C.A., Song, J.: Towards a 4d/2d correspondence for Sicilian quiv-
            ers. JHEP **1110**, 100 (2011)
[HV]   Hinich, V., Vaintrob, A.: Augmented Teichmüller spaces and orbifolds. Selecta Math.
            (N.S.) **16**, 533–629 (2010)
[IOT]   Ito, Y., Okuda, T., Taki, M.: Line operators on $S^1 \times R^3$ and quantization of the Hitchin
            moduli space. JHEP **1204**, 010 (2012)

[Ka1]   Kashaev, R.M.: Quantization of Teichmüller spaces and the quantum dilogarithm. Lett. Math. Phys. **43**, 105–115 (1998)

[Ka4]   Kashaev, R.M.: The quantum dilogarithm and Dehn twists in quantum Teichmüller theory. Integrable structures of exactly solvable two-dimensional models of quantum field theory (Kiev, 2000). Nato Science Series II: Mathematics, Physics and Chemistry, vol. 35, pp. 211–221. Kluwer Academic Publishers, Dordrecht (2001)

[LNS]   Losev, A.S., Nekrasov, N.A., Shatashvili, S.: Testing Seiberg-Witten solution. Strings, branes and dualities (Cargèse, 1997). Nato Science Series C: Mathematical and Physical Sciences, vol. 520, pp. 359–372. Kluwer Academic Publishers, Dordrecht (1999)

[Ma]   Marden, A.: Geometric complex coordinates for Teichmüller space. Mathematical Aspects of String Theory. Adv. Ser. Math. Phys., vol. 1, pp. 341–354. World Scientific, San Diego (1987)

[MNS1]   Moore, G., Nekrasov, N.A., Shatashvili, S.: Integrating over Higgs branches. Commun. Math. Phys. **209**, 97–121 (2000)

[MNS2]   Moore, G., Nekrasov, N.A., Shatashvili, S.: D-particle bound states and generalized instantons. Commun. Math. Phys. **209**, 77–95 (2000)

[MS]   Moore, G., Seiberg, N.: Classical and quantum conformal field theory. Commun. Math. Phys. **123**, 177–254 (1989)

[N]   Nekrasov, N.A.: Seiberg-Witten prepotential from instanton counting. Adv. Theor. Math. Phys. **7**, 831–864 (2003)

[NRS]   Nekrasov, N., Rosly, A., Shatashvili, S.: Darboux coordinates, Yang-Yang functional, and gauge theory. Nucl. Phys. Proc. Suppl. **216**, 69–93 (2011)

[NS04]   Nekrasov, N., Shadchin, S.: ABCD of instantons. Commun. Math. Phys. **252**, 359–391 (2004)

[NT]   Nidaeiev, I., Teschner, J.: On the relation between the modular double of $\mathcal{U}_q(\mathfrak{sl}(2,\mathbb{R}))$ and the quantum Teichmüller theory. Preprint arXiv:1302.3454

[NW]   Nekrasov, N., Witten, E.: The omega deformation, branes, integrability, and Liouville theory. JHEP **1009**, 092 (2010)

[Ok]   Okai, T.: Effects of change of pants decomposition on their Fenchel-Nielsen coordinates. Kobe J. Math. **10**, 215–223 (1993)

[Pe]   Pestun, V.: Localization of gauge theory on a four-sphere and supersymmetric Wilson loops. Commun. Math. Phys. **313**, 71–129 (2012)

[RS]   Robbin, J.W., Salamon, D.A.: A construction of the Deligne-Mumford orbifold. J. Eur. Math. Soc. (JEMS) **8**, 611–699 (2006)

[Ta13]   Tachikawa, Y.: On the 6d origin of discrete additional data of 4d gauge theories. JHEP **1405**, 020 (2014)

[TZ87a]   Takhtajan, L.A., Zograf, P.G.: On the Liouville equation, accessory parameters and the geometry of Teichmüller space for Riemann surfaces of genus 0. Math. USSR-Sb. **60**, 143–161 (1988)

[TZ03]   Takhtajan, L.A., Zograf, P.G.: Hyperbolic 2-spheres with conical singularities, accessory parameters and Kähler metrics on $\mathcal{M}_{0,n}$. Trans. Am. Math. Soc. **355**(5), 1857–1867 (2003)

[T01]   Teschner, J.: Liouville theory revisited. Class. Quant. Grav. **18**, R153–R222 (2001)

[T03a]   Teschner, J.: A lecture on the Liouville vertex operators. Int. J. Mod. Phys. **A19S2**, 436–458 (2004)

[T03b]   Teschner, J.: From Liouville theory to the quantum geometry of Riemann surfaces. Prospects in Mathematical Physics. Contemporary Mathematics, vol. 437, pp. 231–246. American Mathematical Society, Providence (2007)

[T05]   Teschner, J.: An analog of a modular functor from quantized Teichmüller theory. In: Papadopoulos, A. (ed.) Handbook of Teichmüller Theory, vol. I, pp. 685–760. EMS Publishing House, Zürich (2007)

[T10]   Teschner, J.: Quantization of the Hitchin moduli spaces, Liouville theory, and the geometric Langlands correspondence I. Adv. Theor. Math. Phys. **15**, 471–564 (2011)

[TV13]   Teschner, J., Vartanov, G.S.: Supersymmetric gauge theories, quantization of moduli
            spaces of flat connections, and conformal field theory. Adv. Theor. Math. Phys. **19**,
            1–135 (2015)
[ZZ]   Zamolodchikov, A.B., Zamolodchikov, AlB: Structure constants and conformal boot-
            strap in Liouville field theory. Nucl. Phys. **B477**, 577–605 (1996)

Gauge/Vortex Duality and AGT

Mina Aganagic and Shamil Shakirov

Abstract AGT correspondence relates a class of 4d gauge theories in four dimensions to conformal blocks of Liouville CFT. There is a simple proof of the correspondence when the conformal blocks admit a free field representation. In those cases, vortex defects of the gauge theory play a crucial role, extending the correspondence to a triality. This makes use of a duality between 4d gauge theories in a certain background, and the theories on their vortices. The gauge/vortex duality is a physical realization of large N duality of topological string which was conjectured in Dijkgraaf and Vafa (Toda theories, matrix models, topological strings, and N = 2 gauge systems [1]) to provide an explanation for AGT correspondence. This paper is a review of Aganagic et al. (Gauge/Liouville triality [2]), written for the special volume edited by J. Teschner.

1 Introduction

Large N duality plays the central role in understanding dynamics of physical string theory. This duality is inherited by the simpler, topological string with target space a Calabi-Yau three-fold [3–5]. The topological large N duality, like the large N duality of the physical string theory, relates the gauge theory on D-branes to closed topological string on a different background. In the topological string case, the duality is in principle tractable, since topological string is tractable.

In some cases, study of topological string theory is related to studying supersymmetric gauge theory in 4d with $\mathcal{N} = 2$ supersymmetry, see e.g. ([6, 7], [V:14]). It is natural to ask what the large N duality of topological string theory means in gauge theory terms. We will see that the large N duality of topological string becomes a gauge/vortex duality [8–10] which relates a 4d gauge theory in a variant of 2d Ω

A citation of the form [V:x] refers to article number x in this volume.

M. Aganagic (✉) · S. Shakirov
Center for Theoretical Physics, University of California, Berkeley, USA
e-mail: mina@math.berkeley.edu

Department of Mathematics, University of California, Berkeley, USA

© Springer International Publishing Switzerland 2016
J. Teschner (ed.), *New Dualities of Supersymmetric Gauge Theories*,
Mathematical Physics Studies, DOI 10.1007/978-3-319-18769-3_13

background with flux, and the theory living on its vortices.[1] The vortices in the gauge theory play the role of D-branes of the topological string. In fact, the gauge theory duality implies the topological string duality, but not the other way around.

What does this have to do with AGT correspondence [16]? As we will review, [1] conjectured that large N duality of topological string provides a physical explanation for AGT correspondence, under certain conditions: Conformal block should admit free field representation, and Liouville theory should have central charge $c = 1$ to correspond to topological string.

We interpret this purely in the gauge theory language, in the context of the gauge/vortex duality, and show that this leads to a proof of correspondence in a fairly general setting. The partition function of the 4d $\mathcal{N} = 2$ gauge theory associated in [17, 18] to a genus zero Riemann surface with arbitrary number of punctures equals the conformal block of Liouville theory with arbitrary central charge c, on the same surface. The free field representation of conformal blocks implies Coulomb moduli are quantized, but all other parameters remain arbitrary. The crucial role vortices play, extends AGT correspondence to a triality—between the gauge theory, its vortices, and Liouville theory. The striking aspect of this result, which appeared first in [2], is the simplicity of the proof. While in this review we focus on the simplest variant of AGT correspondence, relevant for Liouville theory, same ideas apply for more general Toda CFTs (Liouville theory corresponds to A_1 Toda). The generalization to A_n Toda case can be found in [19].[2]

2 Background

Alday et al. [16] conjectured a correspondence between conformal blocks of Liouville CFT and partition functions of a class of four-dimensional theories, in 4d Ω-background [6]. The 4d theories are conformal field theories with $\mathcal{N} = 2$ supersymmetry defined in [17, 18] (see also [V:2]) in terms of a pair of M5 branes wrapping a Riemann surface C, which we will call the Gaiotto curve. Specifying both the conformal block and the 4d theory $\mathcal{T}_{4d}$ in this class, involves a choice of the curve C with punctures, data at the punctures and pants decomposition. The conjecture is often referred to as 4d/2d correspondence.

2.1 4d Gauge Theory

Let Σ be the Seiberg-Witten curve of $\mathcal{T}_{4d}$,

$$\Sigma : \quad p^2 + \phi^{(2)}(z) = 0. \tag{2.1}$$

---

[1]For early studies leading to [8–10], see [11–15].

[2]Proofs of (some aspects of) AGT correspondence using different ideas appeared in [20–24].

with meromorphic one form $\lambda = p\,dz$. Σ is a double cover of C, z is a local coordinate on C, and $\phi^{(2)}(z)(dz)^2$ is a degree 2 differential on C, whose choice specifies the IR data of the theory (the point on the Coulomb branch). Specifying the UV data of the theory requires fixing the behavior of the Seiberg-Witten differential λ near the punctures.

At a puncture at $z = z_i$, the λ has a pole of order 1, with residues

$$p \sim \pm \frac{\alpha_i}{z - z_i}$$

on the two sheets. These lead to second order poles of $\phi^{(2)}(z)dz^2$. In the gauge theory, α_i's and z_j's are the UV data; the mass parameters and the gauge couplings. Σ also depends on the IR data of the gauge theory, the choice of Coulomb branch moduli. These are associated to the sub-leading behavior of the $\phi^{(2)}(z)$ near the punctures.

Let

$$\mathcal{Z}_{T_{4d}}(\Sigma)$$

be the partition function of the theory, in 4d Ω-background. Given a gauge theory description of T_{4d}, $\mathcal{Z}_{T_{4d}}(\Sigma)$ can be computed using results of Nekrasov in [6] (see also [V:4]). In addition to the geometric parameters entering Σ, $\mathcal{Z}_{T_{4d}}$ depends on

$$\epsilon_1, \quad \epsilon_2,$$

the two parameters of the Ω background [6]. $\mathcal{Z}$ can in principle depend on data beyond the geometry of Σ; different choices of the pants decomposition can lead to different descriptions of the theory with different but related $\mathcal{Z}$'s.

2.2   2d Liouville CFT

The Liouville CFT has a representation in terms of a boson ϕ:

$$S_{Liouv.} = \int dz d\bar{z} \, \sqrt{g} \, [g^{z\bar{z}} \partial_z \phi \, \partial_{\bar{z}} \phi + Q\phi R + e^{2b\phi}].$$

Consider a conformal block on C with insertions of primaries with momenta α_i at points z_i:

$$\mathcal{B}(\alpha, z) = \langle V_{\alpha_0}(z_0) \cdots V_{\alpha_\ell}(z_\ell) V_{\alpha_\infty}(\infty)\rangle,$$

where

$$V_\alpha(z) = \exp\left(-\frac{\alpha}{b}\phi(z)\right)$$

is the vertex operator of a primary with momentum α. Above, Q is the background charge, $Q = b + \frac{1}{b}$; Liouville theory with this background charge has central charge

$c = 1 + 6Q^2$. In addition to momenta and positions of the vertex operators inserted, the conformal block depends on the momenta in the intermediate channels; in denoting the conformal block by $\mathcal{B}(\alpha, z)$ we have suppressed the dependence on the latter.

2.3 The Correspondence

The conjecture of [16] is that the partition function $\mathcal{Z}_{T_{4d}}(\Sigma)$ computes a conformal block of Liouville CFT on C:

$$\mathcal{Z}_{T_{4d}}(\Sigma) = \mathcal{B}(\alpha, z),$$

where b is related to two parameters $\epsilon_{1,2}$ by

$$b = \sqrt{\frac{\epsilon_1}{\epsilon_2}},$$

while the parameters α_i, z_i of Σ map to the corresponding parameters in the conformal block and the Coulomb branch parameters map to the momenta in intermediate channels.

3 AGT and Large N Duality

In [1] Dijkgraaf and Vafa explained the correspondence, in a particular case of the self-dual Ω-background,

$$\epsilon_1 = g_s = -\epsilon_2, \tag{3.1}$$

in terms of a large N duality in topological string theory. The argument of [1] has three parts, which we will now describe. As everywhere else in this review, we will focus on the case when the Gaiotto curve C is genus zero. One can extend the argument more generally [1], as all the ingredients generalize to Σ a double cover of an arbitrary genus g Riemann surface C.

3.1 The Physical and the Topological String

The gauge theory partition function $\mathcal{Z}_{T_{4d}}(\Sigma)$ in the self-dual Ω-background is conjectured in [1] to be the same as the partition function

$$Z(Y_\Sigma)$$

of the topological B-model on a Calabi-Yau manifold Y_Σ, with topological string coupling g_s. The Calabi-Yau Y_Σ is a hyper surface

$$Y_\Sigma : \qquad p^2 + \phi^{(2)}(z) = uv, \qquad (3.2)$$

with holomorphic three-zero form $dudpdz/u$. The geometry of Y_Σ and the Seiberg-Witten curve Σ (2.1) are closely related: the latter is recovered from the former by setting u or v to zero.

This is a consequence of two facts. First, one observes that IIB string theory on Y_Σ is dual to M-theory with an M5 brane wrapping Σ.[3] This gives us another way to obtain the same 4d, $\mathcal{N} = 2$ theory $\mathcal{T}_{4d}$. Second, the partition function of IIB string theory on Y_Σ times the self-dual Ω background is the same as the topological B-model string partition function on Y_Σ [6, 25, 26]. Thus, one can simply identify the physical and the topological string partition functions

$$\mathcal{Z}_{\mathcal{T}_{4d}}(\Sigma) = Z(Y_\Sigma). \qquad (3.3)$$

The power of this observation is that the topological B-model partition function is well defined even when the Nekrasov partition function is not—because for example, the gauge theory lacks a Lagrangian description. It is also important that sometimes one and the same topological string background gives rise to several different Lagrangian descriptions for one and the same theory—for example, $SU(2)^{l-2}$ with four fundamentals vs. $SU(l)$ with $2l$ fundamentals. The former is the theory which is usually associated in the AGT literature to Liouville theory on the sphere with $l + 1$ punctures; the latter is the one that naturally comes out from our approach.

3.2 Large N Duality in Topological String

Next, [1] show that the B-model on Y_Σ has a dual, holographic description, in terms of N topological B-model branes on a different Calabi-Yau, related to Y_Σ, by a geometric transition. Let us first describe the Calabi-Yau that results. Then, we will explain the duality.

3.2.1 A Geometric Transition

By varying Coulomb branch moduli of $\mathcal{T}_{4d}$ we can get the Seiberg-Witten curve Σ to degenerate. Let us call the degenerate curve that results the S-curve:

---

[3]This follows by compactifying M-theory with M5 brane on Σ on a T^2 transverse to the M5 brane. Since the T^2 is transverse to the branes, it does not change the low energy physics. By shrinking one of the cycles of the T^2 first, we go to down to IIA string with an NS5 brane wrapping Σ. T-dualizing on the remaining compact transverse circle, we obtain IIB on Y_Σ.

$$S: \qquad p^2 - (W'(z))^2 = 0. \tag{3.4}$$

Here

$$W'(z) = \sum_{i=0}^{\ell} \frac{\alpha_i}{z - z_i},$$

is determined by keeping the behavior of the Seiberg-Witten differential fixed at the punctures. The S-curve describes the degeneration of the Seiberg-Witten curve to two components, $p \pm W'(z) = 0$. Correspondingly, a single M5 brane wrapping Σ breaks into two branes, wrapping the two components.

The S-curve corresponds to a singular Calabi-Yau Y_S:

$$Y_S: \qquad p^2 - (W'(z))^2 = uv, \tag{3.5}$$

with singularities at u, v, p equal to zero and points in the z-plane where

$$W'(z) = 0.$$

The Calabi-Yau we need is obtained by blowing up the singularities. One can picture this by viewing Y_S as a family of A_1 surfaces, one for each point in the z-plane. At every z there is an S^2 in the A_1 surface whose area is proportional to $|W'(z)|$, The singularity occurs where the S^2 shrinks. After blowing up, we get a family of S^2's of non-zero area, one at each point in the z plane, and all homologous to each other. The minimal area S^2's are where the singularities were—at points in the z plane with $W'(z) = 0$.

The geometric transition trades Y_Σ for the blowup of Y_S. For economy of notations, we will denote Y_S and its blowup in the same way, since their complex structure is the same, given by (3.5).

3.2.2 Large N Duality

The B-model on Y_Σ has a holographic description in terms of B-model on (the blowup of) Y_S with N topological B-model D-branes wrapping the S^2 class. The branes get distributed between the minimal S^2's at points in the z-plane where $W'(z)$ vanishes. This breaks the gauge group from $U(N)$ to $\prod_{i=0}^{\ell} U(N_i)$, with $\sum_i N_i = N$. The Coulomb-branch moduli of Y_Σ get related to t'Hooft couplings $N_i g_s$ in the theory on B-branes. The remaining parameters, α, z and the topological string coupling g_s are the same on both sides. This is the topological B-string version of gauge/gravity duality [4].

The large N duality relates the closed topological string partition function of the B-model on Y_Σ, and thus the partition function $\mathcal{Z}(\Sigma)$, to partition function of the N topological B-branes on (the blowup of) Y_S,

$$Z(Y_\Sigma) = Z(Y_S; N).$$

The right hand side depends not only on the net number of branes, but also how they are split between the different $\mathbb{P}^1$'s.

The partition function of N B-type branes wrapping the S^2 in a Calabi-Yau of the form of (3.5) was found in [4]. It equals

$$\frac{1}{\text{vol}(U(N))} \int d\Phi \, \exp(\text{Tr} W(\Phi)/g_s), \tag{3.6}$$

where $\text{vol}(U(N))$ is the volume of $U(N)$. The integral is a holomorphic integral, over $N \times N$ complex matrices Φ. In evaluating it, one has to pick a contour, ending at a critical point of the potential. In the present case,

$$W(x) = \sum_i \alpha_i \log(x - z_i).$$

Diagonalizing Φ and integrating over the angles, the integral reduces to

$$Z(Y_S; N) = \frac{1}{N!} \int d^N x \prod_{I<J} (x_I - x_J)^2 \prod_{I,i} (x_I - z_i)^{\alpha_i/g_s}. \tag{3.7}$$

Here $N!$ is the order of the Weyl group that remains as a group of gauge symmetries.

The claim is that large N expansion of the integral equals topological B-model partition function on (3.2). At the level of planar diagrams this can be seen as follows. In the matrix integral, define an operator

$$\partial\phi(z) = W'(z) + g_s \sum_I \frac{1}{z - x_I}, \tag{3.8}$$

where x_I are the eigenvalues of Φ. The expectation value of

$$T(z) = (\partial\phi)^2$$

computed in the matrix theory captures the geometry of the underlying Riemann surface by identifying $\phi^{(2)}(z)$ in (2.1) with

$$\phi^{(2)}(z) = \langle T(z)\rangle.$$

There are two limits in which a classical geometry emerges from this. First, by simply sending g_s to zero we recover the S-curve, since then $\langle T(z)\rangle = (W')^2$. But, there is also a new classical geometry that emerges at large N. Letting N_i's go to infinity, keeping $N_i g_s$ fixed we get

$$\langle T(z)\rangle \sim (W'(z))^2 + f(z),$$

with

$$f(z) = \left\langle g_s \sum_I \frac{W'(z) - W'(x_I)}{z - x_I} \right\rangle.$$

From the form of the potential $W(z)$, it follows that $f(z)$ has the form

$$f(x) = \sum_i \frac{\mu_i}{x - z_i}$$

with at most single poles. Thus, the branes deform the geometry of the Calabi-Yau we started with. The resulting Calabi-Yau is exactly of the form Y_Σ (3.2), corresponding to the Seiberg-Witten curve Σ in (2.1) at a generic point of its moduli space.

The large N duality is expected to hold order by order in the $1/N$ expansion; we just gave evidence it holds in the planar limit (the full proof of the correspondence in the planar limit is easy to give along these lines, see [4]). The good variable in the large N limit turns out to be the chiral operator $\phi(z)$ we defined in (3.8). The field $\phi(z)$, is in fact the string field of the B-model.

The B-model string field theory, called Kodaira-Spencer theory of gravity, was constructed in [27], capturing variations of complex structure. For Calabi-Yau manifolds of the form (3.2) the Kodaira-Spencer theory becomes a two dimensional theory on the curve Σ. The theory describes variations of complex structures of Y_Σ, so the Kodaira-Spencer field can be identified with fluctuations of the holomorphic $(3, 0)$ form of the Calabi-Yau. For Y_Σ fluctuations of the $(3, 0)$ form are equivalent to fluctuations of the meromorphic $(1, 0)$ form on Σ:

$$\delta\lambda = \delta p\,dz = \partial\phi\,dz.$$

The Kodaira-Spencer field is a chiral boson ϕ which lives on Σ. When Σ is a double cover of a curve C, a single boson on Σ is really a pair of bosons ϕ_1, ϕ_2 on C, one corresponding to each sheet. The field ϕ that arises in the matrix model in (3.8) can be thought of as off diagonal combination of the two. The diagonal combination is a center of mass degree of freedom and decouples from the dynamics of the branes.[4]

3.3 Topological D-Branes and Liouville Correlators

To complete the argument, [1] observe that the B-brane partition function $Z(Y_S; N)$ equals the Liouville correlator at $c = 1$, when written in the free-field or Dotsenko-Fateev representation [28, 29],

---

[4]The full topological string partition function in the presence of branes is given by the matrix integral in (3.6) and (3.7), describing open strings, times a purely closed topological string partition function of Y_S. This will be relevant later on.

$$Z(Y_S; N) \;=\; \mathcal{B}(\alpha/g_s, z; N)|_{c=1}. \tag{3.9}$$

One treats the Liouville potential as a perturbation and computes the correlator in the free boson CFT

$$\mathcal{B}(\alpha, z; N) = \langle V_{\alpha_1}(z_1)\dots V_{\alpha_\ell}(z_\ell) V_{\alpha_\infty}(\infty) \oint dx_1 S(x_1)\cdots \oint dx_N S(x_N)\rangle_0, \tag{3.10}$$

where we took the chiral half. Here, $S(z)$ is the screening charge

$$S(z) = e^{2b\phi(z)},$$

whose insertions come from bringing down powers of the Liouville potential. It follows that (3.10) vanishes unless

$$\frac{\alpha_\infty}{b} + \sum_{i=0}^{\ell} \frac{\alpha_i}{b} = 2bN + RQ,$$

constraining the net $U(1)$ charge of the vertex operator insertions to be the number of screening charge integrals. This constraint can be found directly from the path integral, by integrating over the zero modes of the bosons [28–30]. We will place a vertex operator at infinity of the x plane, and then the equation determines the momentum of the operator at infinity in terms of the momenta of the $\ell+1$ remaining vertex operators at finite points and numbers of screening charge integrals.

An integral expression for the expectation value of the correlator in (3.10) is easy to obtain, for example, by using the free boson mode expansion

$$b\phi(z) = \phi_0 + h_0 \log z + \sum_{k\neq 0} h_k \frac{z^{-k}}{k},$$

where ϕ_0 is a constant, and h_m satisfy the standard algebra

$$[h_k, h_m] = \frac{-b^2}{2}\, k\, \delta_{k+m,0} \tag{3.11}$$

where $k, m \in \mathbb{Z}$. From this one obtains the two point functions:

$$\langle V_\alpha(z) V_{\alpha'}(z')\rangle = (z - z')^{\frac{-\alpha\alpha'}{2b^2}},$$
$$\langle V_\alpha(z) S(z')\rangle = (z - z')^{\alpha},$$
$$\langle S(z) S(z')\rangle = (z - z')^{-2b^2}.$$

The final result is that (3.10) equals

$$\mathcal{B}(\alpha, z; N) = \frac{r}{N!} \int \prod d^N x \prod_{i,I} (x_I - z_i)^{\alpha_i} \prod_{J \leq I} (x_I - x_J)^{-2b^2},$$

where the integrals are over the position of screening charge insertions and

$$r = \prod_{i,j} (z_i - z_j)^{\frac{-\alpha_i \alpha_j}{2b^2}}$$

is a constant, independent on the integration variables. This is the free-field β-ensemble (with $\beta = -b^2$) reviewed in [V:5].

Setting $\epsilon_1 = -\epsilon_2$ (taking $b^2 = -1$ in Liouville CFT) and rescaling α by g_s, it follows immediately that the free field expression for the conformal block $\mathcal{B}(\alpha/g_s, z; N)$ agrees with the partition function $Z(S; N)$ of B-branes in topological string on Y_S as we claimed in (3.9). Moreover, in the large N limit, the holomorphic part of the Liouville field $\phi(z)$ can be identified with the matrix model operator (3.8). This completes the argument of [1].

3.4 Discussion

The AGT conjecture, for $\epsilon_1 + \epsilon_2 = 0$ can thus be understood as a consequence of a triality relating the closed B-model on Y_Σ, the holographic dual theory of B-branes on the resolution of Y_S and the DF conformal blocks. The first two are conjectured to be related by large N duality[5] in topological string theory, the latter two by the fact that the partition function of B-branes equals the DF block:

$$Z(Y_\Sigma) \stackrel{\text{Large } N}{=} Z(Y_S; N) = \mathcal{B}(\alpha/g_s, z; N)|_{c=1}. \tag{3.12}$$

We also used the embedding of topological string into superstring theory, which implies that the topological string partition function $Z(Y_\Sigma)$ is the same as the physical partition function $\mathcal{Z}_{\mathcal{T}_{4d}}(\Sigma)$.

While this gives an explanation for the AGT correspondence in physical terms, it is by no means a proof: while the partition function of B-branes is manifestly equal to the Liouville conformal block in free field representation, the large N duality is still a conjecture. The exact partition function of the B-model on Y_Σ is not known, so one can only attempt a proof, order by order in the genus expansion. In addition,

---

[5]It may be useful to summarize what the large N asymptotic regime is, on each side of the correspondence. On the B-model side, it is sending g_s to zero while keeping the combination $N g_s$ fixed. On the gauge theory side, it is sending $\epsilon_1 = -\epsilon_2$ to zero while keeping the Coulomb parameters fixed. On the Liouville side, it is sending all the momenta as well as the number N of screening insertions to infinity, while keeping their ratios fixed.

there is a string theory argument, but no proof, that the partition function of the gauge theory $\mathcal{Z}_{\mathcal{T}_{4d}}(\Sigma)$ and topological string partition function $Z(Y_\Sigma)$ agree.

Thirdly, from the perspective of the 4d gauge theory, it is very natural to consider the partition function on general Ω-background, depending on arbitrary ϵ_1, ϵ_2. Topological string on the other hand requires self-dual background, so the argument of [1] can not be extended in this case.[6] In [1], it was suggested to formulate the refinement at the level of B-model string field theory. This remains to be developed better: refinement exists for any Calabi-Yau of the form $F(p, z) = uv$; the predictions from a naive implementation of this idea work for some, but not all choices of $F(p, z)$.

In the rest of the review, we will explain how to solve the last problem, and as it turns out the first two as problem as well, by following a different route.

The relation between topological string and superstring theory suggests one may be able to reformulate [1] in string theory language, replacing topological string branes by branes in string or M-theory. While topological string captures the $\epsilon_1 + \epsilon_2 = 0$ case only, the full superstring or M-theory partition function makes sense for any ϵ_1, ϵ_2. In fact, will will do something simpler yet: We will formulate the *gauge theory analogue of* [1] for any ϵ_1, ϵ_2. We will see that this approach is powerful—in fact it leads to a rigorous yet simple proof that the gauge theory partition function $\mathcal{Z}_{\mathcal{T}_{4d}}(\Sigma)$ agrees with the free field Liouville conformal block for C a sphere with arbitrary number of punctures.

The triality of relations between the 4d gauge theory, its vortices, and Liouville conformal blocks which admit free field representation implies AGT correspondence, however it stops short of the most general case. The restriction to blocks that admit free field representation means, from the 4d perspective, that the Coulomb moduli are quantized to be—arbitrary—integers, which get related to vortex charges on one hand, and numbers of screening charge integrals on the other.

4 Gauge/Vortex Duality

Translated to gauge theory language, the large N duality of topological string theory becomes a duality between the 4d $\mathcal{N} = 2$ gauge theory $\mathcal{T}_{4d}$ and the 2d $\mathcal{N} = (2, 2)$ theory on its vortices; we will denote the later theory $\mathcal{V}_{2d}$. Observations of relations between the two theories go back to [11–14]. Recently [8, 9] proposed that the two theories are dual—indeed this is the "other" 2d/4d relation. On the face of it, the statement is strange at best: to begin with, not even the dimensions of the 4d and the 2d theories match.

In this section we will show that, placed in a certain background, the 4d and the 2d theory describe the same physics, and thus there is good reason why their partition

---

[6]For general $\epsilon_{1,2}$ the background does not simply decouple into a product of a Calabi-Yau manifold times the Ω background where the gauge theory lives. Turning on arbitrary Ω background requires the theory to have an $U(1) \in SU(2)_R$ R-symmetry to preserve supersymmetry. This requires the target Calabi-Yau manifold to admit a $U(1)$ action; this $U(1)$ action is used in constructing the background.

functions agree [10]. The large N duality of [1, 4] becomes a duality between two $d = 2$, $\mathcal{N} = (2, 2)$ theories: the 4d gauge theory $\mathcal{T}_{4d}$ we started with, in a variant of 2d Ω-background with vortex flux turned on, and the 2d theory $\mathcal{V}_{2d}$ on its vortices.

4.1 Higgs to Coulomb Phase Transition and Vortices

In gauge theory language, the geometric transition that relates B-model on a Calabi-Yau Y_Σ, first to a singular Calabi-Yau Y_S and then to a blowup of Y_S, is a Coulomb to Higgs phase transition. This follows from embedding of the B-model into IIB superstring on a Calabi-Yau, and the relation between the string theory and the gauge theory which arises in its low energy limit [31]. The same transition, in the language of M5 branes corresponds to degenerating a single M5 brane wrapping Σ, to a pair of M5 branes wrapping two Riemann surfaces $p \pm W'(z) = 0$ that the S-curve consists of, and then separating these in the transverse directions (these are $x^{7,8,9}$ directions in the language of [32]).

The geometric transition becomes a topological string duality, as opposed to a phase transition, by adding N B-branes on the S^2 in the blowup of Y_S. In terms of IIB string, the N B-branes on the S^2 are N D3 branes wrapping the S^2 and filling 2 of the 4 space-time directions. In terms of M5 branes, the vortices are M2 branes stretching between the M5 brane wrapping $p - W' = 0$ and the one wrapping $p + W' = 0$. In the gauge theory on the Higgs branch, N branes of string/M-theory become N BPS vortices, as explained in [33, 34] and [13, 14].[7]

The vortices in question are non-abelian generalization of Nielsen-Olesen vortex solutions whose BPS tension is set by the value of the FI parameters. These were constructed explicitly in [13, 14]. The net BPS charge of the vortex is $N = \int \mathrm{Tr} F$ where F is the field strength of the corresponding gauge group and the integral is taken in the 2 directions transverse to the vortex.[8]

4.2 Gauge/Vortex Duality

Consider subjecting the 4d $\mathcal{N} = 2$ theory $\mathcal{T}_{4d}$ to a *two*-dimensional Ω-background in the two directions transverse to the vortex. We set $\epsilon_1 = \hbar$ to zero momentarily since the duality we want to claim holds for any $\hbar$. This is the Nekrasov-Shatashvili

---

[7]One should not confuse the vortices here with surface operators in the gauge theory, studied for example in [35–37]. The surface operators are solutions on the Coulomb branch, with infinite tension. From the M5 brane perspective, surface operators are semi-infinite M2 branes ending on M5's.

[8]Usually, the gauge theories on M5 branes wrapping Riemann surfaces are said to be of special unitary type, rather than unitary type. There is no contradiction; the $U(1)$ centers of the gauge groups that arise on branes are typically massive by Green-Schwarz mechanism. This does not affect the BPS tension of the solutions, see e.g. discussion in [38].

background studied in [39]. The 2d Ω-background depends on the one remaining parameter, $\epsilon = \epsilon_2$. (The equivalence of two theories is a stronger statement that the equivalence of their partition functions. The later assumes a specific background, while the former implies equivalence for any background. We will let $\hbar$ be arbitrary once we become interested in the partition functions, as opposed to the theories themselves.)

As in [39], we view this partial Ω-background as a kind of compactification: it results in a 2d theory with infinitely many massive modes, with masses spaced in multiples of ϵ. The background also preserves only 4 out of the 8 supercharges. Under conditions which we will spell out momentarily, the effective 2d $\mathcal{N} = (2, 2)$ theory that we get is equivalent to the theory on its vortices. The condition that is clearly necessary is that we turn on vortex flux. We assume it is also sufficient.

The vortex charge is $\int_D F_i = N_i$ where i labels a $U(1)$ gauge field in the IR, and F_i is the corresponding field strength. Here, D is the cigar, the part of the 4d space time with 2d Ω deformation on it. It is parameterized by one complex coordinate, which we will call w. Without the Ω deformation, turning on $N_i \neq 0$ would be introducing singularities in space-time which one would interpret in terms of surface operator insertions [35]. In Ω background, one can turn on the vortex flux without inserting additional operators—in fact, the only effect of the flux is to shift the effective values of the Coulomb branch moduli. Let us explain this in some detail.

In the Ω background, D gets rotated with rotation parameter ϵ, in such a way that the origin is fixed. The best way to think of the theory that results [39, 40] is in terms of deleting the fixed point of the rotation, and implementing a suitable boundary condition. Because the disk is non-compact, we really need two boundary conditions: one at the origin of the w plane and one at infinity. Turning on flux simply changes the boundary condition we impose at the origin. Without vortices, one imposes the boundary condition [40] that involves setting $A_{i,w} = 0$, where $A_{i,w}$ is the connection of ith $U(1)$ gauge field along D. With N_i units of vortex flux on D, we need instead $A_{i,w} = N_i/w$.

In the Ω-background, the 4d theory in the presence of N_i units of vortex flux $A_{i,w} = N_i/w$ and with Coulomb branch scalar a_i turned on is equivalent to studying the theory without vortices, at $A_{i,w} = 0$, but with a_i shifted by

$$a_i \quad \rightarrow \quad a_i + N_i\epsilon.$$

This comes about because in the Ω background, a_i always appears in the combination [40]

$$a_i + \epsilon w D_{i,w},$$

where $D_w = \partial_w + A_{i,w}$ is the covariant derivative along the w-plane traverse to the vortex. Thus, in the Ω background, at the level of F-terms, turning on vortex flux is indistinguishable from the shift the effective values of the Coulomb branch moduli.[9]

---

[9]In [40] one proves that any flat gauge field on the punctured disk preserves supersymmetry of the Ω background.

The 4d theory placed in 2d Ω-background, with vortex flux turned on has an effective description studied in [39, 40] in terms of the $2d$ theory with $\mathcal{N} = (2, 2)$ supersymmetry with massive modes integrated out. The $(2, 2)$ theory has a non-zero superpotential $\mathcal{W}(a, \epsilon; N) = \mathcal{W}_{NS}(a_i + N_i \epsilon, \epsilon)$, where $\mathcal{W}_{NS}(a_i, \epsilon)$ is the effective superpotential derived in [39], and the shift by $N_i \epsilon$ is due to the flux we turned on. The critical points of the superpotential correspond to supersymmetric vacua of the theory. In the A-type quantization, considered in [39], the vacua are at $\exp(\partial_{a_i} \mathcal{W}_{NS}/\epsilon) = 1$ or, equivalently, at $a_{D,i}/\epsilon = \partial_{a_i} \mathcal{W}_{NS}/\epsilon \in \mathbb{Z}$. In the B-type quantization, they are at $a_i/\epsilon \in \mathbb{Z}$ [8, 9, 41]. Choosing $a_i = 0$, for all i is the vacuum at the intersection of the Higgs and the Coulomb branch. Choosing $a_i = N_i \epsilon$ corresponds to putting the theory at the root of the Higgs branch—but in the background of N_i units of flux.[10]

There is a *second description* of the same system. If we place the theory at the root of the Higgs branch, the 4d theory has vortex solutions of charge N_i even without the Ω-deformation. These are the non-abelian Nielsen-Olsen vortices of [13, 14]. We get a second 2d theory with $\mathcal{N} = (2, 2)$ supersymmetry—this is the theory on vortices themselves. In the theory on the vortex, the only effect of the Ω-deformation is to give the scalar, parameterizing the position of the vortex in the w-plane, twisted mass ϵ. From this perspective, turning on ϵ is necessary since it removes a flat direction (position of vortices in the trasverse space).

Similarity of the two theories at the level of the BPS spectrum was observed in [11–15]. For a class of theories, this duality was first proposed in [8, 9], motivated by study of integrability. The physical explanation for gauge/vortex duality we provided implies the duality should be general, and carry over to many other systems.[11]

4.3 Going up a Dimension

The duality between $\mathcal{T}_{4d}$, in the variant of the 2d Ω-background we described above, and $\mathcal{V}_{2d}$ lifts to a duality in one higher dimension, between a pair of theories, $\mathcal{T}_{5d}$ and $\mathcal{V}_{3d}$, compactified on a circle. We will prove the stronger, higher dimensional version, of the duality. $\mathcal{T}_{4d}$ lifts to a five-dimensional theory $\mathcal{T}_{5d}$ with $\mathcal{N} = 1$ supersymmetry. From 4d perspective, one gets a theory with infinitely many Kaluza-Klein modes. One can view this theory as a deformation of $\mathcal{T}_{4d}$, depending on one parameter, the radius R of the circle. Note that $\mathcal{T}_{5d}$ is not simply placed in a product of 2d Ω-background times a circle—rather the background is a circle fibration

$$(D \times S^1)_t,$$

where as one goes around the S^1 D rotates by t, sending $w \to wt$.[12] Similarly, the 2d theory on the vortex, $\mathcal{V}_{2d}$ lifts to a 3d theory $\mathcal{V}_{3d}$, on a circle of the same radius.

---

[10]We thank Cumrun Vafa for discussion relating to this point.

[11]See [42] for a highly nontrivial example.

[12]This 3d background was used in [6, 25, 40, 43] as a natural path to defining the 2d Ω-background. For a review see [44].

The claim is that the two $d = 2$, $\mathcal{N} = (2, 2)$ theories we get in this way are dual, where the duality holds at least at the level of F-type terms. In the limit when R goes to zero, the KK tower is removed, and we recover the theories we started with.

In the next section we will prove the duality by showing that partition functions of the two theories agree. When we compute the partition function of the 5d theory, we submit it to the full Nekrasov background depending on both ϵ and $\hbar$. This is the background

$$(D \times \mathbb{C} \times S^1)_{q,t}, \tag{4.1}$$

where as one goes around the S^1, we simultaneously rotate D by $t = e^{R\epsilon}$, and $\mathbb{C}$ by $q^{-1} = e^{-R\hbar}$. In the 3d theory on vortices, ϵ is a twisted mass, but $\hbar$ is a parameter of the Ω background along the vortex world volume. The background for $\mathcal{V}_{3d}$ is fixed once we choose the background for $\mathcal{T}_{5d}$, simply by the 5d origin of the vortices. $\mathcal{V}_{3d}$ is compactified on

$$(\mathbb{C} \times S^1)_q. \tag{4.2}$$

As we go around the S^1, $\mathbb{C}$ rotates by q^{-1}, and we turn on a Wilson line t for a global symmetry rotating the adjoint scalar (and thus giving it mass ϵ).

5 Building up Triality

When $\mathcal{T}_{5d}$ is a lift of the M5 brane theory of Sect. 2 to a one higher dimensional theory on a circle of radius R, the gauge/vortex duality extends to a triality. The triality is a correspondence between the 5d gauge theory $\mathcal{T}_{5d}$, the 3d theory on its vortices $\mathcal{V}_{3d}$, both on a circle of radius R and a q-deformation of Liouville conformal block. As R goes to zero, the q deformation goes away and we recover the conformal blocks of Liouville. The q-deformation of the Virasoro algebra was defined in [45, 46], and studied further and as well as extended to W-algebras in [47].

The triality comes about because the partition function of the vortex theory $\mathcal{V}_{3d}$ will turn out to equal the q-deformed Liouville conformal block,

$$\mathcal{Z}_{\mathcal{V}_{3d}} = \mathcal{B}_q, \tag{5.1}$$

analogously to the way the partition function of topological D-branes was the same as the conformal block of Liouville at $b^2 = -1$. The relation between $\mathcal{T}_{5d}$ and $\mathcal{V}_{3d}$ is the gauge/vortex duality. The duality implies that their partition functions are equal,

$$\mathcal{Z}_{\mathcal{T}_{5d}} = \mathcal{Z}_{\mathcal{V}_{3d}}. \tag{5.2}$$

The left hand side is computed on (4.1) and the right hand side, by restriction, on (4.2). Thus, combining the two relations, we get a relation between R-deformation of the partition function of $\mathcal{T}_{4d}$ and the q-deformation of the Liouville conformal block,

$$\mathcal{Z}_{\mathcal{T}_{5d}} = \mathcal{Z}_{\mathcal{V}_{3d}} = \mathcal{B}_q. \tag{5.3}$$

In a limit, both deformations go away and we recover the relation between a partition function of the 4d, $\mathcal{N} = 2$ theory $\mathcal{T}_{4d}$ and the ordinary Liouville conformal block $\mathcal{B}$. We will prove this for the case when C is a sphere with any number of punctures. The equality in (5.2), as we anticipated on physical grounds, holds for special values of Coulomb branch moduli—those corresponding to placing the 5d theory at a point where the Higgs branch and Coulomb branches meet, and turning on fluxes. By taking the large flux limit, where N_i goes to infinity, ϵ goes to zero keeping their product $N_i\epsilon$ fixed, all points of the Coulomb branch and arbitrary conformal blocks get probed in this way.

In the rest of the section we will spell out the details of the theories involved, and their partition functions. Then, in the next section, we will prove their equivalence.

5.1 The 5d Gauge Theory $\mathcal{T}_{5d}$

The 5d $\mathcal{N} = 1$ theory $\mathcal{T}_{5d}$ per definition reduces to, as we send R to zero, the 4d theory $\mathcal{T}_{4d}$ arising from a pair of M5 branes wrapping a genus zero curve C with $\ell + 2$ punctures.

The $\mathcal{T}_{5d}$ theory turns out to be very simple: at low energies it is described by a $U(\ell)$ gauge theory with 2ℓ hyper-multiplets: ℓ hypermultiplets in fundamental representation, ℓ in anti-fundamental, and 5d Chern-Simons level zero.[13] Except for $\ell = 2$, the $U(\ell)$ gauge theory theory is different from the generalized quiver of [17]. This is nothing exotic: there are different ways to take R to zero limit, and different limits can indeed result in inequivalent theories. At finite R, the theory we get is unique, but with possibly more than one description.

The Coulomb branch of the 4d theory $\mathcal{T}_{4d}$ is described by a single M5 brane wrapping the 4d Seiberg-Witten curve (2.1). The Seiberg-Witten curve of $\mathcal{T}_{5d}$ compactified on a circle can be written as

$$\Sigma: \qquad Q_+(e^x)e^p + P(e^x) + Q_-(e^x)e^{-p} = 0, \tag{5.4}$$

with the meromorphic one form equal to $\lambda = pdx$ (see, e.g. [50]). We will denote both the 4d and the 5d Seiberg Witten curves by the same letter, Σ even though the curves are inequivalent; it should be clear from the context which one is meant. Here, $Q_\pm$ are polynomials of degree ℓ in e^x,

---

[13] At very short distances there is a UV fixed point corresponding to it, which is a strongly coupled theory, accessible via its string or M-theory embedding [48, 49].

$$Q_\pm(e^x) = e^{\pm\zeta/2} \prod_{i=1}^{\ell} (1 - e^x/f_{\pm,i}),$$

and $P(x)$ is a polynomial of degree ℓ in x. At points where the Higgs and the Coulomb branch meet, Σ degenerates to:

$$S \ : \ (Q_+(e^x)e^p - Q_-(e^x))(e^{-p} - 1) = 0. \tag{5.5}$$

The 5d Seiberg-Witten curve in (5.4) and the S-curve in (5.5) reduce to the 4d ones in (2.1), and (3.4), by taking the R to zero limit. The limit one needs corresponds to keeping ζ/R and p/R fixed and taking

$$f_{+,i} = z_i, \quad f_{-,i} = z_i \, q^{\alpha_i}. \tag{5.6}$$

Finally, one defines $z = e^x$, and replaces p by pz to get (3.4), the curve with its canonical one form $\lambda = pdz$. Note that one of the punctures we get is automatically placed at $z = 0$.[14]

5.1.1   Partition Function in Ω-background

The 5d Ω-background is defined as a twisted product

$$(\mathbb{C} \times \mathbb{C} \times S^1)_{q,t}, \tag{5.7}$$

where as, one goes around the S^1, one rotates the two complex planes by $q = \exp(R\epsilon_1)$ and $t^{-1} = \exp(R\epsilon_2)$ (the first copy of $\mathbb{C}$ is what we called D before). These are paired together with the 5d $U(1)_R \subset SU(2)_R$ symmetry twist by tq^{-1}, to preserve supersymmetry. The 5d gauge theory partition function in this background is the trace

$$\mathcal{Z}_{T_{5d}}(\Sigma) = \mathrm{Tr}(-1)^F \mathbf{g}_{5d}, \tag{5.8}$$

corresponding to looping around the circle in (5.7). Insertion of $(-1)^F$ turns the partition function of the theory to a supersymmetric partition function. One imposes periodic identifications with a twist by $\mathbf{g}$ where $\mathbf{g}$ is a product of simultaneous rotations: the space-time rotations by q and t^{-1}, the R-symmetry twist, flavor symmetry rotations $f_{i,\pm} = \exp(-Rm_{i,\pm})$, and gauge rotation by $e_i = \exp(Ra_i)$ for the i'th $U(1)$ factor. The latter has the same effect as turning on a Coulomb-branch

---

[14]The second four-dimensional limit gives the 4d $\mathcal{N} = 2$ $U(\ell)$ gauge theory with 2ℓ fundamental hypermultiplets by [17, 32]. In the Seiberg-Witten curve, one writes f_i as $f_i = e^{R\mu_i}$, and takes R to zero keeping x/R, $e^p R$, $e^\zeta R$ and the μ's fixed in the limit. The effect of this is that the 4d curve has the same form as (5.4), but with Q and P replaced by polynomials of the same degree, but in x, rather than e^x.

modulus a_i (see [44] for a review). The partition function of $\mathcal{T}_{5d}$ in this background is computed in [6], using localization. The partition function is a sum

$$\mathcal{Z}_{\mathcal{T}_{5d}}(\Sigma) = r_{5d} \sum_{\vec{R}} I_{\vec{R}}^{5d}, \tag{5.9}$$

over ℓ-touples of 2d partitions

$$\vec{R} = (R_1, \ldots, R_\ell),$$

labeling fixed points in the instanton moduli space. The instanton charge is the net number of boxes $|\vec{R}|$ in the R's. The coefficient r_{5d} contains the perturbative and the one loop contribution to the partition function.

The contribution

$$I_{\vec{R}}^{5d} = q^{\zeta|\vec{R}|} z_{V,\vec{R}} \times z_{H,\vec{R}} \times z_{H^\dagger,\vec{R}}$$

of each fixed point is a product over the contributions of the $U(\ell)$ vector multiplets, the ℓ fundamental and anti-fundamental hypermultiplets $H, H^\dagger$ in $\mathcal{T}_{5d}$. The instanton counting parameter, related to the gauge coupling of the theory, is q^ζ. I^{5d} depends on ℓ Coulomb branch moduli encoded in $\vec{e}$, and the 2ℓ parameters $\vec{f}$ related to the masses of the 2ℓ hypermultiplets. The vector multiplet contributes

$$z_{V,\vec{R}} = \prod_{1 \leq a,b \leq \ell} [N_{R_a R_b}(e_a/e_b)]^{-1}.$$

The ℓ fundamental hypermultiplets contribute

$$z_{H,\vec{R}} = \prod_{1 \leq a \leq \ell} \prod_{1 \leq b \leq \ell} N_{\varnothing R_b}(v f_a/e_b),$$

and the ℓ anti-fundamentals give

$$z_{H^\dagger,\vec{R}} = \prod_{1 \leq a \leq \ell} \prod_{1 \leq b \leq \ell} N_{R_a \varnothing}(v e_a/f_{b+\ell}).$$

The basic building block is the Nekrasov function

$$N_{RP}(Q) = \prod_{i=1}^{\infty} \prod_{j=1}^{\infty} \frac{\varphi(Q q^{R_i - P_j} t^{j-i+1})}{\varphi(Q q^{R_i - P_j} t^{j-i})} \frac{\varphi(Q t^{j-i})}{\varphi(Q t^{j-i+1})},$$

with $\varphi(x) = \prod_{n=0}^{\infty} (1 - q^n x)$ being the quantum dilogarithm [2, 51]. Furthermore, $T_R = (-1)^{|R|} q^{\|R\|/2} t^{-\|R^t\|/2}$, and $v = (q/t)^{1/2}$ as before (we use the conventions of [52]). In what follows, it is good to keep in mind that there is no essential distinction between the fundamental and anti-fundamental hypermultiplets.[15] In keeping with this, it is natural to think of all the 2ℓ matter multiplets at the same footing, and write the partition function, say, in terms of the fundamentals alone, whose masses run over 2ℓ values, f_a, $f_{\ell+a}$, with $a = 1, \ldots, \ell$.

5.2 The Vortex Theory $\mathcal{V}_{3d}$

The non-abelian generalization of Nielsen-Olesen vortices was found in [13, 14]. In particular, starting with a bulk non-abelian gauge theory like $\mathcal{T}_{5d}$, with 8 supercharges, $U(\ell)$ gauge symmetry and 2ℓ hypermultiplets in fundamental representation, they constructed the theories living on its half BPS vortex solutions. The theory on charge N vortices is very simple: it is a $U(N)$ gauge theory with 4 supercharges, with ℓ chiral multiplets in fundamental, and ℓ in anti-fundamental representation, as well as a chiral multiplet in the adjoint representation. The theory has a $U(\ell) \times U(\ell)$ flavor symmetry rotating the chiral and anti-chiral multiplets separately. This symmetry prevents their superpotential couplings. Since $\mathcal{T}_{5d}$ is five dimensional, the theory on its vortices is three dimensional $\mathcal{N} = 2$ theory, which we will denote $\mathcal{V}_{3d}$. Presence of the 2d Ω background transverse to the vortex gives the adjoint chiral field twisted mass ϵ. In addition, the theory is compactified on a circle of radius R. The masses of 2ℓ hypermultiplets of $\mathcal{T}_{5d}$ get related to the 2ℓ twisted masses of the chiral multiplets in $\mathcal{V}_{3d}$. We will see the precise relation momentarily.

5.2.1 Partition Function in Ω-Background

We compactify $\mathcal{V}_{3d}$ on the 3d Ω background:

$$(\mathbb{C} \times S^1)_q.$$

As we go around the S^1 we simultaneously rotate the complex plane by q and twist by the $U(1)_R$-symmetry, to preserve supersymmetry. The partition function of the theory in this background in computes the index

$$\mathcal{Z}_{\mathcal{V}_{3d}}(S; N) = \mathrm{Tr}(-1)^F \mathbf{g}_{3d}, \tag{5.10}$$

---

[15]By varying the Coulomb branch and the mass parameters, the real mass m of the 5d hypermultiplet can go through zero. This exchanges the fundamental hypermultiplet of mass m for an anti-fundamental of mass $-m$, while at the same time the 5d Chern-Simons level jumps by 1 [53]. A relation between the anti-fundamental and the fundamental hypermultiplet contributions to the partition function reflects this, see [2] for details.

where $\mathbf{g}_{3d}$ is a product of space-time rotation by q, an $U(1)_R$ symmetry transformation by q^{-1}, as well as the global symmetry rotation by t. The partition function of the theory can be computed by first viewing the $U(N)$ symmetry as a global symmetry: in this case, since the theory is not gauged, and due to the 3d Ω background, the index in (5.10) is simply a product of contributions from matter fields and the W-bosons, all depending on the N Coulomb branch parameters x_I.

The contribution of the flavor in the fundamental representation is

$$\Phi_F(x) = \prod_{1 \leq I \leq N} \frac{\varphi(e^{Rx_I - Rm_-})}{\varphi(e^{Rx_I - Rm_+})}, \tag{5.11}$$

where $m_\pm$ are the twisted masses. The right hand side is written in terms of Faddeev-Kashaev quantum dilogarithms [2, 51],

$$\varphi(z) = \prod_{n=0}^{\infty}(1 - q^n z).$$

There are different ways to show this, for example, one can reduce the 3d theory down to quantum mechanics on the circle and integrate out a tower of massive states. Alternatively, the index can be obtained by counting holomorphic functions on the target space of the quantum mechanics, see [44]. We can think of the flavor in the fundamental representation in one of two equivalent ways: it is a pair of $\mathcal{N} = 2$ chiral multiplets, one in the fundamental and the other in the anti-fundamental representation. Alternatively, it contains a chiral multiplet and an anti-chiral multiplet, but both transform in the fundamental representation. The above way of writing $\Phi_F(x)$ is adapted to the second viewpoint.

The $\mathcal{N} = 4$ vector multiplet, the adjoint chiral field and the W-bosons, give a universal contribution for any $U(N)$ gauge group:

$$\Phi_V(x) = \prod_{1 \leq I < J \leq N} \frac{\varphi(e^{Rx_I - Rx_J})}{\varphi(t\, e^{Rx_I - Rx_J})}. \tag{5.12}$$

The numerator is due to the W-bosons, and the denominator to the adjoint of mass scalar of mass ϵ. Finally, since the gauge group is gauged, we integrate over x's. This simply projects to gauge invariant functions of the moduli space,

$$\mathcal{Z}_{V_{3d}}(S; N) = \frac{1}{N!} \int d^N x \; \Phi_V(x) \prod_{a=1}^{\ell} \Phi_{F_a}(x) \, e^{\zeta \operatorname{Tr} x/\hbar}. \tag{5.13}$$

The integrand is a product including all contributions of the massive BPS particles in the theory, the W bosons, flavors Φ's, and the adjoint. The exponent contains the classical terms, the FI parameter ζ, and the Chern-Simons level k which is zero in our case. If the gauge symmetry were just a global symmetry, x's would have been parameters of the theory and the partition function of the theory would have been the integrand. Gauging the $U(N)$ symmetry corresponds to simply integrating[16] over x.

We need to determine the contour of integration to fully specify the path integral. The choice of a contour in the matrix model corresponds to the choice of boundary conditions at infinity in the space where the gauge theory lives [65]. At infinity, fields have to approach a vacuum of the theory. For small q and t, the vacua are the critical points of

$$W(x) = \sum_{a=1}^{\ell} \log \frac{\varphi(e^{Rx-Rm_{-,a}})}{\varphi(e^{Rx-Rm_{+,a}})}.$$

There are ℓ vacua of $W(x)$ both before and after the R-deformation. Splitting the N eigenvalues so that N_a of them approach the ath critical point, we break the gauge group,

$$U(N) \quad \rightarrow \quad U(N_1) \times \cdots \times U(N_\ell).$$

We can think of all the quantities appearing in the potential as real; then the integration is along the real x axis. To fully specify the contour of integration, we need to prescribe how we go around the poles in the integrand. The integral can be computed by residues, with slightly different prescriptions for how we go around the poles for the different gauge groups. In this way, we get ℓ distinct contours $\mathcal{C}_{N_1,\dots,N_\ell}$, and with them the partition function,

$$\mathcal{Z}_{\mathcal{V}_{3d}}(S; N) = \frac{1}{\prod_{a=1}^{\ell} N_a!} \oint_{\mathcal{C}_{N_1,\dots,N_\ell}} d^N x \; \Phi_V(x); \prod_{a=1}^{\ell} \Phi_{F_a}(x) \, e^{-\zeta \, \mathrm{Tr} x / \hbar}.$$

Dividing by $N_a!$ corresponds to dividing by the residual gauge symmetry, permuting the N_a eigenvalues in each of the vacua. For $q = t$ this is a topological string partition function of the B-model on Y_S studied in [66], and related to Chern-Simons theory. The $q \neq t$ partition function is the partition function of refined Chern-Simons theory [54], with observables inserted.

We will show that the partition function of $\mathcal{V}_{3d}$ is nothing but the q-deformation of the free-field free field conformal block of the Liouville CFT on a sphere with $\ell + 2$ punctures. Since the q deformation of Liouville CFT might be not familiar, let us review it.

---

[16]This partition function is the index studied in [54–56] with application to knot theory; see also [57]. The index is a chiral building block of the S^3 or $S^2 \times S^1$ partition functions [58–64], deformed by t, the fugacity of a very particular flavor symmetry.

5.3 *q-Liouville*

In this section, we will show that the free field integrals of a q-deformed Liouville conformal field theory [45, 46, 67] have a physical interpretation. They are partition functions of the 3d $\mathcal{N} = 2$ gauge theory, which we will called $\mathcal{V}_{3d}$, in the 3d Ω-background $(\mathbb{C} \times S^1)_q$. The equivalence of the q-Liouville conformal block and the gauge theory partition function is manifest. The screening charge integrals of DF are the integrals over the Coulomb branch of the gauge theory. Inserting the Liouville vertex operators corresponds to coupling the 3d gauge theory to a flavor. The momentum and position of the puncture are given by the real masses of the two chirals within the flavor.

The q-deformed Virasoro algebra is written in terms of the deformed screening charges

$$S(z) = \; : \exp\left(2\phi_0 + 2h_0 \log z + \sum_{k \neq 0} \frac{1 + (t/q)^k}{k} h_k z^{-k}\right) :,$$

where

$$[h_k, h_m] = \frac{1}{1 + (t/q)^k} \frac{1 - t^k}{1 - q^k} m \, \delta_{k+m,0}.$$

The defining property of the generators of the $q-$deformed Virasoro-algebra, is that they commute with the integrals of the screening charges S. The primary vertex operators get deformed as well. The vertex operator carrying momentum α becomes:

$$V_\alpha(z) = \; : \exp\left(-\frac{\alpha}{b^2}\phi_0 - \frac{\alpha}{b^2} h_0 \log z + \sum_{k \neq 0} \frac{1 - q^{-\alpha k}}{k(1 - t^{-k})} h_k z^{-k}\right) :.$$

Note, that these operators manifestly become the usual Liouville operators in the limit where $q = e^{R\epsilon_1}, t = e^{-R\epsilon_2}$ go to 1, by sending R to zero.

Just as before, using these commutation relations, one computes the correlator and obtains the following free field integral:

$$\mathcal{B}_q(\alpha, z; N) = \frac{r}{\prod_{a=1}^{\ell} N_a!} \oint_{C_1,\dots,C_\ell} d^N y \; \Delta_{q,t}^2(y) \prod_{a=0}^{\ell} V_a(y; z_a), \qquad (5.14)$$

where the measure is the q, t-deformed Vandermonde

$$\Delta_{q,t}^2(y) = \prod_{1 \leq I \neq J \leq N} \frac{\varphi(y_I/y_J)}{\varphi(t \, y_I/y_J)},$$

and the potential equals

$$V_a(y; z_a) = \prod_{I=1}^{N} \frac{\varphi(q^{\alpha_a} z_a / y_I)}{\varphi(z_a / y_I)}.$$

In particular, using the properties of the quantum dilogarithm, it is easy to find that $V_0(y; 0) = (y_1 \ldots y_N)^{\alpha_0}$. As in the undeformed case, the relation holds up to a constant of proportionality r. In this paper, we avoid detailed consideration of this normalization constant. The meaning of the constant r, on the Liouville side, is to account for all possible two-point functions between the vertex operators $V_\alpha(z_a)$. Like in the undeformed case, the N eigenvalues are grouped into sets of size N_a, $a = 1, \ldots, \ell$, by the choice of contours they get integrated over.[17]

6 Gauge/Liouville Triality

In what follows, we will prove that there is a triality that relates the 5d and 3d gauge theories T_{5d} and V_{3d}, compactified on a circle, and q-deformation of Liouville conformal blocks. We will show this in two steps.

6.1 q-Liouville and V_{3d}

The first step is to show that q-deformation of the Liouville conformal block (5.14), corresponding to a sphere with $\ell + 2$ punctures equals the partition function of V_{3d}:

$$\mathcal{Z}_{V_{3d}}(S; N) = \mathcal{B}_q(\alpha, z; N).$$

This follows immediately by a simple change of variables that sets

$$z_a = e^{-Rm_{+,a}}, \quad q^{\alpha_a} = e^{Rm_{+,a} - Rm_{-,a}}, \quad y = e^{-Rx}. \tag{6.1}$$

The insertion of a primary vertex operator in Liouville gets related to coupling the 3d gauge theory on the vortex to a flavor: the mass splitting is related to Liouville momentum, the mass itself to the position of the vertex operator. The puncture at $z = 0$ arises from the Fayet-Iliopolous potential, if we set $\alpha_0 = \zeta / \hbar - 1$.

---

[17] The contours of integration *are the same as* in the undeformed case—encircling the segments $[0, z_a]$. The q deformation affects the operators and the algebra, but not the contours. It is important to emphasize that these contours agree with the alternative approach [68] where the free field integrals are replaced by Jackson q-integrals: in our picture, the latter are the residue sums for the former.

6.2 $\mathcal{V}_{3d}$ and $\mathcal{T}_{5d}$: Gauge/Vortex Duality

The second step is to show that the partition function of the 5d gauge theory $\mathcal{T}_{5d}$ and partition function of its vortices, described by the 3d gauge theory $\mathcal{V}_{3d}$ agree

$$\mathcal{Z}_{\mathcal{V}_{3d}}(S, N) = \mathcal{Z}_{\mathcal{T}_{5d}}(\Sigma).$$

For this we place $\mathcal{T}_{5d}$ at the point where the Coulomb and Higgs branches of $\mathcal{T}_{5d}$ meet, $e_a = f_a/v$ with $v = (q/t)^{1/2}$ as before, and Σ degenerates to S. In addition we turn on N_a units of vortex flux.[18] In the Ω-background this is equivalent to not turning on flux and shifting the Coulomb-branch parameters of $\mathcal{T}_{5d}$ so that

$$\mathcal{Z}_{\mathcal{T}_{5d}}(\Sigma) = r_{5d} \sum_{\vec{R}} I_{\vec{R}}^{5d}$$

is evaluated at

$$e_a = t^{N_a} f_a/v, \tag{6.2}$$

where a runs form 1 to ℓ. Here, f_a are the masses of ℓ of the 2ℓ hypermultiplets, and the integer shifts correspond to N_a units of vortex flux turned on. Note that as long as N_a are arbitrary, this is no restriction at all.

To recover $\mathcal{T}_{5d}$ at an arbitrary point of its Coulomb branch, we take the limit $N_a \to \infty$, $\epsilon = \ln(t) \to 0$ keeping the product $N_a \epsilon$ fixed. The gauge/vortex duality is the gauge theory realization of large N duality.

6.2.1 Residues and Instantons

We start by computing the partition function of $\mathcal{V}_{3d}$ by residues. Then we show that the sum over the residues is the instanton sum of the 5d gauge theory $\mathcal{T}_{5d}$. The positions of the poles are labeled by tuples of partitions, and the integrands are equal to Nekrasov summands.

With the change of variables in (6.1), the 3d partition function of $\mathcal{V}_{3d}$ becomes:

$$\mathcal{Z}_{\mathcal{V}_{3d}}(N; S) = \frac{1}{\prod_{a=1}^{\ell} N_a!} \oint_{\mathcal{C}_1, \dots \mathcal{C}_\ell} d^N y \, I^{3d}(y), \tag{6.3}$$

where the integrand $I^{3d}(y)$ equals

$$I^{3d}(y) = V_0(y) \, \Phi_V(y) \prod_{a=1}^{\ell} \Phi_{F_a}(y),$$

---

[18] The shift by v is due to the Ω background. It is natural that the partition function becomes singular at the point where the two branches meet; this determines the shift.

and, in terms of the new variables,

$$\Phi_V(y) = \prod_{1 \le I \ne J \le N} \frac{\varphi(y_J/y_I)}{\varphi(t y_J/y_I)}, \quad \Phi_{F_a}(y) = \prod_{I=1}^{N} \frac{\varphi(q^{\alpha_a} z_a/y_I)}{\varphi(z_a/y_I)}, \quad V_0(y) = \prod_{I=1}^{N} y_I^{\alpha_0}.$$

The ℓ contours $\mathcal{C}_1, \ldots \mathcal{C}_\ell$ run around the intervals in the complex y plane: $\mathcal{C}_a$ circles the interval from $y = 0$ to $y = z_a$, where z_a is the location of a pole in the integral corresponding to a chiral multiplet going massless. The quantum dilogarithm $\varphi(y) = \prod_{n=0}^{\infty} (1 - q^n y)$ [2, 51] has zeros at $y = q^{-n}$, hence the integrand has poles there. The contour is chosen so as to pick up the residues of the poles. For each of the ℓ the groups of eigenvalues we choose the contour that runs from 0 to z_a, circling the poles at

$$y = q^n z_a, \quad n = 0, 1, \ldots.$$

For $|t|, |q| < 1$, the poles interpolate between $y = 0$ and $y = z_a$, and the contours $\mathcal{C}_a$ circle around the interval (this is also where the critical points of the integral are located). However, not all the poles contribute—the numerator in $\Phi_V(y)$ eliminates some: all those for which poles for a pair y_I, y_J coincide up to a q shift. At the same time, the denominator of $\Phi_V(y)$ introduces new poles with y's shifted by t, up to a multiple of q. Up to permutations, the poles that end up contributing are labeled by ℓ-tuples of 2d Young diagrams:

$$\vec{R} = (R_1, \ldots, R_a, \ldots, R_\ell), \tag{6.4}$$

where R_a has at most N_a rows. The poles corresponding to the ath group of variables are at

$$y = y_{\vec{R}},$$

where, up to permutations the components of $y_{\vec{R}}$ equal

$$y_{(N_1 + \cdots + N_{a-1}) + i} = q^{R_{a,i}} t^{N_a - i} z_a, \tag{6.5}$$

where i runs from 1 to N_a and a from 1 to ℓ. The sum over the residues of the integral becomes the sum over the Young diagrams

$$\prod_{a=1}^{\ell} \frac{1}{N_a!} \oint_{\mathcal{C}_1, \ldots \mathcal{C}_\ell} d^N y \quad \rightarrow \quad \sum_{\vec{R}}.$$

While the integrand itself does not make sense at a pole, the ratio of its values at different poles turns out to be finite. This implies that *ratio of the residues* at the poles labeled by $\vec{R}$ and $\vec{\varnothing}$

$$I_{\vec{R}}^{3d} = \mathrm{res}_{\vec{\varnothing}}^{-1} \cdot \mathrm{res}_R \, I^{3d}(y)$$

is simply equal to the *ratio of the integrand* itself at the two poles:

$$I_{\vec{R}}^{3d} = q^{\alpha_0 |\vec{R}|} \cdot \frac{\Phi_V(y_{\vec{R}})}{\Phi_V(y_{\vec{\varnothing}})} \cdot \frac{\prod_{a=1}^{\ell} \Phi_{F_a}(y_{\vec{R}})}{\prod_{a=1}^{\ell} \Phi_{F_a}(y_{\vec{\varnothing}})}. \tag{6.6}$$

Note that $\frac{V_0(y_{\vec{R}})}{V_0(y_{\vec{\varnothing}})} = q^{\alpha_0 |\vec{R}|}$. This makes the sum over residues easy to find:

$$\mathcal{Z}_{V_{3d}}(N; S) = r_{3d} \sum_{\vec{R}} I_{\vec{R}}^{3d}(N, f),$$

where

$$r_{3d} = \mathrm{res}_{\varnothing} I^{3d}(y).$$

The structure of the answer is reminiscent of the 5d partition function $\mathcal{Z}_{T_{5d}}(\Sigma)$, except that the sum in $\mathcal{Z}_{T_{5d}}(\Sigma)$ runs over ℓ-touples of Young diagrams of arbitrary size.

However, from the gauge/vortex duality, we only expect the 3d and the 5d partition functions to equal on the locus (6.2). Restricting to the locus (6.2), the Nekrasov sum truncates to a sum over diagrams R_a with at most N_a rows. Moreover, for every such ℓ-touple, the summand $I_{\vec{R}}^{5d}$ indeed becomes equal to $I_{\vec{R}}^{3d}$. The detailed proof is presented in [2], here we only give a sketch.

Recall

$$I_{\vec{R}}^{5d} = q^{\zeta |R|} \cdot z_{V,\vec{R}} \cdot z_{H,\vec{R}} \cdot z_{H^\dagger,\vec{R}}.$$

The ℓ hypermultiplet contributions $z_{H^\dagger,\vec{R}}$ each contain $N_{R_a \varnothing}(v e_a / f_a)$, as a factor. Restricting this to (6.2) we get $N_{R_a \varnothing}(t^{N_a})$, which, as one can show[19] vanishes if R_a has more than N_a rows. So at this point, $I_{\vec{R}}^{5d}$ is non-zero only for those ℓ-touples of Young diagrams $\vec{R} = (R_1, \ldots, R_a, \ldots R_\ell)$ for which R_a has no more that N_a rows, for each a between 1 and ℓ. Thus, the non-zero fixed point contributions to the instanton sum are the same as the poles of the 3d partition function. Not only does the sum over Young diagrams truncate, but moreover one can prove that the value of the summand in the instanton partition function is exactly $I_{\vec{R}}^{3d}$:

$$I_{\vec{R}}^{3d}(N, f) = I_{\vec{R}}^{5d}(e, f),$$

with identifications

$$e_a / f_a = t^{N_a} / v.$$

Recall we let $f_a = f_{+,a}$ and $f_{a+\ell} = f_{-,a}$ for a running from 1 to ℓ. Finally, we have $q^\zeta = q^{\alpha_0} q$.

---

[19] See [2] for a proof, and [52, 69] for earlier work making use of this.

The vector multiplet contributions in 5d are related to vector multiplet contributions in 3d, and the 5d hypermultiplets to 3d flavors and the instanton counting parameter in 5d to FI term contributions to the potential in 3d. The 5d partition function is actually a product of the instanton sum $I_{\bar{R}}^{5d}$ together with the perturbative and the one loop factors contained in r_{5d}. This equals the partition function of the 5d gauge theory at the root of the Higgs and Coulomb branches in the absence of vortices. On the 3d gauge theory side, one can prove that this is accounted by the product of r_{3d}, the residue at the $y = y_{\bar{\varnothing}}$ pole, together with a contribution that is not captured by the theory on the vortex—this is the partition function of the bulk gauge theory, at the root of the Higgs branch in the absence of vortices. (From the string theory perspective, this contribution is the partition function of Y_S without branes). One can prove that, taking this into account, the full partition functions on the two sides of the duality are equal.

We have thus proven our main claim (5.3) for the case the Gaiotto curve C has genus zero with arbitrary number of punctures. It is elementary to extend this to the case when C is a genus one curve, with arbitrary punctures. We expect the triality to generalize to the case when the Liouville CFT gets replaced by ADE type Toda CFT. The generalization to A_n case will be presented in [19].

References

[1] Dijkgraaf, R., Vafa, C.: Toda Theories, Matrix Models, Topological Strings, and N = 2 Gauge Systems. arXiv:0909.2453 [hep-th]

[2] Aganagic, M., Haouzi, N., Kozcaz, C., Shakirov, S.: Gauge/Liouville Triality. arXiv:1309.1687 [hep-th]

[3] Gopakumar, R., Vafa, C.: On the gauge theory/geometry correspondence. Adv. Theor. Math. Phys. **3**, 1415–1443 (1999). arXiv:hep-th/9811131

[4] Dijkgraaf, R., Vafa, C.: Matrix models, topological strings, and supersymmetric gauge theories. Nucl. Phys. **B644**, 3–20 (2002). doi:10.1016/S0550-3213(02)00766-6. arXiv:hep-th/0206255 [hep-th]

[5] Aganagic, M., Dijkgraaf, R., Klemm, A., Marino, M., Vafa, C.: Topological strings and integrable hierarchies. Commun. Math. Phys. **261**, 451–516 (2006). doi:10.1007/s00220-005-1448-9. arXiv:hep-th/0312085 [hep-th]

[6] Nekrasov, N.A.: Seiberg-Witten prepotential from instanton counting. Adv. Theor. Math. Phys. **7**, 831–864 (2004). arXiv:hep-th/0206161 [hep-th]

[7] Neitzke, A., Vafa, C.: Topological strings and their physical applications. arXiv:hep-th/0410178 [hep-th]

[8] Chen, H.-Y., Dorey, N., Hollowood, T.J., Lee, S.: A New 2d/4d Duality via Integrability. JHEP **1109**, 040 (2011). arXiv:1104.3021 [hep-th]

[9] Dorey, N., Lee, S., Hollowood, T.J.: Quantization of integrable systems and a 2d/4d duality. JHEP **1110**, 077 (2011). arXiv:1103.5726 [hep-th]

[10] Aganagic, M.: M-theory, large N duality and the dynamics of vortices. Talk at 11th Simons Summer Workshop at SCGP (2013)

[11] Dorey, N.: The BPS spectra of two-dimensional supersymmetric gauge theories with twisted mass terms. JHEP **9811**, 005 (1998). arXiv:hep-th/9806056 [hep-th]

[12] Dorey, N., Hollowood T.J., Tong, D.: The BPS spectra of gauge theories in two-dimensions and four-dimensions. JHEP **9905**, 006 (1999). arXiv:hep-th/9902134 [hep-th]

[13] Hanany, A., Tong, D.: Vortices, instantons and branes. JHEP **0307**, 037 (2003). arXiv:hep-th/0306150 [hep-th]

[14] Hanany, A., Tong, D.: Vortex strings and four-dimensional gauge dynamics. JHEP **0404**, 066 (2004). arXiv:hep-th/0403158 [hep-th]

[15] Shifman, M., Yung, A.: NonAbelian string junctions as confined monopoles. Phys. Rev. **D70**, 045004 (2004). arXiv:hep-th/0403149 [hep-th]

[16] Alday, L.F., Gaiotto, D., Tachikawa, Y.: Liouville correlation functions from four-dimensional gauge theories. Lett. Math. Phys. **91**, 167–197 (2010). arXiv:0906.3219 [hep-th]

[17] Gaiotto, D.: N = 2 dualities. JHEP **1208**, 034 (2012). arXiv:0904.2715 [hep-th]

[18] Gaiotto, D., Moore, G.W., Neitzke, A.: Wall-crossing, Hitchin systems, and the WKB approximation. arXiv:0907.3987 [hep-th]

[19] Aganagic, M., Haouzi N., Shakirov, S.: A_n-Triality. arXiv:1403.3657 [hep-th]

[20] Fateev, V., Litvinov, A.: On AGT conjecture. JHEP **1002**, 014 (2010). arXiv:0912.0504 [hep-th]

[21] Alba, V.A., Fateev, V.A., Litvinov, A.V., Tarnopolskiy, G.M.: On combinatorial expansion of the conformal blocks arising from AGT conjecture. Lett. Math. Phys. **98**, 33–64 (2011). arXiv:1012.1312 [hep-th]

[22] Mironov, A., Morozov, A., Shakirov S.: A direct proof of AGT conjecture at beta = 1. JHEP **1102**, 067 (2011). arXiv:1012.3137 [hep-th]

[23] Morozov, A. Smirnov, A.: Finalizing the proof of AGT relations with the help of the generalized Jack polynomials. arXiv:1307.2576

[24] Braverman, A., Finkelberg, M., Nakajima, H.: Instanton moduli spaces and W-algebras. arXiv:1406.2381 [math.QA]

[25] Losev, A.S., Marshakov, A., Nekrasov, N.A.: Small instantons, little strings and free fermions. arXiv:hep-th/0302191 [hep-th]

[26] Hollowood, T.J., Iqbal, A., Vafa, C.: Matrix models, geometric engineering and elliptic genera. JHEP **0803**, 069 (2008). arXiv:hep-th/0310272 [hep-th]

[27] Bershadsky, M., Cecotti, S., Ooguri, H., Vafa, C.: Kodaira-Spencer theory of gravity and exact results for quantum string amplitudes. Commun. Math. Phys. **165**, 311–428 (1994). arXiv:hep-th/9309140 [hep-th]

[28] Dotsenko, V., Fateev, V.: Conformal algebra and multipoint correlation functions in two-dimensional statistical models. Nucl. Phys. B **240**, 312 (1984)

[29] Dotsenko, V., Fateev, V.: Four point correlation functions and the operator algebra in the two-dimensional conformal invariant theories with the central charge c! '1". Nucl. Phys. **B251**, 691 (1985)

[30] Goulian, M., Li, M.: Correlation functions in Liouville theory. Phys. Rev. Lett. **66**, 2051–2055 (1991)

[31] Strominger, A.: Massless black holes and conifolds in string theory. Nucl. Phys. **B451**, 96–108 (1995). arXiv:hep-th/9504090 [hep-th]

[32] Witten, E.: Solutions of four-dimensional field theories via M theory. Nucl. Phys. **B500**, 3–42 (1997). arXiv:hep-th/9703166 [hep-th]

[33] Greene, B.R., Morrison, D.R., Vafa, C.: A geometric realization of confinement. Nucl. Phys. **B481**, 513–538 (1996). arXiv:hep-th/9608039 [hep-th]

[34] Hori, K., Ooguri, H., Vafa, C.: NonAbelian conifold transitions and N = 4 dualities in three-dimensions. Nucl. Phys. **B504**, 147–174 (1997). arXiv:hep-th/9705220 [hep-th]

[35] Gukov, S., Witten, E.: Branes and quantization. Adv. Theor. Math. Phys. **13** (2009). arXiv:0809.0305 [hep-th]

[36] Alday, L.F., Gaiotto, D., Gukov, S., Tachikawa, Y., Verlinde, H.: Loop and surface operators in N = 2 gauge theory and Liouville modular geometry. JHEP **1001**, 113 (2010). arXiv:0909.0945 [hep-th]

[37] Dimofte, T., Gukov S., Hollands, L.: Vortex counting and Lagrangian 3-manifolds. Lett. Math. Phys. **98**, 225–287 (2011). arXiv:1006.0977 [hep-th]

[38] Douglas, M.R., Moore, G.W.: D-branes, quivers, and ALE instantons. arXiv:hep-th/9603167 [hep-th]

[39] Nekrasov, N.A., Shatashvili, S.L.: Quantization of integrable systems and four dimensional gauge theories. arXiv:0908.4052 [hep-th]

[40] Nekrasov, N., Witten, E.: The omega deformation, branes, integrability, and Liouville theory. JHEP **1009**, 092 (2010). arXiv:1002.0888 [hep-th]

[41] Aganagic, M., Cheng, M.C.N., Dijkgraaf, R., Krefl, D., Vafa, C.: Quantum geometry of refined topological strings. arXiv:1105.0630 [hep-th]

[42] Aganagic, M., Shakirov, S.: Refined Chern-Simons theory and topological string. arXiv:1210.2733 [hep-th]

[43] Nekrasov, N., Okounkov, A.: Seiberg-Witten theory and random partitions. arXiv:hep-th/0306238 [hep-th]

[44] Nekrasov, N., Shadchin, S.: ABCD of instantons. Commun. Math. Phys. **252** 359–391 (2004). arXiv:hep-th/0404225 [hep-th]

[45] Shiraishi, J., Kubo, H., Awata, H., Odake, S.: A quantum deformation of the Virasoro algebra and the Macdonald symmetric functions. Lett. Math. Phys. **38**, 33–51 (1996). arXiv:q-alg/9507034 [q-alg]

[46] Awata, H., Kubo, H., Morita, Y., Odake, S., Shiraishi, J.: Vertex operators of the q Virasoro algebra: defining relations, adjoint actions and four point functions. Lett. Math. Phys. **41**, 65–78 (1997). arXiv:q-alg/9604023 [q-alg]

[47] Frenkel, E., Reshetikhin, N.: Deformations of W-algebras associated to simple Lie algebras. In eprint arXiv:q-alg/9708006, p. 8006. August (1997)

[48] Seiberg, N.: Five-dimensional SUSY field theories, nontrivial fixed points and string dynamics. Phys. Lett. **B388**, 753–760 (1996). arXiv:hep-th/9608111 [hep-th]

[49] Intriligator, K.A., Morrison, D.R., Seiberg, N.: Five-dimensional supersymmetric gauge theories and degenerations of Calabi-Yau spaces. Nucl. Phys. **B497**, 56–100 (1997). arXiv:hep-th/9702198 [hep-th]

[50] Nekrasov, N.: Five dimensional gauge theories and relativistic integrable systems. Nucl. Phys. **B531**, 323–344 (1998). arXiv:hep-th/9609219 [hep-th]

[51] Faddeev, L., Kashaev, R.: Quantum dilogarithm. Mod. Phys. Lett. **A9** 427–434 (1994). arXiv:hep-th/9310070 [hep-th]

[52] Awata, H., Kanno, H.: Refined BPS state counting from Nekrasov's formula and Macdonald functions. Int. J. Mod. Phys. **A24**, 2253–2306 (2009). arXiv:0805.0191 [hep-th]

[53] Witten, E.: Phase transitions in M theory and F theory. Nucl. Phys. **B471**, 195–216 (1996). arXiv:hep-th/9603150 [hep-th]

[54] Aganagic, M., Shakirov, S.: Knot homology from refined Chern-Simons theory. arXiv:1105.5117 [hep-th]

[55] Aganagic, M., Shakirov, S.: Refined Chern-Simons theory and knot homology. arXiv:1202.2489 [hep-th]

[56] Aganagic, M., Schaeffer, K.: Orientifolds and the refined topological string. JHEP **1209**, 084 (2012). arXiv:1202.4456 [hep-th]

[57] Fuji, H., Gukov S., Sulkowski, P.: Super-A-polynomial for knots and BPS states. Nucl. Phys. **B867**, 506–546 (2013). arXiv:1205.1515 [hep-th]

[58] Hama, N., Hosomichi, K., Lee, S.: Notes on SUSY Gauge theories on three-sphere. JHEP **1103**, 127 (2011). arXiv:1012.3512 [hep-th]

[59] Kapustin, A., Willett, B.: Generalized superconformal index for three dimensional field theories. arXiv:1106.2484 [hep-th]

[60] Hama, N., Hosomichi, K., Lee, S.: SUSY Gauge theories on squashed three-spheres. JHEP **1105**, 014 (2011). arXiv:1102.4716 [hep-th]

[61] Pasquetti, S.: Factorisation of N = 2 theories on the squashed 3-sphere. JHEP **1204**, 120 (2012). arXiv:1111.6905 [hep-th]

[62] Nieri, F., Pasquetti, S., Passerini, F.: 3d and 5d gauge theory partition functions as q-deformed CFT correlators. arXiv:1303.2626 [hep-th]

[63] Beem, C., Dimofte, T., Pasquetti, S.: Holomorphic blocks in three dimensions. arXiv:1211.1986 [hep-th]

[64] Taki, M.: Holomorphic blocks for 3d non-abelian partition functions. arXiv:1303.5915 [hep-th]

[65] Cheng, M.C.N., Dijkgraaf, R., Vafa, C.: Non-perturbative topological strings and conformal blocks. JHEP **09**, 022 (2011). arXiv:1010.4573 [hep-th]

[66] Aganagic, M., Klemm, A., Marino, M., Vafa, C.: Matrix model as a mirror of Chern-Simons theory. JHEP **0402**, 010 (2004). arXiv:hep-th/0211098 [hep-th]

[67] Awata, H., Yamada, Y.: Five-dimensional AGT relation and the deformed beta-ensemble. Prog. Theor. Phys. **124**, 227–262 (2010). arXiv:1004.5122 [hep-th]

[68] Mironov, A., Morozov, A., Shakirov, S., Smirnov, A.: Proving AGT conjecture as HS duality: extension to five dimensions. Nucl. Phys. **B855**, 128–151 (2012). arXiv:1105.0948 [hep-th]

[69] Dimofte, T., Gukov, S., Hollands, L.: Vortex counting and Lagrangian 3-manifolds. Lett. Math. Phys. **98**, 225–287 (2011). arXiv:1006.0977 [hep-th]

B-Model Approach to Instanton Counting

Daniel Krefl and Johannes Walcher

Abstract The instanton partition function of $\mathcal{N} = 2$ gauge theory in the general Ω-background is, in a suitable analytic continuation, a solution of the holomorphic anomaly equation known from B-model topological strings. The present review of this connection is a contribution to a special volume on recent developments in $\mathcal{N} = 2$ supersymmetric gauge theory and the 2d-4d relation, edited by J. Teschner.

1 Introduction and Key Ideas

The instanton partition function of $\mathcal{N} = 2$ supersymmetric quantum field theories

$$Z^{\text{inst}}(a, \epsilon_1, \epsilon_2; \Lambda) \tag{1.1}$$

is of algebra-geometro-physical interest for at least three different, though related, reasons. First of all, by its very definition, Z^{inst} encapsulates the cohomology of the moduli space of instantons, supersymmetric solutions of the underlying classical field theory, and the algebraic structures on that space (Chapter [V:4]). Secondly, within the 2d-4d correspondences of Alday-Gaiotto-Tachikawa, and Nekrasov-Shatasvhili (Chapter [V:12]), the instanton partition function connects supersymmetric field theories with the world of completely integrable systems and their quantization, specifically Hitchin systems (see Chapter [V:3]). Thirdly, $Z^{\text{inst}}(a, \epsilon_1, \epsilon_2; \Lambda)$ contains information about the structure of the Coulomb branch that goes beyond the weakly coupled description in a Lagrangian field theory. After a suitable analytic continua-

---

A citation of the form [V:x] refers to article number x in this volume.

D. Krefl (✉)
Center for Theoretical Physics, Seoul National University, Seoul, South Korea
e-mail: krefl@snu.ac.kr

J. Walcher
Department of Physics and Department of Mathematics and Statistics, McGill University, Montreal, QC, Canada
e-mail: johannes.walcher@mcgill.ca

© Springer International Publishing Switzerland 2016
J. Teschner (ed.), *New Dualities of Supersymmetric Gauge Theories*,
Mathematical Physics Studies, DOI 10.1007/978-3-319-18769-3_14

tion, it allows to calculate interesting physical quantities everywhere in the moduli space of vacua and marginal couplings, and thereby to study a variety of dualities.

It is in fact, this latter aspect of the instanton partition function that is closest to the approach pioneered by Seiberg and Witten for solving the low-energy dynamics of $\mathcal{N} = 2$ supersymmetric field theories, by exploiting the global constraints on the structure of the moduli space coming from special geometry and modular invariance. The basic ideas are easily explained.

The instanton partition function Z^{inst} calculated via localization (see Chapter [V:4]) is a series

$$Z^{\text{inst}} \sim \sum_n \Lambda^n R_n(a, \epsilon_1, \epsilon_2), \tag{1.2}$$

with rational functions R_n of the Ω-background parameters $\epsilon_{1,2}$, and the Coulomb branch parameters a, that converges well for small instanton counting parameter Λ. As explained by Nekrasov [1], the Seiberg-Witten solution for the low-energy effective action is recovered in the non-equivariant limit $\epsilon_{1,2} \to 0$. Specifically, the $\mathcal{N} = 2$ prepotential is the residue

$$\mathcal{F}^{(0)}(a; \Lambda) = \lim_{\epsilon_1, \epsilon_2 \to 0} \left(\epsilon_1 \epsilon_2 \log Z^{\text{inst}}(a, \epsilon_1, \epsilon_2; \Lambda) \right), \tag{1.3}$$

after the perturbative expansion of the free energy for small Λ (in the asymptotically free case, this is equivalent to the weak-coupling limit $a \to \infty$). It coincides with the prepotential obtained from the Seiberg-Witten effective geometry (the family of hyper-elliptic curves together with the differential), which captures the low-energy dynamics and is the basis for the various embeddings into string theory. Most specifically, the $\mathcal{F}^{(0)}$ appears in the so-called geometric engineering limit of the prepotential governing the compactification of type II string on a (non-compact) Calabi-Yau manifold [2].

It has been a natural question to ask for the analytic and geometric characterization of the terms in Z^{inst} of higher order in ϵ_1, ϵ_2, and their physical interpretation. As anticipated already by Nekrasov [1], the answer is most immediate on the special slice in coupling constant space $\epsilon_1 = -\epsilon_2$. By a detour in one higher dimension, one can see that in general, the Ω-background in the gauge theory arises in the string/M-theory constructions from a vacuum expectation value of the gravi-photon field strength, of the form

$$F = \epsilon_1 dx_1 \wedge dx_2 + \epsilon_2 dx_3 \wedge dx_4 \tag{1.4}$$

(itself a limit of the Melvin background, or "flux-trap" [3] in other duality frames). The specialization $\epsilon_1 = -\epsilon_2$ corresponds to a self-dual gravi-photon background, and the expansion coefficients of the supersymmetric free energy in $\epsilon_1 = -\epsilon_2 =: g_s$ are identified with the higher-derivative F-term couplings $R^2 F^{2g-2}$ in the effective action [4], which can be computed as the *topological* string genus-g free energy [5]. In the geometric engineering limit, one recovers the expansion of the instanton partition function.

$$\log Z^{\text{inst}}(a, g_s, -g_s; \Lambda) = \sum_{g=0}^{\infty} g_s^{2g-2} \mathcal{F}^{(g)}(a) \qquad (1.5)$$

A natural way to test this physical interpretation is to re-calculate the $\mathcal{F}^{(g)}$ using the string theory methods. In the topological B-model, the most universal of these methods is the holomorphic anomaly of BCOV [5].

The basic message of the holomorphic anomaly method is that the higher order corrections $\mathcal{F}^{(g)}(a)$ can still be continued throughout moduli space, in particular any strong coupling regions, but (in distinction to the prepotential $\mathcal{F}^{(0)}$), they are no longer holomorphic functions of a. The physical origin of this non-holomorphicity are the infrared effects, degenerating Riemann surfaces in the perturbative string theory [5], or the distinction between 1 PI and Wilsonian effective action from the point of view of the field theory [6]. Mathematically, the holomorphic anomaly is an expression of the competition between holomorphy and modular invariance [5], and can also be viewed as an embodiment of the wave-function nature of the topological partition function [7].

The holomorphic anomaly *equation* dictates the non-holomorphic dependence of $\mathcal{F}^{(g)}(a, \bar{a})$ recursively in the order of the expansion, $2g - 2$. The meromorphic function on moduli space that is thereby left undetermined at each order is known as the holomorphic ambiguity and can, under favorable circumstances, be determined by imposing appropriate principal parts or "boundary conditions" at the various singular points.

It was shown by Klemm and Huang [8] that the holomorphic anomaly commutes with the geometric engineering limit, and can be used to completely recover the $\mathcal{F}^{(g)}$ in the expansion (1.5). Even though a detailed derivation of the holomorphic anomaly from the gauge theory point of view is missing, the equation itself can be written down based solely on the special geometry data on the moduli space that can be obtained from the prepotential $\mathcal{F}^{(0)}$. And at least in all examples with low-dimensional moduli space, the boundary conditions at the monopole/dyon points are sufficient to completely fix the holomorphic ambiguity, and thereby make the holomorphic anomaly "integrable" in that sense.

From the point of view of instanton counting (1.1), this discussion of the holomorphic anomaly appears as rather tangential. After all, the holomorphy and integrability of the higher order corrections is built into the formalism, while the underlying spectral geometry is completely determined by the first order classical term, *i.e.*, the prepotential. There are nevertheless several very good reasons to explore the connection further, and in particular, to understand the extension of the holomorphic anomaly to the full two-parameter Ω-background, away from the specialization $\epsilon_1 = -\epsilon_2$.

From the gauge theory point of view, the precise role of the holomorphic anomaly, or the wave-function nature of Z^{inst} is not completely understood, for instance in the context of the quantum integrable systems. Moreover, the possible calculation of Z^{inst} as a "sum over instantons" (or some other semi-classical configurations) around other points in moduli space remains to be explored. While conformal field theory in principle provides formal expressions for Z^{inst} in terms of certain contour integrals also elsewhere in moduli space, these have been evaluated explicitly only in a limited number of situations. The continuation of the $\mathcal{F}^{(g)}$ to other points in moduli space via the holomorphic anomaly provides a very welcome benchmark for such calculations.

From the string theory point of view, the existence of the second deformation parameter itself is the most intriguing aspect. Indeed, while the role of $g_s = \sqrt{-\epsilon_1 \epsilon_2}$ as the genus-counting parameter, *i.e.*, the topological string coupling constant, is readily appreciated, the existence of a second "string-coupling like" parameter is much more mysterious. Since, in much more generality than the restricted topological context, string theory does not have any free parameters, the absence of a worldsheet description would be in tremendous tension with the overall picture. To be sure, the role of the second parameter from the macroscopic space-time, or M-theory point of view is completely clear, see [1, 9], as well as the connection with refinement and categorification [10]. What is missing is the *microscopic* explanation.

The main point of the present contribution is to highlight the observation that with the right choice of parameterization of the coupling constants, the deformation away from the special slice $\epsilon_1 = -\epsilon_2$ is indeed as simple as it could be: The higher order corrections for general ϵ_1, ϵ_2 still satisfy the holomorphic anomaly equations, with deformation only in the boundary conditions. In particular, a single infinitesimal coupling constant is sufficient. This was first pointed out in [11–13]. These methods therefore allow the calculation (via "analytic" continuation) of Z^{inst} around points in moduli space other than the weak coupling regime. This constitutes a benchmark for testing the 2d-4d relation this special volume is about away from $\Lambda \to 0$. Coming back to string theory, these observations have allowed the application of the holomorphic anomaly equation for the B-model calculation of refined BPS invariants of local Calabi-Yau manifolds [14, 15]. This can be viewed as further evidence that the second parameter should be lifted to the topological string (not necessarily as a coupling constant, but rather as a deformation parameter), and has been as well applied and interpreted in the context of quantum geometry and quantum integrability [16]. Among the possible stringy explanations of the refinement, we will outline in somewhat more detail an intriguing relation to orbifolds and orientifolds, following [11, 17].

Before closing this introduction, it seems worthwhile to emphasize once again that in this Chapter, we are discussing the instanton partition function from the point of view of the "B-model", meaning the global structure of the moduli space, special geometry and modular invariance. In contrast to (1.2), which is exact in ϵ_1, ϵ_2, but perturbative in Λ, the B-model provides answers that are exact in the instanton expansion, but perturbative in ϵ_1, ϵ_2.

2 Geometric Engineering

Large classes of supersymmetric gauge theories in various dimensions can be systematically obtained from string-, M- and F-theory compactifications. This is usually referred to as geometric engineering, as the geometry of the compactification manifold

X determines the effective gauge theory in the field theory limit. We only give a lightning overview, excellent pedagogical reviews being available in the canonical literature, see for instance, [18, 19].

Any given gauge theory can typically be realized in several ways in string theory. These different constructions are then related by various dualities and limiting procedures. Hence, depending on the gauge theory to be investigated via geometric engineering, and the specific gauge theory property under investigation, a convenient duality frame has to be chosen. A common feature of all geometric engineering approaches is that in order to decouple string and gravity effects, the compactification manifold X has to feature a local singularity, perhaps in the guise of a brane.

We are interested in $\mathcal{N} = 2$ supersymmetric gauge theories in four dimensions, their low-energy effective prepotential $\mathcal{F}^{(0)}(a, m)$, and higher derivative F-term couplings. These are (modulo the holomorphic anomaly) holomorphic functions of the Coulomb moduli a_i and masses of matter fields m_i, and receive their essential contributions from the space-time instantons. This class of theories can be conveniently engineered and investigated in a type IIA/B superstring framework by compactification on a local (non-compact) Calabi-Yau 3-fold, which yields, under certain conditions, a four-dimensional $\mathcal{N} = 2$ supersymmetric gauge theory theory with decoupled gravity.

To be specific, consider type IIA string theory compactified on a Calabi-Yau 3-fold X. Four-dimensional abelian gauge fields arise in the Ramond-Ramond sector by dimensional reduction on the even cohomology of X. But since perturbative string states do not carry Ramond-Ramond charge, in order to obtain interesting non-abelian gauge groups, we must include non-perturbative effects. In particular, D2-branes wrapped on the (compact) 2-cycles of the compactification geometry represent objects electrically charged under the corresponding abelian gauge fields. The masses of these states being proportional to the Kähler class (volume) t_{f_i} of the wrapped 2-cycles, one needs $t_{f_i} \to 0$ in order to have massless charged gauge bosons.

In fact, it is best to view the Calabi-Yau compactification as the dimensional reduction of a K3 compactification near an ADE singularity. The gauge group originates in six dimensions from the (compact) homology of the singularity (of ADE type for ADE gauge group), while further dimensional reduction (on a copy, $B \cong \mathbb{P}^1$, of complex projective space to be specific) leads to an $\mathcal{N} = 2$ gauge theory in four dimensions. In this process, the bare gauge coupling g_{YM} of the four-dimensional theory is proportional to the Kähler class t_B of the 2-dimensional manifold used for the reduction, i.e., $t_B \sim 1/g_{YM}^2$. In order to decouple gravity (and stringy effects) it is sufficient to send the coupling constant to zero, since it pushes the string scale to infinity [2]. This means that we are interested in the limit $t_B \to \infty$.

In order to satisfy the Calabi-Yau condition, the compactification space can not be given by a direct product, but rather must have the structure of a fibration,

$$F \to X \to B, \tag{2.1}$$

where the fiber geometry F (with the ADE singularity) determines the gauge group while the base geometry B the effective gauge coupling in four dimensions. Note that the fibration structure also allows to incorporate matter content via local enhancement of the fiber singularity.

It is important to keep in mind that the limits $t_B \to \infty$ and $t_{f_i} \to 0$ of base and fiber Kähler classes are not independent. This can be illustrated best at hand of a concrete example. Consider the geometric engineering of pure $SU(2)$ along the lines sketched above, as originally discussed in [2]. To obtain the two charged gauge bosons, W^+ and W^-, it is sufficient to fiber a $\mathbb{P}^1$ over the base $\mathbb{P}^1$. [The different ways this can be done are labeled by an integer and the corresponding geometries correspond to the Hirzebruch surfaces. The Calabi-Yau 3-fold itself is the total space of the anti-canonical bundle over this complex surface. All Hirzebruch surfaces give rise to pure $SU(2)$ in four dimensions.] Recall that in the weak coupling regime the running of the gauge coupling is given by

$$\frac{1}{g_{YM}^2} \sim \log \frac{m}{\Lambda}, \qquad (2.2)$$

where m denotes the mass of the W-bosons and Λ the dynamical scale. With the above identifications, we learn that we have to take the limit in a way such that $t_B \sim \log t_f$ holds. The precise proportionality constant can be fixed as follows: We know that at weak coupling the instanton corrections to the bare gauge coupling go in powers of $(\Lambda/a)^4$, with a the Coulomb modulus. Correspondingly, we have to scale $e^{-t_b} \sim \delta^4 \Lambda^4$ and $t_f \sim \delta a$ as $\delta \to 0$, which constitutes the map between the string moduli and the gauge theory parameters, for pure $SU(2)$.

The useful property of the type IIA string construction is that the space-time instanton corrections are mapped to world-sheet instanton corrections. Qualitatively, this is clear from the relation between the 2-cycles and the gauge coupling and gauge bosons sketched above: The Euclidean string worldsheet is wrapped around the 2-cycles of the geometry with worldsheet instanton action $S \sim d_b \, t_b + d_f \, t_f$, where d_b and d_f refer to wrapping numbers. In particular, this means that we do not have to consider the full type IIA string theory to investigate the gauge theory from a string point of view. Rather, the topological sector is sufficient, i.e., the topological string amplitudes which capture world-sheet instanton corrections.

Starting from the topological string tree-level amplitude, taking the above gauge theory limit yields the space-time instanton corrections to the gauge theory prepotential. The higher-genus amplitudes encode the gravitational corrections, as sketched in the introduction. In this way, the string theory provides both a conceptual framework, and a host of computational methods to investigate non-perturbative effects in supersymmetric gauge theories.

There is, however, one important subtlety to keep in mind. Since the geometric engineering limit involves $t_{f_i} \to 0$, the compactification geometry is in fact singular, and we are not expanding the string amplitudes around the large volume point in moduli space. Hence, if we compute the topological string amplitudes using the usual A-model techniques, which are valid at large volume (such as, localization [20] or the topological vertex [21]), these amplitudes have to be analytically continued before we can take the limit. (The necessity of this analytic continuation is quite clear already from the fact that the large volume expansion of the topological amplitudes is a series in the exponentiated Kähler moduli, whereas the gauge theory prepotential at weak coupling is an expansion into negative powers of the Coulomb branch parameter.)

For the topological string amplitudes of the $SU(2)$ engineering geometry sketched above, the analytic continuation can be achieved via a relatively simple resummation [22]. In more complicated examples, one has to switch to the B-model mirror of the type IIA string background in order to perform the analytic continuation. We recall that in general, mirror symmetry maps worldsheet instanton corrections to the expansions of classical geometric quantities. For instance, tree-level worldsheet instantons in type IIA are encoded in the period integrals of the type IIB mirror geometry. As these periods can be calculated as solutions of simple linear differential equations, their analytic continuation all over moduli space is straight-forward.

Under the geometric engineering limit along the lines reviewed above, mirror symmetry may be seen as the stringy origin of the Seiberg-Witten solution of $\mathcal{N} = 2$ gauge theory. That is, the Seiberg-Witten curve and differential arise in the limit of the mirror Calabi-Yau threefold which is mirror dual to the type IIA engineering geometry. In particular, as its B-model parent, the Seiberg-Witten geometry naturally provides a global description of the moduli space.

So far we mainly had in mind spherical world-sheet instantons yielding instanton corrections to the gauge theory prepotential. However, perhaps the most useful property of this stringy construction is that it allows to calculate gravitational corrections to the $\mathcal{N} = 2$ gauge theory, originating from world-sheet instantons of higher genus. In detail, the genus-g topological string amplitude yields $R^2 F^{2g-2}$ corrections to the gauge theory [4], which can be calculated very efficiently via a specific topological string B-model technique, namely the holomorphic anomaly equation, all over the Coulomb moduli space. This is the subject to which we now turn.

3 B-Model

It has been observed some time ago in [8] that the free energy of four dimensional $\mathcal{N} = 2$ supersymmetric gauge theory with gravitational corrections satisfies the holomorphic anomaly equations of [5, 23]. This can be seen as a consequence of the geometric engineering approach to $\mathcal{N} = 2$ gauge theories, where the gauge theory free energy follows as a specific limit of the topological string free energy on a corresponding engineering Calabi-Yau, as outlined in the previous section.

Since the gravitational corrections captured by the topological string are a specific specialization of the Ω-deformed gauge theory, $\epsilon_1 = -\epsilon_2$, it is natural to ask whether the Ω-deformed theory with general equivariant parameters satisfies as well a kind of anomaly equation. The main point of interest is that the holomorphic anomaly equation allows to analytically continue the Ω-deformed partition function over all of the Coulomb moduli space. In contrast, the instanton counting partition function of [1] and as well the CFT calculations for the partition function via the AGT correspondence [24] are most useful only for the asymptotically free theories at a weakly coupled point in Coulomb moduli space (see however [25]).

The complete partition function $Z^{\text{inst}}(\epsilon_1, \epsilon_2)$ obtained via instanton counting is exact in the two equivariant parameters $\epsilon_{1,2}$. In order to get started with the investigation of the anomaly equation, one has to choose a parameterization of the infinitesimal neighborhood of $\epsilon_{1,2} = 0$, and the form of the answer will naively depend on the choice. It turns out that the correct expansion from the topological string point of view is to choose the same parameterization as occurring in the AGT correspondence, used in this context in [11, 13]. Namely, we write

$$\epsilon_1 = \sqrt{\beta} g_s, \quad \epsilon_2 = -\frac{1}{\sqrt{\beta}} g_s, \tag{3.1}$$

with β a fixed constant and g_s being the only infinitesimal expansion parameter. Hence (leaving the Coulomb moduli a implicit) we define the perturbative amplitudes $\mathcal{F}^{(g)}(\beta)$ as the coefficients in the expansion

$$\log Z^{\text{inst}}(\epsilon_1, \epsilon_2) = \mathcal{F}(\epsilon_1, \epsilon_2) = \sum_{g=0}^{\infty} \mathcal{F}^{(g)}(\beta) \, g_s^{2g-2}. \tag{3.2}$$

We note in particular that the Seiberg-Witten prepotential defined via the limit (1.3), is *independent* of β, i.e.,

$$\mathcal{F}(\epsilon_1, \epsilon_2) = \mathcal{F}^{(0)} g_s^{-2} + \mathcal{O}(g_s^0). \tag{3.3}$$

Of course, one might also envisage a double expansion in the two-parameters ϵ_1, ϵ_2, as performed in this context in [12]. However, it is not hard to see that the two-parameter expansion is related via a finite resummation to the one-parameter expansion (3.2). This is related to the fact that the $\mathcal{F}^{(g)}(\beta)$ are *polynomial* in β. As a consequence, an anomaly equation for a two-parameter expansion scheme is algebraically equivalent with the anomaly for the one-parameter expansion (*cf.*, the discussions in [26, 27]). Our results, and specifically, the fact that only the holomorphic ambiguity depends on β, make it clear that the one-parameter expansion (3.2) is most economical and therefore preferred.[1]

Note that with the four-dimensional Lorentz invariance, the expansion (3.2) goes in even powers of g_s only, reflecting the symmetry $\epsilon_{1,2} \to -\epsilon_{1,2}$ of the Ω-background. As noted in [1, 28], localization in the presence of mass parameters in principle can violate this symmetry, and odd powers of g_s will be present as well. However, this odd sector is not fundamental, and can be "gauged away" by a linear shift of the appropriate mass parameters. Notably, this does not apply to the theories in the presence of additional extended objects like surface operators (discussed in Chapter [V:8]). Such a setup breaks 4-d Lorentz invariance and a true odd sector in g_s

---

[1]This is not to say that the microscopic origin of the holomorphic anomaly might not be better explained in the two-parameter scheme, see [27].

will be generated. We will here only consider the even in g_s case, while emphasizing that the general case can be treated as well [11], by appealing to the extended holomorphic anomaly equations of [29, 30].

So let us now explain in detail the method of the holomorphic anomaly for the calculation of the $\mathcal{F}^{(g)}(\beta)$. We begin with the role of $\mathcal{F}^{(0)}$ in special geometry. We denote by u a global coordinate on the moduli space $\mathcal{M}$ of vacua, which is identified with the base space of an appropriate family of complex curves, $\mathcal{C}_u$. (For simplicity, we will write equations only in the case that $\mathcal{M}$ is one-dimensional. The reader might have in mind $SU(2)$ Seiberg-Witten theory with $N_f < 4$ fundamental flavors. Some aspects of the higher-rank theory are discussed in [12, 15].) The family of curves is equipped with a meromorphic one-form λ_{SW}, such that for appropriate choice of one-cycles A and A_D on $\mathcal{C}_u$, the periods

$$a = \oint_A \lambda_{SW}, \qquad a_D = \oint_{A_D} \lambda_{SW}, \qquad (3.4)$$

satisfy the relation

$$a_D = \frac{\partial \mathcal{F}^{(0)}}{\partial a}, \qquad (3.5)$$

after eliminating u from (3.4). We do not need to be explicit about this auxiliary geometric data, for which we refer to chapter [V:2]. However, one should keep in mind, as already mentioned in the previous section, that this auxiliary data originates from the mirror Calabi-Yau geometry of the corresponding geometric engineering geometry.

For expansion in different regions of moduli space, it is most convenient to base the development on the Picard-Fuchs equation, a third order system of linear differential equations,

$$\mathcal{L}\varpi(u) = 0, \qquad (3.6)$$

satisfied by all periods of λ_{SW}. Using a as a local coordinate around $u \to \infty$, the Picard-Fuchs operator takes the form[2]

$$\mathcal{L} = \partial_a \frac{1}{C_{aaa}} \partial_a^2, \qquad (3.7)$$

where

$$C_{aaa} = \partial_a^3 \mathcal{F}^{(0)} = \partial_a^2 a_D(a) = \partial_a \tau(a), \qquad (3.8)$$

is a (meromorphic) rank three symmetric tensor over $\mathcal{M}$, which in the topological string context is referred to as the Yukawa coupling, and $\tau(a)$ corresponds to the complexified effective gauge coupling. In particular, $g \sim \mathrm{Im}\,\tau$ is the Weil-Petersson (or σ-model) metric on $\mathcal{M}$, which plays a central role in special geometry on $\mathcal{M}$.

---

[2]As is now evident, the constant is a third solution of the differential equation. This solution decouples in special cases, such as $SU(2)$ gauge theory with massless hypermultiplets.

Another important feature is the existence of canonical (flat) coordinates [5], which provide a meaningful expansion parameter around any interesting point $u = u_*$ in $\mathcal{M}$. In such a flat coordisnate $t = t(u)$, vanishing at $u = u_*$, the Picard-Fuchs operator takes again the form (3.7) with $a \to t$, i.e.,

$$\mathcal{L} = \partial_t \frac{1}{C_{ttt}} \partial_t^2, \quad C_{ttt} = \left(\frac{\partial u}{\partial t}\right)^3 C_{uuu}. \tag{3.9}$$

We are now ready to write down the holomorphic anomaly equations of [5]. Recall that the specialization of the gauge theory amplitude $\mathcal{F}^{(g)}(\beta)$ to $\beta = 1$, (namely, the self-dual background $\epsilon_1 = -\epsilon_2$) arises via geometric engineering from the genus-g topological string amplitude. The statement of BCOV is that the topological string amplitudes, while holomorphic in the Kähler moduli, are not well-behaved globally over the moduli space. Instead, one should view the topological string amplitudes as a holomorphic limit of non-holomorphic, but globally defined objects. Under the gauge theory limit sketched in the previous section, this translates to the statement that one should view the gauge theory $\mathcal{F}^{(g)}(a)$ (for $g \geq 1$) as the holomorphic limit $\bar{a} \to \infty$ of *non-holomorphic, but globally defined* objects $\mathcal{F}^{(g)}(u, \bar{u})$, arising from the topological string amplitudes in the gauge theory limit. (These are customarily denoted by the same letter, as confusion can not arise.) Similarly, the holomorphic anomaly equation satisfied by the topological string amplitudes translates to a recursive relation for the gauge theory $\mathcal{F}^{(g>1)}(u, \bar{u})$, i.e.,

$$\bar{\partial}_{\bar{u}} \mathcal{F}^{(g)} = \frac{1}{2} \sum_{\substack{g_1 + g_2 = g \\ g_i > 0}} \bar{C}_{\bar{u}}^{uu} \mathcal{F}_u^{(g_1)} \mathcal{F}_u^{(g_2)} + \frac{1}{2} \bar{C}_{\bar{u}}^{uu} \mathcal{F}_{uu}^{(g-1)}, \tag{3.10}$$

where $\mathcal{F}_{uu}^{(g)} = D_u \mathcal{F}_u^{(g)} = D_u^2 \mathcal{F}^{(g)}$, D_u is the covariant derivative over $\mathcal{M}$, and indices are raised and lowered using the Weil-Petersson metric. The holomorphic limit of the connection of the Weil-Petersson metric (entering the covariant derivative) on $\mathcal{M}$ takes the simple form

$$\lim_{\bar{t} \to 0} \Gamma_{uu}^u = \partial_u \log \frac{\partial t(u)}{\partial u}. \tag{3.11}$$

The "one-loop" amplitude satisfies the special equation

$$\bar{\partial}_{\bar{u}} \partial_u \mathcal{F}^{(1)} = \frac{1}{2} \bar{C}_{\bar{u}}^{uu} C_{uuu}. \tag{3.12}$$

At the level of the topological string, the holomorphic anomaly originates from topological anomalies, that is, under coupling to gravity (including integration over moduli of Riemann surfaces), the theory is only "almost" topological, as there are contributions from the boundaries of moduli spaces of genus-g Riemann surfaces to certain topologically trivial correlator insertions. More specifically, a genus-g Riemann surface can degenerate either to a genus-$(g - 1)$ surface with two extra punctures via pinching of a handle, or to two disconnected surfaces with an extra puncture of genus g_a and g_b (with $g = g_a + g_b$) via pinching of a tube. These two boundary contributions are reflected in the holomorphic anomaly Eq. (3.10). Although the topological string origin of the anomaly equation (and in particular of $\mathcal{F}^{(g)}(u, \bar{u})$) is clear, less so is the precise supergravity (or gauge theory) meaning and/or origin thereof. Hence, the main justification of (3.10) at the level of gauge theory comes as a limit of the topological string via geometric engineering. However, an independent justification can be given along the lines of Witten's wave-function interpretation of the topological string partition function [7]. The reasoning leading to this interpretation of the $\mathcal{F}^{(g)}$ and the recursive relations between them relies solely on the special geometry (the holomorphic symplectic structure) of the moduli space (viewed as a clasical phase space). Starting from the Seiberg-Witten geometry, this reasoning can therefore also be applied directly to the gauge theory. We will come back to this interpretation in Sect. 5.

The natural question to ask is how (3.10) should be modified away from $\beta = 1$. The answer provided by [11] is the simplest possible *not at all!* More precisely, in [11], the use of the localization formulas of [1] resulted in the presence of non-vanishing terms of odd order in g_s, suggesting a role for the extended holomorphic anomaly equation of [29, 30], as well as a relation to topological string orientifolds. The extension data (the term at order g_s^{-1}) was also identified in simple geometric terms on the Seiberg-Witten curve. In [13] it was observed that the shift of the mass parameters [28] removes those odd terms, as mentioned above. While it is reassuring to see that the formalism works well with either prescription, we here only present the shifted version, as it is more economical.

The content of the Eqs. (3.10) and (3.12) is that due to the anti-holomorphic derivative, the $\mathcal{F}^{(g)}$ are determined up to holomorphic terms, the so-called holomorphic ambiguity. The standard technique to fix this ambiguity is by taking known characteristics of the $\mathcal{F}^{(g)}$ at specific points in moduli space as boundary conditions into account. For instance, topological string amplitudes expanded near a point in moduli space where the target space develops a conifold singularity show a characteristic "gap" structure [8, 31] (we denote the modulus of the deformation as t_c)

$$\mathcal{F}^{(g>1)} = \Psi^{(2g-2)}(1)\, t_c^{-2g+2} + \mathcal{O}(t_c^0), \tag{3.13}$$

with leading non-vanishing coefficients $\Psi^{(g)}(1)$ of the singular terms given by the free energy of the $c = 1$ string at the self-dual radius $R = 1$ [32] (we use here a normalization different from the one usually used in the CFT context). Knowledge of the conifold expansion (3.13) is usually sufficient to fix the holomorphic ambiguity to very high genus [31], or, even to fix it completely [33], depending on the specific model. The coefficient of the singular term in (3.13), $\Psi^{(n)}(1)$, can be seen as due to integrating out a single massless hypermultiplet in the effective action [34] and is therefore rather universal.[3] In particular, expansion of the gauge theory free energy

---

[3] It that sense, the singularity structure (but not the regular terms) in those strong coupling regions does follow from a field theory computation.

near a point in moduli space with a massless monopole/dyon (hypermultiplet) should show the same behavior, and indeed does [8].

In the generalization to arbitrary $\beta \neq 1$ we then must have that the boundary conditions are not given by integrating out a massless hypermultiplet in an anti-selfdual background, but rather in the Ω-background. Hence, the coefficients $\Psi^{(g)}(1)$ change to β-dependent functions $\Psi^{(n)}(\beta)$ captured by the Schwinger type integral

$$
\mathcal{F}_{c=1}(\epsilon_1, \epsilon_2; t_c) := \int \frac{ds}{s} \frac{e^{-t_c s}}{4 \sinh\left(\frac{\epsilon_1 s}{2}\right) \sinh\left(\frac{\epsilon_2 s}{2}\right)} \sim \cdots + \sum_{n>0} \Psi^{(n)}(\beta) \left(\frac{g_s}{t_c}\right)^n,
$$

under the usage of (3.1). Interestingly, the free energy $\mathcal{F}_{c=1}(\epsilon_1, \epsilon_2; t_c)$ still corresponds to the $c = 1$ string free energy, albeit at general radius $R = \beta$. The corresponding partition function is also known as Gross-Klebanov partition function, and we have for the expansion coefficients the following closed expressions [35]

$$
\Psi^{(0)}(\beta) = -\frac{1}{24}\left(\beta + \frac{1}{\beta}\right),
$$

$$
\Psi^{(n)}(\beta) = (n-1)! \sum_{k=0}^{n+2} (-1)^k \frac{B_k B_{n+2-k}}{k!(n+2-k)!}(2^{1-k}-1)(2^{k-n-1}-1)\,\beta^{k-n/2-1}.
$$
$$
\tag{3.14}
$$

Using the coefficients (3.14) as boundary conditions for general (real) β and analytically continuing back to the weakly coupled regime, somewhat surprisingly reproduces for $SU(2)$ with massless $N_f < 4$ flavors the instanton counting results of [1] (after appropriate choice of gauge of mass parameters, $cf.$, discussion above), as first reported in [11, 13].

Similarly, using (3.14) as boundary conditions for the topological string expanded near a conifold singularity of specific local Calabi-Yau geometries reproduces under analytic continuation the refined free energy defined via the 5d instanton counting. However, there is one important subtlety, which is usually not explicitly mentioned in the literature. Namely, even for a simple Calabi-Yau like local $\mathbb{P}^2$ or $\mathbb{P}^1 \times \mathbb{P}^1$, the boundary conditions (3.14) alone are not sufficient to completely fix the holomorphic ambiguity. The actual difference comes in at 1-loop. Generally, the 1-loop holomorphic ambiguity possesses not only a contribution from the conifold discriminant, but also from the large volume divisor. For example, the 1-loop ambiguity $a^{(1)}(\beta)$ of refined local $\mathbb{P}^2$ reads [16]

$$
a^{(1)}(\beta) = \Psi^{(0)}(\beta) \log \Delta + \kappa(\beta) \log z, \tag{3.15}
$$

with Δ parameterizing the conifold locus, $i.e.$, $\Delta := (1 - 27z)$ and

$$
\kappa(\beta) = -\Psi^{(0)}(\beta) - \frac{2}{3}. \tag{3.16}
$$

Note that in contrast to $\Psi^{(n)}(\beta)$, we do not know how to infer $\kappa(\beta)$ from first principles. Rather $\kappa(\beta)$ has to be manually chosen appropriately to reproduce the desired 1-loop free energy model by model.

4 Refinement Versus Orbifolds

We observed in the previous section that refinement near a conifold point in moduli space can be interpreted as a radius deformation of the $c = 1$ string. The well-known duality between integer radius deformations and orbifolding suggests that at least locally and for integer β, refinement can also be given a more geometric interpretation in terms of a $\mathbb{Z}_\beta$ orbifold. Namely, one may view the refinement in the B-model (for fixed integer β) near a conifold point in moduli space effectively as a replacement of the conifold singularity with an A_β singularity.

There is an apparent puzzle in this proposed orbifold interpretation. In the orbifold case, we have only an anti-selfdual background, so how do the coefficients $\Psi(\beta)$ arise then? Well, the answer is relatively simple. Under the $\mathbb{Z}_\beta$ action, we do not have just one, but β massless hypermultiplets contributing to the leading coefficient of (3.13). This is reflected in the fact that we can decompose the above Schwinger integral representation as (cf., [36])

$$\mathcal{F}_{c=1}(\epsilon_1, \epsilon_2; t_c) = \sum_{n=0}^{\beta-1} \mathcal{F}_{c=1}(\epsilon_1, -\epsilon_1; t_n),$$

with $t_n := t_c - n\epsilon_2 - (\epsilon_1 + \epsilon_2)/2$. Hence, the heart of the proposed orbifold interpretation for integer β lies in the fact that we can trade a single massless hypermultiplet in the corresponding Ω-background for β massless hypermultiplets in an anti-selfdual background.

It is well known that the coefficients $\Psi^{(2n-2)}(1)$ correspond to the virtual Euler characteristic of the moduli space of complex curves of genus n, and one might ask for a similar interpretation for general β. As observed in [37, 38], up to a shift the coefficients for general β in fact match with the parameterized Euler characteristic interpolating between the virtual Euler characteristic of the moduli space of real and complex curves proposed in [39] (see also [40] for a more detailed discussion of this correspondence). The $c = 1$ string orbifold interpretation sketched above now allows us to conjecture a geometric interpretation of this parameterized Euler characteristic for integer β. Namely, it should correspond to the virtual Euler characteristic of genus n curves with a $\mathbb{Z}_\beta$ action.

These local facts lead to the important question whether there exists as well a purely geometric interpretation of the refined partition function at large volume. That is, we should ask if there exists a target space $\widehat{X}_\beta$ such that the refined free energy corresponds to the count of maps

$$\Sigma^{(g)} \to \widehat{X}_\beta. \tag{4.1}$$

Naively, one would suspect that $\widehat{X}_\beta$ corresponds to a free $\mathbb{Z}_\beta$ orbifold of the original Calabi-Yau X (for integer β). In particular, the only visible effect of such a free orbifold on the level of the holomorphic anomaly equations would be a mere change of boundary conditions, as we observed for refinement.

Indeed, one can find for specific models and values of β concrete proof that the refined partition function is dual to the usual topological string on an orbifold of the original geometry. The simplest example has been already given in [11], where it was observed that the quotient of local $\mathbb{P}^1 \times \mathbb{P}^1$ by its obvious $\mathbb{Z}_2$ symmetry equals the refined partition function on the original geometry at $\beta = 2$. A similar observation can be made for orbifolding local $\mathbb{P}^2$ by its cyclic $\mathbb{Z}_3$ symmetry, which corresponds to the refined partition function at $\beta = 3$ [17].

One should note that the a priori undetermined function $\kappa(\beta)$ leaves the freedom for different analytic continuations of the $\mathbb{Z}_\beta$ symmetry of the conifold to large volume. The necessity of such an ambiguity is intuitively clear, as there might exist at large volume differently acting symmetries, which still yield under analytic continuation the same leading singular behavior at the conifold point. For instance, there might be differently acting $\mathbb{Z}_2$ orbifolds at large volume, which, due to the high symmetry of the conifold, all possess the same coefficients $\Psi^{(n)}(2)$ (the massless hypermultiplet does not care which symmetry it feels).

In general, however, it is far from clear how, if at all, the $\mathbb{Z}_\beta$ symmetry of the conifold point in moduli space translates to the large volume regime (*i.e.*, is globally preserved). In the above two examples, the correspondence between the refined topological string for particular values of β to the usual topological string on a different (orbifold) background could be argued to be a consequence of the large global symmetry group, and therefore somewhat accidental. This still leaves open the possibility for the existence of a new classical target space $\widehat{X}_\beta$, which could be obtained for instance in the case of β integer as a suitable partial compactification of an orbifold of the conifold.

5   Wave-Function Interpretation

The special geometry relation (3.5) between the flat coordinate a and the magnetic dual a_D is identical to the relation between canonically conjugate variables (p, q) of a classical integrable system. In particular, comparison with the Hamilton-Jacobi equation $H(q, \frac{\partial S}{\partial q}) = 0$ shows that in this interpretation, the prepotential $\mathcal{F}^{(0)}(a)$ should be identified with Hamilton's principal function (or classical action) effecting the canonical transformation to the action-angle variables.

Consider now the full perturbative partition function Z expanded as a series in g_s, as in (3.2). We have

$$Z = f(a, g_s, \beta) \, e^{\frac{1}{g_s^2} \int a_D \, da}, \tag{5.1}$$

with $f(a, g_s, \beta)$ some regular series in g_s^2. This expansion shows that the partition function should be interpreted as a WKB-type wavefunction in a semi-classical approximation to a quantization of the original hamiltonian system.

In this context, it is important to keep in mind that quantization is intrinsically ambiguous, *i.e.*, in general on cannot associate a unique quantum operator $\hat{H}$ to a classical Hamiltonian H. This is most clearly apparent in the ordering ambiguities that plague the lifting of functions of the phase-space coordinates to quantum-mechanical operators. While the semi-classical terms are universal, the higher order terms in the expansion (5.1) are sensitive to these ambiguities.

The holomorphic anomaly plays in interesting role in this so-called "wavefunction interpretation" of the topological partition function. In fact, as pointed out by Witten [7], the holomorphic anomaly equation simply expresses the change of the wavefunction under a change of polarization of the underlying classical system (*i.e.*, the separation of the canonical coordinates into position and momentum variables). Above, we considered a real polarization, but complex polarizations are natural as well. In the sense of this wave-function interpretation, the topological partition function Z is a representation of the true ground state for a particular choice of polarization.

Witten's original proposal was made for the partition function of topological strings (for which the holomorphic anomaly was first discovered), but given the results of [8, 11] that we have reviewed above, the interpretation is very natural in the context of gauge theory as well (*cf.*, [41]). An interesting consequence of the fact that the holomorphic anomaly equation is insensitive to the deformation parameter β is that the refined partition function corresponds to a family of quantum states with the same semi-classical expansion. On the other hand, it remains unclear how to determine the additional conditions that would select the topological partition function as the unique ground state of the system. To our knowledge, the wavefunction interpretation has not been successfully exploited for fixing the holomorphic ambiguity.

Via the AGT conjecture (for which there is now substantial evidence, as reviewed elsewhere in this volume), quantum Liouville theory provides an answer to the quantization problem that is in principle independent from the relation to topological strings, and has the advantage of being algorithmic. Yet another approach to the quantization problem are the so-called topological recursions of Eynard-Orantin [42]. A detailed comparison between these various schemes remains an interesting avenue for further research.

6 Outlook

We see two major open problems whose solution would constitute significant progress. The first, more technical in nature, is the direct and explicit calculation of the gauge theory partition function at strong coupling either via instanton counting or CFT. The results reviewed here, specifically the holomorphic limit of the $\mathcal{F}^{(g)}$

as $\bar{t}_D = 0$, $t_D \to 0$, provide a benchmark for such a calculation. A first indication that the 2d-4d relation holds beyond weak coupling has been found at hand of an explicit example recently in [25]. However, the general picture is far from clear. The perhaps most closely related work from a CFT point of view is [43]. Ultimately, a simple state like construction, as in [44], for the strongly coupled expansion would be desirable.

The second major open problem is to obtain a better understanding of the deformation parameter β in the topological string context. The core question is whether the deformation really involves a new world-sheet theory (for instance, with a second string coupling, in case the extra parameter is viewed as an infinitesimal coupling constant), or whether it might be sufficient to view the β-degree of freedom entirely as a geometric deformation of the target space of the usual topological string. In this review, we somewhat focussed on the latter point of view.

Though we did not discuss them in this review, it has to be mentioned that there are several explicit proposals in the literature [15, 27, 45, 46], supporting the possibility for an actual world-sheet interpretation with two infinitesimal coupling constants. In the mathematical formulation of the perturbative topological string (Gromov-Witten theory), this might involve a sort of refined count of holomorphic maps [9, 27]

$$\Sigma^{(g_1, g_2)} \to X, \tag{6.1}$$

with worldsheets Σ of genus $g = g_1 + g_2$, carrying an additional $\mathbb{Z}_2$ valued decoration of the handles.

Finally, another proposed interpretation involves replacing the B-model geometry by a sort of "quantum geometry" $\widetilde{Y}_q$ (encoding one of the parameters in a suitable parameterization). In this approach the tree-level special geometry depends explicitly on the extra parameter [16, 47]. This is analogous to the replacement of the spectral curve with a "quantum" spectral curve in the β-deformed matrix models reviewed in chapter [V:5], and the "quantization" of Seiberg-Witten theory outlined in chapter [V:12].

One should note that the quantum geometry $\widetilde{Y}_q$ is not directly related to the wave-function interpretation of the partition function discussed in Sect. 5 (at least the precise relation is not known). While in the former we quantize the underlying curve, in the latter we quantize the periods. In this sense we can understand the refined topological string also as a double quantization. For $\widetilde{Y}_q$ the ordinary special geometry relation is lifted to a so-called quantum special geometry relation and the ordinary periods to quantum periods (depending on the extra parameter). Quantizing similarly as in Sect. 5 the quantum periods, will again lead to the holomorphic anomaly equation, now for the (double) quantum states.

While neither proposal is entirely convincing in the present form, a possible connection between the two proposed target space deformations might even be more tantalizing. Optimistically, it might indicate a new type of classical-quantum duality relating the topological string on $\widehat{X}_\beta$ with that on $\widetilde{Y}_q$. If in addition a refined topological string exists, in the sense of a deformed world-sheet theory, this duality would extend to a triality. This, at least, is the inspiration that we take away.

Acknowledgments We would like to thank J. Teschner for the invitation to participate in this joint review effort, his hard work and patience. We thank all other contributors for their valuable comments and input. The work of D.K. has been supported in part by a Simons fellowship, the Berkeley Center for Theoretical Physics and the National Research Foundation of Korea Grant No. 2012R1A2A2A02046739. The research of J.W. is supported in part by an NSERC discovery grant and a Tier II Canada Research Chair.

References

[1]   Nekrasov, N.A.: Seiberg-Witten prepotential from instanton counting. Adv. Theor. Math. Phys. **7**, 831 (2004). arXiv:hep-th/0206161

[2]   Kachru, S., Vafa, C.: Exact results for N = 2 compactifications of heterotic strings. Nucl. Phys. **B450**, 69 (1995). arXiv:hep-th/9505105. Kachru, S., Klemm, A., Lerche, W., Mayr, P., Vafa, C.: Nonperturbative results on the point particle limit of N=2 heterotic string compactifications. Nucl. Phys. **B459**, 537 (1996). arXiv:hep-th/9508155. Katz, S.H., Klemm, A., Vafa, C.: Geometric engineering of quantum field theories. Nucl. Phys. **B497**, 173 (1997). arXiv:hep-th/9609239. Katz, S., Mayr, P., Vafa, C.: Mirror symmetry and exact solution of 4-D N=2 gauge theories: 1. Adv. Theor. Math. Phys. **1**, 53 (1998). arXiv:hep-th/9706110

[3]   Hellerman, S., Orlando, D., Reffert, S.: String theory of the omega deformation. JHEP **1201**, 148 (2012). arXiv:1106.0279 [hep-th]

[4]   Antoniadis, I., Gava, E., Narain, K.S., Taylor, T.R.: Topological amplitudes in string theory. Nucl. Phys. **B413**, 162 (1994). arXiv:hep-th/9307158

[5]   Bershadsky, M., Cecotti, S., Ooguri, H., Vafa, C.: Kodaira-Spencer theory of gravity and exact results for quantum string amplitudes. Commun. Math. Phys. **165**, 311 (1994). arXiv:hep-th/9309140

[6]   Shifman, M.A. Vainshtein, A.I.: Solution of the anomaly puzzle in SUSY gauge theories and the Wilson operator expansion. Nucl. Phys. **B277**, 456 (1986) [Sov. Phys. JETP **64**, 428 (1986)] [Zh. Eksp. Teor. Fiz. **91**, 723 (1986)]

[7]   Witten, E.: Quantum background independence in string theory. arXiv:hep-th/9306122

[8]   Huang, M.-x., Klemm, A.: Holomorphic anomaly in gauge theories and matrix models. JHEP **0709**, 054 (2007). arXiv:hep-th/0605195

[9]   Hollowood, T.J., Iqbal, A., Vafa, C.: Matrix models, geometric engineering and elliptic genera. JHEP **0803**, 069 (2008). arXiv:hep-th/0310272. Iqbal, A., Kozcaz, C., Vafa, C.: The refined topological vertex. JHEP **0910**, 069 (2009). arXiv:hep-th/0701156

[10]   Gukov, S., Schwarz, A.S., Vafa, C.: Khovanov-Rozansky homology and topological strings. Lett. Math. Phys. **74**, 53 (2005). arXiv:hep-th/0412243

[11]   Krefl, D., Walcher, J.: Extended holomorphic anomaly in gauge theory. Lett. Math. Phys. **95**, 67 (2011). arXiv:1007.0263 [hep-th]

[12]   Huang, M.-x., Klemm, A.: Direct integration for general Ω deformed B-model for rigid $N = 2$ backgrounds. arXiv:1009.1126 [hep-th]

[13]   Krefl, D., Walcher, J.: Shift versus extension in refined partition functions. arXiv:1010.2635 [hep-th]

[14]   Krefl, D., Walcher, J.: Unpublished (2010)

[15]   Huang, M.-x., Kashani-Poor, A.-K., Klemm, A.: The Ω deformed B-model for rigid $N = 2$ theories. Ann. Henri Poincare **14**, 425 (2013). arXiv:1109.5728 [hep-th]

[16]   Aganagic, M., Cheng, M.C.N., Dijkgraaf, R., Krefl, D., Vafa, C.: Quantum geometry of refined topological strings. JHEP **1211**, 019 (2012). arXiv:1105.0630 [hep-th]

[17]   Krefl, D.: unpublished (2012)

[18]   Klemm, A.: On the geometry behind N=2 supersymmetric effective actions in four-dimensions. In: Trieste 1996, High Energy Physics and Cosmology, pp. 120–242. arXiv:hep-th/9705131

[19]   Mayr, P.: Geometric construction of N = 2 gauge theories. Fortsch. Phys. **47**, 39 (1999). arXiv:hep-th/9807096

[20]   Kontsevich, M.: Enumeration of rational curves via torus actions. arXiv:hep-th/9405035

[21]   Aganagic, M., Klemm, A., Marino, M., Vafa, C.: The topological vertex. Commun. Math. Phys. **254**, 425 (2005). arXiv:hep-th/0305132

[22]   Klemm, A., Marino, M., Theisen, S.: Gravitational corrections in supersymmetric gauge theory and matrix models. JHEP **0303**, 051 (2003). arXiv:hep-th/0211216

[23]   Bershadsky, M., Cecotti, S., Ooguri, H., Vafa, C.: Holomorphic anomalies in topological field theories. Nucl. Phys. **B405**, 279 (1993). arXiv:hep-th/9302103

[24]   Alday, L.F., Gaiotto, D., Tachikawa, Y.: Liouville correlation functions from four-dimensional gauge theories. Lett. Math. Phys. **91**, 167 (2010). arXiv:0906.3219 [hep-th]

[25]   Krefl, D.: Penner type ensemble for gauge theories revisited. arXiv:1209.6009 [hep-th]

[26]   Krefl, D., Shih, S.-Y.D.: Holomorphic anomaly in gauge theory on ALE space. arXiv:1112.2718 [hep-th]

[27]   Prudenziati, A.: Double genus expansion for general Ω background. arXiv:1204.2322 [hep-th]

[28]   Okuda, T., Pestun, V.: On the instantons and the hypermultiplet mass of N=2* super Yang-Mills on S^4. JHEP **1203**, 017 (2012). arXiv:1004.1222 [hep-th]

[29]   Walcher, J.: Extended holomorphic anomaly and loop amplitudes in open topological string. Nucl. Phys. **B817**, 167 (2009). arXiv:0705.4098 [hep-th]

[30]   Walcher, J.: Evidence for tadpole cancellation in the topological string. Commun. Num. Theor. Phys. **3**, 111 (2009). arXiv:0712.2775 [hep-th]

[31]   Huang, M.-x., Klemm, A., Quackenbush, S.: Topological string theory on compact Calabi-Yau: modularity and boundary conditions. Lect. Notes Phys. **757**, 45 (2009). arXiv:hep-th/0612125

[32]   Ghoshal, D., Vafa, C.: C = 1 string as the topological theory of the conifold. Nucl. Phys. **B453**, 121 (1995). arXiv:hep-th/9506122

[33]   Haghighat, B., Klemm, A., Rauch, M.: Integrability of the holomorphic anomaly equations. JHEP **0810**, 097 (2008). arXiv:0809.1674 [hep-th]

[34]   Vafa, C.: A stringy test of the fate of the conifold. Nucl. Phys. **B447**, 252 (1995). arXiv:hep-th/9505023

[35]   Gross, D.J., Klebanov, I.R.: One-dimensional string theory on a circle. Nucl. Phys. **B344**, 475 (1990)

[36]   Gopakumar, R., Vafa, C.: Topological gravity as large N topological gauge theory. Adv. Theor. Math. Phys. **2**, 413 (1998). arXiv:hep-th/9802016

[37]   Krefl, D., Walcher, J.: ABCD of beta ensembles and topological strings. JHEP **1211**, 111 (2012). arXiv:1207.1438 [hep-th]

[38]   Krefl, D., Schwarz, A.: Refined Chern-Simons versus Vogel universality. J. Geom. Phys. **74**, 119 (2013). arXiv:1304.7873 [hep-th]

[39]   Goulden, I.P., Harer, J.L., Jackson, D.M.: A geometric parametrization for the virtual Euler characteristic of the moduli space of real and complex algebraic curves. Trans. Am. Math. Soc. **353**, 4405 (2001)

[40]   Chair, N.: Generalized Penner model and the Gaussian beta ensemble. Nucl. Phys. B (in press)

[41]   Aganagic, M., Bouchard, V., Klemm, A.: Topological strings and (almost) modular forms. Commun. Math. Phys. **277**, 771 (2008). arXiv:hep-th/0607100

[42]   Eynard, B., Orantin, N.: Invariants of algebraic curves and topological expansion. Commun. Number Theory Phys. **1**, 347 (2007). arXiv:math-ph/0702045

[43]   Gaiotto, D., Teschner, J.: Irregular singularities in Liouville theory and Argyres-Douglas type gauge theories, I. JHEP **1212**, 050 (2012). arXiv:1203.1052 [hep-th]

[44]   Gaiotto, D.: Asymptotically free N=2 theories and irregular conformal blocks. arXiv:0908.0307 [hep-th]

[45] Antoniadis, I., Hohenegger, S., Narain, K.S., Taylor, T.R.: Deformed topological partition function and Nekrasov backgrounds. Nucl. Phys. **B838**, 253 (2010). arXiv:1003.2832 [hep-th]. Antoniadis, I., Florakis, I., Hohenegger, S., Narain, K.S., Zein Assi, A.: Worldsheet realization of the refined topological string. Nucl. Phys. B**875**, 101 (2013). arXiv:1302.6993 [hep-th]. Antoniadis, I., Florakis, I., Hohenegger, S., Narain, K.S., Zein Assi, A.: Non-perturbative Nekrasov partition function from string theory. arXiv:1309.6688 [hep-th]

[46] Nakayama, Y., Ooguri, H.: Comments on worldsheet description of the omega background. Nucl. Phys. **B856**, 342 (2012). arXiv:1106.5503 [hep-th]

[47] Mironov, A., Morozov, A.: Nekrasov functions and exact Bohr-Zommerfeld integrals. JHEP **1004**, 040 (2010). arXiv:0910.5670 [hep-th]